HIGHEST TRADITIONS

On the morning of 4 December 1942, a flight of three Hudsons of No 2 Squadron limped back to its base at Batchelor from a raid in the Banda Sea. Despite thirty minutes of determined attacks on the formation by an enemy Zero, which killed a gunner and damaged all Hudsons, the Australians repeatedly fought off the fighter and returned to base. The official report on this operation recorded:

... the actions of all members of the three crews were in the highest traditions of the RAAF ...

This book is one of a series of publications produced for the 75th Anniversary of the Royal Australian Air Force.

Royal Australian Air Force

HIGHEST TRADITIONS

The History of No 2 Squadron, RAAF

John Bennett

An AGPS Press publication
Australian Government Publishing Service
Canberra

National Library of Australia Cataloguing-in-Publication data:

Bennett, John (John William), 1949–.
Highest traditions: the history of No. 2 Squadron RAAF.

Bibliography.
Includes index.
ISBN 0 644 35230 2.

1. Australia. Royal Australian Air Force. Squadron, 2. – History. 2. World War, 1939–45 – Aerial operations, Australian. 3. World War, 1939–45 – Regimental histories – Australia. 4. World War, 1914–1918 – Aerial operations, Australian. 5. World War, 1914–1918 – Regimental histories – Australia. I. Australia. Royal Australian Air Force. II. Title.
358.4130994

Design by The AGPS Design Studio
Produced by the Australian Government Publishing Service

FOREWORD

Regrettably, the RAAF has been rather negligent in recording its history. Indeed, it has been quite uncaring in regard to both history and tradition. Customs of the service are given scant regard, as is a conscious effort to build on traditional values. I share the guilt for this. This excellent coverage of the history of Number 2 Squadron should therefore be especially welcomed; hopefully heralding a more responsible attitude to passing on history to the benefit of those who follow.

Number 2 Squadron has served Australia with distinction in two world wars, during the Malayan Emergency, the Vietnam conflict and in peacetime. The sacrifice, at times heavy, was never shirked.

One of the many things that should be noted from John Bennett's fine book is the importance of good leadership—a factor that probably transcends all others. There are several examples, particularly during the two world wars, where a particular commander lifted the Squadron performance to new heights, simply by virtue of his leadership. Such qualities should be developed to the greatest possible extent in service training and never subordinated to academic or other pursuits.

As Chief of the Air Staff I had the melancholy task of authorising the closing down of Number 2 Squadron and its retention on a name only basis. I have no doubt that should Australia again be in peril, the Squadron will reform to serve the nation with the same distinction as has been described in this book.

I commend *Highest Traditions* to all who place value on determination, courage, esprit de corps and leadership.

David Evans
Air Marshal RAAF (Ret)

CONTENTS

Part One—Control of the Air

Part Two—Air Strike

Part Three—Air Support

APPENDICES

ILLUSTRATIONS

Colour Plates

(between pages 168 and 169)

INTRODUCTION

The development of air power over this century has probably been the most notable technological advance in history. Australia has been part of this development as the air age shrank the world. In 1914, soon after the outbreak of war, Australian military aircraft were deployed for overseas operations. At that stage, flying units were formed as part of the army in the Australian Flying Corps. No 2 Squadron was born in Egypt in 1916, and manned by members of the Light Horse and from the first squadron of the Flying Corps that had been sent to the Middle East.

In 1917, after training in England, the unit proceeded to the aerial battlefield in France, operating above the static trench warfare on the Western Front. The scout, or fighter, "machines" of No 2 Squadron gave sterling support to the Royal Flying Corps in gaining the upper hand over the German Air Service in the air campaign. This *control of the air* enabled the freedom of operation for Allied reconnaissance aircraft, and eliminated the threat of air attack to the Allied forces on the ground.

After World War I, Australia was one of the first countries to form an independent air force. Other nations—notably the United States and Canada—restricted air arms to their armies. Through the tough economic period between the wars, Australia was fortunate to maintain the Royal Australian Air Force as a specialised arm to prosecute the many facets of air power. It was this small nucleus that provided the base for the massive expansion required in 1939.

The tactics that developed the forms of air combat for the "machines" in World War I were to hold good for the "kites" in World War II. Now, as a bomber reconnaissance unit, No 2 Squadron took the war to the enemy—repeatedly attacking his bases in occupied territories around the Banda and Arafura Seas. It was this campaign of *air strike* which contributed significantly to wearing down his resources and war-making capability.

From the dark days of 1942, as our nation faced the threat of invasion, No 2 Squadron maintained an offensive air strike campaign in the North-Western Area, and by 1943 the war was swinging against Japan. At the end of the following year, the enemy had lost air superiority, apart from some isolated pockets. One No 2 Squadron airman was to comment: "Twelve months ago it would have been madness for a plane to have passed so close to Koepang's fighter strips in daylight. Today, we hardly gave the place a second glance".

Post World War II, the Squadron was the first Australian operational unit to equip with the Canberra jet bomber. No 2 Squadron's Canberras flew operations in Malaya and, for four years, in Vietnam. The Canberra went through many role changes as a bomber, and by virtue of its accuracy as a weapon platform, and its reliability, the Canberra proved itself in the tactical air campaign by providing *air support* for combat forces. This flexibility not only demonstrated the principle of the flexibility of air power, but the flexibility of the Canberra itself. That was to be further demonstrated by its success in the photographic role during the survey of Papua New Guinea and Indonesia. Unfortunately, the Canberra was to die of old age after almost 30 years of service. With the Canberra's passing, we also lost No 2 Squadron.

No 2 Squadron is unique for Australia in that it has operated in four conflicts. The Squadron is also unique in participating in the three air campaigns—control of the air, air strike, and air support for combat forces. "The Battle of Australia" that No 2 Squadron had waged during World War II was won at high cost—174 members of the Squadron had died. Of those who were captured by an inhumane enemy, only two were to survive the atrocious treatment. This should not be forgotten.

Nor should the lessons that are learnt from each conflict be forgotten. We learn from history that some basic golden rules are sometimes so obvious they can be easily overlooked. Our preparedness with training, readiness with equipment backed by efficient logistics support, the elements of successful leadership, familiarity with where we may have to fight, and the dispersal and protection of our assets, are all factors that the military professional should always be conscious of. History also shows, as borne out in this story, that the persistence of individuals in improving systems—whether they be of a human or technical nature—makes for a more effective organisation.

A military force needs to train in peace the way it will fight in war. Finance, of course, takes the edge off this philosophy. Within the realities of peacetime constraints on our human resources, the effective use of Reserve forces needs always to be considered. Moreover, as the RAAF shrinks and faces civilians filling servicemen's shoes in peace through a commercialisation program, it is necessary to be aware that a civilian cannot replace the specialist and highly trained combatant in war. It would be to our peril if we did not recognise this fact.

It would be wishful thinking to suppose that from our study of history the mistakes of the past will not be repeated. It would be idealistic to think the lessons learnt will place us in good stead for the future. In reality, let us hope that the sacrifices have not been in vain, and that the "highest traditions" are maintained.

JWB
January 1995

ACKNOWLEDGMENTS

This book was written under the auspices of a Chief of Air Staff Fellowship, which amounted to a full-time task for me over 1993 at the RAAF Air Power Studies Centre. The assistance of the Director, Group Captain Gary Waters, and the advice of historian Dr Alan Stephens, kept me, I hope, on track. My colleague there, Wing Commander Al Curr, cast his eye over my staff work. Any errors that remain, however, must reside with me—it would be wrong to "re-course" him at Staff College!

No 2 Squadron's story could not have been written with detail and colour without the help of many past members, and their descendants, who have contributed to making what, I hope, will prove to be an accurate and interesting record of this fine unit. In addition, some of the research facilities we are fortunate in having in Australia have also been generous with their cooperation.

For assistance in researching World War I, I would like to thank John Leeuwin-Clark, Garth Cox and Hubert Billings, all of whose fathers served with No 2 Squadron, Australian Flying Corps. In addition, Wing Commander Mark Lax has provided helpful advice on AFC personnel. The Australian War Memorial is basically the centre in Australia for research into the Great War, and their research facilities are second to none. I also received the assistance of the RAAF Museum at Point Cook, where I would like to thank librarian Jim De Bomfort, and the Museum of Army Aviation and Flying at Oakey, under its curator Brian Reardon. The Australian Archives in Canberra and Melbourne also provided much assistance, with records of early Australian aviation. The United Kingdom Public Records Office at Kew, together with the RAF Museum and the Imperial War Museum in London, provided material that was not available in Australia.

The very active No 2 Squadron Association, based in Melbourne, provided much personal contact for World War II, which I hope has been able to elevate this book above being purely a summary of the facts and figures of history. I thank the Patron, Sir Raymond Garrett, President Peter Helder, and Secretary Rex Freeman. In her role as coordinator for the association, Linda Henry, with the assistance of her husband Al, have made my task so much easier. I thank them both for their time, efforts and patience in bringing this book to fruition.

Also the RAAF Hudson Squadrons Association, based in Adelaide, has enabled an accurate coverage of No 2 Squadron to be attempted. I thank their Patron, Group Captain Tich McFarlane, who commanded the Squadron over what may have been its most notable period—1942. He was generous with his time, and provided valuable advice in ensuring the accuracy of events. In addition, I must thank President John Dorsett and Secretary Murry Lawson. One member, David Vincent, an acknowledged historian and author, was forthcoming with his own research and advice. David is following his previous fine works on the Catalina and Mosquito with a book on the Hudson.

Other members of these two associations include Kym Bonython and Colin Humphris, both of whom allowed me to use their own published works on their involvement in No 2 Squadron. Also Sid Grantham, who has written the history of No 13 Squadron, welcomed me to use his work of No 2's sister unit. Brian Hawthorne, who kept fine diaries of his period during World War II, allowed me full use of his material. Brian's skill makes his writing publishable as stand-alone documents. I would also like to acknowledge the diary of Bruce Wallace, which he had kept up to his capture and subsequent execution. His brother, John Wallace, was understanding in allowing me to include excerpts, and I hope that I have used them properly.

The poem "To The Hudson Men" was written by Evelyn Pitman, sister of a deceased No 2 Squadron member Jack Mawdsley, and I am indebted to her for allowing me to use it for this work. Another former Squadron member who had recorded his story was Sid Wadey, and I thank his widow for allowing me to use his unpublished manuscript. Others who have assisted have been acknowledged with footnotes throughout the book. I thank them all.

One great centre of research for the RAAF in World War II is the RAAF Historical Section, in Canberra, and this has provided access to many wartime records. I thank David Wilson and his staff. For personnel details, I am indebted to the efforts of David Pullen, Dianne Davis and the staff at the RAAF Personnel Records section. The National Library, in Canberra, has also been useful in filling in a few gaps.

Any record of this kind is, of course, incomplete without a full photographic coverage of the life of a unit. To the CO of the RAAF's Central Photographic Establishment at Laverton, Wing Commander Graeme "Wally" Walton, I extend my heartfelt thanks. We had served on No 2 Squadron in Vietnam together, so we had a vested interest in ensuring the finished product was right. Without the help of CPE, I would have felt the book to be inadequate. Other former photographers of No 2 Squadron who have volunteered their works have included Frank Jefferies, Kip Porteus and Aubrey "Strawb" McEgan. Some of the work from their professional lenses have appeared for the first time in print in this book.

For the postwar periods in Australia, Malaya/Malaysia and Vietnam, the many who contributed their assistance have been acknowledged in the footnotes. Air Marshal David Evans, a former No 2 Squadron commander and Chief of Air Staff, provided much of his time and offered advice throughout the writing of the book. I am grateful to him for this assistance, and for agreeing to write the foreword. I also thank Bob Howe for access to his comprehensive Vietnam collection, and to other colleagues who have provided photographs of our time on No 2 Squadron.

Finally, it is reassuring to know that the RAAF is actively pursuing a policy to ensure the traditions of our service remain alive. Group Captain Phil Morrall, of the Chief of Air Staff's office, enabled the work to be put to bed with the publishers. It is the intent to produce a series of works to ensure the RAAF's heritage will not be forgotten. We are all fortunate that the RAAF has such vision.

ABBREVIATIONS

Most of the abbreviations listed below have been used in the biographical footnotes. Generally, abbreviations in the text have been explained in full.

AA	Australian Archives
AA	Air Attache
AA	Anti Aircraft
AAA	Anti Aircraft Artillery
A&AEE	Aircraft and Armament Experimental Establishment
AAF	Allied Air Forces
AAF	Australian Air Force
AAMC	Australian Army Medical Corps
AAS	Air Armament School
AASC	Australian Army Service Corps
AATTV	Australian Army Training Team Vietnam
AAU	Air Ambulance Unit
ABC	Australian Broadcasting Commission
ABDA	American British Dutch Australian Command
AC	Aircraftman
AC	Airfield Construction
ACAUST	Air Commander Australia
ACH	Area Combined Headquarters
ACM	Air Chief Marshal
ACS	Airfield Construction Squadron
AD	Aircraft Depot
ADF	Australian Defence Force
ADFA	Australian Defence Force Academy
ADFWC	ADF Warfare Centre
ADG	Airfield Defence Guard
Adv	Advanced
AE	Army Engineers
AFB	Air Force Base (US)
AFC	Air Force Cross
AFC	Australian Flying Corps
AFM	Air Force Medal
AFTS	Applied Flying Training School
AGA	Australian Garrison Artillery
AHQ	Air Headquarters
AIF	Australian Imperial Force
Air Cdre	Air Commodore
AJWE	Australian Joint Warfare Establishment
ALG	advanced landing ground
AM	Air Marshal
A/M	Air Mechanic
Amb	Amberley
AMP	Air Member for Personnel
AMSE	Air Member for Supply and Equipment
ANS	Air Navigation School
Ant	Antarctic
ANZAC	Australian and New Zealand Army Corps
AO	Order of Australia
AP	Air Publication
AP	Aircraft Park
APU	Aircraft Performance Unit
AOA	Air Officer Administration
AOB	advanced operational base
AOC	Air Officer Commanding
AOS	Air Observer School
ARD	Aircraft Repair Depot
ARDU	Aircraft Research and Development Unit
Armt	Armament
ARS	Aeroplane Repair Section
ARVN	Army of the Republic of Vietnam
ASD	Aeroplane Supply Depot
ASR	Air Sea Rescue
ASRT	Air Support Radar Team
ATC	air traffic control
ATC	Air Training Corps
ATU	Air Trials Unit
AUP	Avionics Update Program
AUW	all up weight
AVM	Air Vice-Marshal
Avn	Aviation
AWM	Australian War Memorial
BAGS	Bombing and Air Gunnery School
BCOF	British Commonwealth Occupation Force

BDA	bomb damage assessment
Bde	Brigade
BE	Bristol Experimental
BEM	British Empire Medal
BFTS	Basic Flying Training School
Bn	Battalion
BSqn	Base Squadron
Bty	Battery
But	Butterworth
BW	Bombardment Wing
CAC	Commonwealth Aircraft Corporation
CAF	Citizen Air Force
CAFM	Chief of Air Force Materiel
CAFOPS	Chief of Air Force Operations
CAFP	Chief of Air Force Personnel
cal	calibre
CAMD	Central Ammunition Depot
Capt	Captain
CAS	Chief of the Air Staff
CAW	College of Air Warfare
CB	Companion of the Bath
CBE	Commander of the British Empire
Cd	Command
CF	Communications Flight
CFI	Chief Flying Instructor
CFS	Central Flying School
CGS	Chief of the General Staff
CGS	Central Gunnery School
CI	Chief Instructor
C-in-C	Commander in Chief
CMG	Companion of St michael and St George
c/n	constructor's number
CO	Commanding Officer
Col	Colonel
comd	commanded
Comdt	Commandant
COP	Close Offensive Patrol
COS	Chief of Staff
COSC	Chiefs of Staff Committee
Coy	Company
CPE	Central Photographic Establishment
Cpl	Corporal
CR	Combat Report
CR	crashed
CR	Central Reserve
CRD	Crash Repair Depot
CRS	Commonwealth Record Series
CSS	Combat Skyspot
CT	Communist Terrorist
CU	Communications Unit
CVO	Companion of the Royal Victorian Order
D	Director
dam	damaged
DAP	Department of Aircraft Production
Dar	Darwin
DASC	Direct Air Support Centre
DCAS	Deputy Chief of Air Staff
DCM	Distinguished Conduct Medal
DD	Deputy Director
Dept	Department
DFC	Distinguished Flying Cross
DFM	Distinguished Flying Medal
DFW	*Deutsche Flugzeug Werke*
DG	Director General
DH	De Havilland
Div	Division
DMZ	Demilitarised Zone
DOD	Department of Defence
DOP	Distant Offensive Patrol
DSO	Distinguished Service Order
E	Eastern
EA	Eastern Area
EA	enemy aircraft
EATS	Empire Air Training Scheme
Ech	Echelon
Edn	Edinburgh
EFTS	Elementary Flying Training School
ELG	emergency landing ground
Equip	Equipment
Esl	East Sale
Exped	Expeditionary
FAC	Forward Air Controller
Fbn	Fairbairn
FBRD	Flying Boat Repair Depot
Fd	Field
FEAF	Far Eastern Air Force
Flg Off (F/O)	Flying Officer
F/Sgt	Flight Sergeant
Flt	Flight
Flt Lt (F/L)	Flight Lieutenant

FSHQ	Fighter Sector Headquarters
Ftr	Fighter
FTS	Flying Training School
Fwd	Forward
GAF	Government Aircraft Factories
GCB	Knight Grand Cross of the Bath
GCVO	Knight Grand Cross of Royal Victorian Order
Gen	General
GHQ	General Headquarters
GM	George Medal
GOC	General Officer Commanding
GP	General Purpose
Gp	Group
GR	General Reconnaissance
HE	high explosive
HMAS	His/Her Majesty's Australian Ship
HMAT	His Majesty's Australian Transport
HMS	His/Her Majesty's Ship
hp	horsepower
HQ	Headquarters
HQOC	Headquarters Operational Command
HQHC	Headquarters Home Command
HQSC	Headquarters Support Command
HQTC	Headquarters Training Command
IFF	Identification Friend or Foe
ITS	Initial Training School
IWM	Imperial War Museum
JG	*Jagdgeschwarde*, Fighter Wing
KBE	Knight Commander of the British Empire
KCB	Knight Commander of the Bath
KCSI	Knight Commander of the Star of India
kg	kilogram
LAC	Leading Aircraftman
Lav	Laverton
lb	pound
LCH	Landing Craft Heavy
Lieut	Lieutenant
LO	Liaison Officer
LRN	Long Range Navigation
Lt	Lieutenant
Lt	Light
LVG	*Luft-Verkehrs Gesellschaft*
MACV	Military Assistance Command Vietnam
Maj	Major
MBE	Member of the British Empire
MC	Military Cross
mk	mark
Met	Meteorological
mg	machine-gun
MHR	Member of the House of Representatives
MID	Mentioned in Dispatches
MM	Military Medal
MSM	Meritorious Service Medal
MT	Motor Transport
MUR	Melbourne University Rifles
MW	Maintenance Works
NCO	Non Commissioned Officer
NE	North Eastern
NEA	North Eastern Area
NEI	Netherlands East Indies
NR	Navigation Reconnaissance
NSW	New South Wales
NT	Northern Territory
NTW	Navigation Training Wing
NVA	North Vietnamese Army
NW	North Western
NWA	North Western Area
NZ	New Zealand
OBE	Order of the British Empire
OBU	Operational Base Unit
OC	Officer Commanding
OCU	Operational Conversion Unit
Offr	Officer
OG	Operational Group
OOC	out of control
Opnl	Operational
Ops	Operations
OR	Operational Requirements
Org	Organisation
OSG	Operational Support Group
OSU	Operational Support Unit
OTS	Officer Training School
OTU	Operational Training Unit
Para	Parachute

Pea	Pearce
Plt Off (P/O)	Pilot Officer
POW	prisoner of war
PRO	Public Record Office
PTS	Primary Training School
QFI	qualified flying instructor
Qld	Queensland
RAAF	Royal Australian Air Force
RAAFM	RAAF Museum
RAAFV	RAAF Vietnam
RAAFSC	RAAF Staff College
RAF	Royal Air Force
RAFA	Royal Australian Field Artillery
RCAF	Royal Canadian Air Force
RDF	Radio Direction Finding
RE	Reconnaissance Experimental
Regt	Regiment
Rep	Representative
Reqs	Requirements
RFC	Royal Flying Corps
Ric	Richmond
RMAF	Royal Malaysian Air Force
RMS	Royal Mail Ship
RSU	Repair and Salvage Unit
S	Southern
SA	South Australia
SA	Southern Area
SAN	School of Air Navigation
SAP	semi-armour piercing
SASO	Senior Air Staff Officer
SD	Stores Depot
SDSC	Strategic Defence Studies Centre
SE	Scouting Experimental
SEATO	South East Asian Treaty Organisation
SFTS	Service Flying Training School
SHQ	Station Headquarters
Sgt	Sergeant
Sig	Signals
SO	Staff Officer
SOA	Staff Officer Administration
SPAD	*Societe anonyme pour l'Aviation et ses Derives*
Sqn	Squadron
Sqn Ldr (S/L)	Squadron Leader
SRO	Senior Recruiting Officer
SS	Steamship
S/Sgt	Staff Sergeant
STT	School of Technical Training
SWPA	South West Pacific Area
TACAN	tactical air navigation
TAF	Tactical Air Force
Tas	Tasmania
TDP	Target Direction Post
TFS	Tactical Fighter Squadron
TFTS	Tactical Fighter Training Squadron
TFW	Tactical Fighter Wing
TG	Training Group
TI	Target Illuminator
Torp	Torpedo
Trg	Training
TRS	Tactical Reconnaissance Squadron
TSF	Transport Support Flight
TU	Transport Unit
UHF	ultra high frequency
UK	United Kingdom
US	United States
USAAC	United States Army Air Corps
USAAF	United States Army Air Force
USAF	United States Air Force
USS	United States Ship/ Submarine
VBM	visual bombing mission
VC	Viet Cong
Vic	Victoria
VNAF	Vietnamese Air Force
W	Western
WA	Western Australia
WAAAF	Women's Australian Auxiliary Air Force
WAG	wireless operator-air gunner
WAGS	Wireless Air Gunners School
WFS	withdrawn from service
Wg	Wing
Wlm	Williamtown
WO	Warrant Officer
w-o	written-off
WOff	Warrant Officer
W/T	wireless telegraphy

PART ONE

CONTROL OF THE AIR

CHAPTER 1
THOSE EARLY DAYS

We flew only at dawn and at sunset, when there was no wind. Our labouring box-kite, capable of only forty-five miles per hour, was provided with no instruments other than a barometer ... The senses took the place of instruments. One's ears did duty as engine counters; the rush of air in the face told whether the climb or glide was at the right angle ...

Tom White[1]

Military aviation in Australia was a natural consequence of aircraft development in the first decade of the twentieth century. Australia's vast distances suited air travel, and the young and developing nation fostered the new air-mindedness. Australia's geographic insulation had not isolated the island continent from the dream that man could fly.

The use of the aeroplane for defence purposes followed. This new technology was recognised as having military application, although the roles initially remained undefined. In September 1909, the Commonwealth Government offered a £5,000 ($10,000) prize for the successful design of a "flying machine ... for military purposes".[2] During 1910, plans for an air arm were submitted to the Department of Defence, and the following year the Defence Minister, Senator G F Pearce, was able to gazette the requirement for personnel to create a service flying school. The formation of a Flying Corps was approved in October 1912.[3]

The site chosen that year for Australia's first military aerodrome was Point Cook, to the west of Melbourne on Port Phillip Bay. Two pilots, Henry Petre[4] and Eric Harrison,[5] who had gained their licences in September 1911 in England, had been appointed to run the flying school. Orders were also placed for training aircraft. A part-time Australian Army officer on attachment in the United Kingdom, Captain Oswald Watt,[6] from the New South Wales Reserve of Officers, inspected British aeroplane manufacturers, and four aircraft were ordered for the Defence Department—two BE.2a tractor biplanes and two Deperdussin tractor monoplanes.

---

1 T W White, *Guests of the Unspeakable*, Little Hills Press, Sydney, 1990, p.13.

2 "Wanted—A Flying Machine For Military Purposes", *RAAF News*, Vol 2, No 4, May 1962.

3 Military Order 570, of 22 Oct 1912.

4 Maj H A Petre DSO MC. CFS; comd 1st Half-Flight AFC 1915–16; No 30 Sqn RFC; comd 29 Trg Sqn/5 Sqn AFC 1917–18. Aviator of Melbourne; b Ingatestone, Essex, Eng, 12 Jun 1884.

5 Gp Capt E Harrison. Comd CFS 1914–18; AIF 1918; later served RAAF. Aviator of Melbourne; b Castlemaine, Vic, 10 Aug 1886.

6 Lt Col W O Watt OBE. NSW Scottish Rifles, French Service d'Aviation Militaire, No 1 Sqn AFC; comd 68 Sqn RFC/2 Sqn AFC 1916–18, 1st Austn Trg Wg 1918–19. Merchant and grazier of Sydney; b. Bournemouth, Eng, 11 Feb 1878.

(Australia's fifth military aeroplane—a Bristol Boxkite pusher biplane—was ordered later in 1912.) Watt reported on 30 August 1912 to the Director of Military Training at Victoria Barracks, Melbourne:

> The arrangements made on behalf of the Corps are in my opinion excellent. Both the Deperdussin monoplane and the Factory (BE) engine-in-front biplane are unsurpassed by any.[7]

Captain Watt's involvement in military aviation at this early stage had stemmed from his personal interest in flying. By obtaining a pilot's licence while on duty in England during 1911,[8] Watt had become the first Australian serviceman to be trained as a pilot. He subsequently bought his own Bleriot XI monoplane, which he flew in Egypt over 1913–14, and then in Paris, where he was working for Bleriot at the outbreak of war.

Lieutenant Eric Harrison prepares to fly a Bristol Boxkite at Point Cook's Central Flying School. RAAF

Meanwhile the rudimentary aeroplanes of Australia's embryo Central Flying School (CFS) were shipped to Sydney in 1913, and stored until the Point Cook[9] school was established. CFS was part of the Army, and its newly appointed personnel

7 AA CRS MP84/1 1954/23/144 letter, of 30 Aug 1912.

8 Royal Aero Club licence No.112, awarded at the Bristol School, Salisbury Plain, on 1 Aug 1911.

9 Named after John Murray Cooke, mate on *HMS Rattlesnake* during 1837 survey of Port Phillip Bay. Admiralty Charts and Land Survey Maps were to erase the "e" from Cooke, and although recommendations were made in 1937 to rectify the error, the base was to remain Point Cook. AA CRS A705/1 24/2/143, of 30 Nov 1937. The site was also referred to as Laverton and Werribee.

formed the Aviation Instructional Staff. The function of CFS was "the training of the personnel of the Australian Flying Corps (Citizen Forces) and Flying Corps Reserve (Permanent and Citizen Forces) ... Four officers of the Permanent Forces and eight officers of the Citizen Forces will be trained annually".[10] The first Australian military aircraft took to the air on 1 March 1914, when Harrison flew the Boxkite, and then Petre a Deperdussin. Point Cook had become the birthplace of Australian air power.

The first training course for CFS commenced on 17 August 1914. The students were four soldiers—Captain Tommy White,[11] and Lieutenants Richard Williams,[12] George Merz[13] and David Manwell.[14] Training was conducted "dual" on the Boxkite, meaning the student leant over the instructor to reach the controls. When the student then had some time in the air, these positions were reversed, until he was assessed as safe to solo. Williams soloed after 2 3/4 hours instruction.

As calm weather was needed to fly the Boxkite, it was the orderly officer's duty to test the air at daybreak. Williams recalled "he did so by holding his handkerchief in the air by one corner; if it did not hang still there was no flying that morning".[15] The course finished that November, the students each completing under eight hours flying time.

With the advent of the Great War in August 1914, the services of the Australian pilots for operations were soon required. On 30 November the first

Oswald Watt, in France at the outbreak of war, flew for the French forces, and is shown here with a Bleriot monoplane at Lesges during February 1915. RAAF

10 Military Order 381, of 14 Jul 1914.

11 Gp Capt Hon Sir Thomas White KBE DFC, 250875. 60th Bn, CFS, 1st Half-Flight AFC; comd No 1 ITS 1940–41, Liaison Offr Trg Cd RAF 1942; Minister for Air & Civil Avn 1949–51; Austn High Commissioner UK 1951–56. Reserve army offr of Melbourne; b Melbourne 26 Apr 1888.

12 AM Sir Richard Williams KBE CB DSO. Army Instr Staff, CFS, No 1 Sqn AFC; comd 67 Sqn RFC/1 Sqn AFC 1917–18, 40 Wg RAF 1918. RAAF CAS 1921–38; AOA Coastal Cd RAF 1939; AOC RAAF O'seas HQ 1941–42; RAAF Rep Joint Chiefs of Staff Wash 1942–46; DG Civil Avn 1946–55. Regular army offr of Adelaide; b Moonta, SA, 3 Aug 1890.

13 Lt G P Merz. MUR, CFS, 1st Half-Flight AFC. Medical Practitioner of Melbourne; b Melbourne 10 Oct 1891. Killed in action 30 Jul 1915.

14 Lt D T W Manwell MBE. 16th Lt Horse, CFS, No 1 Sqn AFC/67 Sqn RFC, SO Equip AFC London 1918. Commission agent; b Queenscliff, Vic, 23 Aug 1890.

15 Air Marshal Sir Richard Williams, *These Are Facts*, AWM, Canberra, 1977, p.26.

Australian aviation unit for active service left Melbourne for German New Guinea.[16] Germany had annexed north-east New Guinea in 1884, and this outpost of the German Empire held the potential for hostilities, bordering to its south the British Empire. The unit was commanded by Harrison, with Merz as the other pilot. They sailed on *HMAS Una* with four mechanics and two aircraft, a BE.2a and Farman Hydroplane, donated by Sydney businessman Lebbeus Hordern. By the end of 1914, all German posts in New Guinea had been occupied, and German New Guinea was placed under an Australian military administration. Although the detachment had sailed to Wilhelmshafen (now Madang) and Petershafen, on Witu Island north of New Britain, the aircraft remained unpacked, and arrived back from Rabaul in early 1915.[17]

Following this first expedition, a cable message received from the Viceroy of India on 8 February 1915 requested Australian air assistance further afield:

> Could you provide any trained aviators for service in Tigris Valley? All our trained officers are in Egypt and England.[18]

The request amplified that aircraft could be provided from England. So, on 20 April 1915, one half of a flight, known as the 1st Half-Flight Australian Flying Corps—consisting of four officers, 42 other ranks, and no aircraft—was sent to serve in Mesopotamia (now Iraq). Sailing on the *RMS Morea* to Bombay, and then on the *SS Bankura* to the Persian Gulf, the Australians arrived at Basra on 25 May.[19] Under the command of Captain Petre, the other pilots were White and Merz, from the first CFS course, and Lieutenant William Treloar,[20] who had trained in England before the war. They were all members of the Australian Flying Corps (AFC) and, being on operational service, were part of the Australian Imperial Force (AIF). Later, a further 10 groundcrew were to sail from Sydney on the *RMS Persia* on 10 August 1915, to join up with the original group.[21]

In Mesopotamia the 1st Half-Flight was absorbed into the Indian forces, and flew Maurice Farman Shorthorn and Caudron G.III biplanes on reconnaissance missions. Although the aircraft were unarmed, the observer could drop small bombs from the cockpit by hand. On one sortie in July 1915, Merz and his New Zealand observer, Lieutenant William Burn, forced-landed. They were attacked by Arabs and killed. By the end of the year both White and Treloar had been captured, and Petre continued flying in 1916 with the Royal Flying Corps (RFC). The remaining mechanics who had survived death or capture eventually reached the AIF in Egypt.

---

16 Gp Capt K Isaacs, *Military Aircraft of Australia 1909–1918,* AWM, Canberra, 1971, p.17.

17 AA CRS A2023/1 items A38/3/130, of Dec 1914, and A38/3/128, of Dec 1914 and Jan 1915, cite the movements of the New Guinea expedition.

18 F M Cutlack, *The Official History of Australia in the War of 1914–1918, Vol VIII, The Australian Flying Corps,* AWM, Canberra, 1923, p.422.

19 AWM8 Embarkation Roll 8/15/1, of 20 Apr 1915; and AWM224 MSS 513 255/3/39a, of 18 Jul 1919.

20 Lieut W H Treloar. 70th Bn, 1st Half-Flight AFC. Motor mechanic of Melbourne; b Melbourne 8 Aug 1889.

21 AWM8 Embarkation Roll 34/13/1, of 10 Aug 1915.

The first pilot training course at Richmond in August 1916, in front of a Curtiss JN3 Jenny trainer: Front Row: N Clark, G Taylor, possibly L M Samson, Sutherland, R Smallwood, N B Love, Williams, Burgess, R L Clark, A Gregory. Rear Row: C Dagg, J R Faviell, J H Weingarth, J Francis, unknown, W Stutt, D R Williams, B F Weaver, C R Burton, Cleary, E Coleman, G H Potts. H Chester, engineer, standing on plane at rear J LEEUWIN-CLARK

A student from the first Richmond course, Robert Clark, in the cockpit of a Curtiss Jenny trainer J LEEUWIN-CLARK

Later a second half-flight was offered to India. Lieutenant Williams would have commanded this unit, but it was not required. After the disasters that had befallen the first contingent, Williams felt fortunate that the offer had not been accepted.[22]

While the primary activity associated with military aviation in Australia was centred on Point Cook, pilot training was also being conducted at Richmond, in New South Wales. Here the New South Wales Government had established the New South Wales State Aviation School in 1916 under Billy Stutt.[23] Stutt, trained in England before the war, served as the chief instructor at Richmond, and in November 1917 he was to fly one of the school's Curtiss Jenny biplanes to Point Cook. This was, most probably, the first Sydney to Melbourne flight. He was later to join the staff at CFS. The first course commenced at Richmond in September 1916, and 19 pilots graduated in December—10 joined the AFC, one remained on staff to instruct, and the others were accepted into Britain's Royal Flying Corps. The types of machines flown at the school were the French Caudron and the American Curtiss Jenny. During 1917 the New South Wales Department of Education took responsibility for the school to widen its scope for "general after-the-war purposes".[24] By the end of the war, six courses had been completed.[25]

This CFS Shorthorn trainer, known as the "Rumpety", was extensively damaged in a crash near Point Cook in 1917. RAAF

22 Williams, p.39.

23 Capt W J Stutt. RFC; comd NSW Avn School 1916–19, ARS Austn Air Corps 1919–20. Aviator of Melbourne; b Melbourne 11 Dec 1889. Killed in flying accident 23 Sep 1920.

24 "State Aviation School", *The Sydney Morning Herald*, 20 Jul 1917.

25 D Roylance, *Air Base Richmond*, RAAF Base Richmond, 1991, p.12.

The output of students at CFS also built up, with over 20 students starting on both the Sixth Course in 1916 and the Seventh Course in 1917.[26] There was to be a total of 11 courses at CFS during the war. The earlier aircraft were supplemented by a French Bleriot and Caudron, and later by the British FE-2b, Bristol Scout and Grahame-White Boxkite types. More substantial numbers of each aeroplane type were not operated until Maurice Farman Shorthorns, and then DH-6s, were received in limited quantities.

The decision to constitute a complete flying unit in Australia was made in late 1915. Britain had previously requested personnel for service in the Royal Flying Corps, but on 24 November revised this by stating that no further groundcrew ("mechanics") could be accepted. However,

> Above does not apply to men joining any complete aviation unit raised and paid for by your Government.[27]

On 27 December, the Australian Government advised its intention "to organise a squadron, flying corps, 28 officers and 181 other ranks, for dispatch in February".[28] No 1 Squadron AFC was duly formed in January 1916, and embarked on 16 March to Egypt.

Australia thereby became the first dominion to form a flying corps and dispatch flying units overseas. No 1 Squadron was to fly with distinction in the Middle East for the remainder of World War I. But more importantly to this story, it was to give birth to a second unit—No 2 Squadron.

26 Cutlack, p.426.

27 Cutlack, p.423.

28 Cutlack, p.423.

CHAPTER 2

No 68 (Australian) Squadron

The whole value of the RFC lies in its cooperation with the other arms and in the assistance it can give to the conduct of operations on the ground. Fighting in the air is not an end in itself, but under present conditions it is necessary to enable this assistance to be given.

Policy in the Air[1]

The first Australian squadron joined the Egyptian Expeditionary Force on arrival in Egypt on 14 April 1916. After training with the RFC and equipping with British aircraft, No 1 Squadron AFC flew reconnaissance sorties in support of the Army defending the Suez Canal. This was the vital link in the line of communications for the British Empire. The AFC unit then went through a confusing change of identity as it was briefly referred to as No 68 (Australian) Squadron, Royal Flying Corps.

The raising of another Australian flying unit was then proposed. On 27 July 1916 the Department of the Navy, in reply to a request from the Chief of General Staff in Melbourne, advised that transport for the dispatch of a flying unit of 13 officers and 186 other ranks could be arranged for November. But at this stage it was unsure whether the new squadron would be required in England or Egypt:

> If the destination is the United Kingdom, arrangements could probably be made to dispatch the unit complete.
>
> If the troops are to be forwarded to Egypt, it would be necessary for the majority of the officers to proceed by mail steamer. It is understood that a number of the rank and file could be despatched to Egypt in October if required, and also, that only personal equipment will be taken by the unit.[2]

As these plans were being made for a new squadron to be shipped from Australia, plans were also afoot to prepare a further unit in Egypt for service on the Western Front. In London the War Office advised in early September its intention of forming the new unit, No 67 Squadron, in Egypt during November 1916.[3] The date of formation was soon brought forward, as authority was received on 15 September to form this second squadron:

> From: WAR OFFICE. LONDON.
> To: Aeronautics, Egypforce.
> 22842. CIPHER A.0.1. 15/9/16
>
> Your A.M.958. Formation of Australian Squadron RFC is authorised. Will be known as 67th (Australian) Sqdn.[4]

---

1 AWM25 81/24 OB/1837 "Policy in the Air", HQ 1 ANZAC 16/187, undated.

2 AA CRS A2023/1 A38/8/816 Navy Office M.T.O. 2408, of 27 Jul 1916.

3 PRO AIR1/2086/207/6/1 "Air Org Memos Jun 1916–Oct 1918" War Office AO/173, of 5 Sep 1916.

4 AWM22 AIF HQ 31/2/2005 War Office 22842 Cipher A.0.1, of 15 Sep 1916.

Five days later the War Office advised the RFC (most probably for reasons of seniority in squadron numbering) that the new unit should instead be No 68 Squadron:

Air Board
GOC, RFC in the Field
GOC, RFC Middle East Brigade

. . . .

The following notification is issued concerning the Australian Squadrons:-
1. The present Australian Squadron in Egypt, known as No 68 (Aus) Squadron will in future be:-

No 67 (Aus) Squadron.

2. A second Australian Squadron, to be composed of personnel recruited from other Australian units serving in the East, has been authorised, and, when formed, will be known as:-

No 68 (Aus) Squadron.

B C H Drew
Major
General Staff
for Director of Air Organisation

War Office
20.9.16[5]

The confusion created by the War Office was not to end there. In December it advised that No 67 Squadron, referring to the newly formed second Australian unit, would become a fighter squadron at Harlaxton in England, with its first three aircraft planned for delivery on 15 February 1917.[6] Apparently the edict of this unit being No 68 Squadron had temporarily been forgotten.

So, surrounded by some confusion, the second Australian squadron—No 68—was formed, at Kantara in Egypt, from personnel from No 67 (Australian) Squadron and the Light Horse Regiments of the AIF, on Wednesday 20 September 1916.[7] Extra personnel were requested from Australia, and headquarters in Melbourne advised the AIF HQ in Egypt that 14 air mechanics would embark on the *Barunga* on 20 October.[8] The main shortfall of ground staff remained with wireless operators, illustrated by this signal from London in December:

From: Troopers London
To: Perardua Cairo
No 19862 A.O.2 14 December 1916
Your T.605, wire as soon as possible exact date you expect mechanics to complete 68 Sqn would be obtained from Anzac Mounted Division as it is proposed to complete this Squadron as early as possible. If you cannot obtain these men, arrangements will be made here; we will collect 26 Wireless Operators, 10 riggers, 4 photographers from Australian Forces in England.[9]

5 PRO AIR1/2086/207/6/1 War Office AO/199, of 20 Sep 1916, and AWM22 AIF HQ 31/2/2005, received 6 Oct 1916.

6 PRO AIR1/2086/207/6/1 War Office AO/279, of 22 Dec 1916.

7 AWM224 MSS 517 Part 1.

8 AWM22 AIF HQ 31/2/2005 Signal WY 75, of 13 Oct 1916.

9 AWM22 AIF HQ 31/2/2005 Signal 19862, of 14 Dec 1916.

The Commanding Officer of No 68 (Australian) Squadron was Captain Oswald Watt. Having flown with the French *Service d'Aviation Militaire* since the outbreak of war in 1914, "Toby" Watt had gained combat experience in the air that no other Australian shared. Watt had initially flown Bleriot single-seat monoplanes, which were soon found to be unsuitable for active service and considered bad machines for war. If left in the open overnight, the fabric would soon become flabby, with a consequent loss of performance and manoeuvrability. By the end of 1915, the unarmed Bleriot was, for all active service purposes, a relic of the past.[10]

Oswald Watt, centre, with his French Shorthorn, while flying with the Service d'Aviation Militaire, *marked appropriately with a kangaroo and "Advance Australia"*

RAAF

Soon Watt had advanced to flying the pusher Maurice Farman Shorthorn, with an observer standing in the front armed with a rifle to fire at the enemy.[11] Watt named his two-seater "Advance Australia". During 1915, the French had awarded him both the *Legion d' Honneur* and the *Croix de Guerre*. However, as Watt was not a French citizen, he had been unable to hold a command, so he transferred back to the Australian forces in 1916. He joined No 1 Squadron as the "B" Flight Commander in Egypt during May.[12]

The confusion of the squadron numbering of the new unit had not been restricted only to the Royal Flying Corps. Australia contributed by forming another "second Australian squadron" at Point Cook. A draft Military Order from the General Staff in Melbourne in August 1916 had stated:

AUSTRALIAN MILITARY FORCE
No 2 Flying Squadron

1. Approval has been given for the raising and dispatch of No.2 Australian Flying Squadron for active service abroad. NCOs will only be appointed temporarily for the voyage.
2. Table showing allotment and distribution of this unit is issued separately.
CGS
No 412/1916
31/8/16[13]

10 PRO AIR1/727/152/6 "The Bleriot Machines 1906–1915", p.5.

11 C G Grey, "Oswald Watt the Airman", in S Ure Smith, B Stevens & E Watt, *Oswald Watt*, Art in Australia Ltd, Sydney, 1921, p.29.

12 Cutlack, p.41; Williams, p.43.

13 AA CRS A2023/1 A38/8/857 CGS No 412/1916, of 31 Aug 1916.

Consequently, the eleven graduates of the Fifth Course at CFS, conducted from 1 August to 1 October 1916, were appointed as Second Lieutenants in No 2 Squadron Flying Corps on 1 October.[14] One of these pilots, Lieutenant Henry Wrigley,[15] related that:

> The Squadron was assembled and organised in the AIF camp adjacent to CFS at Point Cook and, by early October, issues of personal equipment had been completed and personnel were granted leave prior to embarkation, which was provisionally fixed for the end of the month.[16]

The Squadron marched out of Point Cook for Melbourne on 25 October, proceeded by road and rail to Port Melbourne, and departed on the *SS Ulysses,* bound for England. This "No 2 Squadron" anchored in Devonport Harbour, near Plymouth, on 28 December. The unit was subsequently designated No 69 (Australian) Squadron, and ultimately became No 3 Squadron AFC in 1918.

The confusion of this "other No 2 Squadron" has confounded historians since. Even Cutlack, the official historian of the AFC in World War I, got this wrong[17] and the No 2 Squadron Battle Honours board recorded the Squadron as being formed under Watt at Point Cook!

Table 2-1: Formation of the Operational Squadrons of the Australian Flying Corps

AFC Squadron	*Formation*	*RFC Title*	*Implementation on Unit of Authority of 4 Jan 1918*
No 1 Squadron	1st Squadron at Point Cook 6 Jan 1916	68 Sqn RFC 14 Sep 1916, 67 Sqn RFC 20 Sep 1916	No 1 Sqn AFC 13 Feb 1918
No 2 Squadron	2nd Squadron at Kantara, Egypt 20 Sep 1916	67 Sqn RFC 20 Sep 1916, 68 Sqn RFC 20 Sep 1916	No 2 Sqn AFC 15 Jan 1918
No 3 Squadron	2nd Squadron at Point Cook 19 Sep 1916	69 Sqn RFC 28 Dec 1916	No 3 Sqn AFC 20 Jan 1918
No 4 Squadron	4th Squadron at Point Cook 25 Oct 1916	71 Sqn RFC 27 Mar 1917	No 4 Sqn AFC 18 Jan 1918

14 AA CRS A2023/1 A38/8/969 Draft Military Order CFS 5th Course of Instruction; AA CRS A2023/1 A38/8/969 DOD Minute "Appointment of Officers to No 2 Squadron AFC" of 25 Sep 1916: AA CRS 2023/1 A38/8/969 DOD letter CFS.425, of 20 Sep 1916.

15 AVM H N Wrigley CBE DFC AFC. CFS, No 2 Sqn AFC, 69 Sqn RFC/3 Sqn AFC; comd 1 Trg Gp 1939–40; AOC HQSA 1940; AMP 1940–42; AOC O'seas HQ 1942–46. Regular army offr of Melbourne; b Melbourne 21 Apr 1892.

16 Wg Cdr H N Wrigley, *The Battle Below*, Errol G Knox, Sydney, 1935, p.20.

17 Cutlack, p.424.

While most of the tradesmen had been transferred from No 67 Squadron, other candidates for air mechanics and pilots had been recruited from the Light Horse. The methods used by Watt in obtaining applicants was somewhat unorthodox. One recruit was 2nd class Air Mechanic Fergus Cox.[18] He recollected that AFC personnel in three Leyland trucks arrived at the camp calling out for "any trades, any mechanics, any engineers". When queried if the volunteers should be processed through the orderly room, the response was: "No, we have to snatch them."[19]

Private Fergus Cox, of the 1st Light Horse Regiment, prior to joining No 68 Squadron in Egypt GARTH COX

The trade volunteers—and there was no shortage from the Light Horse—were tested at No 67 Squadron; those selected were absorbed into the new Squadron at the Kantara camp in Alexandria, those not required were returned to their units. The new Flying Corps held a fascination to those in the Light Horse. It was, perhaps, as 67 Squadron pilot L W Sutherland explained:

> Aeroplanes are like horses. Some are docile, reliable, well-mouthed, and comfortable to ride. Others, like polo ponies, are sharp and snappy on the turns ... lastly, the outlaw, which fights man's mastery up to the last; in its record there is sure to be at least one victory, involving maiming or death for the vanquished.[20]

Forming the nucleus of groundcrew for the new unit were 13 of the 18 AFC mechanics who had survived from the 1st Half-Flight in Mesopotamia.[21] Among this

---

18 2A/M F R Cox, A.1544. 11th Lt Horse, No 68 Sqn RFC/2 Sqn AFC. Farm hand of Oakey, Qld; b Toowoomba, Qld, 17 Oct 1895.

19 F R Cox, The First World War Memoirs of Fergus Cox, 2 Squadron AFC, unpublished manuscript courtesy of Garth Cox and the Museum of Army Aviation and Flying, Oakey, p.2.

20 L W Sutherland, *Aces and Kings,* Hamilton, London, p.15.

21 No 68 Sqn Routine Orders No.1, of 25 Nov 1916; and AWM22 31/4/2004, of 10 May 1917.
The following members of the Mesopotamian Half-Flight were absorbed into No 68 Squadron:

No	Rank	Name	Trade	No	Rank	Name	Trade
3	Sgt Maj	Garling S W	Sgt Maj	2	S/Sgt	Heath C V	Technical WO
43	F/Sgt	Mackinolty G J W	Rigger	14	F/Sgt	Abdy C E	Rigger
6	F/Sgt	Wardell C	Carpenter	8	Sgt	Cowper G H	Driver MT
17	Sgt	Clayton S G	Rigger	13	Cpl	Bissett H	Driver MT
20	Cpl	Lonsdale R	Carpenter	46	A/Sgt	Dobney W E	Fitter & T
41	2/AM	Lewis O	Batman	40	2/AM	Carvell H	W/T Operator
439	2/AM	Anlezark J	Batman				

TELEPHONE No. W'mstown.290

R.A.

COMMONWEALTH OF AUSTRALIA.

DEPARTMENT OF DEFENCE.

CENTRAL FLYING SCHOOL, LAVERTON,

No. CFS.425 20th. September, 1916.

Secretary,

M.O.3

In reply to your 77822 of 9th. instant relative to officers attending 5th. Course of Instruction recommended for appointment as 2nd. Lieutenants in No. 2 Squadron A.F.C., the following is the order of merit based upon six weekly progress examinations in theoretical subjects, practical flying and apparent capabilities as officers :-

1. Lieut. Griggs ✓
2. ,, Kitson ✓
3. ,, Morrison ✓
4. ,, Garrett ✓
5. ,, Mowle ✓
6. ,, Wrigley ✓
7. ,, Petschler ✓
8. ,, Tregilles ✓
9. ,, Murray
10. ,, Walker ✓
11. ,, Taylor ✓

Note. – Exclusive of Lt. Sleeman who headed the list but is selected by Capt. Harrison for assistant Instructor vide C.F.S. 426 of 20/9/16 L.P.

The necessary statements in writing and authorized forms A.22 are enclosed, and the appointment of these officers to No. 2 Squadron A.F.C. is recommended.

E Harrison
Captain,
O.C., C.F.S.

C.F.S. 20/9/16

CFS letter of 20 September 1916 RAAF

team was Flight Sergeant George Mackinolty,[22] a rigger, who was appointed the Flight Sergeant of "A" Flight. He was later to join the stores and accounting branch of the new RAAF, and rose to the rank of Air Vice Marshal as the Air Member for Supply and Equipment.

When complete, No 68 Squadron under Major Watt, who had been promoted on 23 December,[23] proceeded to Alexandria to board ship for England. The Squadron departed without pilots or observers[24]—they would be trained in Britain—and set sail on the *Kingstonian* on Saturday 13 January 1917, berthing at Malta, and then arriving in Marseilles Harbour on 24 January.

George Mackinolty, in CFS uniform at Point Cook, saw overseas service with the Mesopotamian Half-Flight before becoming a founding member of No 68 Squadron. RAAF

The journey just to Malta was not without incident, with two ships in the convoy being torpedoed. Disembarking in Marseilles, the Australians boarded the crowded train for a three-day trip north through France, circumventing Paris and disembarking at Le Havre to await shipping for England. The Australians from the desert could not believe the cold. 1st Class Air Mechanic Hubert Billings,[25] a wireless mechanic recruit from the Light Horse, recalled that icicles were hanging under the engine where leaking water had frozen.[26] At Le Havre personnel were billeted in tents which offered little protection from the freezing conditions. "I got a tin of hot water from the cook for shaving, and ice was forming on the sides before I finished".[27] Fergus Cox recalled: "it was the coldest night for 80 years" and two of their British Army sentries died—"frozen on guard duty".[28] 2nd Class Air Mechanic Leslie Ward,[29] another

22 AVM G J W Mackinolty OBE. 1st Half-Flight AFC, No 68 Sqn RFC/2 Sqn AFC; Director of Supply 1940–42; AMSE 1942–50. Of Melbourne; b Leongatha, Vic, 24 Mar 1895.

23 Dept of the Army 706/R2/14, of 29 Dec 1966.

24 Williams, p.54.

25 1 A/M H D Billings A.14. 1st Lt Horse Signals, No 68 Sqn RFC/2 Sqn AFC. Railways clerk of Melbourne; b Avenel, Vic, 29 Oct 1894.

26 H D Billings, *My Own Story*, Glen Iris, 29 Oct 1989, p.68, held at RAAF Museum, Point Cook.

27 Billings, p.68.

28 Cox, p.6.

29 Lt L N Ward. 3rd & 1st Lt Horse, No 68 Sqn RFC. Clerk of Walkerville, SA; b Walkerville 29 Jan 1893. POW 20 Nov 1917.

recruit from the Light Horse who was to become a pilot on the Squadron, spent his birthday here and recorded: "we stayed and shivered all day at this camp".[30]

From this makeshift camp, No 68 Squadron sailed for Southampton on the *Donegal*, arriving at daybreak on Tuesday 30 January 1917. The Australians caught a train to London, and then another to Grantham in Lincolnshire, arriving that night. Here No 68 Squadron was to be based at the Royal Flying Corps aerodrome at Harlaxton, three kilometres south-west of Grantham, near the Harlaxton Manor. The Australians were to share this aerodrome with No 44 (Reserve) Squadron. Both units were part of the 24th Training Wing, headquartered at nearby Spittlegate.[31]

30 AWM PR 83/230 Personal Diary of 2/AM Ward, of 29 Jan 1917.

31 The 24th Wing had been formed on 25 Sep 1916 and, in addition to No 68 Sqn, consisted of HQ and No 49 (Reserve) Sqn at Spittlegate, No 31 (R) Sqn and No 65 Sqn at Wyton, and No 44 (R) Sqn at Harlaxton. PRO AIR1/2086/207/6/1 AO/187, of 13 Sep 1916, and AO/280, of 29 Dec 1916.

ROUTINE ORDERS
by
Capt. W.O.Watt, Commanding No. 68 (Australian) Squadron R.F.C.
KANTARA Saturday 25th November 1916. No. 1.

---

1. MOVEMENTS 2/Lieut C.De.C. Matulich, 2/Lieut. A.O. McPhillamy and 2/Lieut N. Steele proceeded to CAIRO on leave for 2 days 24.11.'16.

PART II.

1. APPOINTMENT (Extract from 5th Wing R.O. 329 of 28.10.'16)
Capt. W.O. Watt, No. 67 (Australian) Squadron, R.F.C. as acting Squadron Commander 27.10.'16.

2. POSTINGS The following Officers, NCOs and men have been posted to No. 68 (Australian) Squadron, RFC on dates shown, and are taken on the strength accordingly:-

2/Lieut C.De.C. Matulich	24.11.'16.
2/Lieut A.O. McPhillamy	24.11.'16.
2/Lieut N. Steele	24.11.'16.

(Authority: A & NZ.M'td.Div.R.O. No.2 of 21.11.'16)

No. 235 A/F.Sgt Reid E.F.	22.9.'16.

(Authority: No. 67 (Australian) Squadron R.O. No. 218)

No. 462	2/A.M.	Jenyns R.	20.9.'16.
No. 464	"	Mitchell R.	"
No. 466	"	McLeod R.	"
No. 447	"	Arnold A.	"
No. 458	"	Hansen H.	"
No. 465	"	McRae J.K.	"
No. 467	"	Roberts F.L.	"
No. 469	"	Pound W.J.	"

(Authority: No.67 (Australian) Sqn R.O.215,20/9/16)

No. A 45	Cpl	Dobney W.E.	10.10.'16.

(Authority: M.E.B R.F.C Memo No.B/20a of 8.10.'16)

No. A 9	Sgt	Head	15.10.'16.

(Authority: M.E.B. B/33a of 15.10.'16)

No. PA.619	2/A.M.	Collins H.E.	8.11.'16.
No. PA.2465	"	Eaton J.W.	"
No. PA.1512	"	Scott J.E.	"

(Authority: No. 67 (Australian) Squadron R.O.260)

No. A 2	S/Sgt.	Heath C.V	3.11.'16.
No. A 8	Sgt.	Cowper G.H.	"
No. A 17	Cpl.	Clayton S.G.	"
No. A 41	2/A.M.	Lewis O.	"
No. A 20	Cpl.	Lonsdale R.	"
No. A 43	F/Sgt.	Makinolty G.	30.11.'16.
No. A 6	Sgt.	Wardell C.	"
No. A 13	Cpl.	Bissett H.	"
No. A 40	2/A.M.	Carvell H.	"

(Authority: 5th Wing R.O. No. 345 of 13.11.'16)

No. A 393	2/A.M.	Moore N.	9.11.'16.

(Authority: 5th Wing R.O. No. 346 of 14.11.'16)

No. 2415	Tpr.	Higgins E.C.	21.11.'16.

(Authority: A.I.F. 307/17/2020 of 21.11.'16)

(Signed) W.O.Watt, Capt.,
Commanding No.68 (Australian) Squadron,
Royal Flying Corps.

CHAPTER 3

To the Battlefield

It must seem strange that the sky, too, is about to become another battlefield no less important than the battlefields on land and sea. But from now on we had better get accustomed to this idea and prepare ourselves for the new conflicts to come ...

Major Giulio Douhet, 1909[1]

No 68 Squadron was now to begin preparing for its role as a "service" squadron—a unit at the Front. It would be trained as a scout squadron and support the ground forces in fighting for control of the air over the battlefield. For its groundcrew, this meant adapting their technical expertise, mainly gained before joining the Light Horse, to the new technologies of aviation. For the pilots, many of whom were recruited from horsemen, this involved learning to fly.

Technical training for the Australian mechanics was conducted at different RFC depots. Some were detached to Catterick, near York, for training with the 8th Training Wing on aeroplane rotary engines, and before returning to Harlaxton more advanced tuition was given at different factories. Manufacturers, such as Rolls-Royce and Smiths Instruments, hosted the Australians and were impressed by the standard and eagerness of the students.

Trainee pilots receiving technical instruction at one of the RFC's Schools of Military Aeronautics.
RAAF

1 E M Emme, *The Impact of Air Power*, Van Nostrand, Princeton, 1959, p.5.

For those who had been selected as pilots, training was to commence at the RFC's Schools of Military Aeronautics at Oxford and Reading. Here a month's technical groundschool was undertaken before selection for preliminary flying training. 2nd Class Air Mechanic Leslie Ward was taken by surprise by his sudden selection for pilot training, as his diary indicated:

> Thursday 15 March. Waiting on parade this morning, two of us were suddenly ordered to pack up and proceed to Reading on a course of Aeronautics, in other words to be trained as pilots. We were almost stunned as it was more than we could realise, however, we were soon packed.[2]

The Aeronautics course covered the technical aspects of aeroplane engines and rigging, the theory of flight and map reading. Specialist training was given in wireless signalling and receiving, the care of machine-guns, the operation of aerial cameras and, by the use of models, the observation of artillery fire.[3]

RFC riggers undergoing training on a BE-2 airframe RAAF

Basic flying training at this stage was normally conducted on the Maurice Farman Shorthorn, known as the "Rumpety", with advanced training on the Avro 504. Trainees went to RFC aerodromes around Lincoln for their instruction before returning to Harlaxton. Tuition here with the 24th Training Wing continued on the Avro 504 and the Sopwith Pup. Also flown by the Australians was the Sopwith Trainer, known as the "1 ½ Strutter" because of its splayed centre-section wing struts. The Strutter had, by this stage, been relegated to the training role from the Western Front, where in 1916 it had been the first RFC aeroplane to be fitted with machine-gun interrupter gear.[4] It had, therefore, been considered as the first real British fighting machine.[5]

Avro 504s comprised 68 Squadron's "A" Flight, and "B" Flight consisted of two Pups and four Strutters.[6] This latter type was adorned under Watt's guidance with kangaroo markings, similar to that which had adorned his Maurice Farman in

2 AWM PR 83/230 Personal diary of 2AM Ward, 15 Mar 1917.

3 "Training the Military Flyer", *Flying*, Aeronautical Society, London, 14 Feb 1917, p.87.

4 O G Thetford & E J Riding, *Aircraft of the 1914–1918 War*, Harleyford Publications, Marlow, 1954, p.34.

5 PRO AIR1/2301/215/5 "Arming of Aircraft", paper prepared by Capt Batsford, Mar 1919, p.2.

6 A H Cobby, *High Adventure*, Kookaburra, Melbourne, 1981, pp.27, 29.

France, to promote the new Squadron's Australian identity. "C" Flight flew Pups and Avros.

The key to the control of the air was fast, armed fighting scouts flown by good pilots. Up until December 1916, the total flying experience of a replacement pilot arriving in France had averaged 15 hours. Instruction was then increased to 22 hours, but the heavy losses to the RFC during the "Bloody April" of 1917 forced a cut back to 17 hours.[7] The typical standard of training was described by the famous Australian aviator P G Taylor. He met his instructor immediately before his first flight and received no briefing. The instructor:

> ... walked off towards a Maurice Farman standing on the tarmac and I followed him in a somewhat confused state of mind. This wasn't exactly what I had expected. I had pictured some sort of orderly approach; a talk with the instructor before my flight, and some idea of the theory behind controlling an aeroplane. Instead, I climbed up after my instructor into the rear seat ...
>
> The engine was throttled back, the nose went down, and a shout came from the figure in the front seat, "You land it!" The shock of this remark caused an immediate physical reaction. My hands shot out to the scissors handles in front of me and my feet onto the rudder bar on the floor. I imagined that unless I took the controls instantly the aeroplane might fall out of control to the ground. I knew nothing. Nothing beyond the absurdity of the demand that I should land the aeroplane. I checked my panic. The instructor was still in the aeroplane with me. Even if he were mad enough to expect me to land it without instruction he would probably be sane enough to stop me crashing it ... The engine came on, the controls were snatched out of my hands and a savage shout came back to me from the instructor, "Bloody awful!".[8]

Fortunately, this standard of instruction was not to persist; most probably the RFC had lost too many students for that. Pilot training was then to become systemised and comprehensive through the efforts of Major Robert Smith-Barry[9] who established the flying school at Gosport, near Portsmouth in Hampshire. Under the commander of Training Division, Major-General John Salmond,[10] the Gosport school developed into the School of Special Flying.[11] Here instructors were taught how to teach, and aerobatic manoeuvres became part of the basic pilot training curriculum. This concentrated training program resulted in pilots going to France by 1918 with an average of 50 hours instruction.[12] "Smith-B", as he was always called, was to ensure that the "Gosport System" of training would ultimately form the basis of flying instruction as we know it today.

---

7 M Paris, *Winged Warfare*, Manchester University Press, Manchester, 1992, p.219.

8 Sir Gordon Taylor, *Sopwith Scout 7309*, Cassell, London, 1968, pp.17–19.

9 Lt Col R R Smith-Barry. No 2 Sqn RFC, 5 Sqn, 60 Sqn; comd 60 Sqn 1916, 1 Reserve Sqn 1917, School of Special Flying 1917–18. Officer of RFC Special Reserve of Stowell Park, Wiltshire, Eng; b London 4 Apr 1886.

10 Marshal of RAF Sir John Salmond GCB CMG CVO DSO. Comd RFC and RAF in Field 1918–19; CAS 1930–33; D Armt Prodn, and DG Flg Control and ASR 1939–45. Of London; b 17 Jul 1881.

11 H A Jones, *The War in The Air*, Vol V, Clarendon Press, Oxford, 1935, pp.429–32.

12 Paris, pp.220–1.

One pilot to train with No 68 Squadron at Harlaxton was the future ace, Lieutenant Harry Cobby.[13] A graduate of the Sixth Course at CFS, Cobby had sailed with No 4 Squadron, arriving in England in March 1917. He was to become the highest scoring pilot of the Australian Flying Corps. Cobby had soloed on the Shorthorn, and then flew the Avro 504 on 68 Squadron.

> There was a vast difference in the manner in which we were treated at Harlaxton, to that meted out to us at Royal Flying Corps schools. We were Australians with Australians and no longer gentlemen visitors from the Antipodes and instruction and comment was direct and to the point. The senior instructors were Guilfoyle, Muir and Matthews and they just put us through the hoops. We were started off on Avros, something in our experience that was more like a proper aeroplane than anything we had touched. There were no frills about the place at all. A few hours with an instructor, a few "straights", that is a fast taxy and then lift the machine off and put it down again, then one dual flight right away, and you went "solo". Not only did we have to follow a fixed sequence of movements in the air, in sight of our instructors on the ground, but we were all under a moral obligation to loop on our first solo. Somebody had started the habit some little time before and although nothing was ever said until after you had done it, even amongst the pupils, it had to be gone through otherwise you lost caste. It did not matter how many times you looped the loop later on, it had to be done on the first solo.[14]

At the beginning of the war, in 1914, most aircraft had been two-seaters, as the role of the aeroplane had been observation and reconnaissance. Information was needed about the enemy army, its disposition and strength, and its movements. Aircraft crews were the "eyes of the army"—mapping trench systems, reconnoitring the battle area and directing artillery barrages. Machine-guns were not carried as their weight restricted the rate of climb. But it soon became evident that the air force with the best scouts, or fighters, could control the air for its own observation machines.

Pusher scouts, which were developed for fighter squadrons in 1915, had soon become obsolete. The pusher normally had a crew of two—a pilot, and an observer in the front cockpit to shoot a machine-gun. This, however, had a restrictive field of fire, and the crew had blind spots which covered an approaching enemy. The replacement of the pusher was, therefore, not due solely to its relatively poor performance, but to the adoption of the synchronised machine-gun to fire forward through the propeller arc.[15] This innovation had been pioneered by the French in February 1915. By fitting metal plates to the wooden propeller blades, bullets fired by a forward-firing machine-gun could be deflected.[16] But when this aircraft was forced down behind German lines in April, the idea was effectively developed by Fokker in the form of an interrupter gear.[17] This was a synchronised mechanical device which stopped firing

13 AirCdre A H Cobby CBE DSO DFC* GM. 47th Bn, CFS, No 4 Sqn AFC, 68 Sqn RFC, 71 Sqn RFC/ 4 Sqn AFC, 5 Sqn; AOC NE Area 1942, Comdt RAAF Staff School 1943–44; AOC 10 Gp 1944; AOC 1st TAF 1944–45. Regular army offr of Melbourne; b Melbourne 26 Aug 1894.

14 Cobby, op cit, pp.28–9.

15 "The DH-5", *Flight*, 9 Jan 1919, London.

16 E H Sims, *Fighter Tactics and Strategy 1914–1970*, Cassell, London, 1972, p.9.

17 B Robertson, *Air Aces of the 1914–18 War*, Harleyford, Letchworth, Hertfordshire, 1964, p.15.

when the blade obstructed the path of fire. The new design flew in the Fokker E-1 *Eindekker* (itself modelled on the French Morane-Saulnier monoplane) and in turn was captured by the Allies during the summer of 1915.

The British initially copied Fokker's system with the Scarff-Dibrovski cam interrupter. Then, with the assistance of a Romanian engineer, Georges Constantinesco, the Allies were able to develop a gun which fired efficiently through the propeller by way of an hydraulic system. The advantage of this system over mechanical gears was that it was easy to adapt to all classes of engines and aircraft. The connection between the generator on the engine and the trigger motor on the gun was by means of a simple pipe instead of complicated push rods and bell cranks. This enabled a fairly constant and high rate of fire.[18]

The "Fokker Scourge" — the Fokker E-1 Eindecker *monoplane with the forward-firing machine-gun* *RAAF*

As the Germans had used their development to gain control of the air during this "Fokker Scourge", in turn the Allies by 1916 were now able to gain air superiority. Gun synchronisation—"gear that synchronised miraculously the absence of a blade with the presence of a bullet"[19]—in addition to a scout's speed and manoeuvrability, was of major importance. The resultant tactic of getting onto an adversary's tail, with minimum deflection or angle-off, is one which has remained to this day. To use this to advantage, a scout pilot could use superior height for concealment in the sun, and then pounce to achieve success by accurate close-in shooting.

18 PRO AIR1/2301/215/5, p.5.

19 V M Yeates, *Winged Victory*, Mayflower, St Albans, Herts, 1975, p.26.

Control of the air then see-sawed during 1916 as performance of the opposing scouts improved. Firepower, too, was developed and the German Albatros D.I was the first scout to appear with twin synchronised guns.[20] With this machine, and the Halberstadt scout, the German Air Service was able to gain the advantage by late 1916. Field Marshal Haig, the Commander-in-Chief of British Armies in France, in a letter to the Chief of the General Staff in early 1917, advised:

> The position as regards fighting squadrons in particular is most serious. Our fighting machines will almost certainly be inferior in number and quite certainly in performance to those of the enemy. In view, therefore, of the marked increase in the number and efficiency of the German aeroplanes it appears that we cannot expect to gain supremacy in the air in April, and it is even possible that it may pass to the enemy.[21]

Haig's prediction was quite correct. The German scouts did maintain their superiority, and during "Bloody April" the RFC was to lose no less than 316 aircrew.[22] In spite of this, the RFC was to maintain offensive operations.

A tactical development, pioneered by the Germans, was the flying of aeroplanes in formations. This had started from a pair, to provide mutual support, developed into flights of four, then into massed formations of close-flying aircraft. The official RFC tactical manual explained:

> The development of aerial fighting has shown that certain fundamental maxims which govern fighting on land and sea are equally applicable in the air. Among these are concentration, mutual co-operation and support, and a well organised system of command, under which no individual has more than a limited number of units under his immediate control. The adoption of formation flying has followed as an inevitable result. Any mission which has fighting for its object, or for the accomplishment of which fighting may normally be expected, must usually therefore, be carried out by a number of machines, the number depending on the amount of opposition likely to be encountered.[23]

Selected German fighter squadrons (*Jagdstaffeln*, abbreviated to *Jasta*) were grouped into wings (*Jagdgeschwadern—JG*), or circuses, for large-scale operations. The most renowned became JG I, commanded by Baron Manfred von Richthofen. The German strategy was to engage the RFC over the German lines. This gave them several advantages. The prevailing westerly wind on the Continent drifted the ensuing melee over German territory. Allied fighters then had the handicap of heading back home into wind, while the Germans were closer to their airfields, and therefore needed less fuel. Also aircraft forced down in these dogfights would, more often than not, fall into German territory. This meant the Allies became prisoners of war while the Germans could be returned, wiser and more experienced, to their squadrons.

---

20 An Albatros D.Va scout, number 5390/17, is on display at the Australian War Memorial in Canberra. This was forced down by a 3 Sqn AFC RE-8 crew, Lt Sandy and Sgt Hughes, on 7 Dec 1917. In addition, the AWM holds a Pfalz D.XII scout, 2600/18, supposedly forced down by 4 Sqn AFC pilots during Oct 1918; however, this is not confirmed by Squadron records.

21 PRO AIR8/2/ID2/137 letter OB/1837, of 13 Feb 1917.

22 C Shores, N Franks & R Guest, *Above the Trenches*, Grub Street, London, 1990, p.17.

23 PRO AIR10/324 AP302 "Fighting in the Air", March 1917, pp.8–9.

The operational aeroplane for which the Australian novice flyers were training was the De Havilland DH-5 single-seat scout. De Havilland designed the DH-5 to combine the better tractor performance with the good forward and upward view of the pusher, resulting in its negative wing stagger—the bottom wing was forward of the upper wing. It was most probably the first aircraft specifically designed for ground attack, as the Air Ministry "required a machine for 'trench-strafing' and low flying, and instructions were issued to designers to produce a type which would give the pilot a good view overhead as well as below and in front".[24]

The DH-5's performance was reasonable for the power offered by its 110 hp Le Rhone engine and it was not unduly difficult to handle. It had a maximum speed of 105 mph at 6,500 feet and could climb to 10,000 feet in 12 minutes. But this performance was inferior to contemporary German scouts, and the DH-5 was generally considered the least successful of De Havilland's 10 wartime designs.

Captain Stan Muir seated in A9197, one of No 68 Squadron's first DH-5 scouts, at Harlaxton during September 1917 — the inscription indicates it was the 15th aircraft funded by donations from New South Wales, in this case from the upper Hunter region AWM C1862

While its backward-staggered wing (with the pilot sitting ahead of the upper wing leading edge) did provide good forward visibility, the rear visibility was fatal for hunting "the Hun in the sun". Again P G Taylor recollected:

24 Cutlack, p.402.

> Apart from the DH-5's performance, which was inadequate for the period of its service, its pilot's vital rearward view was almost entirely obscured by the upper wing. And any pilot in France who didn't, or couldn't, look behind to watch for the enemy on his tail, didn't live very long.[25]

The DH-5 was armed with a synchronised Vickers machine-gun mounted above the fuselage in front of the cockpit, slightly to starboard. Gun aiming was by means of ring and bead sight, and also an Aldis telescopic sight was fitted. The Aldis sight was described by Australian pilot L W Sutherland:

> The Aldis is a clever and efficient job. This is it—this tube about twenty inches long, and two inches in diameter ... inside are three lenses at equal distances apart. On one of them is painted a ring with a dot in the centre. When you look through the tube, this ring appears to be hung in space. At 200 yards range it gives a diameter of approximately forty feet; that is the average wing-span of the Hun machines. The pilot's job is to get the figure of the Hun pilot cutting the rim of this circle, and heading towards the black dot. Then he presses the trigger. Theoretically, the Hun pilot and your bullets should arrive at the black dot together. When the theory works out, as it mostly does, you have some interesting information for your 'Combat in the Air Report'.[26]

Fighting tactics had evolved since the first recorded aerial combat on 22 August 1914, when a German two-seater was shot down by rifle from an RFC aircraft.[27] Pilots grew in experience of air warfare, and distinctive classes of aeroplanes were developed. The single-seat scout was born to provide air superiority over and behind enemy lines to safeguard the slow-flying observation machines. These offensive fighter tactics against hostile aeroplanes then led to the growing realisation that the aircraft was highly effective at harassing enemy troops and communications. This was now being developed into bombing and strafing—the beginning of ground attack.

These were the concepts that No 68 Squadron's pilots were to learn during their stay at Harlaxton during 1917. They soloed on Shorthorns at basic training units, typically taking two to three hours to achieve this, then flew another 10 hours solo on Rumpeties to qualify for advanced instruction.[28] This next stage of training was conducted on one of the Australian units preparing for the Front—Nos 68, 69 or 71 Squadrons—or on an RFC Reserve Squadron.[29] After this, pilots were qualified to wear their "wings",[30] and No 68 Squadron was able to concentrate on combat training. Several of the pilots were fortunate to receive extra training at the School of Gunnery at Turnberry, to the south of Glasgow in Scotland, and on the Fighting Course at Spittlegate, near Harlaxton.[31] These were considered the "finishing schools" for a combat pilot.

---

25 Taylor, p.55.

26 Sutherland, pp.3–4.

27 The claim was credited to Sgt-Maj Tillings of No 2 Sqn RFC. Sims, p.7.

28 IWM 73/235/1 and IWM 88/7/1 refer.

29 In May 1917, "Reserve" Squadrons were retitled "Training" Squadrons.

30 PRO AIR1/2038/204/356/1.

31 Pilots who later joined the Squadron also flew at the School of Aerial Fighting at Ayr in Scotland.

An ingenious development in the training for aerial fighting was the camera gun, described in a contemporary *Flying* magazine:

> We now have a Lewis gun, which, instead of being fitted to fire bullets, is fitted to take photographs, and aerial battles in which one aeroplane endeavours to catch and photograph another with this photographic gun are of daily occurrence. The value of such training is obvious. In addition to this a school of gunnery has been developed, and is going on developing, in which practice from the air takes place against every sort of target, both on the ground and in the air. Targets are now towed by aeroplanes without any danger or difficulty, and these are used for actually counting the hits made from an aeroplane in the air.[32]

Several of the Australian instructors with No 68 Squadron at Harlaxton had come from No 67 Squadron in Egypt, including Captains William Guilfoyle,[33] Stan Muir[34] and John Bell.[35] Guilfoyle, a Scot, had gone to war with the Light Horse in August 1914, transferring to the Royal Artillery the following year. Late in 1915 he commenced flying training with the RFC, serving first on No 19 Squadron, then back to the desert in July 1916 to rejoin Australians in No 1 Squadron. He was then a founding member of No 68 Squadron on the flying staff at Harlaxton in January 1917, and in August ceased his Australian ties to command a number of RFC training squadrons. He later served with the Royal Air Force.

Over that summer, the Commanding Officer and most of the pilots were able to spend at least a month in France to gain operational experience on attachment to RFC squadrons at the Front. The first weeks over the lines were the most dangerous for an inexperienced pilot—the novices were easy prey, as they did not sense approaching danger. So building up experience levels before proceeding to the Front en masse proved invaluable. Most of the Australian pilots flew the DH-5 on their attachments, but several flew the Camel and other types.[36]

One of the Australian pilots, Lieutenant Richard Howard,[37] was attached from May until July to No 57 Squadron RFC at Boisdinghem, near St Omer, flying DH-4 two-seaters. Another, the adjutant of No 68 Squadron, Captain Roy Phillipps,[38] had applied for pilot training, and by July was flying the DH-5 at Harlaxton. At the

32 "Training the Military Flyer" p.89.

33 Gp Capt W J Y Guilfoyle OBE MC. 4th Lt Horse, No 19 Sqn RFC, 67 Sqn , 68 Sqn; comd RFC trg sqns 1917–18, 28 Sqn RAF 1918. Later RAF. Agricultural student of Melbourne; b Edinburgh, Scotland, 1890.

34 Capt S K Muir MC. 4th Lt Horse, RFC, No 1 Sqn AFC/67 Sqn RFC, 68 Sqn. Station overseer of Mathoura, NSW; b Melbourne 6 Apr 1892. Killed in flying accident 12 Sep 1917.

35 Capt J Bell. 1st Lt Horse, RFC, No 1 Sqn AFC/67 Sqn RFC, 68 Sqn. Grazier of Rokewood, Vic; b Melbourne 7 Oct 1885. Died of wounds in France 27 Dec 1917.

36 PRO AIR1/1044/204/5/1506 "AFC in France Jul 1917–Oct 1918", CRFC 2305/1(A), of 5 Aug 1917.

37 Capt R W Howard MC. Army Engineers, No 68 Sqn RFC, 57 Sqn, 68 Sqn/2 Sqn AFC. Engineering student of Hamilton, NSW; b Sydney 9 Oct 1896. Died of wounds 22 Mar 1918.

38 Sqn Ldr R C Phillipps MC* DFC, 860. 28th Bn, No 68 Sqn RFC, 32 Sqn, 68Sqn/2 Sqn AFC; comd 6 Sqn AFC 1918–19; comd 2EFTS 1940–41. Accountant of Perth; b Sydney 1 Mar 1892. Killed in flying accident 21 May 1941.

beginning of August, he was attached to No 32 Squadron RFC at Droglandt, in Flanders, to build up his DH-5 experience. On 6 August, he was hit by groundfire and crashed near Ypres. He was unharmed and rejoined No 68 Squadron in September.

Lieutenant George Matthews[39] and two future flight commanders of the Squadron—Lieutenants Gordon Wilson[40] and Henry Forrest[41]—were also sent to the Front, the latter with 43 and 32 Squadrons flying Strutters, then DH-5s. On 5 August, flying DH-5 A9380 with No 32 Squadron, Forrest crashed soon after take-off, completely wrecking his aircraft and injuring himself over the eye.[42] He was able to rejoin No 68 Squadron in November. Lieutenant Victor Norvill[43] was attached to No 29 Squadron RFC to fly Nieuport 17s at Poperinghe on reconnaissance patrols prior to the Third Battle of Ypres (31 July–10 November). In an engagement on 29 July between seven British machines and 25 German scouts, while flying Nieuport B1677, he was shot down, wounded and taken prisoner.[44]

Among the junior pilot ranks were two from CFS at Point Cook. Lieutenants Douglas Morrison[45] and Albert Griggs,[46] an American living in Hobart, had been students on the Fifth Course at Point Cook, and had been posted to the second unit—No 3 Squadron—to be shipped from Australia. Morrison then served for four months on No 49 Reserve Squadron to complete his training, and in July went for a month at the Front with No 24 Squadron at Baizieux on the DH-5. Griggs finished his training with No 48 Reserve Squadron and joined No 68 Squadron in June. Unfortunately, they both died from wounds in the Squadron's first two months of combat.

Most of the pilots had returned from their attachments to France by 18 August. Sadly, Stan Muir, the "B" Flight Commander, was killed in an accident on 12 September while demonstrating aerobatics at 500 feet over the aerodrome in DH-5 A9197. He over-stressed the wings during inverted flight, they folded, and he died instantly as the wreckage hit the ground. No 68 Squadron was to dig its first grave—he was buried with full Military Honours in Harlaxton Cemetery. Muir had been popular with the groundcrew: one day he had overheard an RFC training pilot being abrupt with one of the 68 Squadron riggers. Muir explained to the pilot: "Don't make trouble here, that laddie can do ten times as much for you as you could do for him, and civility is free."[47]

---

39 Capt G C Matthews AFC. 9th Lt Horse, No 68 Sqn RFC, 71 Sqn RFC/4Sqn AFC,1st Trg Wg AFC. Master mariner of Adelaide; b Stranraer, Scotland, 25 Jul 1883.

40 Capt G C Wilson MC AFC DCM. Army Engineers, No 68 Sqn RFC/2 Sqn AFC, comd 8 Sqn AFC 1919. Pattern maker of Newcastle, NSW; b Minmi, NSW, 6 Oct 1895.

41 Capt H G Forrest DFC. 23rd and 46th Bns, No 68 Sqn RFC, 43 Sqn, 32 Sqn, 68 Sqn/2 Sqn AFC. Clerk of Melbourne; b Melbourne 5 Dec 1895.

42 PRO AIR1/970/204/5/1103 "Record of Casualties of Australian Personnel 21 Jun 1917–14 Nov 1918", p.59.

43 Lt V A Norvill. 82nd Bn, CFS, No 68 Sqn RFC, 29 Sqn. Mechanical engineer of Adelaide; b Melbourne 31 May 1895. POW 29 Jul 1917.

44 Cutlack, p.177.

45 Lt D G Morrison. 64th Bn, CFS, No 24 Sqn RFC, 68 Sqn. Orchardist of Melbourne; b Kilmore, Vic, 8 May 1895. Died of wounds 29 Oct 1917.

46 Lt A Griggs. Army Engineers, CFS, No 68 Sqn RFC. Civil engineer of Hobart; b Meridian, Miss, USA, 23 Sep 1887. Died of wounds 23 Nov 1917.

47 Cox, p.9.

In anticipation of the move to the Front, No 68 Squadron had been established along the RFC lines of having three flying Flights—"A", "B" and "C"—and an administrative Headquarters Flight. Each Flight was formed with six aeroplanes and pilots and its own establishment of groundcrew. At this stage in the RFC, the Squadron Commanding Officer was not permitted to fly on operations. It was the Flight Commander who led aircraft into combat. In addition to the daring and initiative required of an aviator, the patrol leader had literally to see everything in the sky so as never to lead his flight into a bad position. After Stan Muir's death, Gordon Wilson was appointed to command "B" Flight, and was subsequently promoted to captain the following month.

The Australian Squadron about to depart England for France, 21 September 1917 RAAF

On 16 September, the Squadron's advance ground party with transports left Harlaxton for France, under command of the Equipment Officer, Lieutenant Eric Tooth.[48] They proceeded by road to Portsmouth, arriving on the 19th, and then sailed to Le Havre. Setting off then by road to Amiens, the ground personnel arrived at Baizieux on the evening of 26 September. The remainder of the groundcrew, including spare pilots, departed Harlaxton for Southampton by rail on 21 September, to provide the Squadron with a total strength of 170 ground personnel.[49]

The move of No 68 Squadron's aircraft from England on Friday 21 September 1917 was the first time that a whole unit of the Royal Flying Corps was able to deploy overseas in one day. Leaving Harlaxton at 9.30 a.m., Major Oswald Watt had led the Squadron to Lympne in Kent and then, with his "A" Flight Commander, Captain

48 Capt E N Tooth. 2nd Lt Horse, 1 Fd Sqn Engnrs, No 68 Sqn RFC/2 Sqn AFC, 1st Trg Wg AFC.

49 The groundcrew establishment for the Squadron was normally around 158 personnel. AWM 25 911/13 Strength Returns, No 68 Sqn/No 2 Sqn, 1917–1919.

Wilfred McClaughry,[50] across on a 40-minute over-water leg to France. It had been a beautiful day for the crossing, and two Flights lunched at St Omer, with the last of the 15 aircraft arriving at 5 p.m. The following day they flew on to Warloy, and then to their new base at Baizieux, near Albert, on 23 September. The one-day deployment to France is believed to be a record they held to the conclusion of hostilities.[51]

DEPLOYMENT OF NO 68 SQUADRON TO FRANCE

The following 15 pilots flew from Harlaxton, via Lympne in Kent, to St Omer in France on 21 September 1917. Although established for a complement of 18 DH-5 aircraft, No 68 Squadron possessed 15, and was issued with a further three on arrival at Baizieux from No 2 Aeroplane Supply Depot (2ASD). The depot then resupplied aircraft to make good attrition and maintain squadron strength at 18 aeroplanes.

Major W O Watt
Captain J Bell
Captain W A McClaughry
2/Lieutenant G C Wilson
Captain R C Phillipps
Lieutenant G C Matthews
Lieutenant A Griggs
2/Lieutenant C H James
2/Lieutenant L H Holden
2/Lieutenant R W Howard
2/Lieutenant W A Robertson
2/Lieutenant D G Morrison
2/Lieutenant F G Huxley
2/Lieutenant A J Pratt
2/Lieutenant H Taylor

The following officers proceeded with ground personnel, leaving Harlaxton on 16 and 21 September, arriving at Baizieux on 25 and 26 September.

Lieutenant E N Tooth	Equipment Officer
2/Lieutenant W A Turner	Recording Officer
2/Lieutenant L F Loder	Armament Officer
2/Lieutenant J R Y Bartlam	Flying Officer
2/Lieutenant R W McKenzie	"
2/Lieutenant I C F Agnew	"
2/Lieutenant L N Ward	"
2/Lieutenant S W Ayers	"

Another pilot to join No 68 Squadron on its arrival at Baizieux was 2/Lieutenant C C Sands.

50 AVM W A McClaughry CB DSO MC DFC. 9th Lt Horse, RFC, No 68 Sqn RFC; comd 71 Sqn RFC/4 Sqn AFC 1917–18, 8 Sqn 1918–19. Later RAF. Law student of Adelaide; b Knightsbridge, SA, 26 Nov 1894.

51 H C Brinsmead, "In France with the 68th (2nd Squadron) AFC", in Ure Smith, Stevens and Watt, p.25.

CHAPTER 4

THEY WERE SPLENDID

Each Battleplane was equivalent in fighting value to 2000 men …

Lord Kitchener[1]

At Baizieux No 68 Squadron formed part of the RFC's 13th (Army) Wing in support of the British Third Army. The aerodrome was 40 kilometres behind the lines. With the arrival of the groundcrew on 26 September, there was minimal time for settling in. Having prepared and tested the aeroplanes, the Squadron was able to commence area familiarisation flying on the 28th. Also, the No 68 Squadron identification marking was painted on the sides of the fuselages, in the form of a single white band immediately in front of the tail plane.[2] In addition, a letter or number was applied, so that individual machines and pilots could be recognised from a distance. The side letters A to F indicated "A" Flight, 1 to 6 "B" Flight, and U to Z "C" Flight. Easy identification would be crucial as they entered the fray beyond the lines, over "Hunland".

To further assist identification, the Flight Commander would stream a pennant from a wing strut to be distinguishable to the other members of the patrol. As Squadron Commanders were expressly forbidden to fly on operations, Watt was virtually desk-bound. Each pilot normally flew his own aircraft, and the constitution of the three flights was:

"A" Flight		**"B" Flight**		**"C" Flight**	
A Howard	A9284	1 Wilson	A9464	U Griggs	A9469
B Bartlam	A9273	2 Pratt	A9265	V Bell	A9473
C Holden	A9245	3 Phillipps	A9288	W Morrison	A9242
D McClaughry	A9459	4 Huxley	A9462	X Matthews	A9457
E James	A9226	5 Taylor	A9224	Y Robertson	A9483
F Ward	A9399	6 Sands	A9263	Z McKenzie	B377

(In addition Agnew was allotted A9271, and joined "A" Flight. By early October, Huxley's machine had been replaced by A9461 and Morrison's by A9277.)

Operations for the Squadron involved flying Close Offensive Patrols (COP) to control the heights from the front-line trenches to about 10 kilometres beyond, keeping the skies clear of enemy activity. The formal role of the Offensive Patrol was defined as:

> … whose sole mission is to find and defeat the enemy's aeroplanes. The further such patrols penetrate behind the hostile front the greater will be the moral effect of the success they gain, and the more they will interrupt the work of the enemy's machines, while enabling ours to accomplish their missions without interference.[3]

---

1 AA CRS A2 series 1919/362, cited from letter, of 15 Jul 1918.

2 PRO AIR1/867/204/5/523 CRFC 1693.G, of 19 Sep 1917.

3 PRO AIR10/324 AP302 "Fighting in The Air", Mar 1917, p.3.

The first air engagement by an Australian squadron in France occurred on Tuesday 2 October 1917. A patrol of four aircraft from "A" Flight, led by Captain Wilfred McClaughry, was returning over St Quentin, south-east of Baizieux, at 10,000 feet when an enemy two-seater was spotted. They dived on the enemy aeroplane (EA), but in spite of their superior height, they were unable to catch the faster German. Fifteen minutes later they encountered another two-seater, which Lieutenants Les Holden[4] and Richard Howard attacked, but again the EA had superior speed. "A" Flight returned home without Lieutenant Ivor Agnew[5] (flying A9271), who had been forced to land behind enemy lines, due presumably to battle damage. The Germans dropped a message over the lines stating that Agnew was an unwounded prisoner of war.[6] He was the first loss in battle for the Australian squadrons in France.

With the aerial war constantly being conducted over Hunland, the German groundfire was able to get plenty of practice in shooting at Allied aircraft. On No 68 Squadron's patrols, not only was groundfire from machine-guns and small arms to exact their toll, but enemy artillery—"Archie"—was another hazard. As patrols were always carried out over enemy territory, with the German Air Service rarely venturing west across the lines, the black smudges of German Archie had to be avoided.[7] On the Western Front, German anti-aircraft gunners were to claim 1,588 Allied aircraft.[8] Typical of these COP sorties was that recorded by "A" Flight pilot Leslie Ward:

> October 10th. We were on dawn patrol this morning and it lasted for two hours. We went over the lines as far as Cambrai and Archie was very hot. He seems to have some new batteries and they are very good shots.[9]

Returning from a patrol on 13 October, two aircraft of "C" Flight, flown by Lieutenants Douglas Morrison (A9277) and Robert McKenzie[10] (B377), were bounced by four Albatros scouts. McKenzie got in one burst at the enemy leader, but had to withdraw with an engine problem. Morrison was shot down near Queant, and crashed between the lines in No Man's Land, severely wounded. He was rescued by men of the 13th London Regiment while the remains of his wrecked machine were shelled by the enemy. Morrison died from his wounds on 29 October and was buried in Grevillers Cemetery.

---

4 Capt L H Holden MC AFC. 2nd Div AIF, No 68 Sqn RFC/2 Sqn AFC, 6 Sqn AFC. Assistant manager of Sydney; b Adelaide 6 Mar 1895.

5 Lt I C F Agnew. Artillery 17th Bty, RFC, No 68 Sqn RFC. Pastoralist of Epping Forest, Tas; b Oatlands, Tas, 17 Feb 1893. POW 2 Oct 1917.

6 PRO AIR1/970/204/5/1103 "Record of Casualties", p.58.

7 Archie could also be used to the advantage of the flyers. While the German smoke was black, Allied shells exploded with white smoke. The enemy would use his black barrage to warn his machines of advancing Allied scouts. Also, if the enemy's Archie suddenly stopped harassing a patrol, it was a sure sign EA were about to swoop.

8 K P Werrell, *Archie, Flak, AAA and SAM*, Air University Press, Maxwell AFB, Alabama, 1988, p.1.

9 AWM PR 83/230 Personal diary of 2AM Ward, 10 Oct 1917.

10 Lt R W McKenzie MC. AAMC, No 68 Sqn RFC, 41 Sqn, 68 Sqn/2 Sqn AFC, 8 Sqn AFC. Chemist of Melbourne; b Adelaide Jun 1895.

Lieutenant Douglas Morrison, in the cockpit of A9432, was shot down in No-Man's Land on 13 October 1917 — Captain Heath of the 13th London Regiment was awarded a bar to his MC for rescuing Morrison from his wrecked machine. AWM C2800

Two new pilots then joined the Squadron to replace these losses. Lieutenant Robert Clark[11] replaced Agnew in "A" Flight, and Lieutenant Clive Johnson[12] joined "C" Flight. Clark had been trained at Richmond on the First Course in 1916 and, like some others on that course and members of the Sixth CFS Course, had sailed from Australia as the Fourth Australian Squadron. He had then trained in England as a member of No 71 Squadron (which later became No 4 Squadron AFC), but soon found himself required at the Front as a replacement. His experience was typical of the pilots who then joined the Squadron in France. Clark disembarked at Boulogne on 6 October, and was held at No 2 Aeroplane Supply Depot (2ASD), located 30 kilometres back from Baizieux at Fienvillers. The depot's role was to supply aircraft and pilots to the frontline units to ensure they were always at strength. Clark was able to join No 68 Squadron on 14 October, but later pilots were held in the depot pool for periods of up to two months. One RFC pilot described his holding period at an ASD by these frustrated lines: "I'm still in this God forsaken hole. I wish they would hurry up and post me to a squadron, as this place is as dull as ditchwater."[13]

11 Lt R L Clark. No 4 Sqn AFC/71 Sqn RFC, 68 Sqn, 6 Sqn AFC. Mining engineer of Sydney; b Sydney 4 Sep 1889.

12 Lt C L Johnson. 1st Lt Horse, 4th Camel Regt, No 68 Sqn RFC/2 Sqn AFC. Accountant of Sydney; b Sydney 13 Aug 1887.

13 IWM 73/235/1, letter from Lt Y E S Kirkpatrick RFC, of 7 May 1918.

As the Squadron's flying hours mounted over the Front, the DH-5 was plagued with engine problems—magneto failures and broken tappet rods being common. During October alone, the Squadron had eight crashes not due to enemy action, resulting either in write-off of the machines or their return to depot for repair. This necessitated a continuous stream of replacement machines from No 2ASD's repair and issue sections which, with the nearby 2 Aircraft Depot (2AD) at Candas, were responsible for the aircraft supply system in the southern area of the Western Front.

On 2 November, McClaughry's patrol fought off an attack above Gouy at 10,000 feet by eight Albatros scouts of the "black-tailed circus". Again the enemy had the advantage of superior speed. In this engagement 2nd Lieutenant Howard's machine (A9284) was damaged with about 25 bullet holes through all four main wing struts.[14] The aircraft was unrepairable and was struck off strength. In addition to the Albatros, another formidable adversary was the Fokker Dr.I triplane, known to the Germans as the *Dreidecker*, which had first appeared over the front in early September. It was very manoeuvrable and could out-turn and outclimb any of the Allied scouts.

Throughout the previous three months, the British Army had been continuously on the offensive in Flanders—the third Battle of Ypres—with the aim of capturing the German-held Belgian ports. Scout squadrons were coordinated with the infantry advance by attacking enemy infantry columns and targets of opportunity with light bombs and machine-guns. For the first time large-scale ground attack operations were conducted, with devastating effects. This proved a suitable task for the DH-5 scout squadrons. The ground attack role was dangerous and even though they suffered heavy casualties from groundfire, by September the Allies were able to gain air superiority.[15]

The offensive petered out at the end of autumn. The British had managed to secure the Passchendaele Ridge, but at enormous cost. Their forces suffered over 300,000 casualties, the Germans less than 200,000. With German forces concentrated in Flanders, an innovative plan was being developed for an assault some 60 kilometres to the south, at Cambrai.

Traditionally, offensives had been opened by massive artillery bombardments, quite often with devastating effects on the enemy. But the bombardment also destroyed the roads and made the landscape virtually impenetrable for a speedy advance. The British Third Army wanted to break enemy defences by use of modern technology ahead of the assault. This would be the first employment of massed tanks and air power and at Cambrai, as at Third Ypres, the successful ground attacks by bombing and strafing would become a significant factor. The DH-5 scouts of Nos 64 and 68 Squadrons were tasked with this role. In great secrecy, the tanks had been assembled, and with these the Third Army would attempt to breach the fortifications of the Hindenburg Line at Flesquieres, 10 kilometres south-west of Cambrai.

In preparation for this British offensive—the Battle of Cambrai (20 November–7 December)—the Australian DH-5s had been training for a new role in ground attack.

---

14 F R Cox, personal diary, of 16 Oct 1917, Museum of Army Aviation and Flying, Oakey.

15 In aerial combat on 25 Sep, the RFC lost only one aircraft to the German loss of 19.

Two of the experienced pilots had returned to England during the month to help form the next Australian scout squadron—No 71 Squadron. McClaughry was promoted to Major and took command of this new unit on 11 November, and Lieutenant George Matthews left "C" Flight to take command of 71 Squadron's "A" Flight. Captain Roy Phillipps then became the new Flight Commander, and he commenced training and developing tactics by practice bombing and strafing at a weapons range near Baizieux. The DH-5 was fitted with bomb racks, which could carry four 25 lb Cooper bombs.

Some of the 68 Squadron pilots that returned to England as instructors: Rear, from left, Capt L H Holden, Maj R C Phillipps, Capt W A Robertson. Front, Lt H Clark, Lt Col W O Watt, Lt R L Clark RAAF

The Australians were then to concentrate their tactics on low-flying to deliver their weapons. It was essential to be below 1,500 feet to make out the details of trenches and dug-outs, and below 1,000 feet to distinguish the grey uniforms of the enemy from the khaki of the Allies.[16] This was quite often the way that the position of the Front could be verified, and necessitated formation flying at low level so that targets could be identified. Attacks would be made from the Front, to as far as 25 kilometres beyond the German lines, and these became known as Distant Offensive Patrols (DOP).

On the misty morning of Tuesday 20 November, the Third Army launched a heavy attack on the Front near Cambrai. As a maximum effort was required, No 68 Squadron took off in three flights of six. It had not been the practice to fly in sections of this size, and when a flight of six was necessary, it would normally be flown in two vics, a vee formation of three machines. The second vic was usually on the port side

16 AWM25 85/16 Part 2, GHQ OB/1656, of 24 Dec 1916.

slightly to the rear. But on this day the visibility in the dense fog and drizzle was too bad to maintain formation, so pilots had to acquire and attack their targets individually. Bell led "C" Flight off first, followed an hour later by Wilson with "B" Flight and Phillipps with "A" Flight. The whole of the Squadron had entered the fray, bombing and strafing from an altitude of 100 feet.

"C" Flight attacked German forces near Cambrai and immediately took losses from the hostile groundfire. The leader, Captain John Bell (A9473), was shot through the chest, and crashed beyond the lines. He was rescued but finally succumbed to his injuries on 27 December and was buried in Tincourt Cemetery. Another member of Bell's flight, Lieutenant William Robertson[17] (A9483), was badly damaged in an engagement with EA, but was able to land at Bapaume. Here, 30 kilometres forward of Baizieux towards Cambrai, 68 Squadron's groundcrew were manning an advanced landing ground (ALG). Being closer to the battle, only 12 kilometres from the Front, this forward position was to expedite operations.

While Wilson's "B" Flight was attacking enemy infantry and batteries, Lieutenant Harry Taylor[18] (A9378) was brought down in No Man's Land. He climbed from the wreckage and picked up an enemy rifle to fire at snipers. He met up with an advanced British patrol collecting the wounded, then came across Bell's damaged machine. He was unsuccessful in getting it started as the fuel tank had been holed. He then assisted at a casualty dressing station, and was eventually able to get a lift back to the aerodrome.

Phillipps' "A" Flight made successful attacks on enemy communication trenches. Lieutenants Les Holden and Robert Clark then fired into the confusion caused by the direct hits of their bombs. Holden (flying A9278) was then attacked by an enemy fighter and, like Robertson, was able to coax his badly damaged machine back to the Bapaume ALG. Also flying with "A" Flight Lieutenant Leslie Ward (A9399) attacked enemy infantry near Marcoing to the south-west of Cambrai when he was shot down. He later described his experience:

> I was brought down by machine-gun fire, my leg was broken in the crash. I was carried by Germans to a dressing station where they attended to me. I stayed there for a couple of hours and was then carried for about one mile on a stretcher.[19]

Ward was to spend the remainder of the war as a POW. The first authorities knew of his condition was after his sister received a letter explaining how he had sustained a broken leg in the crash and was taken prisoner.[20]

One of the tactics used by the DH-5 pilots had been to crater the roads and then strafe the stranded German traffic. The boggy ground added to the congestion and the enemy's frustration. Two of "C" Flight's pilots landed at the Bapaume ALG,

---

17 Capt W A Robertson. Army Engineers, No 68 Sqn RFC/2 Sqn AFC. Engineering student of Melbourne; b Melbourne 30 Mar 1896.

18 Lt H Taylor MC MM. AASC, No 68 Sqn RFC/2 Sqn AFC, 1st Austn Trg Wg. Mechanic of Sydney, b Birmingham, England, 1889. Killed in flying accident 18 Aug 1918.

19 AWM PR 83/230 Personal diary of 2AM Ward, 20 Nov 1917.

20 PRO AIR1/970/204/5/1103 "Record of Casualties", p.56, citing 3rd Bde letter 3b/57/A, of 14 Jan 1918.

replenished for further attacks, and were airborne at noon. Lieutenant Frederick Sheppard[21] (A9457), who had been at the Front for three weeks, was shot down and wounded. Lieutenant Robert McKenzie (B377) crashed when his fuel tank burst due to enemy groundfire near Queant, but he was safe.

Throughout this first day of the battle the pilots returned to the advanced landing ground, reloaded with petrol, bombs and ammunition, and disappeared again into the mist. The weather had been so bad as to mostly keep the German Air Service on the ground. Of the 18 aircraft flown by the Australians, seven had been destroyed or severely damaged, together with three pilots as casualties. The losses of the low-flying squadrons were to average 30 per cent on the days they were used throughout the battle.[22] The dangers of flying ground attack at tree-top height into enemy groundfire were obvious. Recognising these hazards, the General Officer Commanding the RFC in France, Major-General Hugh "Boom" Trenchard,[23] cabled Watt with the encouraging words: "Congratulate all pilots on their gallant work under impossible conditions".

The battle was fiercest on 22 and 23 November, and three more pilots were lost to groundfire. On the 22nd, Lieutenant David Clark[24] (A9477), who had been on the Squadron for only two weeks, failed to return from a morning mission. He had last been seen heading east over Bourlon Wood, a heavily defended enemy stronghold seven kilometres west of Cambrai. The possession of this wood, a nest of machine-guns on the lofty Bourlon ridge, was vital to provide observation of German defences south of the Scarpe and Sensee Rivers. Lieutenant Archie Pratt[25] (A9265) bombed and strafed a battalion headquarters in the north-west corner of Bourlon Wood before being wounded. He was able to crash to safety near the frontline trenches.

That morning, in the persistently bad weather, Lieutenant Frederick Huxley[26] (A9461) scored the first enemy aircraft destroyed by the Australians. He was at 700 feet over Marquion, north of Bourlon Wood, having just attacked enemy infantry, and saw an Albatros scout below and in front of him. He pounced onto the German's tail, firing 30 rounds from 15 metres, and saw it nose-dive into the ground. "It was a gift", he said.[27] An hour later Lieutenant Howard (A9294) drove down two others, while Captain Phillipps (A9288) drove down a third into a garden in Cambrai.

Ground attack sorties continued on the 23rd. Lieutenant Robert Clark's attack report for his morning sortie to Bourlon typified the Squadron's efforts:

---

21 Lt F H Sheppard. No 69 Sqn RFC, No 68 Sqn. Photographer of Melbourne; b 1893.

22 H A Jones, *The War in the Air*, Vol IV, Clarendon Press, Oxford, 1934, p.239.

23 Marshal of RAF Viscount Trenchard GCB GCVO DSO. GOC RFC in Field 1915–17; CAS RAF 1918–1929. Of London; b 3 Feb 1873.

24 Lt D G Clark. No 68 Sqn RFC. Grazier of Sydney; b Picton, NSW, 1891. Killed in action 22 Nov 1917.

25 Lt A J Pratt. Army Engineers, No 68 Sqn RFC. Engineer of Melbourne; b Melbourne 1893.

26 Lt F G Huxley MC. 26th & 52nd Bns, No 68 Sqn RFC/2 Sqn AFC. Shopkeeper of Currie, King Island, Tas; b King Island 8 Aug 1891.

27 Gp Capt A D J Garrisson OBE, "Number Two Has Notable Firsts", *RAAF News*, Vol 1, No 2, Feb 1960, p.4.

> 2 bombs from 100 feet on troops in the Wood advancing to reinforce front line as tanks were advancing on SW edge of Wood. 200 rounds on troops opposing tanks' advance and 200 rds on trenches SW of Fontaine which were holding up our infantry.[28]

68 Squadron pilot — Lieutenant Robert Clark
J LEEUWIN-CLARK

However, during the course of the morning's operations, Lieutenant Sydney Ayers[29] (A9263) was shot down by groundfire over the heavily defended wood. Although severely wounded, he was able to crash back near the lines, but died from his wounds the next day. His brother, Private Charles Ayers of the 55th Battalion, had been killed at the battle in nearby Polygon Wood at Ypres, only two months previously.

To wrest back control of the air, German reinforcements were rushed to the Front. The four squadrons of Richthofen's Circus[30] arrived in the skies over Cambrai from Courtrai, in Flanders. In the heavy aerial fighting that developed, Lieutenant Les Holden's machine (A9326) was badly damaged near Bourlon Wood but, unhurt, he was able to safely return.

From his attack sorties Captain Roy Phillipps (A9288) was able to pass on the following intelligence report:

> Saw tanks NW of Bourlon valley and NE of Bourlon Wood. Fontaine Notre Dame appears to be completely ours, as British troops were seen on Eastern outskirts. Big explosion seen in Cambrai, apparently in centre, and heard above the noise of engine when 1500 yards away. 6 EA ("Black tail circus") seen at 6000 feet over Cambrai.[31]

On the west of Bourlon Wood an enemy strongpoint was holding up the Allied advance. On the afternoon of the 23rd, Lieutenant Griggs (A9428) overflew the position of the 10th Royal Irish Rifles, and made repeated strafing attacks on the German defenders. Captain Wilson saw an enemy scout attack Griggs' machine, but Griggs fought him off. He continued his strafing attacks, but was met with heavy fire, ultimately being shot down and killed. However, the support he had provided

28 AWM4 8/5/1, No 68 Squadron Ground Attack Report No 9, of 23 Nov 1917.

29 Lt S W Ayers. No 68 Sqn RFC. Mechanic of Cootamundra, NSW; b Cootamundra 1893. Died of wounds 24 Nov 1917.

30 The Circus was comprised of *Jastas* 4, 6, 10 and 11. Jones, p.245.

31 AWM4 8/5/1 Ground Attack Report No 12, of 23 Nov 1917.

endured long after the battle had ended. In the "In Memorial" column of *The Times* in 1918 appeared:

> To an UNKNOWN AIRMAN, shot down 23rd November 1917, whilst attacking a German strong-point south-west of Bourlon Wood, in an effort to help out a Company of the Royal Irish Rifles, when other help had failed.[32]

It was a testimony to the efforts of the ground attack pilots that the advance achieved the ground it did. Tank personnel and infantry acknowledged that the ground attack aircraft had made advance possible, when attacking troops would otherwise have been pinned down.[33]

On 26 November, Lieutenant Taylor (A9336) was flying over Bourlon Wood when he spotted an enemy DFW reconnaissance two-seater.[34] He dived on the EA and opened fire at 200 metres, to see the two-seater crash to the ground. Three days later in the same area, Lieutenant Howard (A9517) was attacked by a similar two-seater. He manoeuvred onto the enemy's tail and fired 40 rounds from about 20 metres. As the EA climbed away Howard was able to score some hits from underneath, then fired another 30 rounds from abeam. This was evidently enough to fatally wound the observer, who slumped over the rear cockpit. The EA dived to the ground and was able to land intact north of Cambrai. This combat, although decisive and confirmed by another pilot, Les Holden, emphasised the difficulty in claiming enemy machines in World War I. As this EA was not destroyed, or "forced down out of control" (which was termed as a "probable" in World War II), the system did not allow for this victory to be added to the pilot's score. Instead, this type of inconclusive result was referred to as "driven down".

The German counter-attack at Cambrai started on the misty morning of the 30th. The pilots flew back time and again to the battle area to bomb and strafe the German onslaught. Captain Wilson (A9449) was leading "B" Flight in an attack on enemy infantry when his petrol tank was hit by groundfire. He forced-landed near an anti-aircraft battery, but was soon able to be airborne again. Over Gonnelieu, south of Cambrai, he attacked an enemy DFW two-seater. He shot the observer with his first burst, but was unable to inflict more damage as he chased his target east towards Bantouzelle. He was then surprised by a further DFW, so turned on his attacker and fired a burst which put the EA into a dive. The DFW overturned as it hit the ground, and Wilson then successfully bombed the wreckage. Returning to base he was set upon by a third two-seater and, having exhausted his ammunition, was forced to feint an attack. This was enough for the German, who turned immediately and headed east. It had been a busy sortie for the "B" Flight Commander.

One pilot reported "an absolute melee of aircraft around Bourlon Wood—the air was thick with DH-5s".[35] Often during that day, there were 50 or more RFC aeroplanes over the eight kilometre front south of Bourlon Wood, with as many of the

32 Jones, p.246.

33 Jones, p.247.

34 Machine produced by *Deutsche Flugzeug Werke.*

35 Jones, p.254.

enemy. Great difficulty was being encountered in crossing the lines owing to the presence of enemy scouts.

Richthofen's Circus was active, and found a solitary DH-5 coming from Cambrai. British army troops witnessed two Germans chasing it, firing at the DH-5 as low down as 20 feet, but the DH-5 was able to zigzag along the road to Bapaume.[36] Perhaps this had been Lieutenant Harold Cornell,[37] a new pilot in "A" Flight, who had replaced Leslie Ward, missing since the 20th. Cornell (A9532) had departed that morning as part of an eight aircraft formation with "A" and "C" Flights on a ground attack mission at Bourlon Wood, but became separated after attacking his target. He was then attacked by enemy aircraft, and was shot down near the lines. After spending an exciting 24 hours in a heavily shelled position, he was able to make his way back to the aerodrome.

During the afternoon, a further 18 sorties were flown over the lines attacking the enemy army near Bourlon, and then about 15 kilometres to the south at Gonnelieu. A complete reconnaissance of the area was then passed to the Corps Headquarters, which enabled Allied artillery to be brought into action onto the indicated objectives.

December opened with concentrated ground attack activity south of Cambrai. "B" Flight departed at dawn and, flying low in the mist at 800 feet, Lieutenant Fred Huxley (A9461) spotted an enemy reconnaissance machine over Bourlon Wood. He closed to within 50 metres of the blue and green Aviatik two-seater, opening fire and sending it down. On hitting the ground the machine overturned and, after the formation replenished at the ALG, Huxley returned and bombed the crowd that had gathered to inspect his wreck.

Later that morning Henry Forrest (A9255) led "A" Flight, four-strong, to bomb enemy troop concentrations at Villers Guislain, 15 kilometres south of Cambrai. With him were Lieutenants Robert McKenzie (A9541), Bill Robertson (A9466) and a newcomer with two weeks on the Squadron, Lawrence Benjamin[38] (A9341). Robertson's machine was badly hit by groundfire on crossing the lines, but Robertson was able to limp back over friendly territory to land near Bapaume. On delivering their weapons the others encountered an enemy patrol. Forrest recorded:

> Coming out of a dive after dropping bomb, I found 5 scouts above me, apparently escorting 3-seater. I attacked 1 scout who turned and fired a few rounds and then climbed up into the clouds. I fired 30 rounds without apparent effect till I lost him. I then dived on trenches to machine gun Infantry, and on coming out of dive, I saw the 3-seater about 400 feet above me on my left with the 2 gunners standing up and firing at me. I zoomed up under him and fired 20 rounds at him and he turned East climbing.[39]

---

36 Sims, p.49.

37 Lt H G Cornell. No 69 Sqn RFC, No 68 Sqn. Electrical engineer of Ballarat, Vic; b Melbourne 10 Aug 1891. Killed in action 11 Dec 1917.

38 Lt L Benjamin. AAMC, No 68 Sqn RFC/2 Sqn AFC. Student of Melbourne; b Melbourne 23 Aug 1896.

39 AWM4 8/5/1 Combat in the Air Report No 21, of 1 Dec 1917.

McKenzie was more successful in the engagement with another Albatros:

> I saw 5 EA coming towards and 400 feet above me. One was lagging slightly and I zoomed up and got in a stalling burst at 50 yards range, when he immediately turned East and started a steep glide and I was unable to follow him on account of the other 4 EA, but I saw him try to land in most unfavourable country and go nose first into a shell hole where he remained with his tail in the air.[40]

McKenzie was credited with the Albatros destroyed, but during the engagement Benjamin's machine was hit. With his fuel tank shot through, he was able to reach the lines safely and deposit his damaged aircraft on an ALG near Wagholien.

Concentrated ground attack and reconnaissance sorties were maintained over the next days. With tank and air support the Third Army was able to break through German defences, but without offensive reserves was unable to exploit its advantage and was driven back from Bourlon Wood.

In recognition of the air support the Australians were providing, Major-General Trenchard wrote to General Birdwood, the commander of the ANZAC Corps:

ADV HEADQUARTERS,
Royal Flying Corps,
22nd November, 1917.

Dear General,

I have just been to see the Australian Fighting Squadron, No.68 for the second time in the last week, and I have talked to some of the pilots who carried out the great work on the 20th, 21st and today. Their work was really magnificent and their machines, I am afraid were very much shot about, but they only lost one officer and machine missing, though several were wounded.

These pilots came down low and fairly straffed the Hun. They bombed him and attacked him with machine gun fire from 50 feet flying amongst the tree tops; they apparently revelled in this work, which was of great value.

You might like to let some of your people know that I think them really great men, and I am certain in the summer next year, they will all give a very fine account of themselves. They are splendid.

I hope to be able to come and see you after I have been home, but I have to go home as soon as I can get away.

Yours Sincerely,
(Signed)
H. Trenchard.

General Sir W.R. Birdwood,
KCB, KCSI, etc.etc,
Headquarters,
ANZAC Corps.[41]

Trenchard had left his Advanced Headquarters at Fienvillers for London on 27 November, but hastily returned to France on 1 December on the news of the surprise German counter-attack at Cambrai which had threatened the whole of the

40 AWM4 8/5/1 Combat in the Air Report No 22, of 1 Dec 1917.

41 No 68 Squadron AFC, Operations Record Book, AWM4 8/5/1, of Nov 1917.

Third Army. The German thrust was held, and by 4 December the British line was reformed. That night the British Minister of Munitions, Winston Churchill, stayed at RFC Headquarters, predicting a long war. "Of course", Churchill said, "if we gave in we could have peace tomorrow".[42]

Two days later, 6 December, strong enemy attacks south of Bourlon Wood were repulsed as No 68 Squadron flew wave after wave of ground attack missions. "B" and "C" Flights flew sorties in the morning around Bourlon. After Forrest's formation dropped its bombs, the DH-5s climbed to carry out a Close Offensive Patrol to the south over Marcoing. At 7,000 feet they encountered several formations of Albatros scouts. After some inconclusive skirmishes, Lieutenant Robert McKenzie's new machine (A9544) had its aileron controls shot away. With other damage to his machine's wings and fuselage, he was able to make the lines and crashed safely at Fremicourt. Lieutenant Johnson's aircraft (A9279) was also found on return to be badly damaged.

The remains of Lieutenant Robert McKenzie's machine A9544, shot down in the lines at Fremlicourt on 6 December, and returned the next day to Baizieux IWM E(AUS) 1483

That afternoon, eight aircraft of "A" and "B" Flights attacked the enemy on the III Corps Front near Flesquieres, south-west of Cambrai. Lieutenant Howard (A9517) recorded:

> 2 bombs OK on cross roads E of Bantouzelle which was congested with traffic. Returned to ALG for 2 new plugs. On return 100 rounds on enemy infantry in small parties close to Flesquieres, then too dark to fire.[43]

---

42 M Baring, *Flying Corps Headquarters 1914–1918*, Buchan & Enright, London, 1985, p.260.

43 AWM4 8/5/1 Ground Attack Report No 34, of 6 Dec 1917.

Lieutenant Huxley (A9461) bombed the crossroads at nearby Lateau Wood then strafed enemy troops. As he climbed away to the south through 3,000 feet, he saw through the haze below two enemy DFW reconnaissance two-seaters. He immediately dived on one:

> As soon as I opened fire the enemy Observer opened fire also but after I had fired 30 rounds at 50 feet range he dived steeply and on striking the ground burst into flames.[44]

The next day, 7 December, the Battle of Cambrai was over. This had been the first time the new technologies of warfare—armour and air—were used to support a ground offensive. Because of the deficiencies in the design of the DH-5, the Australian scouts had been re-roled from control of the air to a new development, that of specialised ground attack. However, when the situation had warranted, the DH-5 flown by the tenacious Australians did have limited success in countering the superior Albatros scouts.

The Germans also now adopted ground attack by so-called "battle flights", the *Schlachtstaffeln* (abbreviated as *Schlasta*). Each *Schlasta* consisted of eight machines, whose duty was to assist the advance of the infantry by support with machine-gun fire, bombs and hand grenades.[45] They were allotted to Armies, Corps and Divisions as circumstances demanded, and led to a considerable expansion of the German Air Service over the winter of 1917–1918. Air power had been used to effect, and exploiting new roles had demonstrated its flexibility.

44 AWM4 8/5/1 Combat in the Air Report No 25, of 6 Dec 1917.

45 PRO AIR10/111 "Handbook of German Military and Naval Aviation (War) 1914–1918", AP No 71, Air Ministry Directorate of Air Intelligence, London, Oct 1918, p.38.

CHAPTER 5

NEW MACHINES

Once you have brought down your first or even your second or third, you suddenly realise "Now that's the way to do it".

Baron von Richthofen[1]

On Thursday 13 December 1917 an announcement brought much joy and pride to the Squadron. For their role in the Battle of Cambrai, six of the Australian pilots were awarded the Military Cross.

> Tonight news came through of 6 MCs being awarded to 68 Sqn officers—Lt Howard, Capt Phillipps, Lt Holden, Lt Taylor, Capt Wilson, Lt Huxley. Lt Howard came up to us and thanked George [Kompton] and myself for the way we had kept his machine up and said it was due greatly to its efficiency that he had been so successful.[2]

In addition to the six pilots being decorated, the Military Medal was also awarded to four groundcrew for bravery in salvaging aircraft wrecks under fire on the front line. Those recognised for their hazardous work were Sergeants Bertram Jones and Richard Lonsdale, Corporal William Campbell and 1st Class Air Mechanic Herbert Raphael.[3]

The editor of *The Aeroplane* magazine, C G Grey, wrote from London to Major Eric Harrison at Point Cook with some interesting comments on what he had seen of the men of the Australian Flying Corps. He observed:

> I hear great things about your Australian aviators. The week before Christmas I was in France staying at Flying Corps Headquarters, and the General told me that 68 Squadron, commanded by Oswald Watt, had won more Military Crosses in one day in the Cambrai battle than had been won in a day by any squadron in the Corps. I think they raked in six of them, out of 18 pilots.
>
> All the Australian squadrons have a very high reputation in the Corps, and I have never seen anything to equal the keenness of the Australian mechanics. Whenever a ship load of them arrives in this country they besiege this office with calls and letters, asking for information about where they can get the latest books on aeroplanes and engines so as to bring themselves thoroughly up to date.
>
> Of course on the average they are individually better class men than our own mechanics, because many of them are properly trained engineers who in this country would probably be Equipment Officers and not mechanics, but even their keenness is extraordinary, and apparently the work which they do on active service is as well and as carefully done as if they were working in a comfortable factory at home. It is not surprising that a race of sportsmen like the Australians should make good pilots, but it is rather extraordinary and

---

1 PRO AIR1/1/4/26/4 "Extracts from translation of von Richthofen's book *Der Rote Kampfflieger*", 29 Sep 1917, p.4.

2 F R Cox, personal diary, of 13 Dec 1917.

3 See Appendix 3.

certainly highly gratifying that they should make such excellent mechanics. You and all the people associated with you at the School have every reason to be proud of yourselves.[4]

Australia's official war historian, C E W (Charles) Bean, visited the Squadron during December just as the DH-5 was being withdrawn from service. His unpublished notes give an interesting insight into Watt and the Squadron's daily routine.

Officers of the No 68 Squadron at Baizieux, 7 December 1917:
Kneeling, from left: Lt W A Turner, Capt F G Huxley. Front: Lt L S Truscott, Lt F A Power, Lt P H Lawson, Lt D C Allardice, Capt H G Forrest, Maj W O Watt, Lt R L Clark, Capt L H Holden, Capt R W Howard, Lt L Benjamin, Capt W A Robertson, Lt A Pratt AWM E1434

7 Dec. I went ... to the 68th Squadron AFC at Baizieux. Oswald Watt, their major, was away lecturing at the Army HQ at Albert to 200 officers and NCOs on the cooperation of aircraft and infantry. "Little" Turner,[5] their adjutant, persuaded me to stay the night. Watt had given his youngsters a night off in Amiens.

8 Dec. They are winning themselves a magnificent name, this first Australian fighting Squadron. They had a quiet time from September 21 to November 20 and then suddenly they plunged into bad flying weather into this very heavy fight. Their great doing I have made a note of in a record book.

It is Watt who has worked them up to this remarkably high level of conduct and general tone—as Australian boys can always be worked. Little Turner, who was shellshocked in the infantry, told me: "I hope to God we don't lose him through his overwork; for believe me we shall never get another like him!"

---

4 AA CRS A1952/1 E524/10/408, letter from C G Grey, editor of *The Aeroplane*, to Maj E Harrison, CFS Point Cook, of 7 Feb 1918.

5 Lt W A Turner. 29th Bn, No 68 Sqn RFC/2 Sqn AFC. Engineering student of Melbourne; b Melbourne 20 Jun 1893.

He was in Europe at the start of the war and at once entered the French Flying Corps and flew for a time over Verdun and Nancy. Then he was sent to command the Australian squadron forming in Egypt; and after flying there came to England and lastly flew over with his Squadron in September.

His plan is, in order to give his boys a complete night's rest, to send them to bed about 10; and not to let any of them know overnight what work, if any, they have to do in the morning. With this view he himself, with Little Turner, stays up till the orders come in about 12 or 1AM from the Wing. He then gets out the order for work of the morning and the men who are to do it. He makes out a list of the youngsters to stand by; then a list of their sergeant mechanics; then a list of their batmen. The batmen are the first to give the officers their orders of the work they are to do when they call them in the morning. They may be called at 5. Breakfast is ordered for 5.30. While the officer is having breakfast the mechanic is going through the machine. By the time breakfast is over everything is ready to start. By order all lights are turned out at night; when breakfast is on the lights are turned on again. So that the CO knows, if he sees no lights at 5AM, that they are late.

For the CO this has meant an exceedingly heavy strain—up writing till 3, sleep till 5—up again to see them off punctually. Watt himself told me that he could always get odd hours during the day but he is very worn and he fell asleep after dinner at night over the fire. It was not a cold night, but he was shivering. His little adjutant told me of the strain—he did not mention it himself.

The one strain he did mention was that of one peculiar trial of the last 3 weeks heavy fighting. He said that the wing commander knew him well and trusted very much to him to decide if the weather were suitable for flying. During the critical times in the Cambrai battle or any battle it was necessary to go out sometimes and carry out flights on days when flying was exceedingly dangerous and almost impossible. It was left to him to say if it were possible or not. He had often had to send four of his youngsters out on days when we knew that flying was perilous. He had to make the order for these friends to go. It was not as if they were a battalion or even a company say of 120 men. They were four boys who sat at the same table every day for 6 months and had become exceedingly well loved friends. "You can just imagine the feelings with which I used to sit there during two hours waiting for them to come back", he said. Five were lost in those two weeks.

One day one youngster was brought down at Coutaing and as he did not return hour after hour, he was posted missing. 14 hours later came a wire from him to say he was on his way in—he had been brought down, had got into Coutaing where he found no one of either side. He lay down in a house and went to sleep and while he slept the town was heavily shelled and the house was hit. He got out and managed to get through his message—and Watt, sitting at home in the office there threw up his cap till it hit the ceiling.

Of 20 officers who went out in September, 10 are still there. Of the missing only 2 are known to be killed. They may hear of the others any day—from a French hospital, or from Germany. Often the first thing they hear of a British officer prisoner is the passing of his cheque through Cox's Bank.

Watt is careful to encourage in every way good relations between the AFC and the RFC—insisting that the Australian Corps is really part of the RFC and making them proud of it. It makes the work better and smoother and the boys more generously disposed to those who are on the same big work. It makes one a little ashamed of having taken a narrower view. The youngsters certainly <u>are</u> proud of their connection with the RFC, and like to think that they carry out its orders, and can go home and train its recruits when their rest time comes, and so forth.

I wrote an article on them at the Hotel du Rhin.[6]

6 C E W Bean, personal diary Nov–Dec 1917, AWM38 3DRL Item 94 [1], pp.84–91.

One of the Squadron's pilots, Lieutenant Lawrence Benjamin, also remembered the CO's long days:

> Every morning, no matter how early the dawn patrol left the aerodrome, "Toby" would be up to see them off, even if his work had kept him up late the night before. His one thought was for us, and we in our turn idolised him.[7]

Now in the lull after Cambrai the DH-5 was phased out of service and replaced by the SE-5A in all RFC squadrons in December 1917. No 68 Squadron had received its first SE-5A (A4856) back in late October to begin pilot conversions. Within two days, 18 pilots on the Squadron had flown the new type. Deliveries commenced after the Cambrai Battle on 7 December.

On the afternoon of 11 December, 2nd Lieutenant Harold Cornell took off from Baizieux in one of the Squadron's remaining DH-5s (A9324) when he stalled and nose-dived into the ground from 50 feet. He was killed on impact, and buried in Dernancourt Cemetery.

The final DH-5 ground attack missions were flown on 12 December. A line patrol training sortie was flown for the benefit of some new pilots on 15 December, then the last of the DH-5s were transferred to No 2 Aeroplane Supply Depot four days later. Now the Squadron was properly equipped and could concentrate on training with its new machines.

One of 68 Squadron's first SE-5A scouts, B55, photographed by war correspondent Charles Bean during his visit to Baizieux on 7 December 1917 IWM E(AUS) 1447

7 Letter from Lt L Benjamin, of 22 Sep 1921, in Ure Smith, Stevens and Watt, pp.43–4.

The Scout Experimental-5 (SE-5), a product of Britain's Royal Aircraft Factory, was introduced to beat the latest German Albatros scout that had proved so effective at the Front.[8] It was the first Allied scout into service to be armed with two guns—a drum-fed Lewis mounted on the top wing centre section, and a synchronised Vickers in front. The synchronised Vickers had a rate of fire of 1,000 rounds per minute.[9] The Lewis had been adapted for aerial use and improved to fire at 600 rounds per minute, with a magazine capacity of 97 rounds.[10]

The SE-5 had begun operations with No 56 Squadron in April 1917, and its 150 hp Hispano-Suiza engine gave a top speed of 120 mph and an excellent rate of climb. The improved SE-5A, powered by a 200 hp Hispano-Suiza followed in service in June, and by the end of 1917 seven RFC squadrons, including No 68, had equipped with this very capable and popular machine.

Cockpit of the SE-5A, showing the Aldis sight and Vickers gun IWM Q67071

These new SE-5As could fly three-hour sorties, and this endurance was to become of great importance to the attacker. Its versatility extended No 68 Squadron's effectiveness in delivering air power—it now had superior speed to chase hostile aircraft from the sky, to defend its own reconnaissance vehicles and to check enemy

8 PRO AIR1/2132/207/137/1 "Development of Airplanes During the World War", paper by Lt Col V E Clark US Army, Feb 1919, p.13.

9 PRO AIR1/2301/215/1 "Aerial Armament 1914–1918".

10 PRO AIR1/2301/215/2 "Aircraft Armament". The Lewis was manufactured by the Birmingham Small Arms Co and in France by *La Societe Francaise des Armes Lewis*, IWM 73/183/1.

movements. Its performance would allow a true multi-role capability. While controlling the air, this new scout would also effectively conduct the ground attack role against enemy troops, communications and supply lines that it had pioneered within the limitations of the lumbering DH-5.

The Squadron now began an intensive training program in aerial combat to exploit its advantage over the German scouts. The pilots' experience gained already at the Front had taught them some tricks of the trade. Approaching from the rear and below in the blind spot of the slow two-seaters was quite effective, enabling an attack to be pressed home at close range. However, against a fast scout a quick single firing pass in a dive was often enough to send down an adversary. Deflection shooting, when at a large angle off an adversary's tail, was also being pioneered, but with inaccurate results.

The overwing Lewis gun fitted to the SE-5A RAAF

Another effective tactic in combat was to rudder turn. This meant the machine skidded around a level turn, which was difficult for an adversary to detect, and threw off his aim. Good shooting was the critical factor in aerial fighting, and pilots needed the nerve to get in and shoot at point blank range, or at least within 20 metres.

The SE not only provided superior speed, but a good climbing performance over the enemy. Rate of climb to attain altitude quickly was used to get the tactical advantage from height and had become an important requirement for the fighter. By this closing year of the war fighters were flying up to the cold heights of 20,000 feet without oxygen. These altitudes were sought, in spite of the intense cold and the effects of hypoxia, for safety and the advantage of bouncing the enemy by surprise with the additional speed from the dive. Rarely would combat be initiated from the same level or below, unless into an enemy's blind spot. Control of the air was won by the side exploiting these advantages.

Training in handling the new machines, formation flying and weaponry at the nearby Warloy Aerial Range continued throughout December and into the new year. The flexibility of the SE meant that the Squadron could effectively be dual tasked. In Close Offensive Patrols, high level patrolling to maintain control of the air, the SE-5A was more than a match for the German scouts and, if required, the Squadron could be thrown into the ground attack role it had helped to develop. In these final years of the war air power had become a significant factor in the success of major ground offensives.

The bad weather of January 1918 had not been suitable for offensive air operations on the Western Front, and the month passed quietly for the Australian Squadron. Training continued on the new mounts, and accidents, particularly landing, prevailed. Although pilots were full of praise for the SE's fine flying qualities and performance, it was a tricky machine to land. Unless a three-point touchdown could be mastered, a bounce was assured. Also the undercarriage was not as strong as the contemporary Sopwith Camel, so throughout the SE squadrons a smashed undercarriage became the expected thing. If these difficulties were not enough, the SE also tended to swing on touchdown.

On 27 December 2nd Lieutenant Alex Clark[11] had written-off one machine (B64) when he ran into a snow drift after landing. Lieutenant Robert Clark came to grief touching down at Baizieux on 3 January, and penned to the back of this photo of his crashed machine B4859:

> "At this game—those who live, learn—and those who don't, teach others by their mistakes."
> —Oswald Watt

"At this game — those who live, learn — and those who don't, teach others by their mistakes" — Oswald Watt J LEEUWIN-CLARK

11 Capt A G Clark. 12th Lt Horse, No 68 Sqn RFC/2 Sqn AFC, 6 Sqn. Grazier of Sydney; b Sydney 11 Aug 1896.

Lieutenant Taylor (B74) was also to crash on landing that afternoon, but was uninjured. In his accident Robert Clark had broken his ankle and was repatriated to England, where he later instructed with fellow 68 Squadron pilots on the Australian SE-5 training unit—No 6 (Training) Squadron at Minchinhampton.

Significant for the Australians at this stage was the change of title for the unit. From Friday 4 January, its old identity was regained, and No 68 (Australian) Squadron Royal Flying Corps became No 2 Squadron Australian Flying Corps.[12]

Orders were then received to deploy to join the 10th Wing of the First Army at the Auchel/Lozinghem aerodrome, 60 kilometres north of Baizieux, near Bethune. On 6 January the first of the Squadron's machines left for their new base. 2nd Lieutenant Percy Lawson[13] (B55) crashed on the approach to land at Auchel. His machine flat turned into a spin from 400 feet, caught fire on impact, and he was burnt to death. He had been on the Squadron for six weeks. The remainder of the aircraft deployed from Baizieux three days later. Lieutenant Lewis Truscott,[14] another new arrival, was diverted by bad weather to Yvrench. He crash landed, damaging his machine (A8906), but was uninjured.

Another Squadron move was carried out later in the month to Savy, an aerodrome 20 kilometres to the south, near Arras. On 22 January, Les Holden wrecked his aircraft (A8936) taking off from Auchel for Savy. On the same detail 2nd Lieutenant Frank Power[15] (C5312) crashed on landing at the new airfield. He survived another incident two days later when his new machine (B504) suffered engine problems, and he crashed during his forced landing. If minor damage could be repaired within 36 hours, this work was done on the Squadron. Otherwise, more extensive damage ensured that No 2 Aircraft Depot, at Candas, was kept busy with SE-5 repair work.

The first operational flying with the SEs was flown by "C" Flight on 24 January, led by the new Flight Commander Henry Forrest, escorting photographic reconnaissance machines of No 2 Squadron RFC. Offensive patrols were mounted for the remainder of the month, but these were uneventful, with no engagements with the enemy.

Air operations in February were again severely hampered by bad weather. On 5 February, four enemy scouts were engaged near Lille with no result. Heavy fogs over the next 10 days restricted operations to weaponry practice and general flying. On Saturday 16 February, command of No 2 Squadron passed from Oswald Watt to Major William Sheldon.[16] A graduate of the Second Course at Point Cook, Sheldon

---

12 Official records refer at various times to No 68 (Australian) Sqn RFC, and to No 68 Sqn AFC. With the new identity, the unit was alternatively called No 2 Sqn AFC or the 2nd Sqn AFC. AWM10 4301/10/35, of 31 Mar 1917; AWM25 225/6 Pt 1 180, of 13 Nov 1917; AWM25 225/6 Pt 1 204, of 11 Dec 1917; AWM22 509/1/1, of 4 Jan 1918; PRO AIR1/1044/204/5/1506, of 11 Jan, 27 Jan and 21 Aug 1918; and PRO AIR1/2086/207/6/1 AO/1005, of 21 Aug 1918 all refer.

13 Lt P H Lawson. No 68 Sqn RFC. Engine driver of Adelaide; b Largs Bay, SA, 1897. Killed in flying accident 6 Jan 1918.

14 Lt L S Truscott. No 68 Sqn RFC/2 Sqn AFC. Motor mechanic of Sydney; b Newcastle, NSW, 2 Dec 1895.

15 Lt F A Power. 12th Lt Horse, No 68 Sqn RFC/2 Sqn AFC. Agriculturalist of Melbourne; b Melbourne 9 Sep 1891.

16 Maj W Sheldon. RAFA, CFS, No 1 Sqn AFC, 14 Sqn RFC, 67 Sqn; comd No 71 Sqn RFC 1917, No 2 Sqn AFC 1918. Army officer of Melbourne; b Singapore 20 Aug 1889.

had been the "A" Flight Commander in Egypt with No 67 Squadron, and then had led the other Australian fighter unit, No 71 Squadron RFC, destined to become No 4 Squadron AFC, for its training period in England during 1917. Watt returned to England to organise and lead the new Australian Training Wing.

Fred Huxley (B4895) did have some success on the 18th to open the account with the Squadron's new machines. He was leading a pair, and with 2nd Lieutenant Anthony Paxton[17] (C5347) encountered four Albatros scouts over Lille. Huxley attacked the leading machine which had a red nose, green fuselage with yellow stripes and blue tail. He was able to drive this gaudy machine down out of control:

> I had superior height and dived on the leader who turned on his wing tip and fell out of control after I had fired about 50 rounds at 150 yards range. I then attacked another machine of hostile formation, with no result. The EA then flew East of Lille. One enemy machine attacked me whilst I engaged the leader. 2/Lt Paxton engaged one of the enemy machines and fired a burst of 20 rounds with Lewis gun with no result.[18]

Continuing their patrol over the lines, the pair then encountered a DFW two-seater, which Huxley pursued, exhausting his ammunition as the enemy machine dived to the east. Later that morning Richard Howard (D212) and Les Holden (D214) departed on an offensive patrol east of the lines towards Lille, and near Carvin met six red-tailed Albatros scouts. Howard dived his pair into the attack, sending one Albatros into a vertical spin, with the others heading off east to land at Lille.

The following morning, Lieutenants McKenzie (C9539) and Benjamin (B579) engaged three Albatros scouts near Pont a Vendin, north of Lens. Although McKenzie's Vickers gun had failed, he was able to successfully engage one aircraft with his Lewis gun, and he saw it fall out of control. The British Army observed the enemy machine crashing, and McKenzie was credited with its destruction.

Six RFC pilots had been attached to the Squadron on 10 January to gain experience at the Front, and two were to feature in the operations of 21 February—2nd Lieutenants Ronald Lang[19] and George Logan.[20] Lieutenant Alex Clark (B4895) led four machines off on a morning patrol, with Benjamin (C1060), Lang (B535) and another new arrival, a Canadian, 2nd Lieutenant William Adams[21] (C9496). Flying at 15,000 feet, Clark sighted an enemy formation of 10 red-nosed Albatros scouts near Brebieres, south-west of Douai:

> One hostile machine dived at me. I fired both Lewis & Vickers into the nose of this machine, getting a burst of about 50 rounds into it at a range of 10 yards. The machine fell over on its left wing and dropped vertically in an uncontrolled spin.

---

17 Wg Cdr A L Paxton DFC RAF. No 88 Sqn RFC, 2 Sqn AFC, later RAF. Rancher of Sonora, Mexico; b Sydney 7 Aug 1896.

18 AWM4 8/5/1 Combat in the Air Report No 27, of 18 Feb 1918.

19 Lt R Lang MM. Royal Engineers, No 88 Sqn RFC, 2 Sqn AFC. Of London; b 14 May 1895.

20 Lt G C Logan. RFC, 2 Sqn AFC. Of Peterborough, Ontario, Canada. POW 21 Feb 1918.

21 Lt W Q Adams RAF. Can Exped Force, No 65 Sqn RFC, 2 Sqn AFC. Of Crystal City, Manitoba, Canada; b Edinburgh, North Dakota, USA, 6 Sep 1893.

> Lieut Benjamin engaged an enemy machine which dived across his front. He fired a burst of about 50 rounds from his Lewis gun at about 30 yards range, the machine went down out of control, finally getting into a steep fast spin and was not observed to come out of it 2000 feet lower.[22]

Ronald Lang observed one of the enemy machines crashing, but in the engagement he was badly wounded. He succeeeded in making it back to land at No 5 Squadron's aerodrome at Acq, between the lines and Savy; however, his wounds necessitated the amputation of his leg. Later in the morning, 2nd Lieutenants Paxton (C5347) and Logan (B619) departed for an offensive patrol. Over the German lines they were engaged by groundfire, with Logan being shot down and taken prisoner.

Bad weather brought a quiet period for the rest of February and early March and the enemy squadrons used this respite to nurse their strength for the imminent great offensive. A resurgence of air activity on 8 March saw Captain Howard (D212) drive down two black two-seaters near Lens, sharing one with Paxton, and then an Albatros to the south at Brebieres. That afternoon Les Holden (D214) also scored an Albatros.

The Squadron was successful above the Front near La Bassee over the next two days with Richard Howard (D212) driving down two Albatros scouts on 9 March, and another on the 10th. Over the next days, both Les Holden and Robert McKenzie were then to score several enemy machines each, with Howard scoring again on the 18th, destroying a Rumpler two-seater in flames over Haubourdin, near Lille:

> I first noticed this machine flying East from Lines towards Lille gradually losing height. Our formation approached unobserved and I dived and fired about 60 rounds into enemy machine; smoke immediately issued from fuselage. Lieut Holden and Lieut Rackett saw the machine diving steeply out of control and in flames.[23]

Perhaps the most significant development at the end of 1917 had been the Armistice between Germany and the revolutionary Bolshevik regime in Russia. Finally, in early 1918, all was quiet on the Eastern Front.

In France the Allies once again had the control of the air, but now hundreds of thousands of German troops with large numbers of aircraft that had been fighting against Tsarist Russia could be released to the Western Front. Now, as the weather cleared, a final massive offensive was expected to smash the Allied armies once and for all.

22 AWM4 8/5/1 Combat in the Air Report No 35, of 21 Feb 1918.

23 AWM4 8/5/1 Combat in the Air Report No 47, of 18 Mar 1918.

CHAPTER 6

THE GERMAN SPRING OFFENSIVES

No history can be expected to furnish a full record of all acts of prowess that were performed in the air during the long course of the war. Many of the best of them can never be known; the Victoria Cross has surely been earned, over and over again, by pilots and observers who went east, and lie in unvisited graves.

Sir Walter Raleigh[1]

The Armistice with Russia allowed Germany to strengthen its forces in the west in early 1918. With America's entry into the war, General Ludendorff, commander of the German Army, needed to strike before the arrival of any substantial numbers of United States forces, but any offensive could not be planned before the end of winter.

During March, with the German assault imminent, No 2 Squadron's offensive patrols, flying normally at around 15,000 feet, had been reporting on all enemy movements. As anticipation of the German attack grew, the strength of each offensive patrol was increased to two flights, numbering up to 10 aircraft. The weight of the German offensive was to fall on the fronts held by the British Third and Fifth Armies, to the south of Arras. Enemy air activity had intensified, although it was still restricted by increasingly bad weather.

On the misty morning of Thursday 21 March, the Germans began their push, which was to become their last great offensive, with an immense assault of 56 divisions (the First Battle of the Somme 21 March–5 April). This was directed against a 100 kilometre front, between the Rivers Scarpe and Oise. The RFC had 31 squadrons in the area of the Third and Fifth Armies, a total of 579 serviceable machines of which 261 were single-seat fighters. Opposing these the Germans had 730 aircraft, including 326 single-seat fighters.[2] When the fog cleared, every machine that could carry bombs was thrown into the attack. The heroic bombing and strafing attacks by Allied aircraft helped prevent the British retreat from turning into a rout. By 1918, air power had indeed become a major factor in the war on the ground.

No 2 Squadron, attached to the First Army, was diverted with other fighting squadrons to cover the Third Army front. Here, to the south of Lille around Douai, the Germans had concentrated their best units, with their brightly painted Albatros scouts and Fokker triplanes. The Australians were required to operate on this front for the remainder of the battle. Large formations of scouts would engage daily in dogfights, preventing opponents from attacking and creating havoc amongst the low-flying ground support flights.

Tactics were now changed with the Squadron flying across the lines at around 1,000 feet to protect the dedicated ground attack aircraft and carry out reconnaissance

1 Sir Walter Raleigh, *The War in The Air*, Vol I, Clarendon Press, Oxford, 1922, pp.6–7.

2 Jones, p.273.

of the enemy's movements. The SEs were also briefed to engage targets of opportunity, such as troop concentrations and transport; anything to hamper the enemy's advance. Over the last 10 days of the month, No 2 Squadron's SEs fired 14,000 rounds at ground targets. This changeover to low flying was not to change the overall situation as the Germans, too, had left the upper air.

Officers of No 2 Squadron at Savy, 25 March 1918:
Front (from left): Lt A G Clark, Capt L H Holden, Capt H G Forrest, CO Maj W Sheldon, Capt R C Phillipps, Capt F R Smith, Lt L S Truscott. Middle: Lt A C Cameron, Lt W Q Adams, Lt L Benjamin, Capt R W McKenzie, Lt F W Sexton, Lt C F Felstead, Lt L J Primrose, Lt L F Loder, Capt R L Manuel. Back: Lt T Hosking, Capt E D Cummings, Lt F W Follett, Lt O T Flight, Capt G H Blaxland, Lt G Brettingham-Moore, Lt H E Hamilton, Lt A R Rackett AWM E1883

The effect that Allied air power had on the enemy in their assaults was described by the war diary of the German 73rd Regiment:

> The English got valuable support from their aircraft which attacked regardless of consequences. The squadrons, flying very low, found profitable targets for bomb and machine-gun in the thickly concentrated masses of the 111th and 2nd Guards Reserve Divisions. Our own aircraft were absent.[3]

On 22 March, the British line was broken west of St Quentin, enabling the Germans to push towards the River Somme. Ten SE-5As of No 2 Squadron took off to rush the 90 kilometres south-east to St Quentin, on the Fifth Army front. Two turned back with engine problems, the other eight encountered five two-seater machines escorted by a number of single-seaters. Captain Forrest (C9539) sent a two-seater down in flames,

3 War Diary, German 73rd Regiment, of 22 Mar 1918. Cited in Jones, p.301.

and Lieutenant McKenzie (C5382) forced down an Albatros D.V out of control. They headed north to Bourlon Wood, where Captain Phillipps (C9541) attacked the leader of a flight of Fokker triplanes, sending it down into a spin. On the return to Savy more Albatros scouts were engaged over the lines at Bullecourt and Forrest sent down another two, with Lieutenant Les Holden (D214) scoring a third.

No 2 Squadron's "C" Flight at Savy during March 1918 IWM E(AUS) 1881

On an evening patrol, Lieutenant Archie Rackett[4] (B70) drove down a two-seater at Honnecourt, near Bapaume. However, 22 March was not to prove totally successful for the Australians. From this formation the Squadron lost one of its experienced flight leaders. Captain Richard Howard (D212) was last seen at 3,000 feet about three kilometres over the lines near Epehy, south of Cambrai. Then, in an engagement 15 kilometres to the south near Vermand, he was shot down by Lieutenant Boehning of Jasta 79, a 17-victory ace, and died of his wounds that night.[5] He had come to France with the Squadron in December, and had been awarded the Military Cross for conspicuous gallantry and devotion to duty during the Battle of Cambrai. In February, he had been promoted to Flight Commander and during March he had created a record on the Squadron by accounting for seven enemy aircraft. His place as "B" Flight Commander was taken by Les Holden.

By the 23rd, the Allies were in retreat, with the Germans making substantial ground on the Front from Cambrai down to St Quentin. The roads on the Allied side were full of traffic, as 10 Australian SEs went out on a morning offensive patrol along the Third Army Front between Arras and Douai. McKenzie (C5382) dived on a two-

4 Lt A R Rackett. Army Signals, No 2 Sqn AFC. Painter of Mt Gambier, SA; b Adelaide 18 Feb 1896. POW 1 Jun 1918.

5 Shores, Franks & Guest, p.200.

seater between Brebieres and Vitry and destroyed it from close range. Ten minutes later, flying south about 15 kilometres behind the lines, they encountered a formation of eight enemy scouts. Phillipps (C9541) was successful in driving down a red-nosed Albatros north of Bourlon Wood. That afternoon, Forrest (C9539) destroyed a two-seater near Bapaume, and Phillipps (C9541) drove down another, killing the observer, before it made the safety of the enemy's lines. No 2 Squadron flew 65 hours over the lines this day.

The following day, another two enemy aircraft were claimed by the Squadron in addition to its continual harassment of the German advance, which was now threatening Bapaume and pushing towards Albert. With all Allied fighter squadrons thrown into ground attack, air activity was having quite an effect on German morale. A German soldier of the German 8th Grenadier Regiment recorded:

> As we were moving forward after crossing the Somme, there suddenly appeared before us some twenty British aeroplanes which dived to a height of about 100 metres and then, continuing to within 2–3 metres of the ground, attacked us with their machine-guns. Several Tommies flew so low that the wheels of their aeroplanes touched the ground. My company commander, Lieutenant Nocke, had to fling himself flat on the ground, but for all that he was struck on the back by the wheels of one machine, thus being literally run over. Not far from me an aeroplane appeared about one metre above the ground making straight for me and for the moment I did not know in what direction to throw myself: the pilot appeared determined to run over me.[6]

By 25 March, the position on the Front was grave, with large concentrations of enemy infantry making further progress between Arras and Peronne, and forming to the east of Bapaume. Stubborn resistance by the Third Army was successful but the Fifth Army defending Amiens, in spite of French reinforcement, was forced to withdraw, and some airfields had to be evacuated. Allied ground attackers were taking heavy losses from groundfire during these near suicidal missions. A flight of Australian Camels from No 4 Squadron was bounced by enemy scouts during a strafing run, but was able to shoot down three of its attackers. However, the Camels lost two pilots, with another three wounded, all to ground fire.

That evening, Major-General John Salmond, the new GOC of the RFC in the Field, wrote to his predecessor, Trenchard:

> We managed to concentrate 100 machines on the threatened line of the 3rd Army. They had orders to low fly and to take every risk; nothing was to count in carrying out their duties. I had news from the 1st Brigade that our machines were so thick over this point that there was every danger of collision in the air, so it looks as if we must have had a great effect against the enemy attack.[7]

The next morning, as the Allied Front was driven back towards Arras and Albert, the Squadron was tasked with escorting the low-flying Camels, and joined them in strafing the enemy advance. Lieutenant Lawrence Benjamin (C5382) was over the lines, as enemy forces were approaching Albert, when he spotted an Albatros scout:

---

6 War Diary, German 8th Grenadier Regiment, cited in Jones, p.316.

7 PRO AIR1/475/15/312/201, General Salmond's letter to General Trenchard, JS 2/1/48A, of 25 Mar 1918.

THE FRONT - 23 MARCH 1918

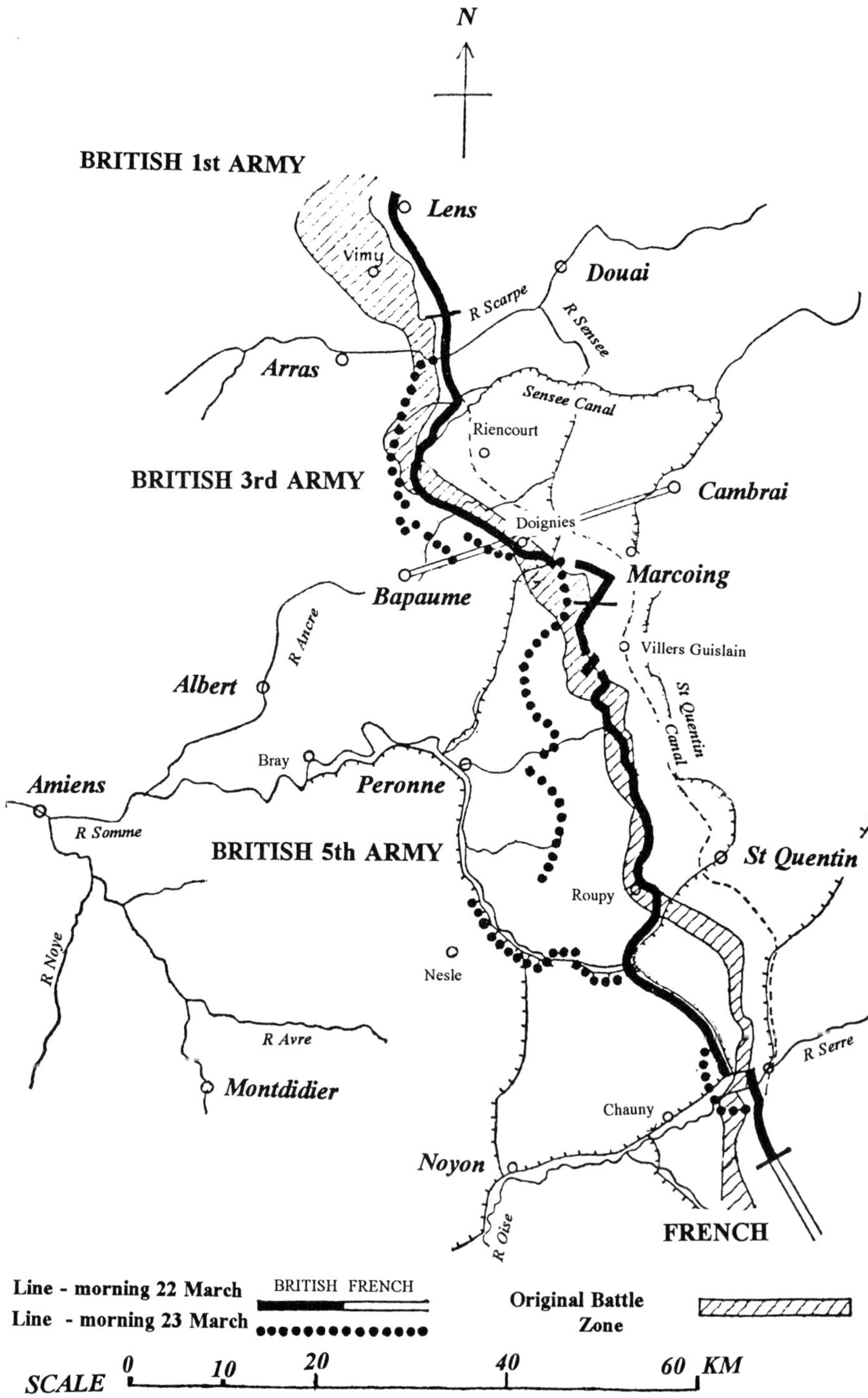

The Front — 23 March 1918 RAAF

> I observed this machine near Pozieres flying NW towards Arras. I dived on it, firing a continuous burst with both guns until within 20 yards of it; it rolled over on its back, burst into a cloud of smoke and went down in a vertical nose dive.[8]

By the 27th, the rapid German advance over the week had succeeded in the capture of Albert. No 2 Squadron's offensive patrols, in addition to protecting the low-flying Camels, strafed the enemy's trenches and transport. At noon, Phillipps (C9541) reported:

> When near Albert, our formation was attacked by 6 triplanes. I zoomed up into the mist and came out over Pozieres. I got onto the tail of a triplane and caught it over Suzanne and when at close range I fired about 50 rounds and the enemy machine burst into flames, and fell slowly to earth.[9]

Twenty minutes later, returning to patrol Albert, he spotted a new type enemy scout:

> I was following one of our machines across the line, when I saw an enemy machine below me about 200 feet, firing at our troops. I dived on it, and, when at close range, fired about 100 rounds and machine rolled over and crashed, apparently on our side of lines near Meault.[10]

These were the fifth and sixth victories for the "A" Flight Commander. In the same formation 2nd Lieutenant Thomas Hosking[11] (A8906) destroyed a Fokker triplane at Fricourt, five kilometres east of Albert.

On the morning of Thursday 28 March, the Germans launched a major assault on Arras. Lieutenant Les Holden led a formation, 12 strong, on an offensive patrol escorting low-flying Camels of No 4 Squadron AFC, which was now also part of the 10th Wing. From this mission Thomas Hosking (A8913) failed to return. He had only been on the Squadron a month, and had just claimed a kill the previous day. Another newcomer was also lost that afternoon, when Lieutenant Oscar Flight[12] (B102) departed on a similar sortie. He became separated from his formation and was eventually bounced and shot down behind enemy lines at Etaing, near Arras. He later described the details of this, his last flight.

> Four of us left the Aerodrome to escort No 4 Australian Squadron and also for trench strafing. Just after getting over the enemy's lines we were split up on diving through clouds. After this I sighted one of No 4's machines and followed him about ½ hour. At the end of this time the machine I had been following went home. This was about 12.45pm. Then I went to Arras and strafed along Arras-Cambrai Road. I was there about 15 minutes under a good deal of AA and machine gun fire when all of a sudden AA fire stopped. Very soon after this,

8 AWM4 8/5/1 Combat in the Air Report No 59, of 26 Mar 1918.

9 AWM4 8/5/1 Combat in the Air Report No 60, of 27 Mar 1918.

10 AWM4 8/5/1 Combat in the Air Report No 61, of 27 Mar 1918. The enemy scout could possibly have been a Siemens-Schuckert D.III.

11 Lt T Hosking. 8th & 39th Bns, No 71 Sqn RFC/4 Sqn AFC, 2 Sqn. Electrician of Melbourne; b London 1896. Killed in action 28 Mar 1918.

12 Lt O T Flight. Army Engineers, No 2 Sqn AFC. Student of Bendigo, Vic; b Bendigo 1895. POW 28 Mar 1918.

shots came at me from the rear. On looking round I saw three enemy Albatros machines. I pulled round in a climbing turn, but two of the machines had too much height on me. The fire was more or less incessant for a minute or so. Up to that time my machine had not been hit. About this time four enemy triplanes joined in the fight and gradually forced me down to within 150 feet of the ground. Bullets were coming from all directions. All at once one machine got a big burst into my right plane and rear strut, also severing a flying wire. A few seconds later my rear strut flew out; the bottom socket had been blown out. The machine immediately heeled over and all attempts to right it failed. I switched off and held onto the front of the cockpit. The machine then dived into the ground tearing down a number of enemy telephone wires. I was rendered unconscious.[13]

When Flight regained consciousness he was able to set fire to the wreckage with his Verey pistol before being captured. He was sent to Landshut POW camp, in Bavaria. During September, he tried to tunnel out of the prison with another AFC pilot, Lieutenant Cecil Feez of 4 Squadron, who coincidently had been shot down and captured on the same day in March. They were informed upon by an Italian prisoner, and remained POWs until the Armistice. In December 1918, he was repatriated to England.

The German thrust to the south-east of No 2 Squadron's aerodrome at Savy had pushed the Third Army front back to a line joining Arras and Albert, but here they were brought to a standstill. On Good Friday, 29 March, the enemy was then pushed back in retreat, with heavy German losses. Indications were that the tension on the Third and Fifth Army sectors had ended, and the First Army to the north might be attacked any time in early April. Consequently, the Squadron deployed on 2 April south to Bertangles, north of Amiens, with the 22nd Wing RAF. Then, two days later, the Squadron was grouped into the 51st Wing of IX Brigade at La Bellevue. This was 20 kilometres from Arras, which the Allies considered would be the vital point for the next attack on the threatened northern front.

A German Halberstadt Cl.IV two-seater forced down into the Australian lines AWM E3046

13 AWM30 B3.6 Statement by Lt O Flight, Repatriated POW, of 27 Dec 1918.

The start of the month—Easter Monday 1 April 1918—had seen the birth of the first independent air arm with the formation of the Royal Air Force by the amalgamation of the Army's RFC and the Royal Naval Air Service. There had been considerable inter-service wrangling over the creation of this new service, but Lieutenant-General Jan Smuts' committee had won the day, and with foresight in 1917 they had pointed out:

> The day may not be far off when aerial operations with their devastation of enemy lands and destruction of industrial and populous centres on a vast scale may become the principal operations of war, to which the older forms of military and naval operations may become secondary and subordinate.[14]

The first chief of the new force was Major-General Sir Hugh Trenchard. It was a significant time for the birth of the world's first air force. The German advances had been halted and air power was being decisive in ground battles, preventing an enemy breakthrough to the Channel coast.

Because of enemy air inactivity, most of early April for No 2 Squadron was spent on low-level work, dropping bombs and strafing enemy trenches and strongpoints. The German troops were exhausted and, with transport hampered, any further offensive against Amiens was frustrated. Now the focus of German attacks was switched against the Second Army, further north in the Ypres-Armentieres sector, on the plain of the River Lys. It was here, between La Bassee and Armentieres, that Ludendorff decided to strike in the German Lys offensive (9 April–8 May).

The second week of April began with pouring rain, restricting the Squadron's flying effort until the 11th. On that evening, an offensive patrol of 10 SEs met a large formation of enemy Fokker triplanes and Albatros scouts, and drove them back to the east. The next morning, another large effort was mounted, this time with Captain Alex Clark (B4895), now the "B" Flight Commander, driving down an Albatros over Lorgies, between Bethune and Lille. That afternoon, Captain Forrest (C9539), the "C" Flight Commander, destroyed a two-seater beyond the lines over Vieille Chapelle, north of Bethune.

The German attacks had led to rapid advances on the First and Second Army fronts between Bethune and Ypres, but by the end of the month the Allies had consolidated into a strong defence. To the south, the German commander Ludendorff was determined that his offensive would not again develop into what he described as a "battle of exhaustion".[15] His last push on the Somme was made on 24 April, when the Germans captured Villers-Bretonneux and Hangard. However, the following night, these towns were retaken by Australian troops, thus ending the last German chance to take Amiens.

For his leadership during the spring offensives, Captain Roy Phillipps was awarded a bar to his Military Cross during April. His award read:

---

14 PRO AIR8 2/ID2/132 "Committee on Air Organisation and Home Defence Against Air Raids (2nd Report)", paper 5, GT.1658, of 17 Aug 1917, p.4.

15 General Ludendorff, *My War Memories*, p.600, cited in Jones, p.367.

Conspicuous gallantry and devotion to duty. When engaged with hostile aircraft during the recent operations, he has destroyed three enemy machines, and has in addition brought down two hostile planes out of control. He has also brought back accurate and valuable information regarding hostile movements under the most adverse conditions, particularly on one occasion, when he flew through a heavy barrage at a low altitude behind enemy lines. He has shown conspicuous skill and determination when escorting low flying bombing patrols.[16]

A new pilot to the Squadron, who was to follow in Phillipps' footsteps in becoming one of its most successful aces, was Lieutenant Eric Cummings.[17] The press later reported his first combat of 3 May 1918 with much excitement:

One of the most popular pilots in No 2 Squadron was Captain Cummings DFC of Hobart. During the spring of 1918 the Hun was doing a lot of damage with his tri-planes—facetiously known to all our men as "Tripes"—which machines were then being flown by Richthofen and his famous Circus. Encounters with these machines at about 17,000 feet were a daily occurrence, and some terrific "dog-fights" resulted. Whilst doing a patrol Captain Cummings—then a Lieutenant, by the way, and almost new to the Front—spotted a Hun two-seater a few thousand feet below. Sticking the nose of his SE-5 down he dived after the Hun, and opening fire when about 100 feet away, had the satisfaction of seeing the two-seater topple over and fall after a few bursts. Turning to climb back again, Cummings saw one or two suspicious rips appear in his wings, and the next second a bullet whizzed past his head into the windscreen. He looked round to see the ugly shapes of half-a-dozen tri-planes diving on to him, and a sky that was full of little phosphorous paths made by their tracer bullets. Cut off from any assistance his position was desperate, but he decided on his tactics immediately. Turning the nose of his machine towards the nearest Hun he flew straight at him with both guns blazing tracer bullets. It was a great nerve test, for both machines were racing nose-on towards one another, apparently to a certain collision. The nerve of the Hun gave out. He pushed his joy-stick forward and dived under the very nose of Cummings' machine. Instantly the Australian dived after him; the couple of seconds that he was on the Hun's tail were sufficient, for a well-placed burst hit the machine in a vital spot and it went spinning down to earth. But the other enemy machines followed Captain Cummings down, and the bullets were ripping his tailplane to shreds. The dashboard in front of him was perforated with bullet-holes, and the cockpit filled with fumes from the compass and the thermometer as one after another of his instruments was smashed.

It was hopeless to try and fight the Germans off, so he put his machine into a terrific dive in the hope of getting clear. The Huns followed down with an unceasing stream of bullets but the speed forced them to pull out of the dive for safety one by one. With a "conked" engine and a machine shattered by bullets, and the strain of over diving, the Pilot gingerly steadied up, glided just over "No Man's Land" and crashed into our support trenches, while the Hun made frantic efforts with machine and field guns to hit him. Beyond a shaking up from the crash he was unhurt.[18]

In this engagement over Meteren, on the Second Army front south of Ypres, Cummings (B188) was credited with a Fokker Dr.I destroyed, and it is believed he was

16 AWM4 8/5/1 Summary of War Diary, for month ending 30 Apr 1918.

17 Capt E D Cummings DFC. AASC, No 2 Sqn AFC; later Austn Air Corps, then RAF. Student of Franklin, Tas; b Braidwood, NSW, 13 Apr 1896.

18 "Knights of the Air", *Smith's Weekly*, Sydney, Saturday 29 Mar 1919, p.13.

shot down by Lieutenant Karl Bolle of Jasta "Boelcke".[19] He crashed beyond the lines near Merville, and was rescued by Australian troops while German guns destroyed what remained of his wrecked SE.

The German Air Service commander, General von Hoeppner, had considered that both the Somme and the Lys Battles had similar characteristics.[20] In the first two days of each offensive, the Germans had held air superiority. Then, with the arrival of Allied aerial reinforcements, control of the air had passed back to the Allies.

Now the Allies were developing new tactics to maintain that control of the air—the German doctrine of "Circus" formations was being adopted. From the Allies' perspective, the difference was that the Germans were always unwilling to attack, unless from a winning position with surprise, superiority in numbers, and an advantage in height. Indeed, they were not very enterprising, and did not hesitate to flee from a scrap if these factors were not in their favour. On the other hand, the Allies would always maintain the offensive in their quest for control of the air.

19 AWM4 8/5/1 Combat in the Air Report No 69, of 3 May 1918; and Shores, Franks & Guest, p.128.

20 General von Hoeppner, *Deutschlands Krieg in der Luft*, pp.157–8, cited in Jones, p.403.

CHAPTER 7

SWEEPING THE SKIES

I do not care who the man is, or how stout of heart, his early days of flying over enemy country are characterised by fear, and it is the sort of fear that robs a man of his initiative and determination, the two greatest factors in the character of a successful fighter.

Harry Cobby[1]

The move south for No 2 Squadron in April had also provided a change in operations for the unit, as it had been selected to roam the Front in large numbers, forming a Circus. This was a stage in the development of air power on the Western Front as the RAF commenced flying large formations, maintaining continuous patrols, to seek out and destroy enemy aircraft. These fighting formations, in close support of one another and hitherto considered as unwieldy, were then to become fairly common.[2]

The amalgamation of the Australian squadron into such a wing was coincident with the death of the greatest of the Circus exponents—Richthofen. The German ace, credited with 80 victories, was shot down on 21 April. The psychological impact of his death, according to the commander of the German Army, General Ludendorff, was equivalent to the loss of 30 divisions.[3] Richthofen had held some interesting views on the attributes of different nationalities in air fighting:

> In my opinion the whole thing is a question of initiative, and that is what we Germans possess. Thanks to it we shall always maintain the supremacy of the air. The Frenchman likes to lie in ambush and pounce upon his prey. This will not work in the air. The Englishman now and then shows a trace of his Germanic blood.[4]

By the beginning of May, the German offensive on the Lys had been contained. The focus of operations now reverted to the south as the Squadron took on its new role as a Circus unit. The RAF's adoption of German Circus tactics saw No 2 Squadron flying with two Royal Air Force Camel units—Nos 43 and 80 Squadrons—forming the 51st Wing. The Circus operated with the Australians at 16,000 feet, above the Camel squadrons, which were staggered below at 2,000 foot intervals. The Circus was to sweep the sky of enemy machines, roaming beyond the lines of the Third Army front around Bapaume. But activity was quiet and not many enemy patrols were encountered. The Squadron felt that perhaps the large formations were acting as a

---

1 Capt Harry Cobby, "Aerial Fighting", cited in E J Richards, *Australian Airmen*, Bruce & Co, Melbourne, p.61.

2 Jones, Vol IV, p.351.

3 R Jackson, *Aces' Twilight*, Sphere Books, London, 1988, p.65.

4 PRO AIR1/1/4/26/4 "Extracts from translation of von Richthofen's book *Der Rote Kampfflieger*", 20 Sep 1917, p.7.

deterrent which, while keeping the upper air clear of enemy machines, was having the detrimental effect of not being able to engage and destroy them. The philosophy in No 2 Squadron was:

> There are two methods of working scouts, viz:—
> (a) In large organised formations capable of dealing with similar formations of the enemy, and
> (b) small patrols of anything up to five machines, whose primary object is to destroy two seaters, balloons and small enemy scout patrols who are liable to molest our Artillery machines.
> The advantages of these methods are as follows:—
> (a) Pilots are encouraged to work with one another and to sacrifice individual results for those of the patrol.
> (b) Much more is given to the individual pilot so that he may take advantage of the fleeting chance which often offers itself.
> The disadvantages are:—
> (a) The initiative of the pilot is apt to be cramped, and
> (b) no experience is gained of the organised methods of Aerial Warfare which become so necessary during Offensive Operations.[5]

A good landing is one you can walk away from — only minor damage was sustained by Captain H G Forrest's SE-5A C9539 from this mishap at Savy in March 1918, and the aircraft was repaired, eventually to be withdrawn from service as "war worn" in June 1918. AWM E1882

Therefore, the suitability of Circus tactics for control of the air was being questioned. The Squadron considered the tactics viable against large enemy formations at height, but during May they had only served to frighten any enemy

5 AWM4 8/5/1 Summary of War Diary, for month ending 31 May 1918.

machines off to the east to avoid combat. Among the few scorers for the remainder of the month were Lieutenants James Adam,[6] Frank Smith[7] and Gregory Blaxland.[8]

On an evening patrol on 17 May, Captain Murray Jones[9] (C9496) was at 16,000 feet over Bapaume, when he pounced onto a red Fokker triplane:

> This machine was one of a formation of 8 enemy triplanes. I followed one down and dived on it, firing about 180 rounds. The machine fell out of control, going down in a vertical nose dive in which it remained until my attention was attracted by three other machines which were attacking me.[10]

In a Circus patrol in the same area on 30 May, 2nd Lieutenant George Cox[11] (D3431), flying behind Captain Forrest, opened his account when they dived on a formation of six Pfalz scouts:

> I singled out one machine and fired a burst of about 50 rounds at a range of 200 yards without effect. I zoomed up and dived a second time, reserving my fire till I was about 50 yards from the EA, then firing a long burst into it. The EA turned over sideways on its back. I zoomed up to engage another EA and did not observe what became of the first machine.
>
> Capt Forrest states:— When diving on this formation, I observed one EA Pfalz scout turn over on its back and burst into flames just as the Camels came up to engage.
>
> This machine is also confirmed by a patrol of 43 Squadron.[12]

An unfortunate accident, common in the fog of war, had occurred to the Squadron earlier in the month. On 4 May Lieutenant Blaxland had mistakenly shot down a French SPAD[13] killing its pilot. This was not an isolated instance in confusion between the Allies. Another SPAD was brought down near Hazebrouck four days later by a British patrol, which reported:

> The French machines were apparently single-seaters coloured yellow and red, the latter predominating, and making the national markings hard to distinguish.[14]

Three days later, on 11 May, a flight of five French SPADs was attacked by another RAF patrol. Salmond noted that the similarity of the colouring of French aircraft to Albatros scouts was causing this misidentification. Accordingly, the *Commandant l'Aeronautique Francaise* was requested to adopt a single colour, one toned similar to the RAF scheme, when French aircraft were operating over the British lines.

---

6 Capt J A Adam RAF. Canadian Exped Force, No 85 Sqn RFC, 2 Sqn AFC, later RAF. Dentist of Dunedin, NZ; b Timaru, NZ, 8 Apr 1888.

7 Capt F R Smith MC DFC. 31st Bn, No 2 Sqn AFC. Clerk of Brisbane; b Brisbane 1896.

8 Capt G H Blaxland. 10th Lt Horse, 16th Bn, No 2 Sqn AFC, 8 Sqn, 2 Sqn. Electrical engineering student of Fremantle, WA; b Broken Hill, NSW, 10 Mar 1896.

9 Maj A M Jones MC DFC*. 46th Bn, CFS, No 1 Sqn AFC, 17 Sqn RFC, 67 Sqn; comd 2 Sqn AFC 1918–19. Pharmacist of Melbourne; b Melbourne 25 Feb 1895.

10 AWM4 8/5/1 Combat in the Air Report No 75, of 17 May 1918.

11 Lt G Cox. No 4 Sqn AFC/71 Sqn RFC, 2 Sqn AFC, 1st Trg Wg. Cabinet maker of Melbourne; b Melbourne 17 Jul 1894.

12 AWM4 8/5/1 Combat in the Air Report No 76, of 30 May 1918.

13 French single-seat fighter manufactured by *Societe anonyme pour l'Aviation et ses Derives.*

14 PRO AIR1/533/16/12/122, letter CRAF 1730.G from Salmond to GHQ, of 21 May 1918.

No 2 Squadron had attempted to cover up this incident, but had been unsuccessful. The following week, Blaxland was posted back to England for instructional duties, and the CO was replaced.[15] Murray Jones was promoted to Major on the 22nd, and took over command. Like other COs of flying units, the squadron commander hitherto had chiefly administrative duties, "and his influence on morale of the squadron had to be exerted from his chair in the squadron office rather from the seat of an aeroplane".[16] Now, as the CO was allowed to fly in combat, Jones became No 2 Squadron's first commander to lead his unit on operations.

The aircraft establishment had also been raised to 19 machines: six for each Flight and one for the CO. During May the Squadron began receiving SE-5As with different power plants. The geared 200 hp Hispano engine, driving a four-bladed propeller, had proved unreliable. As a result machines powered by the licensed-built Wolseley variant, the 180 hp Viper with direct drive to a two-bladed airscrew, began arriving.[17]

In late May, reports of large-scale troop movements on the River Aisne, in the Champagne district near Reims, were being made by the reconnaissance crews. These were ignored, and on 27 May the Germans made a deep advance through the Allied front. The RAF C-in-C in France, Major-General John Salmond, vowed that such an error would not happen again. Air power would guarantee no more such unpreparedness: "the responsibility that the British Army is not surprised is on the Royal Air Force".[18]

By 30 May, the Germans had pushed past Reims and had reached the River Marne, where the Allies were able to hold them. The Germans had failed to achieve control of the air to assure the success of their spring offensives on the Somme and Lys. Perhaps, if their new fighters had become operational earlier, they may have wrested back air superiority, and the result of the German spring offensives could then have been devastatingly different.

During the first week of June on the Western Front, the RAF claimed 56 enemy machines destroyed and a further 26 driven-down out of control. Of this total, 16 were credited to No 2 Squadron within the first two days of the month.

On a dawn Circus patrol on Saturday 1 June to the south of Bapaume, Lieutenants Eric Cummings (B195) and William Adams (D3429) both destroyed Pfalz scouts. That evening, Lieutenant Archie Rackett (B525) departed in the first flight of 12 Squadron aircraft for an offensive patrol 20 kilometres east of Villers-Brettoneux, which now lay in ruins. Flying at 13,500 feet, the Australians encountered eight Pfalz and Triplane scouts. In Rackett's Flight, Captain Adrian "King" Cole[19] destroyed a Fokker, and

15 Shores, Franks & Guest, p.79.

16 Jones, Vol IV, p.287.

17 By August 1918, all Hispano Suiza powered machines had been replaced by Vipers.

18 Jones, Vol VI, p.405.

19 AVM A L T Cole CBE DSO MC DFC. No 67 Sqn RFC/1 Sqn AFC, 2 Sqn; comd Central and Southern Areas 1939–41, 235 Wg RAF 1941–42; Fwd Air Controller Dieppe Raid 1942; AOC RAF Northern Ireland 1942–43; comd NWA 1943–44; RAAF Liaison Offr SEAC 1945. Regular army offr of Melbourne; b Melbourne 19 Jun 1895.

Lieutenant Leslie Primrose[20] sent down a Pfalz out of control. From the other flight, Captain Forrest destroyed a triplane, and Lieutenants George Cox and Roby Manuel[21] each scored a Pfalz out of control. Rackett, however, was shot down and later related the engagement:

> We engaged a number of enemy scouts over and behind Villers-Bretonneux. After several minutes fighting I hit an Albatros[22] which burst into flames and smoke. I also hit a Triplane which withdrew from the fight, starting to spin and got out of control towards the ground. At the same time my engine cut out and I was forced down, landing at Bourchoir (between Roze and Amiens), turning over among some disused trenches. Two Albatros scouts kept hovering over me while unarmed German infantry came from nearby and pounced upon me. These German soldiers proceeded to violently assault me, hitting me about the head and kicking me. Several German Officers appeared and intervened, one of them being a Flying Officer. I was taken by motor to a village nearby. There I was closely interrogated regarding the make and power of the machine, my squadron number and work, and the situation of the aerodrome. Of course they gained no information.[23]

Rackett was sent to Landshut POW camp, and then with Lieutenant Leslie Ward, who had been shot down in his DH-5 in November, was sent to Stralsund until the Armistice.

On another Circus the next morning, 2 June, six Pfalz scouts were attacked near Albert. In the ensuing melee, No 2 Squadron brought down five, with two being credited to Roby Manuel.[24] Another was brought down by William Adams and two by Henry Forrest (C9539). Forrest's tally now stood at 11, all scored in the same machine. To commemorate Forrest's record in his aircraft C9539, the Australian War Memorial in Canberra has restored an ex-RAAF SE-5A, serial number A2-4 (ex C1916) to represent Forrest's 2 Squadron mount.

An evening patrol at 16,000 feet over Peronne, towards the unknown territory around St Quentin, encountered a formation of 10 triplanes and Pfalz scouts. The leader, Lieutenant James Adam (D3962), drove one Pfalz down:

---

20 Lt L J Primrose. CFS, No 2 Sqn AFC. Lecturer in science and mathematics of Ballarat, Vic; b Ballarat 14 May 1890. Killed in action 4 Jun 1818.

21 Capt R L Manuel DFC*, 43rd Bn, No 2 Sqn AFC, 6 Sqn. Later RAAF. Farmer of Kerang, Vic; b Kerang 7 Oct 1895.

22 The Pfalz and the Albatros did have a similar appearance. The Pfalz D.III had only recently been encountered by No 2 Squadron, and was intended as the successor to the Albatros D.V. Although slower, it was a lot more manoeuvrable than the D.V, and could be distinguished by the different tail assembly and wing shape. The RAF had recently evaluated the Pfalz, as one had been brought down in British lines in February. It was assessed as handling better than the Albatros, with an excellent view from the cockpit, but with a top speed of only 102 mph. PRO AIR1/1977/204/273/59 "Report on EA Pfalz G.141"

23 AWM30 B3.11, Statement by Lt A R Rackett, Repatriated POW, of 31 Dec 1918.

24 Confusion exists over 2 Squadron's records for 1 and 2 Jun 1918. This is due to the Squadron Record Book confusingly recording a day's activity from 4 p.m. one day to 4 p.m. the next—an RAF requirement for daily statistics. As a result Shores, Franks & Guest claim Manuel's score of 1 Jun (Combat Rpt no 79) was on the same sortie as his two on 2 Jun (CR 80). The Squadron Record Book clearly shows that these were separate patrols.

> I dived on the highest machine firing a short burst at 100 yards range without apparent result. I dived on another of the formation and fired a burst of about 200 rounds at about 80 yards. This machine immediately turned over on its side and fell down out of control.[25]

This patrol was to send down a further three Pfalz scouts. Captain King Cole (D3429) and Lieutenant Hubert Hamilton[26] (B4895) each scored, with the third being shared between the formation.

The Squadron Record Book made the following observations on the operations of the German Air Service:

> The Squadron accounted for 23 EA during the month. It will be noted that all were shot down in four days, namely 1st, 2nd, 11th and 12th. This may be accounted for by the way the enemy makes use of his aircraft: his usual method being—while concentrating his troops etc, for an offensive, he keeps his machines on the ground relying on concealment and night movements, and not on his chaser squadrons to prevent our reconnaissance machines from obtaining information, then just before his attack commences, he uses all available machines, and aerial activity becomes most marked. As the attack progresses his activity becomes less and less marked until it returns to the normal. Thus, just prior to the Montdidier-Noyon offensive, which commenced on the 6th June, his activity was very accentuated in the Albert sector, and this would account for the number of machines destroyed on the 1st and 2nd.[27]

By 2 June, it had been apparent where the next German attack would be, and the French Commander-in-Chief, General Foch, requested the assistance of British fighter squadrons.[28] The 51st Wing was dispatched the following day and No 2 Squadron ground personnel moved to south of Amiens, to the aerodrome at Fouquerolles, 60 kilometres north of Paris. The aircraft followed the next morning. In this new sector the Circus remained under the RAF's IX Brigade, but provided support for the Fifth Army and the French forces facing the expected offensive on the Montdidier-Noyon front. That afternoon, Lieutenant Leslie Primrose (C5441), who had commenced flying training at Point Cook on the Sixth Course with Harry Cobby, was killed when he flew into the ground low flying near Clermont. Shortly afterwards, Lieutenant Hamilton (B70) was taking off from the new aerodrome when his engine cut out. He crash landed and his wrecked machine burnt, his injuries necessitating his repatriation from active service.

By the second week of June, German fighters were once again swarming over the Somme. A new German scout just making its appearance was the Fokker D.VII. Powered by a 160hp Mercedes engine and armed with twin Spandau synchronised guns, its top speed was 118mph and it could attain 20,000 feet.[29] As it equipped German squadrons through late spring and summer, this new scout was to prove a

---

25 AWM4 8/5/1 Combat in the Air Report No 86, of 2 Jun 1918.

26 Lt H E Hamilton. 24th Sig Coy AE, No 2 Sqn AFC. Electrical mechanic of Sydney; b 14 May 1894.

27 AWM4 8/5/1 Squadron Record Book, Record of Service, of Jun 1918.

28 Jones, Vol VI, p.401.

29 The Fokker D.VII's Mercedes D.III engine was soon replaced by the 185hp BMW IIIa which further improved performance, giving a top speed of 125mph.

formidable opponent, with its rapid climb and lightning recovery from a dive. It had the ability to appear to hang from its propeller and retain control at low flying speeds. But, as with its predecessors, through the doctrine of fighting over German territory, it could remain airborne for only 1 ½ hours—a lot less than its adversaries. The SE-5A was unable to turn with the new Fokker, but because of its speed, rate of climb and endurance, it was able to regain control of the air—with the Sopwith Camel—during 1918. The superiority of these two fighters would not change for the remainder of the war.

The Fokker D.VII was possibly the best scout to be operated by the Germans during the war.
RAAF

The Germans launched their attack in the south early on the morning of Sunday 9 June. They achieved initial success along the River Matz, and during the three days of the main battle, No 2 Squadron was employed on low-flying attacks. Targets of troops and transports were plentiful for the bombing and strafing attacks, which attempted to halt the enemy's advance.

By the 11th, the German offensive had been repelled, and the Squadron's high offensive patrols for control of the air recommenced. The 51st Wing's Circus modus operandi was for six of 2 Squadron's SEs to work on high patrol at 16,000 to 18,000 feet, and 12 Camels from Nos 43 or 80 Squadrons flying 2,000 to 3,000 feet below. The Camel was more manoeuvrable, and always flew at a lower altitude, while height suited the diving and zooming tactics of the SE.

The Squadron claimed its first of the new Fokker D.VII scouts on 12 June, with Captain Roy Phillipps (D6860) having a most successful sortie that morning. Leading an offensive patrol of six SEs over Ribecourt, 10 kilometres south of Noyon on the River Oise, he surprised a formation of six triplanes at 13,000 feet attempting to attack the lower Camels of the Circus:

> I dived down on the highest which did not appear to see me and fired a burst of 50 rounds at very close range into it; this machine side slipped and fell out of control, crashing near Gury.
>
> Zooming up into the fight again, I attacked another triplane head on and fired another burst of 50 rounds into it; this machine turned over on its back and went down vertically, crashing into a wood south of Cannectancourt.[30]

Twenty minutes later, Phillipps spotted a pair of enemy observation aircraft near Gournay, 15 kilometres south of Montdidier. He plunged onto an LVG two-seater,[31] firing inside 100 metres to send it crashing into Allied lines near Marqeglise. Thirty minutes later, continuing their sweep, a flight of 10 of the new Fokker biplanes dived

30 AWM4 8/5/1 Combat in the Air Report No 91, of 12 Jun 1918.

31 Two-seat reconnaissance machine made by *Luft-Verkehrs Gesellschaft.*

on to the Camels near Moreuil, between Amiens and Montdidier. Phillipps led the SEs down again, and selecting the nearest scout, shot it down in a spin. All of these claims were confirmed by No 43 Squadron's Camels, raising Phillipps' personal score to 11 victories. However, on his return, he was advised that Lieutenant Themotre Hammond[32] (D3960) had failed to return from a morning patrol. Hammond had served three weeks at the Front.

An Australian mechanic with the lower engine cowl dropped to allow access to the SE-5A's Wolseley engine AWM D439

Bad weather for the next week again kept things quiet, and as the immediate danger to Compiegne from the Montdidier-Noyon sector had passed, the Circus was moved north again. On 21 June, the Squadron joined the RAF's 10th Brigade at Liettres aerodrome, about two kilometres from the village of Estree-Blanche, 20 kilometres south of St Omer. Bad weather and lull in enemy activity provided a break from operations. Circus activity was to remain the role of the Squadron as it moved again on 1 July from Liettres to Reclinghem, 20 kilometres to the south-west. Here a new wing was formed, with No 4 Squadron AFC and No 46 Squadron RAF, both flying Camels, and No 103 Squadron RAF with DH-9 bombers. This new 80th Wing was to provide support in the north for the Fifth Army.

Ludendorff's forces regrouped to attack again on 15 July. This next offensive was on the 100-kilometre front against the French forces around Reims. The American army was also involved in the defence, having entered the war the previous year. The main thrust of this last German offensive of the war was across the Marne, and initially their advance was rapid. However, the Allied counter-attack from Villers Cotterets Forest, south of Noyon, forced the enemy back over the Marne in a confused retreat.[33]

32 Lt T J Hammond. 13th Lt Horse, No 2 Sqn AFC. Grazier of Adavale, Qld; b Sydney 1893. Killed in action 12 Jun 1918.

33 Montgomery, p.5.

The unseasonal weather during July meant that only occasional hostile aircraft were encountered, and these rarely accepted combat. Six enemy aircraft were destroyed or driven down until the 31st, when overcast and rain gave way to sunny conditions. This was to be a day suitable for aerial work. As the visibility improved that Wednesday morning, all the Squadron's aircraft were launched. From the 17 machines airborne on offensive patrols, over 100 enemy aircraft were seen between Lille and Armentieres.[34]

Captain Roby Manuel (C1948), who had replaced Henry Forrest as the "C" Flight Commander, dived on an Albatros scout over Estaires, 10 kilometres south-west of Armentieres. He fired at 150 metres, sending it down out of control. "The EA turned over and fell like a leaf. He was still going down out of control … when my attention was attracted by other EA diving from above."[35]

A little over an hour later, an "A" Flight patrol led by Roy Phillipps (D6860), flying above No 4 Squadron's Camels, encountered eight enemy LVG two-seaters near Laventie, between Armentieres and Bethune. Above them were their escorts—nine Fokker D.VII scouts. In this engagement, Eric Cummings (C6473) destroyed one two-seater and sent another down out of control. Lieutenant Follett[36] (C1125) destroyed one of the escorting Fokkers. "I fired a burst of 60 rounds at under 15 yards range; the EA turned up vertically and dived steeply down with black smoke issuing from fuselage."[37]

Then another pilot on a "B" Flight patrol led by Cole, flying at 20,000 feet over Armentieres, Lieutenant James Wellwood[38] (D6913), dived on a Rumpler two-seater. In an inconclusive engagement, he fired 30 rounds inside 100 metres range, then continued to dive below it and fired another 50 rounds in the zoom up:

> On completion of the zoom I was still above the EA and saw that there was apparently no one in the observer's cockpit. I got on its tail again and fired a short burst of about 40 rounds after which both guns jammed. The EA was by this time over the southern edge of Lille.[39]

Roy Phillipps (D6860) on an afternoon air test of his aircraft, west of La Bassee, spotted a DFW two-seater. He fired 100 rounds into it from under its tail, which was sufficient to send it down to crash. He was then attacked by a Fokker D.VII, which he sent down out of control.

> I fought it for about five minutes, finally getting a good burst of about 50 rounds side on from about 30 yards. The EA rolled over and kept side slipping and rolling towards the ground.

---

34 AWM4 8/5/1 War Diary Intelligence Summary, of 31 Jul 1918.

35 AWM4 8/5/1 Combat in the Air Report No 102, of 31 Mar 1918.

36 Lt F W Follett. Army Engineers, No 2 Sqn AFC, 6 Sqn. Civil engineer of Sydney; b Sydney 27 Mar 1892.

37 AWM4 8/5/1 Combat in the Air Report No 100, of 31 Jul 1918.

38 Lt J J Wellwood DFC. Artillery, No 2 Sqn AFC. Motor engineer of Drouin, Vic; b Drouin 15 Oct 1892.

39 AWM4 8/5/1 Combat in the Air Report No 103, of 31 Jul 1918.

> I was unable to watch this machine as I was immediately attacked by seven Fokker biplanes, but saw it last about 500 feet above the ground when it was still apparently out of control. The seven EA chased me back to the lines and shot my machine about on the way.[40]

By autumn 1918, every German scout squadron on the Western Front was armed with the new Fokker D.VII. The Germans tried in vain to establish air superiority over the battlefront. This was to prove impossible against the SE-5As, Camels and newly arrived Sopwith Snipes.

40 AWM4 8/5/1 Combat in the Air Report No 104, of 31 Jul 1918.

CHAPTER 8

The Battles of the Hundred Days

August 8th was the black day of the German Army in the history of the War.

General Ludendorff[1]

By the beginning of August, after four years of battle on the Western Front, the opposing forces still confronted one another with no appreciable gain to either side. This was finally about to change.

The Somme River rises near St Quentin, and flows by Amiens and Abbeville to the Channel. The British Somme offensive was to commence on Thursday 8 August, with the Fourth Army, which had relieved the Fifth Army, striking along the river, east of Amiens—the Battle of Amiens (8–12 August). From this initial assault, the offensive would be joined by the other British Armies to the north, and the French to the south. From this Allied offensive, the losses during August would prove to be the heaviest since the great battles of the 1917 spring.

For the preceding week, Allied air forces made concentrated attacks on enemy airfields, particularly those of the ground attack *Schlachtstaffeln* "battle flights". On the opening day of the Allied assault, fighter squadrons were tasked with supporting the bombers in continuing attacks against enemy aerodromes on the Fourth Army front, and to standby ready to counter any enemy air activity.[2]

The weather initially proved unfit for flying, with bad visibility in mist and rain. As conditions improved during the morning, aircraft were able to join the battle, and found the German retreat offered lucrative targets. However, the losses were high with the RAF losing, that day alone, 45 aircraft in combat and a further 52 wrecked on landing.[3] This was an overall wastage rate of 13 per cent for the day. Many of the losses were from attacks on German forces fleeing over the bridges across the Somme, and from these low-flying sorties the attrition rate was 23 per cent. These bridges were almost in sight of the German airfields, and presented the enemy with an easy solution of defending what was a series of pinpoint targets.

With the focus of activity initially to the south, fighter squadrons were able to be released from the northern flank. On 11 August, the day after His Majesty King George V and staff visited to inspect the Squadron, the SEs were deployed south to Fienvillers, for operations in support of the French forces and the Fourth Army.

The Squadron moved further south the next morning to Allonville, near Amiens, and while escorting two flights of Camels at 15,000 feet over Licourt, engaged seven Fokker biplanes. "B" Flight's Lieutenant James Wellwood (D6968) chased one off the

---

1 General Ludendorff, *My War Memories 1914–1918,* Vol II, p.679, quoted from Maj-Gen Sir Archibald Montgomery, *The Story of the Fourth Army,* Hodder & Stoughton, London, 1919, p.8.

2 Jones, Vol VI, pp.433–4.

3 Jones, Vol VI, p.445.

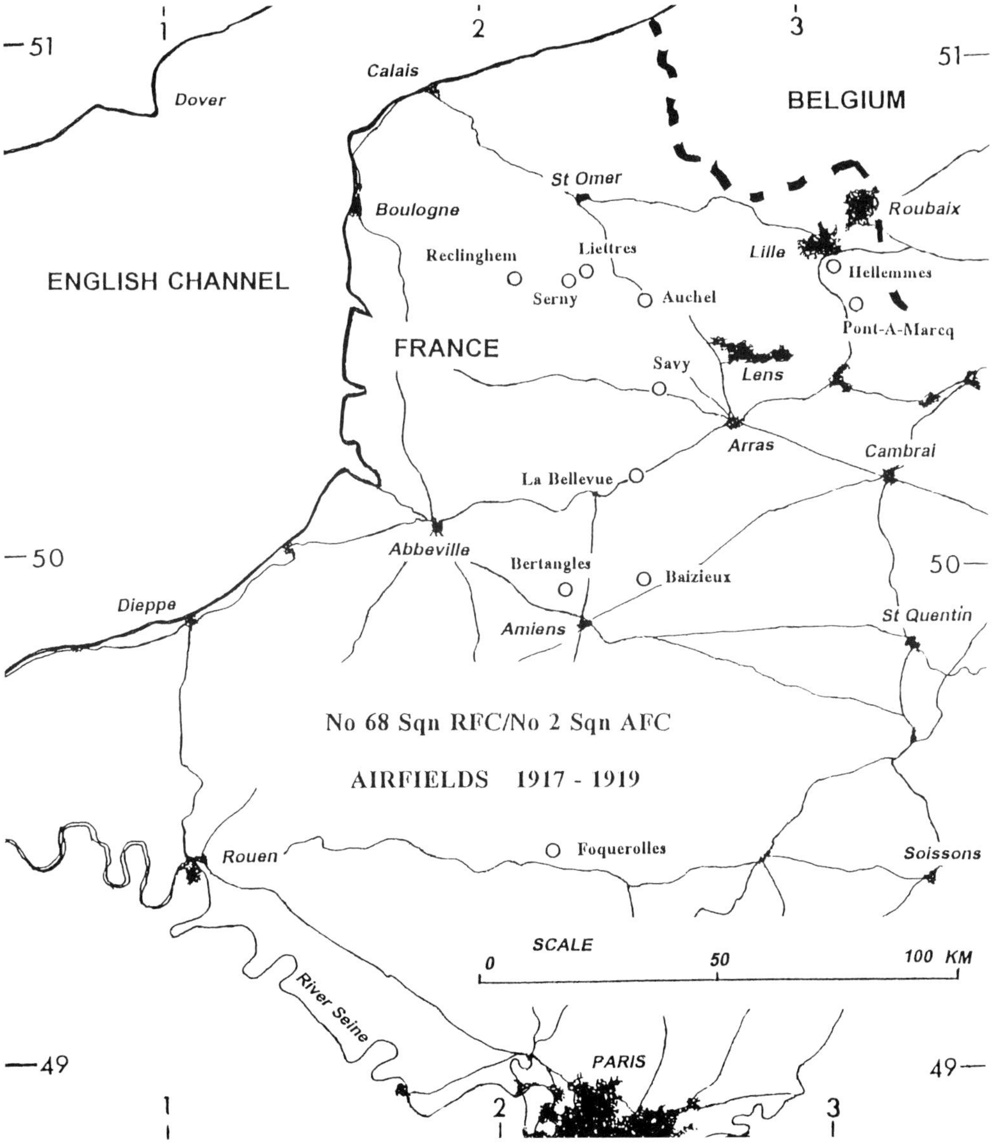

No 68 Sqn RFC/No 2 Sqn AFC Airfields 1917–1919 RAAF

tail of a Camel, and it spun into the ground. Then "A" Flight joined the fray, with Captain Roy Phillipps (D6860) destroying another, which as it slowly twisted round in a vertical dive was seen to disintegrate. Lieutenant Frederic Markham-Mills[4] (E5989) sent down a third out of control. Lieutenant Frank Alberry[5] (D6948) had been forced to retire with engine trouble, and attempted to land in the lines. His machine was wrecked as it crashed in the trenches, and Alberry received minor injuries. It was just

4 Lt F C Markham-Mills. AASC, No 2 Sqn AFC. Engineering student of Sydney; b Tuena, NSW, 20 Oct 1893. Died of illness 18 Dec 1918.

5 Lt F Alberry DCM. 8th Bn, No 2 Sqn AFC. Later RAAF. Sleeper hewer of Hobart; b Hobart 29 Sep 1892.

as well he was not more seriously injured, as Alberry—the "one-legged ace"—was to go on and score seven victories on the Squadron. Alberry had lost a leg while serving with the infantry and, fitted with an artificial leg, had personally obtained a letter from the King recommending his acceptance for flying training.

The following morning, 13 August, two more Fokker D.VIIs fell to the Squadron at 12,000 feet over Erchen, between Montdidier and St Quentin. One of these was again to Markham-Mills (E5989), the other to Lieutenant Follett (C1125). After this, the SEs returned to Reclinghem, and the Squadron's efforts were again directed against the northern fronts.

SE-5A D6995 OF "C" Flight, showing the revised squadron marking of a white bar below the cockpit, carried from mid-1918 RAAF

Success continued the next morning when, near Armentieres, three more enemy machines were destroyed. Markham-Mills (E5989) was on a morning patrol with "A" Flight when he spotted a new type of biplane scout:

> Whilst over Steenwerck, our patrol encountered some EA chasing Camels westward at 12,000 feet. I dived on one and at close range (under 50 yards) fired 150 rounds at him, sweeping the machine from front to back; it immediately turned over and fell out of control. I followed it down to 5000 feet and saw it crash near Nieppe.[6]

This scout downed by Mills could possibly have been a Siemens-Schuckert D.IV or a Pfalz D.XII, both of which had appeared in August.[7] That evening, the "C" Flight Commander, Roby Manuel (C1948), sent down a Fokker out of control, followed by "B" Flight having a strange engagement with a two-seater. The patrol leader, King Cole, reported:

> When returning from Tournai with a formation of DH-4s, a two-seater was seen flying east over Wavrin at 12,000 feet. On the approach of our patrol this machine, without a shot being

6 AWM4 8/5/1 Combat in the Air Report No 112, of 14 Aug 1918.

7 Following this, in October 1918, a new enemy scout—the Fokker D.VIII monoplane—arrived at the Front in small numbers, but it had little opportunity to prove itself in action.

> fired at it, commenced to spin down. It was noticed to spin for about 3000 feet, and then one of its wings fell away in the air.[8]

In addition to its offensive patrols during August, No 2 Squadron flew two successful daylight bombing raids on the enemy airfields at Harbourdin and Lomme. On 16 August, the raid against Haubourdin Aerodrome was carried out by Nos 88 (Bristol Fighter) and 92 (SE-5A) Squadrons RAF, and Nos 2 (SE-5A) and 4 (Camel) Squadrons AFC. The Squadron provided 19 out of the 65 machines taking part in the raid. Each SE dropped four 25 lb bombs, then emptied its guns on the various targets around the aerodrome. Other aircraft of the Wing dropped 60 25 lb and six 40 lb bombs, which succeeded in setting fire to hangars and buildings, and blowing up the main fuel dump. When Lieutenant Follett (C1125) was about to dive on the aerodrome, he spotted an Albatros two-seater:

> Turning sharply from the formation, I engaged the EA from the rear and fired about 70 rounds from 50 yards range; the EA turned south-east and glided away very steeply.[9]

Captain Harry Cobby, from 4 Squadron's formation, saw Follett's victim crash.

This success was repeated the next day when 14 SEs of the Squadron participated in another Wing raid against Lomme Aerodrome. The Squadron Intelligence Summary recorded:

> Our machines dropped 44 (25-lb) bombs from an average height of 200 feet on hangars, workshops and quarters with good results. Two hangars were observed to be burning furiously. 2000 rounds were fired at various ground targets, including machine guns, one of which was silenced. Some days after the raid, information was received per medium of a prisoner that the raid was very successful, 17 Fokker biplanes being totally destroyed, besides other material damage being done[10]

On the 19th, the Squadron sent a patrol back to Haubourdin, where the leader, Captain Cole (D6948), sent two Fokker D.VII scouts down. Four days later, one of his "B" Flight pilots, Lieutenant Claude Ebeling,[11] was visiting No 1 Aeroplane Supply Depot's aerodrome at Marquise, between Calais and Boulogne. Taking off from the Depot, Ebeling (D6913) crashed, with his injuries proving fatal a few hours afterwards. Ebeling had been an original Squadron member in Egypt, serving as a sergeant fitter with "C" Flight over 1916–17. After pilot training in 1918, he had flown as a ferry pilot at 1 Aeroplane Supply Depot for a few months before rejoining the Squadron in July.

The Fourth Army's victory in front of Amiens caused Ludendorff to tender his resignation, and on 14 August the Kaiser had directed Secretary of State von Hinze to open peace negotiations.[12] As a result of the Battle of Amiens, the Allies realised that,

---

8 AWM4 8/5/1 Combat in the Air Report No 114, of 14 Aug 1918.

9 AWM4 8/5/1 Combat in the Air Report No 115, of 16 Aug 1918.

10 AWM4 8/5/1 War Diary Intelligence Summary, of 17 Aug 1918.

11 Lt C R Ebeling. No 1 Sqn AFC, 14 Sqn RFC, 68 Sqn/2 Sqn AFC. Engineer and draughtsman of Melbourne; b Melbourne 10 Mar 1893. Died of injuries 23 Aug 1918.

12 General Lord Rawlinson, cited in Montgomery, p.viii.

in the face of declining enemy resistance, it was time to mount a final offensive. The British C-in-C, Sir Douglas Haig, decided this would be between Albert and Arras, towards the strategic objective of the Front between Cambrai and St Quentin.[13]

On 21 August, when the advance of the Fourth Army had reached level with Albert, the Third Army widened the Front to the north, resulting in the successful Battle of Bapaume (21–31 August), which forced the enemy to the east bank of the Somme. This led to a push on the whole northern Somme defences in the Second Battle of the Somme, as the victories to the south had given the First Army, on the left of the Third Army, the opportunity to join the rout in the successful Battle of Arras (26 August–3 September). In these final offensives, air power continued to play a crucial part.

Flying at 17,000 feet above the low cloud and rain on 27 August, Lieutenant George Cox (E5965) was on patrol between Arras and Cambrai. Over Sains, 10 kilometres west of Cambrai, his flight encountered 30 Fokker biplanes. He sent one down in flames, then zoomed to dive on another which he fired at from inside 150 metres, to send it straight down. He then climbed south-east to find a Pfalz scout:

> I dived on him putting in a long burst from 150 to 50 yards range. He turned over and went down out of control in a slow spiral.[14]

By the end of August, the Australian forces were successful in their attacks against Mont St Quentin, which commanded the Somme crossings, and then at Peronne. This was further evidence that victory over the Germans was finally within sight, as the enemy dug in near St Quentin. This marked the end of the second phase of the great British offensive on the Somme. The third stage began on 2 September in heavy rain when the formidable Drocourt-Queant line, west of Cambrai, was stormed, breaching the enemy defences towards the Hindenburg Line and throwing the enemy into retreat. As they withdrew, the German troops and transport were continuously harassed by low-flying fighters.

No 2 Squadron was now tasked with providing Circus offensive patrols at lower levels on the Arras front, north of Cambrai. On an evening patrol on 6 September, led by the CO, flying with 4 Squadron's Camels and 88 Squadron's Bristol Fighters, the Squadron attacked seven Fokker D.VIIs between Douai and Lille. Lieutenant Frank Smith (C6403) destroyed one, and Lieutenant James Wellwood (D6968) sent down another. During the engagement Lieutenant James Ross[15] (D8392) was wounded. (Postwar, Ross and another member of the AFC, Lieutenant R Douglas, attempted to fly from England to Australia. Shortly after leaving Hounslow aerodrome, their single-engined Alliance crashed at Surbiton, in Surrey, killing both pilots.[16])

After a week of appallingly bad weather of gale force winds and heavy rainstorms, the Squadron was to achieve one of its best engagements of the war in a sweep over

13 Jones, Vol VI, p.469.

14 AWM4 8/5/1 Combat in the Air Report No 119, of 27 Aug 1918.

15 Lt J S L Ross. No 2 Sqn AFC. Telegraphist of Moruya, NSW; b Moruya 20 Oct 1895. Killed in flying accident 13 Nov 1919.

16 Cutlack, p.387.

Lille. Captain Roby Manuel (C1948) was leading two flights on the morning patrol, when they encountered 15 Fokker biplanes. Manuel sent one down out of control, as did Lieutenants Charles Copp[17] (D6869) and George Holroyde[18] (D6903), while Frank Smith (C6403) scored two, and Frank Alberry (D6995) sent down another in flames. On the return, Manuel spotted another Fokker near La Bassee heading north-west towards St Omer, and with only 20 rounds remaining, he engaged in a long chase:

> I pursued the EA a fair distance until he reached St Omer, when he turned north-east thus enabling me to catch up. I fired a burst in front of EA making him put his nose down. I followed him down to within 100 feet of the ground at last firing a burst of ten rounds at 20 yards range. The EA dived straight into the ground about 1½ miles east of Droglandt—totally wrecking the machine. The pilot, who was a Sergt-Major, being seriously injured.[19]

Lt Frank Currie with SE-5A B8392 of "B" Flight, at Reclinghem in September 1918 RAAF

Five days later, 21 September—the anniversary of the Squadron arriving in France—Lieutenant George Cox (E5965) failed to return from patrol. He had last been seen over Armentieres flying very low. He later recounted his experience:

17 Capt C H Copp. 5th & 46th Bns, No 2 Sqn AFC. Consulting optician of Melbourne; b Melbourne 29 Mar 1893.

18 Lt G E Holroyde. 6th Lt Horse, 46th Bn, No 2 Sqn AFC. Station hand; b Brisbane 3 Aug 1894.

19 AWM4 8/5/1 Combat in the Air Report No 132, of 16 Sep 1918.

> While over Lille at about 1500 feet encountered EA. I immediately chased one EA eastwards. I did not catch him, however. As I was about to dive on him, I collided with something which knocked my engine out of its bearers.[20] I turned west and made for home, but after a couple of minutes my plane got out of control and I went down through the clouds for about 1000 feet. At about 800 feet I regained partial control and finally came to earth amongst a system of German trenches.[21]

It is believed that Cox's machine had been hit by groundfire. Captured, George Cox saw out the rest of the war in Bavaria, at the prison camps in Landshut and Ingolstadt. In December, he left for Switzerland, and arrived at Dover two days before Christmas 1918.

For the tactical employment of fighters supporting forces on the Third Army front, the following instructions were issued on 22 September:

> The duties of the various sub-formations of a patrol are as follows:
> (i) Top sub-formation will watch the upper air and is responsible that the formation is not surprised.
> (ii) The lowest sub-formation is responsible for initiating attacks and will search for hostile formations below or on the same level.
> (iii) Intermediate sub-formations will ensure that no hostile machines dive below them without being attacked.[22]

This type of tactical coordination between flights proved itself two days later. On 24 September, as the rain and low cloud gave way to a spell of fine weather, No 2 Squadron had considerable success in one combat, destroying five Fokker D.VIIs, and driving another three down out of control. A squadron-strength evening patrol led by the "B" Flight Commander, Captain Adrian Cole (D6964), engaged four German formations of 25 Fokker D.VII and Pfalz D.III scouts between Armentieres and Lille. Captain Eric Simonson[23] (D406) destroyed one, sending another down out of control, while Cole and Lieutenants Smith (C6403), Wellwood (D6968) and Franks (F5465) each destroyed one. Lieutenant Fred Markham-Mills (E5989) and one of the Squadron's original air mechanics, Lieutenant Frank Roberts[24] (D6995), each sent down a Fokker out of control. The Australians suffered no casualties.

With the successful Allied push in the north, the Germans were slowly retreating on the Ypres-Arras front, losing all the ground they had gained during the spring. By the third week of September, the Fourth, Third and First Armies were confronted by the last of the enemy's defences—the formidable defences of the Hindenburg Line. On Friday 27 September, the Allies began to assault the fortifications of the Hindenburg Line, with attacks against the main defences between Bellenglise and Vendhuille, on the St Quentin Canal east of Peronne, being launched two days later. By nightfall on

20 Cox's machine was evidently hit by heavy anti-aircraft fire.

21 AWM30 B3.5, Statement by Lt G Cox, Repatriated POW, of 14 Jan 1919.

22 Jones, Vol VI, pp.506–7.

23 Capt E L Simonson. 8th & 14th Bns, No 2 Sqn AFC, 6 Sqn. Engineering student of Melbourne; b Melbourne 23 Jan 1894.

24 Lt F L Roberts. No 1 Sqn AFC, 68 Sqn RFC, 71 Sqn/4 Sqn AFC, 2 Sqn. Engineer of Bendigo, Vic; b Bendigo 18 Aug 1896.

the 30th, Allied forces had taken the canal and almost all the Hindenburg Line defences had been overwhelmed, opening the way for the rapid and complete defeat of the enemy.[25]

At the end of the month, the Front was approaching a line Lens-Cambrai-St Quentin, so No 2 Squadron prepared to move closer to the action, 20 kilometres east to Serny. From here the Wing would support the Fifth Army, recently reconstituted and deployed on the northern flank. The move to this large airfield was carried out on Tuesday 1 October, with No 2 Squadron again being with No 4 (Camel) Squadron AFC, and Nos 88 (Bristol Fighter) and 103 (DH-9) Squadrons RAF. At Serny, No 4 Squadron began to exchange its Camels for the new Sopwith fighter, the Snipe.

October saw the resumption of low bombing operations for the Australians as the Allied air effort was devoted to hampering the enemy's withdrawal, attacking road junctions, railway stations and other bottlenecks. Both Nos 2 and 4 Squadrons also carried out very low altitude attack sorties against enemy airfields. October also saw some of the greatest air battles of the war, and No 2 Squadron accounted for 32 enemy machines during the month, a record. Of this total, 13 were encountered during two patrols on Monday 14 October led by Lieutenant Frank Smith, who personally accounted for two on each flight.

That morning, as the Allies successfully attacked in the Lys area, Smith led his formation on a bombing raid to Fretin, 10 kilometres south-east of Lille, escorted by "A" Flight. Dropping their bombs, they were attacked by 15 Fokker biplanes, which were successfully engaged by the escort. Captain Eric Cummings (C6473), who had replaced Phillipps as "A" Flight Commander, crashed one near Cysoing and drove down another two, while Lieutenant Greg Blaxland (C1125) shot down one in flames, sending another out of control. While the escorts were occupying the enemy, Smith (C6403) led his low flight climbing out to the west and encountered a further formation of Fokker D.VIIs, and he attacked and destroyed two. His wingman, Lieutenant Len Franks (F5465), closed on another and, firing 60 rounds from 15 metres, saw his victim crash near Mouveaux, north of Lille.

On an afternoon offensive patrol north of Tournai, a dozen Fokker biplanes were engaged in a long dog fight resulting in more success. Both Smith (C6403) and Lieutenant Charles Stone[26] (D6919) each destroyed one and sent down another out of control, while Lieutenant Roberts (C1125) also scored. However, from the 10 No 2 Squadron aircraft airborne that afternoon, two failed to return from patrol.

Lieutenant John McKeown (D6968) was shot down and killed south of Tournai. Captain Edward Cornish[27] (E5989), who was a graduate of the Gosport Special School of Flying and had served as "A" Flight Commander on No 6 (Training) Squadron AFC, was shot down and wounded. He had been on the Squadron for only nine days, and became No 2 Squadron's last POW:

---

25 Montgomery, p.9.

26 Lt C O Stone. No 69 Sqn RFC, 2 Sqn AFC. Chauffeur of Geelong, Vic; b Chingford, Essex, Eng Sep 1893.

27 Capt E W Cornish MC. 13th & 45th Bns, No 6 Sqn AFC, 2 Sqn. Clerk of Sydney; b Sydney 16 Jul 1897. POW 14 Oct 1918.

> We were flying in formation of five at 1700 feet with three other planes flying higher up. While manoeuvring for a scrap with a party of twelve Huns, we were surprised by five others. When the leader of the last party of Huns dived I managed to get on his tail. After firing several bursts the Hun went straight down, apparently under control. I, myself being wounded in the back, dived away from the Huns and was again hit in the arm. I went straight to the ground and was taken prisoner.[28]

Replacement pilots for No 2 Squadron came from No 6 Squadron, at the AFC Training Wing's base at Minchinhampton AWM D421

Cornish was hospitalised until the end of the war, and arrived back in England in December. Of the seven No 2 Squadron pilots taken prisoner of war, all were to be repatriated. The focus of the Squadron's attention remained in this area for the next days, and on 18 October the defence at Lille fell to the Allies.

For the efforts of the Wing against targets around Tournai, the Commander of the Fifth Army sent the following congratulatory words to the Commander of the RAF's Tenth Brigade:

> I am to request you to convey to the 2nd and 4th Squadrons AFC and the 88th and 103rd Squadrons RAF, the Army Commander's thanks for the excellent work carried out by these Squadrons in their low altitude bombarding attack on Tournai and neighbourhood. The results achieved bear testimony to the skill and dash with which all taking part in the raid executed their tasks.[29]

On a patrol on 27 October, Lieutenant Charles Copp (E5659) drove down two Fokker biplanes over Tournai but Lieutenant Francis Howard[30] (D6950) of "C" Flight was shot down and killed in the action. The following morning, Captain Cole (D6964)

28 AWM30 B3.4 Statement by Capt E W Cornish, Repatriated POW, of 17 Dec 1918.

29 AWM4 8/5/1 Letter from Commander Fifth Army GA 186/9 to Commander Xth Brigade RAF, of 20 Oct 1918.

30 Lt F Howard. Army Engineers, No 2 Sqn AFC. Railway engineering assistant of Maryborough, Vic; b Clunes, Vic, 17 Apr 1894. Killed in action 27 Oct 1918.

and Lieutenants Blaxland (F5457) and Stone (D6919) all sent down Fokker biplanes, while Frank Alberry (D6995) scored a further two. Adrian Cole saw his victim jump from his stricken machine by parachute, the first time the Australians had seen this.[31]

That afternoon, No 4 Squadron Snipes escorted twelve SE-5As, led by the CO, on a bombing raid on Lessines, north of Ath, on the Dendre River. The SEs bombed then climbed from low level to join the Snipes which were engaging about 30 Fokkers. Captain Eric Simonson (C1125) destroyed a Fokker D.VII from a Snipe's tail and Lieutenant James Egan[32] (F860) destroyed an LVG reconnaissance two-seater.

On 30 October, Australian pilots had reported great air activity around the German aerodrome at Rebaix, north of Ath. All squadrons of the 80th Wing were immediately tasked for a raid. The bomb-carrying SEs of No 2 Squadron, led by Major Murray Jones and escorted by 11 No 4 Squadron Snipes, joined DH-9 bombers in a low-level attack on the airfield. Bombing down to a height of 20 feet, they destroyed hangars, buildings and several LVG two-seaters. The Australians' Wing Leader, a veteran aviator since the RFC's early days in France in 1914, Lieutenant-Colonel Louis Strange,[33] reported the whole attack:

> Four 230-lb bombs, six 112-lb bombs, and 115 25-lb bombs were dropped from under 1000 feet. 5400 rounds were fired at ground targets. Machines of No 4 AFC Squadron and No 88 Squadron ably carried out escort for bombing machines. Machines of No 103, No 54 and No 2 AFC Squadrons carried out the bombing, as a result of which three hangars were completely destroyed and one partially. Many bombs were observed bursting all round the remaining hangars, and it is almost certain that every machine on the aerodrome must have been destroyed. Two machines on the ground were totally wrecked. On the return journey the Scouts "contour chased" along the roads causing great havoc to motor transport, horse transport, and troops, especially in the villages Ligne and Leuze. A staff car was shot up, ran into a ditch and overturned. Enemy aircraft were generally active throughout the attack, and as a result of numerous combats from varying heights—2000 feet and upwards—nine enemy aircraft were destroyed, four driven down out of control, and two driven down.[34]

Not only were the Allies now able to inflict heavy losses on enemy ground targets at will, but in the air that day 67 German aircraft were destroyed on the British front. King George V sent the following message to the RAF:

> I offer you and the Royal Air Force my warmest congratulations on the successful results of air fighting on October 30th, and on beating all previous records. Such achievements testify to the spirit which animates all ranks in their determination to maintain our mastery in the air and cannot fail materially to assist the steady advance of my Armies in the field.[35]

---

31 A German pilot had first been seen to bale out on 11 August 1918 from his Pfalz scout. The Germans had begun issuing parachutes that summer, much to the frustration of the Allies who lacked them.

32 Lt J A Egan. No 4 Sqn AFC/71 Sqn RFC, 2 Sqn AFC. Bank clerk of Melbourne; b Bendigo, Vic, Aug 1892.

33 Lt-Col L A Strange DSO MC DFC RAF. Nos 5, 6, 12, 16, 23, 41 Sqns RFC; Asst Comt CFS 1917; comd 23rd Wg 1918, 80th Wg 1918–19, 51st Wg 1919. Farmer & aviator of Blandford, Dorset, Eng; b Tarrant, Keynstone, Eng, 27 Jul 1891.

34 Jones, Vol VI, p.549.

35 Telegram to the Secretary of State for the RAF, cited in Jones, Vol VI, p.548.

Indeed, the Allies were clearly demonstrating their mastery of the sky by the adaptability with which fighters could bomb, and bombers reverted to fighters. The SEs of No 2 Squadron had certainly proved themselves through the versatility of air power.

RECOMMENDATION FOR AN AWARD
to
Capt R. L. Manuel DFC

The name of Captain ROBY LEWIS MANUEL, DFC, is submitted for immediate Award for Conspicuous Gallantry under the following conditions:—

On the morning of the 16th September 1918, when returning from an Offensive Patrol, he encountered a Fokker Biplane flying over our side of the lines, North from LA BASSEE, at 9000 feet. Notwithstanding the fact that he had used practically all his ammunition in an encounter with 15 EA, of which by his able leadership his Patrol had destroyed one in flames, and driven down five out of control, he attacked this machine, keeping to the East of it, and firing a short burst in front, thus endeavouring to get the EA to turn towards our aerodrome. The EA put its nose down and flew North, closely followed by Capt. Manuel, who owing to his shortage of ammunition, could not fire until he made absolutely sure of hitting it. The Fokker continued diving until he was within 100 feet of the ground and then turned East to contour chase home. Getting close to it Captain Manuel fired his remaining few rounds (about 10) and the EA crashed into the ground near DROGLANDT, the pilot being found shot through the head.

This Officer has proved himself to be a very capable Flight Commander as well as a well-balanced Patrol leader. The efficiency of his flight has never been higher, his keenness communicating itself to all his subordinates.

As a leader he has shown marked judgment, leading his Patrol on all occasions with skill and dash, as evidenced by the following:—

On 27th August, while with his patrol of five machines, he attacked 30 Fokker Biplanes, with such skill as to permit one of his Flying Officers to shoot down one EA in flames, and two out of control.

In addition to the above he has previously destroyed four EA, and shot down six out of control. He has also taken part in low bombing, and low shooting patrols during MONTDIDIER-NOYON offensive in June last, and in the successful recent low bombing raids on Harbourdin and Lomme Aerodromes.

Major
Commanding 2nd Squadron
Australian Flying Corps

In the field
16.9.18

This was the recommendation to Captain Roby Manuel's bar to his DFC.

CHAPTER 9

PEACE AFTER STRIFE

He did but do his duty
simply bravely
and in the doing died.[1]

A spell of dull weather was broken on 4 November, and as the rain and mist cleared activity in the air picked up. Although the end was near for the German Air Service, it was still to fight with determination. On a morning offensive patrol, all three flights were airborne. Now venturing further over the lines to Renaix, 25 kilometres north-east of Tournai near Ath, Lieutenant Ernest Davies[2] (E5765) destroyed an LVG two-seater. Then the Australians swooped on seven Fokker biplanes, with three being destroyed by Wellwood (F5611), Simonson (C1129) and Stone (D6919), with Davies driving down a fourth.

That afternoon, the CO led the Squadron with a wing-strength attacking force that carried out a highly successful raid on Wattines airfield, east of Tournai. Escorted by No 4 Squadron and Bristol Fighters of No 88 Squadron, the Squadron shot down six Fokkers in the furious fighting, but three No 4 Squadron pilots were also shot down and killed. After releasing their bombs, No 2 Squadron's pilots joined in the combat to assist the Snipes, and drove down a further seven Fokkers, which one by one fell spinning into the low cloud.[3]

After 4 November, the German withdrawal became a rout. "The enemy was falling back without coherent plan in widespread disorder and confusion", the British Commander-in-Chief, Haig, reported in his dispatch.[4] In the days that followed, both Australian squadrons were engaged in bomb and machine-gun attacks on enemy columns near Ghislenghien, rolling stock at Enghien and on Croisette airfield. The Snipes came down to join the SEs in strafing attacks as there was now no air opposition. Only unfavourable weather saved the retreating Germans from further slaughter, as the Allies now controlled the skies. In this last week of the war, 4–11 November, the RAF claimed 68 enemy machines destroyed, and a further 24 driven down out of control, for a loss of 60.

The last big raid flown by the Australians was on 9 November. The attacking force comprised DH-9 bombers of No 103 Squadron, Camels of No 54 Squadron, and the

---

1 This epitaph, and "Peace After Strife", are inscriptions on Australian graves on French battlefields, cited in J Laffin, *Guide to Australian Battlefields of the Western Front 1916–1918*, Kangaroo Press, Sydney, 1992, pp.176, 178.

2 Capt E E Davies DFC. Lt Horse, No 2 Sqn AFC. Barrister and solicitor of Swan Hill, Vic; b Kerang, Vic, 16 Mar 1890.

3 Cutlack, p.381.

4 Montgomery, p.10.

SEs of No 2 Squadron. The escorting force was Bristol Fighters of No 88 Squadron and Australian Snipes of No 4 Squadron. The leader, CO of 54 Squadron Major R S Maxwell, led the force past Ath to Enghien and later reported:

> There, there was great congestion of troops and transport of all descriptions on the roads, trains on the railways and in the station, also two aerodromes with machines on the ground. On one aerodrome a hangar was completely destroyed and one machine completely wrecked, and bombs seen bursting among other machines which must have been badly damaged. On another aerodrome, one hangar and a machine on the aerodrome were destroyed in flames, and direct hits obtained on machines and hangars. Three large bombs and many 25-lb bombs were seen to burst on troops and transport on the main Ath-Enghiem road between Bassilly and Enghien, which was particularly congested, others on the roads north and south of main road, where there were many targets of troops, mechanical transport, and heavy transport. Lorries were seen to collide, one being set on fire, many others being destroyed by direct hits and others ditched. Horse transport was seen stampeding in all directions, and in numerous cases troops endeavouring to get into houses for cover were shot at and many casualties caused. In the station and junction at Enghien, no less than twenty direct hits were observed on trains. One train was set on fire from end to end, and was still burning furiously when the raid left, sheds and buildings in the station catching fire from it. A direct hit was obtained between some mechanical transport and a train where troops were entraining at a siding just east of Bassilly and a 230-lb bomb scored a direct hit on trains in Bassilly. Escorting machines of No 88 Squadron meeting with no opposition came down and joined in the destruction being caused on the ground. The ground targets were so obvious and numerous that every pilot and observer kept firing until stoppages or lack of ammunition compelled him to cease. The damage done and confusion caused was almost indescribable and impossible to give in detail. It must have been very great, every one agreeing that such an opportunity had never before been met with.[5]

The formation's only loss was Captain Frank Smith (C6403), leader of the 15 No 2 Squadron SEs, who was shot down by groundfire. Smith made it back over the lines by foot a few days later. The German capitulation, by the signing of the Armistice on "the eleventh hour of the eleventh day of the eleventh month", was all that averted the complete destruction of the German Army.

No 2 Squadron, with the other units of the Australian Flying Corps and the Royal Air Force, had continued an offensive policy of air power which was synonymous with the name of "Boom" Trenchard. This policy had resulted in continuously expanding demands for aircraft cooperation and in the last year of war the German Air Service never prevented the Allies from exploiting their control of the air. A fitting description of Trenchard was that he:

> ... saw with clear vision that the aeroplane is essentially an offensive weapon, and although it was in the squadrons that the offensive spirit was cherished, it was the dominant and inspiring personality of Major-General Trenchard which fired and coloured that spirit.[6]

Peacetime flying now involved patrols to monitor that there was no enemy activity, aerial mail runs and a small amount of currency training. An interesting task allocated to the Squadron was the impounding and evaluation of German aeroplanes.

5 Jones, Vol VI, pp.551–2.

6 Jones, Vol VI, pp.551–2.

In groups of about 20, under the command of one of the Squadron's officers, the mechanics were dispatched to several aerodromes in Belgium and Germany. 2nd Class Air Mechanic Fergus Cox was with a group that went to an airfield just to the east of Brussels:

> We had to take stock of all aeroplanes, get signatures for the airworthiness, petrol and oil, and finally put a German pilot up in it with enough petrol to do a good test, but not enough to abscond. In one notable case the German pilot declared the machine was OK, still he would not go up in it. The Teuton ack-emmas[7] stated it was in good condition. Finally he was told it was part of the Peace Agreement, and must be carried out. The pilot detailed how the flying wires were cut almost through with the lock-nut backed off.
>
> Some petrol tanks had sugar in them, which meant complete removal and wash. Loosened nuts on petrol pipes were plentiful. Once a machine was declared ready, it was isolated and guarded by Aussies. The aeroplanes were taken back to coastal France and handed over to a pool of new staff who flew them out of France.[8]

"A" Flight aircraft, D379 and E5765, at Pont-A-Marcq, near Lille, at the war's end, 16 November 1918 IWM E(AUS)3735

At least five of the Squadron's pilots were attached to Rely, south of St Omer, for ferrying these captured machines back to England. In addition, a Fokker D.VII was provided to the Squadron for training.[9] However, before it could be brought home as

7 Ack-emma = AM, aeroplane mechanic.

8 Cox, pp.39–42.

9 AWM4 8/5/1 War Diary, of 25 Jan 1919, records the serial number of this aircraft as A179/18. It most probably should read 4179/18.

a war trophy, it was crashed on 3 February by Lieutenant Egan, while flying between Pont a Marcq and Hellemmes.

With the cessation of hostilities, it was only a matter of time before demobilisation of the active units for their repatriation to Australia. While this continued during December, one of the Squadron's pilots, Lieutenant Fred Markham-Mills, died of illness on the 18th, and was buried in the City of Paris (Pantin) Cemetery.

Lieutenant E Dibbs in "C" Flight's E5782, immediately after the armistice AWM P355/46/29

The New Year saw a Squadron move to Hellemmes aerodrome, east of Lille, on 10 January, and on 1 February a return to Serny to join the 89th Wing in preparation for demobilisation. At Serny, the aircraft were handed over to No 32 Squadron RAF over 18–19 February 1919, and No 2 Squadron came under the administration of the 91st Wing in preparation for its return to England. During the month, two members of the Squadron died of pneumonia and influenza.[10] The move back to United Kingdom commenced on 27 February, and by 4 March personnel were camped with the 3rd Training Brigade at Hurdcott, near Havant, in Wiltshire, to await their return home. The waiting was to last through April.

On Anzac Day, many of No 2 Squadron's pilots, and former members of the Squadron, flew over London. It was their last sortie and as the CO, Major Murray Jones, described, "ably informed the London populace that there was such a unit as the Australian Flying Corps".[11] On 6 May the Squadron, together with all other units

10 Lt Gerald H Cowan died of pneumonia on 22 Feb 1919 at No 4 Stationary Hospital and was buried with Military Honours at St Omer. 2/AM George L Taylor died of influenza on 11 Feb 1919 at 39th Stationary Hospital Lille. AWM4 8/5/1 Monthly Summary, of Feb 1919.

11 AWM4 8/5/1 Summary of Officer Commanding, for month ending 30 Apr 1919.

Australian airmen embarked on the Kaisar-i-Hind *at Southampton, 6 May 1919* RAAF

of the Australian Flying Corps in England, proceeded from Fovent railway station to Southampton to board *HMAT Kaisar-i-Hind* for the return voyage. Oswald Watt, as the senior AFC officer in Europe, was in command of all Corps members.

One member returning was Captain Henry Wrigley, who had sailed with the "other" No 2 Squadron from Melbourne in 1916 and was to serve with distinction with No 3 Squadron. He recorded:

> At 6am on the 7th May the *Kaisar-i-Hind* weighed anchor and the homeward voyage commenced. The normal routine during the voyage was of the simplest form, a physical training parade, the object of which was to keep personnel fit, being held each morning, after which all ranks were free to organise their own recreations. Numerous competitions were organised and concerts, cinema entertainments, etc, arranged for the evenings. The voyage proved uneventful, and, on the 16th May, the *Kaisar-i-Hind* anchored off Port Said.
>
> After leaving Aden some rather boisterous weather was encountered, but this served to relieve the monotony of the voyage, and Columbo was reached about midday on the 29th May. This was the last port of call before reaching Australia, and, early on the morning of the 1st of June, the *Kaisar-i-Hind* weighed anchor.[12]

The voyage across the Indian Ocean was marked by everyone's impatience for the first sight of the Australian coast.

> Very few people on board failed to see the sun rise on the morning of June 9th; for with the dawn came the first glimpse of Australia. Naturally, after two or three years of wandering around Europe, there was much excitement amongst all ranks at the first sight of their native shores.[13]

Those from the West disembarked in Fremantle, and on 10 June the transport resumed her voyage to Sydney. Arriving at the Adelaide's Outer Harbour on 14 June, then Port Melbourne two days later, the *Kaisar-i-Hind* then completed the final leg of the journey to Sydney for the final disbandment of Squadron members on 18 June.

12 Wrigley, p.179.

13 Richards, p.50.

Homeward bound, No 2 Squadron personnel onboard for their return to Australia RAAF

No 2 Squadron had been instrumental in the development of air power and the successes of the Royal Flying Corps and the Royal Air Force. The Australian successes in aerial combat had led to its selection for Circus operations during the final year of the war, moving around the Front to where the action was, to ensure control of the air. Moreover, scouting machines could now be used for other roles to inflict the Allied will on an enemy.

Strafing had been used by the Australians effectively at Cambrai in 1917; it was decisive in avoiding a rout of the Allied armies in the spring of 1918, and it was to play a major role disrupting the German retreat later in 1918. That was now history, and peace offered a relief from the horrors they had endured.

> They had their time in the air tense with exhilaration and danger, but when they turned their backs upon the battlefield the war was left behind.[14]

Unfortunately, the skills they had pioneered, and the lessons they had learnt in cooperation with the army, were to be forgotten in the inter-war years. It was not until 1943, that the Allies were to rediscover and adopt the effective use of air power in assisting ground forces. It had been obvious to one of those early 2 Squadron pilots. In 1940 Robert Clark, on seeing the results of the German *Blitzkrieg* through Europe wrote:

> Battle of Cambrai was first experiment (and very successful) of dive-bombing in fighters being fitted with light bomb racks, and diving on trenches, troops, artillery and transport concentrations. We seem to have forgotten it, and the "Hun" developed it.[15]

---

14 Jones, Vol VI, pp.557–8.

15 Written by Robert Leeuwin Clark in 1940, before his capture in 1942 by the Japanese at Rabaul and subsequent death as a POW on the *Montevideo Maru* off the Philippines on 1 Jul 1942. Courtesy of John Leeuwin-Clark.

PART TWO

AIR STRIKE

CHAPTER 10

A New Air Force

I always found it useful in making recommendations for increases in the Service to be able to say, "This was recommended by Sir John Salmond".

Air Marshal Sir Richard Williams[1]

At the end of World War I, Australia possessed many military aircraft which had been acquired from a variety of sources. There were those that had been obtained for CFS over the war years, and those that were ordered at the war's end, which included 20 Avro 504s and 12 Sopwith Pups, intended to finally equip CFS with adequate training machines. In addition, six Fairey IIID seaplanes were being procured for the Royal Australian Navy.

The British Government had also given an undertaking to present an Imperial Gift of 100 aircraft to any dominion desiring to establish an air arm. In Australia's case, this comprised 30 DH-9A bombers, 35 SE-5A fighters, and a further 35 Avro 504K trainers. In return for the aircraft Australia had donated to the Imperial Government during the war, Britain further agreed to the presentation of another 28 aeroplanes—DH-9 bombers—making a total of 128 Gift aircraft.[2] Lieutenant-Colonel Williams, who was to become the first Chief of Air Staff, then negotiated the supply of spares and tools, mechanical transport and hangars, and with this equipment the Australian Air Force was established on 31 March 1921.[3] The birth date was three years after the formation of the first independent air force, the RAF, and conveniently avoided that ignominious anniversary of 1 April.

For its first year, it was planned that the new service would develop along the following lines:

In Victoria:

Air Force Headquarters
1 Flying Training School
1 Aircraft Depot
1 Recruit & Non-Technical Training Depot
1 Wing Headquarters, comprising:
1 Corps Reconnaissance Squadron
1 Fighter Squadron
1 Flying Boat Squadron.

In New South Wales:

1 Wing Headquarters, comprising:

1 Williams, p.185.

2 Williams, p.117.

3 The "Royal" prefix was added to the AAF on 13 Aug 1921.

1 Corps Reconnaissance Squadron
1 Fighter Squadron
1 Ships Seaplane Squadron.

In Western Australia:

The necessary land and buildings for a station to house one Squadron.[4]

Subsequently, an additional squadron of flying boats was planned and the need for the training depot was negated by the training of RAAF recruits at the Army Central Training School. The proposed organisation for 1921–22, therefore, became:

	Organisation	*Aircraft*
Victoria	Headquarters	
	1st Wing, comprising:	
	No 1 Squadron (Corps Reconnaissance)	12 DH-9
	No 2 Squadron (Single-Seat Fighters)	12 SE-5A
	No 6 Squadron (Flying Boats)	6 P.5[5]
	No 1 Flying Training School	Avro & Pup
	No 1 Aircraft Depot	reserves
NSW	2nd Wing, comprising:	
	No 3 Squadron (Corps Reconnaissance)	12 DH-9
	No 4 Squadron (Single-Seat Fighter)	12 SE-5A
	No 5 Squadron (Fleet Co-op Seaplanes)	6 Fairey 3D
	No 7 Squadron (Flying Boats)[6]	6 P.5

This proposal intended that for each service squadron, one flight would be manned by permanent personnel, the other two flights by members of the Citizen Forces. It had also been decided that the traditional roles of each squadron from the Australian Flying Corps be retained, hence Nos 1 and 3 Squadrons would be Corps Reconnaissance with army cooperation duties, with Nos 2 and 4 Squadron retaining their fighter role.

Accordingly, Nos 1 to 5 Squadrons were formed on a cadre basis with a nucleus of personnel on 10 January 1922. This was a start towards a plan of expansion, and the allotment of personnel of the various units at Point Cook was:

4 Air Board Agendum No 6, approved by the Air Council on 23 Dec 1920.

5 The P.5 flying boat proposed in this instruction was evidently a reference to the Felixstowe F.5, a type which was considered, but not acquired, by the RAAF. Eventually the Seagull III and Southampton I were to assume the flying boat role.

6 Air Board Agendum No 98, of 13 Jul 1921.

	Officers	*Airmen*
No 1 AD	11	88
No 1 FTS	3	12
No 1 Sqn	2	5
No 2 Sqn	2	4
No 3 Sqn	2	27
No 4 Sqn	1	7
No 5 Sqn	1	17[7]

In the event, only two pilots were posted to No 2 Squadron: Flight Lieutenant Harry Cobby and Flying Officer Eric Cummings,[8] and in effect this organisation of No 2 Squadron was little more than a fighter flight within No 1 Flying Training School (No 1 FTS), with only four SE-5As allocated for its operation.[9]

Bad news was soon forthcoming, as funding for the new Air Force was soon cut, with the vote for 1922–23 being just £250,000 ($500,000)—half that of the previous year.[10] Plans for the expansion to five squadrons were shelved, and on 1 July 1922 all squadrons (with the exception of No 1 Squadron, which remained in name only) were disbanded, with the sole flying activity remaining at Point Cook with No 1 FTS. This single conglomerate unit then comprised: "A" and "B" Flights with Avro 504s and Sopwith Pups for training, "C" Flight with SE-5As, "D" Flight with DH-9s, "E" Flight with DH-9As, and "F" Flight with 504 and Fairey IIID floatplanes.[11]

No 1 Squadron, being retained on a cadre basis until reforming as an active unit, finally commenced flying with No 3 Squadron in July 1925. Both units, No 1 at Laverton and No 3 at Richmond, were composed of three flights—fighter, bomber and army cooperation—with a mixture of weary aircraft types from the Imperial Gift. Again to minimise expenditure, there was a heavy reliance on part-time Citizen Air Force (CAF) personnel.

For presentation at the Imperial Conference in 1926, Group Captain Williams prepared a memorandum laying out his planned program of development, stating No 2 (Fighter) Squadron would be formed by 1 July 1929, and based in Victoria.[12] The Squadron was to be equipped with 12 operational aircraft, and a further 12 in reserve. Manning was planned for eight officers, with 22 from the Citizen Forces, and 62 airmen, with a further 114 of the Citizen Forces.

7 AWM54 81/4/165, "Units Located at Point Cook", undated, p.9.

8 Air Board Agendum No 182, of 20 Dec 1921, and AWM38 3DRL 7953 Item 9, undated, appears prepared circa Apr 1922. Other assignments included S/L A M Jones to command No 1 Station (as Point Cook was to be known); F/L S G Brearley to 1 Sqn; S/L R S Brown, F/L F W F Lukis and F/O P E Kenny to 3 Sqn; F/O G Jones to 4 Sqn; F/Os W A Murphy, F C Hawley and A E Mustard to 5 Sqn; and F/O G J W Mackinolty to 1AD.

9 C D Coulthard-Clark, *The Third Brother,* Allen & Unwin, Sydney, 1991, p.42. The four SE-5As being flown at Point Cook during 1922 were A2-2 (ex D370), A2-6 (E3167), A2-13 (D8476) and A2-19 (D8490).

10 This was 5.4% of the Defence vote. Williams, p.149.

11 Coulthard-Clark, p.186.

12 RAF Museum MFC 76/1/293, Memorandum of CAS, of 27 Jul 1926.

Before any heavy financial commitment was to be made by the government, the assistance of the RAF was sought on shaping the organisation into a balanced air force. Consequently, in July 1928, Air Marshal Sir John Salmond, who had commanded the RFC and RAF in the field in 1918, arrived in Melbourne. His findings, the Salmond Report, recommended the establishment of new squadrons, organised along the following lines:

Victoria	—	one fighter flight
		one flying boat flight
		one bomber-recon squadron
NSW	—	one fighter flight
		one flying boat flight
		one bomber-recon squadron
ACT	—	one army co-op squadron
WA	—	one bomber squadron (CAF)[13]

Hawker Demon fighter A1-38 was one of No 2 Squadron's first aircraft on reforming at Laverton in May 1937. K PORTEUS

The only immediate response to Salmond's recommendations was the ordering of a handful of Bristol Bulldog fighters in 1929. Again, the plans for further expansion of the RAAF into an effective fighting force were delayed by the grim economic outlook and its minuscule proportion of the defence vote.[14] Indeed, the very survival of the

13 Gillison, p.33.

14 From 1921 to 1930, the first nine years of the RAAF's existence, the total defence vote had been £57.545m ($115.09m), of which the RAAF had only received £3.635m ($7.27m)—a mere 7 ½ %. Williams, p.200.

RAAF as a separate service was in jeopardy. Member of Parliament Mr Thomas White, a veteran of the AFC's Half-Flight, told the House of Representatives that if the force was split by army and naval rivalry, air power would be set back a decade in its development.[15] Fortunately, the autonomy of the RAAF was retained, and in 1934 Hawker Demon fighters and Supermarine Seagull seaplanes were purchased to carry out more of Salmond's plan. This was how, in April 1936, Nos 21 and 22 Squadrons came to be formed as CAF units with a permanent core.

However, it was not until the following year that the RAAF was in a position for the further expansion that had been envisaged during the intervening years since its formation. In March 1937 the following plan was announced:

> FORMATION OF No 2 (General Reconnaissance) SQUADRON:
> No 2 (General Reconnaissance) Squadron is to form at RAAF Station, Laverton, on 1st May 1937.
> FORMATION OF No 4 (General Reconnaissance) SQUADRON:
> No 4 (General Reconnaissance) Squadron is to form at RAAF Station, Richmond, on 1st May 1937.
> FORMATION OF No 23 (CITY OF PERTH) (General Purpose) SQUADRON:
> No 23 (City of Perth) (General Purpose) Squadron (Permanent Cadre) is to form at RAAF Station, Laverton, on 1st May 1937. This squadron will move at a later date to Pearce Aerodrome, Western Australia.[16]

The formation of No 2 Squadron ultimately took place on 3 May 1937, with three officers and 38 airmen. The Commanding Officer was Squadron Leader Johnny Summers,[17] with Flight Lieutenant Peter Lavarack[18] as Adjutant, and Pilot Officer Arthur Hocking[19] as Equipment Officer. The initial equipment was made up of two Hawker Demon biplanes, sharing a hangar at Laverton with No 23 Squadron, beside that of No 21 Squadron. By the end of August, the Squadron strength had grown to three officers and 71 airmen.[20]

The Demon was a pretty, multi-purpose biplane with the distinctive lineage of the Hawker Hart family. Since its acquisition by the RAAF, the Demon had provided the backbone of the service, equipping the front-line service squadrons and then the CAF units. The first two Demons for No 2 Squadron (serialled A1-19 and A1-38) were to provide the only equipment until the receipt of Bristol Bulldog fighters and an Avro Anson bomber later in the year.

---

15 *Commonwealth Debates,* Vol 122, p.1230, cited in Gillison, p.36.

16 RAAF Historical 4/4/105, HQ RO 1046, RAAF HQ Routine Orders 1937, of 24 Mar 1937.

17 AirCdre J H Summers OBE, 12. RFC/RAF ; comd No 1 Sqn 1928–29 , Ftr Sqn 1932–35, 1 Sqn 1935–36, 21 Sqn 1936–37, 3 Sqn 1937, 2 Sqn 1937–1939, 1FTS 1939–40, 1SFTS 1940, 3BAGS 1942; AOC EA 1943, AOC NEA 1943–46. Regular air force offr of Sydney; b Sydney 12 Aug 1895.

18 Flt Lt P O Lavarack, 103. Nos 1, 2 & 21 Sqns; comd 9 Sqn 1942; 36 & 24 Sqns, 4CF. Student of Melbourne; b Camberley, Eng, 25 Aug 1915.

19 Wg Cdr A N Hocking, 1061/O24271. Nos 2 & 12 Sqns, SHQ Darwin, RAAFLO Wash. Accountant of Sydney; b Melbourne 27 May 1913.

20 No 2 Sqn Operations Record Book (A50), of 31 Aug 1937.

No 2 Squadron Bristol Bulldog, photographed from a Demon near Laverton K PORTEUS

Among the new pilots posted in at the year's end were Pilot Officers Red Green[21] and Keith Hampshire,[22] both of whom were to have notable careers and rise to high rank in the RAAF. By early 1938 sufficient of the modern Ansons were being received from England to equip a flight and provide the foundation for future Squadron expansion. "A" Flight was equipped with the Anson, "B" Flight with the Demon and "C" Flight with the Bulldog and a new type from the United States—the North American NA-16 (the forerunner of the Australian Wirraway). As the RAAF's prime bomber reconnaissance aircraft, the Anson had several innovative features. It was the RAAF's first monoplane, some were fitted for training with dual controls, and it was the first military aircraft in Australia with retractable undercarriage. This, however, was an arduous procedure—it took 165 turns to wind the gear up!

The organisation of the Squadron, comprising seven officers, three sergeant pilots, and 70 other ranks, was made up as follows:

21 AirCdre R F M Green CBE DSO, 146. No 2 Sqn, SA HQ; comd 4 FSHQ 1942, 43 OBU 1942–43, 11 Sqn 1943–44, GR Sch 1944, Pearce 1944–45; SASO RAAF Cd 1945. Regular air force offr of Pingelly, WA; b Pingelly 4 Feb 1916.

22 Gp Capt K McD Hampshire DSO* DFC, 147. No 2 Sqn, SHQ Darwin; comd 6 Sqn 1941–42, 23 Sqn 1942, 22 Sqn 1942–43, 456 Sqn 1943–44, 280 Wg RAF 1945. Regular air force offr of Perth; b Port Macquarie, NSW, 10 Sep 1914.

"A" Flight		"B" Flight		"C" Flight	
Flg Off Eric Read[23]		Plt Off Red Green		Flt Lt Hugh Berry[24]	
Anson	A4-29	Demon	A1-19	Bulldog	A12-1
	A4-33		A1-38		A12-2
	A4-34		A1-58		A12-3
	A4-37		A1-59		A12-6
				NA-16	A20-1

Although this hard landing for Bulldog A12-6 at Laverton on 2 April 1938 looks like the aircraft was a write-off, it was repaired and returned to No 2 Squadron within three months. VIA E PERKINS

The Squadron at this stage possessed more aircraft than pilots and with their experience level fairly low, this meant that pilots from other units at Laverton often assisted with flying; in fact the "C" Flight Commander was on the strength of Station Headquarters. This was the composition (less A12-6 which was damaged in a landing mishap on 2 April) as the unit prepared for its first major exercises, air pageants at Flemington Racecourse in Melbourne and the RAAF Base at Richmond New South Wales in April 1938. For these events, Squadron Leader Fred Scherger,[25] a test pilot

23 Wg Cdr E V Read AFC. Nos 1, 21, 2 & 23 Sqns, AHQ Far East, RAAF HQ; comd 200 Flt 1945–46. Power station operator of Melbourne; b Hobart 9 Feb 1915.

24 Wg Cdr H W Berry OBE, 46. No 101 Flt, 3 & 1 Sqns, SHQ Laverton; comd 22 Sqn 1939–40; AHQ Iraq, 242 Gp RAF. Orchardist of Melbourne; b 19 Jan 1905.

25 ACM Sir Frederick R W Scherger KBE CB DSO AFC. D Trg 1938–40; comd RAAF Darwin 1941–42; AOC 10 Gp 1943–44, 1TAF 1945; AMP 1945,1955–56; DCAS 1947–51; AOC RAF Malaya 1953–55; CAS 1957–61; Chman COSC 1961–66. Regular air force offr; b Ararat, Vic, 18 May 1904.

from RAAF Headquarters in Melbourne, was attached to the Squadron to fly aerobatic demonstrations with the newly arrived NA-33. This aircraft was borrowed from the Commonwealth Aircraft Corporation, where it was serving as the pattern airframe for the Wirraway program. The Victorian RAAF Display, on 9 April, attracted 70,000 spectators at the Racecourse as the Squadron was joined by aircraft from Nos 1, 3, 4 and 21 Squadrons for aerobatic demonstrations and massed flypasts, the largest ever attempted by the RAAF.

Pilot Officer Lance Sutherland was the first casualty on No 2 Squadron RAAF, when he crashed his Bulldog, A12-2, at Richmond on 22 April 1938. RAAF

After the successful completion of this event in Melbourne, the Squadron prepared for the deployment to Richmond for a similar display, departing Laverton on 17 April. The display on 23 April was, however, marred by the fatal crash of a 2 Squadron Bulldog during practice the previous day. Pilot Officer Lance Sutherland[26] (flying Bulldog A12-2) was killed instantly when his starboard wing hit the ground while practising a slow roll over the Richmond aerodrome. He was the first fatal air casualty of the Squadron since its reformation.

On return to Laverton, the Squadron focused on flying training on Anson around-the-clock operations, with emphasis on night operations and instrument flying. During May 1938 the airfields at Cressy, Hamilton and Mt Gambier were inspected for their suitability for night use. This was to prove significant for future Squadron operations, as was the adoption of Yanakie, on Wilson's Promontory, later that year, as an advanced base for maritime patrol operations.

Navigation Reconnaissance Courses[27] were also conducted for the RAAF by No 2 Squadron over this period, the first commencing in July 1938. Training of this type was necessary in the RAAF as at this stage there were no navigators, this aircrew function being performed by pilots. These intense 19-week courses covered Dead-Reckoning Navigation, Reconnaissance, Meteorology, Search and Patrol, Photography,

26 Plt Off L R Sutherland. No 2 Sqn. Jackeroo of Adelaide; b Minlaton, SA, 14 Feb 1915. Killed in flying accident 22 Apr 1938.

27 No 1 NR Course was run from 18 Jul until 25 Nov 1938, and No 2 NR Course from 20 Feb to 16 Jun 1939. An ad hoc third course was commenced by 2 Sqn in Apr 1940, with members completing the syllabus on their respective squadrons. These courses were distinct from the Navigation Specialist Course and the General Reconn Couse, both conducted at Point Cook.

Signalling and Coding, Ship Recognition and Mathematics. This first course was marred by a fatal accident on the morning of 10 August 1938, when Anson A4-29 crashed. The aircraft was one of five in a stream involved in a navigational exercise which had been recalled due to the deteriorating weather.[28] In low cloud near Dromana, Victoria, A4-29 flew into a hill, known as Arthur's Seat, killing four on board.[29]

Future navigation reconnaissance courses were conducted under the guidance of Flight Lieutenant Frank Headlam.[30] Among some of the pilots who undertook this specialised training over 1938 to 1940, when courses were run by No 2 Squadron, were Pilot Officers "Ginty" Lush,[31] Bill Pedrina,[32] Jim Hepburn,[33] "Spud" Spurgeon,[34] Herb Plenty,[35] Wilbur Wackett,[36] John Ramshaw[37] and John Lockwood.[38]

By February 1939, the delivery of sufficient Ansons allowed the Squadron's older aircraft to be pensioned off: "B" Flight's Demons were transferred to a new unit forming at Laverton, No 12 Squadron. Several 2 Squadron personnel were also posted to this new squadron, including equipment officer Flying Officer Arthur Hocking and photographer LAC Kip Porteus.[39]

---

28 Coulthard-Clark, p.329.

29 Killed were pilots Plt Off R S Symonds (also acting as navigator) and Sgt J M Gillespie, W/T Operator AC1 K McKerrow, and fitter AC1 R Mawson. AC1 J Glover, a new recruit on a joy flight in the rear turret, was the sole survivor.

30 AVM F Headlam CB CBE, 70/O331. No 2 Sqn, SHQ Lav, RAAF HQ; comd No 2 Sqn 1941–42, HQ NWA; comd 2 ANS 1942–43, 2 AOS 1943; HQ NWA; comd 90 Wg 1950–53; SASO HQHC, DepAir; AOC HQOC 1961–62; AOC 224 Gp 1962–65; DCAS 1965–66; AOC HQSC 1966–68; Def Staff London 1968–71. Regular air force offr of Hobart; b Launceston, Tas, 15 Jul 1914.

31 AVM J F Lush CBE, 254/O369. Nos 2, 12, 13, 10 & 6 Sqns; comd 32 Sqn 1943; SO Ops HQ NWA 1943–44, RAAF Cd 1944–45, SASO 11 Gp RAF 1945; comd 86 Wg 1949–50, 38 Sqn 1950, Laverton 1957–59; AA Wash 1963–65, COMRAAFV 1967–68, AOC HQSC 1969–70. Regular air force offr of Melbourne; b Melbourne 17 Sep 1915.

32 Sqn Ldr W A Pedrina DFC, 264. Nos 2, 12, 13, 6 & 32 Sqns, 1 OTU. Regular air force offr of Hamilton, Vic; b Hamilton 20 May 1919. Killed in action 15 Dec 1942.

33 Wg Cdr J A Hepburn DFC AFC, 247. Nos 12, 13, 2, 14 & 7 Sqns, 1 OTU; comd 100 Sqn 1943; 1 OTU; comd 1 OTU 1945. Regular air force offr of Warrnambool, Vic; b Melbourne 24 Jan 1919.

34 AirCdre C H Spurgeon CBE DFC, 569/O399. Nos 21, 2 & 8 Sqns, POW Jan 1942. SAN; comd 1 Sqn 1955–56; RAAFSC, RAAF Wash; comd 1 Sqn 1962–63; RAAF Wash 1964–66; comd 82 Wg 1967–69; HQOC; COMRAAFV 1970–71; comd Amberley 1971–74. Regular air force offr of Melbourne; b Melbourne 18 Apr 1920.

35 Gp Capt H C Plenty DFC*, 582/O3103. Nos 21, 2, 8, 14 & 100 Sqns, 1 OTU, 14 Sqn, HQ NEA, DepAir, 1BFTS, HQTC, 1 AFTS; comd CFS 1962–64, 78 Wg 1964–66, Laverton 1966–71; DepAir. Regular air force offr of Warnertown, SA; b Port Pirie, SA, 2 Feb 1921.

36 Sqn Ldr W L Wackett, 588. Nos 2, 24, 75 & 31 Sqns. Regular air force offr of Melbourne; b Townsville, Qld, 19 Feb 1921. Killed in action 24 Sep 1944.

37 Flt Lt J C Ramshaw, 552. Nos 21, 2 & 1 Sqns. Regular air force offr of Melbourne; b Bangalore, India, 18 Oct 1914. Killed in action 8 Dec 1941.

38 Flt Lt J A H Lockwood, 554. Nos 21, 2 & 1 Sqns. Regular air force offr of Geelong, Vic; b Geelong 23 Dec 1919. Killed in action 14 Feb 1942.

39 Sqn Ldr K L Porteus, 2683. Surv Flt, Nos 2, 12 & 13 Sqns, 9 OG, RAAF Wash 1944–46; BCOF 1946–47; comd Phot Sch 1952–54, CPE 1954–56. Photographer of Sydney; b Port Macquarie, NSW, 15 Feb 1917.

The Laverton tarmac in 1938, with Ansons, Seagull amphibians, Demons, a Magister and a Moth
F JEFFERIES

In May, the Bulldogs passed to Communications Flight; however, the unit's association with biplanes was not quite over, as two Avro Cadet trainers were flown until mid-year. At the same time, the following appointments were made for Flight Commanders: "A" Flight—Flight Lieutenant Frank Headlam, "B" Flight—Pilot Officer Reg Burrage,[40] "C" Flight—Flight Lieutenant Fred "Jock" Wittscheibe.[41] This was followed on 1 March by Wing Commander Alan Charlesworth[42] taking command of the Squadron.

40 AirCdre R B Burrage OBE DFC, 212/O365. Nos 2 , 21 & 10 Sqns; comd 461 Sqn 1942; 11 Sqn; comd 76 Wg 1944–45, Pearce 1945, Pt Cook 1945–47; SA HQ; comd 34 Sqn 1951, Mallala 1951–54; RAAF HQ; comd BSqn Ric 1956–59. Regular air force offr of Melbourne; b Melbourne 13 Sep 1917.

41 Gp Capt F A Wittscheibe, 89. Nos 1, 21 & 2 Sqns;comd 23 Sqn 1940, 4SFTS 1941–42, 1ANS 1942–43, 2AOS 1943–45, 13 Sqn 1945. Regular air force offr of Bendigo, Vic; b Melbourne 17 Aug 1913.

42 AVM A M Charlesworth CBE AFC, O37. Nos 1, 3 & 1 Sqns, 1FTS, RAAF HQ; comd 1 Sqn 1937, 2 Sqn 1939, Pearce 1940–41, 2BAGS 1942–43; AOC EA 1943–44; AOC NWA 1944–46; AOC EA 1946; comd SLAW 1947–49, Williamtown 1949; CoS BCOF 1949–51; AOC SA 1951–53; comd O'seas HQ London 1954–55. Regular air force offr of Laverton, Vic; b Lottah, Tas, 17 Sep 1903.

The Anson would ultimately equip the Squadron for almost three years on coastal patrol duties. One exercise, on 26 August 1939, involved a search off Port Phillip for the overdue cargo-passenger vessel *Straasfurt*. The Squadron diary recorded: "The search revealed nothing. The most outstanding and regrettable feature apparent was the very limited range of Anson aircraft".[43]

A week later, 3 September 1939, German hostile intentions in Europe dragged the world once again into war. No 2 Squadron was ordered to mobilise for operations at 7 a.m. on 4 September.

43 No 2 Sqn A50, of 31 Aug 1939.

CHAPTER 11

GUARDIANS OF THE TRADE ROUTES[1]

... during World War II there were 40 documented attacks against Allied shipping by Japanese and German submarines in Australian waters. 20 Allied ships were lost.[2]

Two of the Flight Commanders in this opening month of the war were Flight Lieutenants Bob Hitchcock[3] and Ray Garrett.[4] Garrett had been a pilot with the CAF between the wars, having trained on the SE-5A in 1927 and becoming the first Qualified Flying Instructor (QFI) while on duty with the Citizen Forces. He was called up for active service on the outbreak of war. The third Flight Commander was Squadron Leader Fred Thomas[5] who then took command of the Squadron in December 1939. He was replaced as "B" Flight Commander by Flight Lieutenant John Ryland,[6] another CAF officer who had been flying Lockheed 10 airliners with Ansett Airways. (Ryland was later to become General Manager of Trans Australia Airlines.) At this stage Laverton was also home to No 1 Squadron flying Demons, and No 21 Squadron with Wirraways.

No 2 Squadron's prime activity was the vitally important seaward patrols, searching for enemy raiders to ensure the protection of convoys. For this role the Anson was hardly ideal, but it was the best the RAAF had. The Anson had a range of 1,340 kilometres (790 miles), cruising at 250 km/hr (158 mph). On these patrols, the Ansons were each armed with two 112 lb (48 kg) bombs and other stores, such as flares and markers.

Perhaps the Squadron's most crucial operation was the protection of the first convoys carrying the 2nd AIF to the Middle East during early 1940. Navigation on these overwater patrols was a problem, as there was still no navigator category in the RAAF. Flight Lieutenant Ray Garrett described patrol flying, and the necessity of accurate search patterns:

---

1 The title of this chapter comes from a movie made in 1941 by 2 Squadron pilot, Flg Off Kym Bonython. K Bonython, *Ladies' Legs and Lemonade*, Rigby, Adelaide, 1979, p.42.

2 Wg Cdr I MacFarling, "Airborne Anti-Submarine Warfare", in A W Stephens, *Defending the Air/Sea Gap*, ADSC, Canberra, 1992, p.103.

3 Flt Lt R E Hitchcock, 115. Nos 1 & 2 Sqns. Regular air force officer of Melbourne; b Fremantle, WA, 19 Jun 1912. Killed in flying accident 13 Aug 1940.

4 Gp Capt Sir Raymond W Garrett AFC, 250118. Nos 1 & 2 Sqns; comd 4EFTS 1941, 2EFTS 1942, 2 OTU 1943–44. President Vic Legislative Council 1968–76. Commercial pilot of Melbourne; b Melbourne 19 Oct 1900.

5 Gp Capt Sir Frederick W Thomas, 250097. Comd No 2 Sqn 1939–41, RAAF Townsville 1942; SASO NEA 1942–43; SASO RAAF Mission USA 1944; D Opnl Reqs 1945. Lord Mayor of Melbourne 1957–59. Flour miller of Melbourne; b Nathalia, Vic, 27 Jun 1906.

6 Gp Capt J P Ryland CBE DFC, 250188. Nos 21 & 2 Sqns; comd AOB Namlea 1941, 13 Sqn 1941–42; CFI 1 OTU 1943; comd 1 OTU 1944, 79 Wg 1944–45. Commercial pilot of Melbourne; b Melbourne 26 Jul 1911.

With the commencement of hostilities, No 2 Squadron's Ansons adopted "warpaint". F JEFFERIES

These aeroplanes were equipped with bombsights in the nose, so that by lying down prone in the aircraft and looking through a perspex panel on the lower portion of the nose, the bombsight could be used to measure the drift. Then one would go back to the little desk behind the pilot's seat and plot the course. It was, of course, necessary that while taking this reading of drift, the aircraft be flown straight and level. There was a class of pilot used in the service then called a "driver pilot".[7] These were men with rank of sergeant and they were taught to fly by instruments, that is to say they could fly straight and level only ... they acted actually as a human "George". When you wanted to take drift readings, you got into the prone position at the bombsight and let the driver-pilot keep the aircraft flying straight and level.

The aircraft were largely used to carry out patrols ahead of the convoys to ensure that there were no submarines lurking in the vicinity, and searches were done in various ways. The most usual search was what was called the Diverging Search. It would start off from a spot, called the datum point, usually somewhere on the south coast, most cases a lighthouse, and aircraft would fly fanned out in a diverging manner. In Ansons we used to fly 350 miles out to sea,[8] fifty miles along, and 350 miles back to the datum point. In that manner, by using about 9 or 10 aircraft, a very large area could be covered to ensure there were no submarines in the vicinity.

Another type of search was a Parallel Track Search. In this case one aircraft would fly perhaps 5 or 10 miles out from the shore, another one a few miles further out parallel to it, and another one further out again. Another type was the Anti-Submarine Search, and the aircraft flew at a height of only about 500 feet, and passed along the convoy out about 5 or 10 miles ahead of it, turn and flew back again along the line of ships.[9]

---

7 The first two courses had been conducted by 2 Sqn in May 1939. No 2 Sqn A50, of 1 Jun 1939.

8 This increased radius of action for the Anson was made possible by auxiliary fuel tanks in the cabin, fed by a foot pump operated by a crewman.

9 Tapes of recollections, Sir Raymond Garrett. No 2 Squadron Association, Melbourne.

Some merchant vessel convoys would be picked up by No 2 Squadron Ansons at Sydney Heads, and escorted around the coast to Western Australia. Aerodromes such as Mallacoota, Yanakie, Pat's River (on Flinders Island), Cressy, Parafield and Ceduna became advanced operating bases away from Laverton. Bombing whales became a consequence of the Squadron's attacks on suspected submarines. Ray Garrett later recounted: "A whale from the air looks very much like a submarine … I don't know how many whales I sunk!"[10]

2nd AIF convoy off Wilson's Promontory, Victoria F JEFFERIES

Most of the weather in Bass Strait was cold and wet. When flying in these foul conditions, there was the constant fear of icing: the collection of ice along the wing leading edges and propellers. Any sign of icing meant having to descend to low level to avoid any dangerous accumulations. These concerns were heightened by the possible loss of an engine, as the Anson needed both operating to maintain height. For over six months the Squadron operated under these conditions, and did not lose one aircraft. Ray Garrett attributes this:

> … to the efficiency of our groundstaff, who kept the engines in the best order that they possibly could despite, in some cases, a shortage of spares.[11]

10 Garrett tapes.

11 Garrett tapes.

Now that the Squadron was flying operationally, during 1940 the previously silver Ansons were camouflaged. The joke around Laverton became: "Come to the hangar and feel how many aircraft we've got". One of the gunners on the Ansons was Corporal Wally Scott,[12] who had been a founding member of the Squadron three years previously. He remembered an amusing incident in June 1940, when Italy entered the war on the side of the Germans:

> The day after Italy came into the war we were lined up in flights on the road in front of the No 2 Squadron hangar at Laverton for the CO's parade. The Squadron WO was calling the roll. He calls: "Cpl Scascighini". No answer. He calls again: "Cpl Scascighini!". I nudge 'Scaza' and say quietly "Wake up, Scaza". He says "Not me, mate". The WO says "Is Cpl Scascighini on parade?". Scaza takes one step forward and states clearly: "The name is Smith, sir, as of this morning!".[13]

With the Lockheed Hudson now arriving in Australia in quantity to equip the RAAF general reconnaissance squadrons, Flight Lieutenant John Ryland, as one of 2 Squadron's QFIs, proceeded to Richmond to undergo instruction on the new type in February 1940.[14] Here he was converted to the Hudson by the Lockheed instructor who had come out from California, Mr L D "Swede" Parker. Ryland completed his conversion at the end of March, to return to flying Ansons on 2 Squadron, but occasionally flew No 1 Squadron's Hudsons at Laverton for continuity.[15]

Anson A4-44, displaying the "B" marking to indicate its 2 Squadron ownership, soon after the arrival of the first Hudsons to Laverton in mid-1940 F JEFFERIES

12 WOff W R Scott, 2491. No 2 Sqn, 26 RSU, 11 RSU, CFS, RAAF HQ. Diesel engineer of Melbourne; b Cobden 8 Jun 1911.

13 Wally Scott, letter to 2 Squadron Association, Melbourne, of 22 Oct 1992.

14 John Ryland's Flying Logbook, Manuscript Collection MS 5020 Box 1, National Library, Canberra.

15 At Laverton over Apr–Jun 1940 Ryland flew 1 Sqn's A16-19, -20, -21, -23 and -26.

The Hudson was a comfortable aircraft to fly, with a heated cabin and hydraulically retracted undercarriage, and its electrically started engines meant that the hand cranking required by the Ansons became a thing of the past. The weapons-carrying capability of the Hudson was a vast improvement over the Anson. Whereas two 48 kg bombs had been carried in the Anson, the Hudson was able to carry almost 500 kg of stores. With its two powerful Twin Wasp engines, the Hudson had plenty of power, and it flew well on a single engine. However, Garrett learned to treat the Hudson with utmost respect. When heavy, it was necessary to keep the speed up, as it was more critical near the stall than its contemporaries; pilots did not take liberties near the ground.

The first Hudsons were ferried down to Laverton from Richmond in June 1940. Crew familiarisation and conversions began immediately, with Ryland flying the first sorties on 15 June. By this stage Ray Garrett had also been converted by Parker at Richmond, finishing off his training with Ryland in July. Over the next months, Flight Lieutenants John Ryland and Ray Garrett converted 2 Squadron's pilots to the Hudson, with Garrett passing on his old adage: "The only way to learn to fly is by flying".[16] The intensity of the flying for the instructors was high—in the first six months Ryland flew over 500 hours, mainly instructional.[17] This was a lot of flying, but the wartime urgency warranted such efforts. One of the first new pilots to undergo conversion to Hudsons, and begin a long association with No 2 Squadron, was Pilot Officer Bob Dalkin.[18] He recalled his early experiences at Laverton:

> I arrived at 2 Squadron, Laverton (after completing an SFTS on Demons), on Monday 12 August 1940, with three others for conversion to Hudsons. Bill White[19] was one, David Campbell[20] another, the third being Dick Wiesener[21] who had one ride in a Hudson as a passenger on that day; 24 hours later he was dead in the right seat of A16-97, in the accident which killed the crew and so many high officials at Canberra the next day, Tuesday 13 August.
>
> Conversions thereafter were interrupted by patrols from Richmond, Moruya, Mallacoota, etc. I did some second piloting and nav to Fred Thomas (squadron commander and later, Sir Fred) in A16-34 & -80, Ray Garrett (later Sir Ray) in A16-78 and a short time later, one of my own early second pilots was Bob Law-Smith[22] (yes—of course, also Sir Bob) in A16-18, as

16 Garrett tapes.

17 Ryland's Flying Logbook.

18 AirCdre R N Dalkin DFC, 260612. Nos 2, 7 & 13 Sqns, HQ NWA; CI GR Sch 1943–44; Bomber Cd RAF 1944–45; comd 2 Sqn 1949. Administrator Norfolk Island 1968–72. Regular air force offr of Salamaua, TPNG; b Northumberland, Eng, 21 Feb 1914. AirCdre Dalkin became the founding Patron of the RAAF Hudson Squadrons Association.

19 Flt Lt W V D White DFC, 260624. Kenya Defence Force, Nos 3 & 2 Sqns. Company manager of Sydney; b Inverell, NSW, 25 Oct 1913. Executed by Japanese 20 Feb 1942.

20 Wg Cdr D W I Campbell DFC, 250528. Nos 2, 24 & 13 Sqns; comd 1 Sqn 1943–44, 2 Sqn 1944, 32 Sqn 1945. Clerk of Melbourne; b Melbourne 16 Jul 1915.

21 Plt Off R F Wiesener, 607. No 2 Sqn. Of Sydney; b Sydney 31 Dec 1920. Killed in flying accident 13 Aug 1940.

22 Sqn Ldr R R Law-Smith AFC, 280733. Nos 2, 100, 36 & 34 Sqns; comd 37 Sqn 1944–45; DD Postings 1945–46. Solicitor of Adelaide; b Adelaide 9 Jul 1914.

well as our irrepressible member, Kym Bonython.[23] John Ryland was flight commander, but I soon started my conversions on A16-12, -6 and -9 under that terrific man and great pilot, Dallas Scott,[24] later to be executed at Ambon.

In 2 Sqn in those days we were given our own Hudsons to fly, look after, polish, lock up and boast about, with the aid of the ground staff who were the salt of the earth. We did not, of course, manage to fly our own aircraft exclusively. I still have the aluminium key tag of A16-32, made up in the Squadron, which I carried around on my key ring for over 40 years before I retired it to the trinket box. It certainly brought me luck. I flew 138 sorties at 2 Squadron as captain of A16-32, and when I left the Squadron for a couple of months to join 7 Squadron, and was again posted north to 13 Squadron as Flight Commander, I was soon, to my joy, in command of 32 again, this time at the "shooting war".[25]

Patrol flying continued with the new aircraft; however, with the Hudson being so much more capable, the flying tended to become more mundane. "Negative enemy reports" was the normal postflight result, and obviously prompted Ray Garrett's remark: "All this flying was pretty boring." Nevertheless, Australia was dependent on merchant vessels for vital supplies and munitions from overseas, and the transport of our troops overseas meant patrol flying was crucial to ensure safe passage.[26]

A16-6, one of the first Hudsons delivered to No 2 Squadron in June 1940, shows the lack of dorsal gun turret when these aircraft were originally delivered to the RAAF. F JEFFERIES

One episode which broke up the monotony of patrol flying was when Sweden capitulated to the Germans. Sweden wanted the return of her ships before they could be commandeered by the Allies:

23 Sqn Ldr H R Bonython AC DFC AFC, 280778. Nos 2 & 100 Sqns, 6 Torpedo Trg Unit; comd Torp Dev't Flt 1944; 87 Sqn. Trainee chartered accountant of Adelaide; b. Adelaide 15 Sep 1920.

24 Wg Cdr E D Scott AFC, 250101. No 1 Sqn; comd 7 Sqn 1940, 6 Sqn 1940–41, ACH Halong 1941–42. Commercial pilot of Melbourne; b Melbourne 5 Oct 1913. Executed by Japanese 6 Feb 1942.

25 Bob Dalkin, "Love Affairs with A16-12 and Other A16s", *RAAF Hudson Squadrons Association Newsletter*, No 13, Feb 1990, Adelaide, pp.1–2.

26 The importance of coastal patrols was further highlighted by Japan's entry into the war. Only the fact that Japanese submarines lacked any significant strategic mining capability prevented further losses to Allied shipping. LCdr D M Stevens, "The War Cruise of I-6", *Australian Defence Force Journal*, No 102, Sep/Oct 1993, pp.39–46, refers.

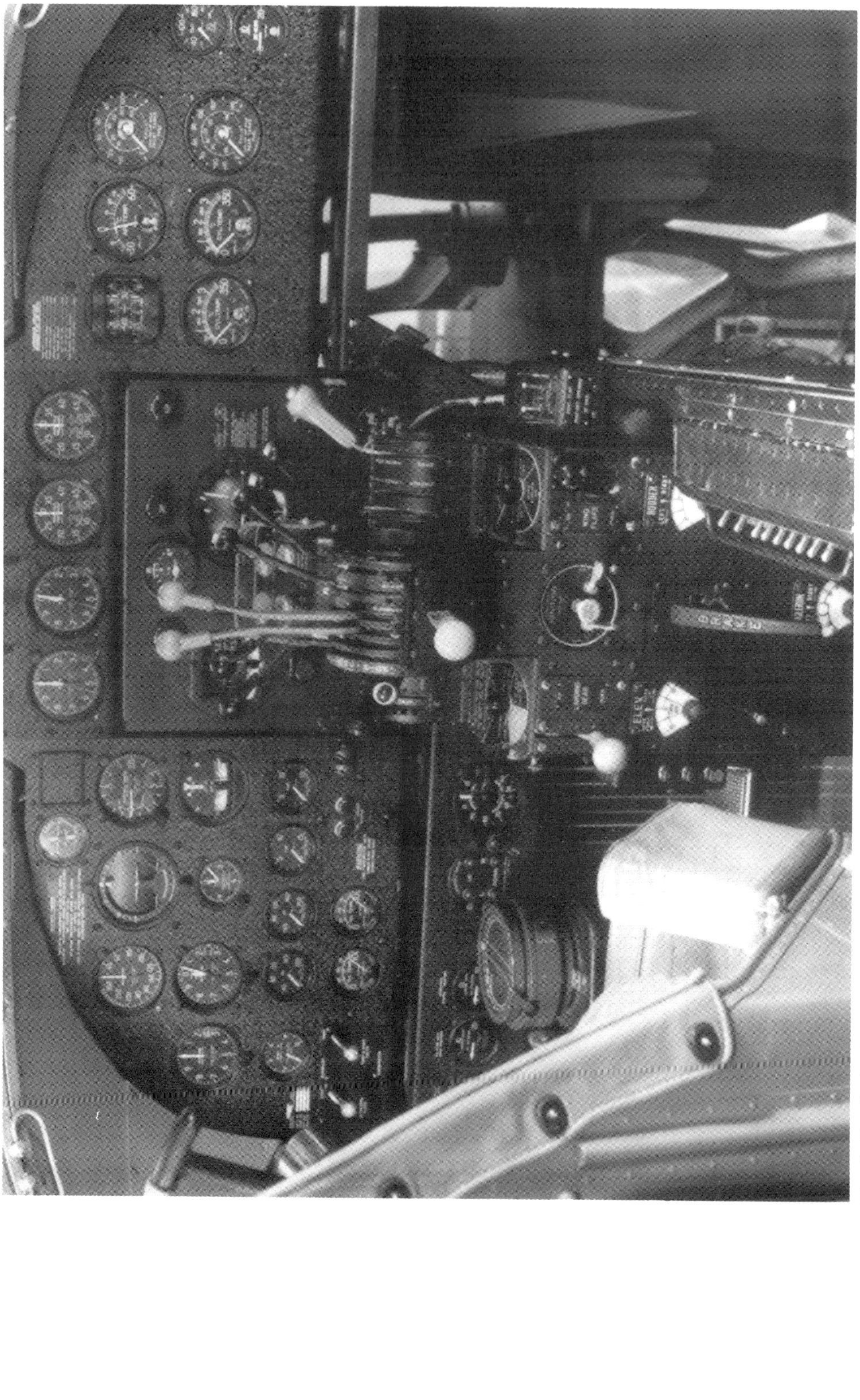

F JEFFERIES

The "office" of a brand new 2 Squadron Hudson at Laverton in 1940

> There was a number of these Swedish ships in Australian waters, and 2 Squadron was given the job of rounding these up. This meant locating a ship, then directing it by signal to proceed to the nearest port. Sometimes they would obey the signal and do so, but more often than not, they would try to get away. We would dive on them and put a burst of gunfire, from our front guns, across their bows, whereupon they would promptly turn around and go to the nearest port. We rounded up quite a number in this way, and got a lot of fun out of it![27]

The Squadron continued to conduct anti-submarine and convoy escort patrols from advanced operational bases at Moruya, Mallacoota, Bairnsdale, Cambridge, and Mount Gambier ("a horrible place to work from ... the ground was undulating and the strip consisted of a series of small hills ... one got bumped into the air"[28]). Operations often involved night take-offs from these deployed bases, in order to locate and rendezvous with convoys at first light. One such mission was to escort the largest convoy to leave Australia, which consisted of the *Queen Mary*, the *Mauritania*, and many ships up to the 10,000 ton class. The convoy was lined up in two parallel lines, with seven or eight ships in each line. The escort was the battleship *HMS Repulse*, which was to be lost off Malaya at the end of 1941.

Pilots of "B" Flight, No 2 Squadron, at Laverton, August 1940:
From left: Robert Law-Smith, Jock Webb, John Ryland, Peter Gorrie, Jim Harper and Bill White
NATIONAL LIBRARY OF AUSTRALIA

The tragedy of the first Hudson loss struck the Squadron on the morning of Tuesday 13 August 1940. Flight Lieutenant Bob Hitchcock (A16-97) crashed, killing the 2 Squadron crew[29] and the six passengers on board, while making an approach into Canberra. The RAAF Base at Canberra was subsequently renamed after one of those who perished, the Honourable James Fairbairn, Minister of State for Air. The other passengers were Brigadier the Honourable Geoffrey Street (Minister of State for

27 Garrett tapes.

28 Garrett tapes.

29 In addition to Hitchcock, the following 2 Squadron members were killed: Plt Off Richard Wiesener (607), Cpl John Palmer (2130), and AC Charles Crosdale (6673).

the Army), the Honourable Sir Henry Gullett (Vice-President of the Executive Council), General Sir Brudenell White (Chief of the General Staff), Lieutenant Colonel Francis Thornthwaite and Mr Richard Elford. No 2 Squadron photographer LAC Frank Jefferies[30] had been tasked on this flight:

> I was to have flown with F/L Hitchcock to Canberra on the 13th August 1940, and on return to go on patrol for two days, and then to do a photographic survey at Woodside, SA.
>
> On the 12th August I was told that my seat on the aircraft would be taken by Mr Elford, Private Secretary to the Minister for Air.
>
> On the morning of the 13th, as I was preparing my cameras, I was told by F/L Garrett to put my gear in his aircraft as A16-97 had crashed and that all had been killed.[31]

Officers of No 2 Squadron, Laverton, August 1940:
Front row: (from left) Dunne, Hemsworth, Chapman, Garrett, Wg Cdt Thomas (CO), Ryland, Rongar (Adjutant), Lavery (Equipment Officer), Milne. Rear row: Fowler, Prowd, White, Law-Smith, Eddison, Dalkin, Cowan, Wackett, Hughes F JEFFERIES

The aircraft had stalled on its approach into Canberra.[32] The Hudson's short, stumpy fuselage made its centre of gravity critical. With undercarriage and flaps down for an approach, and with power applied, the Hudson could depart from flight

30 Flt Lt E F Jefferies, 205742. Nos 21 & 2 Sqns, 1 OTU, HQ NWA, 9 & 10 Op Gps. Paint maker of Melbourne; b Melbourne 15 Aug 1912.

31 Statement "The Crash of A16-97 at Canberra 13th August 1940", by Frank Jefferies, Jun 1993. No 2 Sqn Association, Melbourne.

32 AA CRS A705, 32/10/2729 "Court of Inquiry of Hudson Aircraft A16-97 at Canberra on 13th August 1940", Finding of RAAF Inquiry, of 18 Aug 1940, p.42.

severely into a sudden stall if airspeed was too slow. The Judicial Court of Inquiry found:

> This accident ... emphasise[s] the necessity of impressing on pilots of machines of the type that crashed at Canberra the stalling characteristics of this type of aircraft and the necessity of approaching landing grounds with an ample safety factor of speed to avoid the danger of stalling at a height at which the pilot can have only the remotest chance of regaining control of the machine. This is a warning which seems particularly advisable where the landing is to be made at an aerodrome surrounded by hills the presence of which may set up air currents which may affect the equilibrium of the machine and raise the stalling speed.[33]

Although it has subsequently been suggested that Fairbairn himself may have been flying on that fatal approach,[34] this has never been supported by any evidence.[35]

With the loss of Hitchcock, the make-up of the Squadron now became Flight Lieutenant Rob Burns Cuming[36] as "A" Flight Commander, Flight Lieutenant John Ryland with "B" Flight, and Squadron Leader Allen Love[37] with "C" Flight. Among the pilots with Bob Dalkin, Bill White and David Campbell, were Neville Hemsworth[38] and Kym Bonython who related that the main operational function with Hudsons:

> ... was to give air cover to the troopships sailing westwards with 2nd AIF men for the Middle East. No 6 Squadron, based at Richmond near Sydney, gave coverage from north of Sydney down to Mallacoota on the south-eastern tip of Australia. We took over from there and patrolled almost as far west as Adelaide, watching for lurking U-boats or minelayers.[39]

Flight Lieutenant Bob Hitchcock F JEFFERIES

The wreckage of A16-97 after the tragic crash at Canberra on 13 August 1940 F JEFFERIES

---

33 AA CRS A705, 32/10/2729, Findings of Judicial Court of Inquiry, of 5 Sep 1940, p.12.

34 F Cranston, "Fairbairn Hoped to Pilot New Bomber", *The Canberra Times*, Sunday 12 Aug 1990.

35 Judicial Inquiry, p.103.

36 Flt Lt R W B Cuming, 200540. No 2 Sqn. Chemical engineer of Adelaide; b Adelaide 5 Mar 1911. Killed in action 20 Jan 1942.

37 Gp Capt A R Love, 250185. Nos 1 & 2 Sqns, RAAF HQ. Architect of Melbourne; b Melbourne 27 Jun 1909.

38 Wg Cdr N G Hemsworth DFC, 260417. No 2 Sqn, 1 OTU, 34 Sqn; comd 37 Sqn 1943–44, 12 Sqn 1945. Commercial pilot of Sydney; b Sydney 31 Aug 1917.

39 Bonython, p.41.

No 2 Squadron's flight line at Laverton in 1940 F JEFFERIES

A16-80 in 2 Squadron's hangar at Laverton in 1941 as gun turrets were being fitted to the Hudson fleet F JEFFERIES

The evidence of enemy merchant raiders was brought home, at the end of 1940, when mines were discovered in Australian waters. The loss of the *Cambridge* off Wilson's Promontory and the *City of Rayville* the next day off Cape Otway, both due to mines, led to the closing of Bass Strait for a week to clear the sailing routes.[40] Other vessels were soon lost or damaged to mines off New South Wales and South Australia, but the Germans ensured that the activities of their surface-attack raiders were beyond the limited range of shore-based aircraft. Just in case No 2 Squadron found any raiders, the Hudsons began practice bombing.

One of the air gunners on the Squadron at this stage was Sergeant George Wiburd,[41] a bank clerk and member of the CAF, who had been called up at the outbreak of war.

40 R C Sturtivant, *The Anson File*, Air-Britain, Tonbridge, Kent, 1988, p.161.

41 Flt Lt G Wiburd AFM, 205759. Nos 21 & 2 Sqns, 11 SU, 31 Sqn; t/comd 1 OBU 1944–45. Bank clerk of Melbourne; b Corowa, NSW, 18 Apr 1917.

His keenness in flying as a wireless operator in Ansons and Hudsons resulted in such a high number of flying hours that he was awarded the Air Force Medal.[42]

On 15 April 1941, Squadron Leader Frank Headlam assumed command of No 2 Squadron. By this stage, Japan's expansionism was recognised as the threat in the region. The delivery of Hudsons to the RAAF was increased through the "lend-lease" arrangement between the United States and the British Commonwealth. This agreement acknowledged United States neutrality by supplying Hudsons that were unarmed, the RAAF temporarily fitting a single gun mounted in a low fairing. British Boulton Paul powered gun turrets, fitted with twin Browning machine-guns, were subsequently retrofitted to RAAF Hudsons. With the United States suppliers bending the rules on the supply of armaments in this manner, the Commonwealth air forces received badly needed equipment, and secured the place of the United States as the "arsenal of democracy".[43]

The 1921 Washington Treaty, which was signed by the British Empire, France, Japan and the United States, recognised the island possessions in the Pacific, declaring "it is firmly resolved to respect the rights of the Netherlands in relation to her insular possessions in the region of the Pacific Ocean".[44] But Japan's quest for oil was to override its earlier commitment, and the following assessment was made:

> Japan is, of course, regarded as the great potential danger to the NEI, and the despatch of an expedition to South China in the autumn of 1938 and the subsequent capture of Canton convinced many that Japan's southward expansion was only deferred until a suitable opportunity should arise.[45]

Australia had agreed, in early 1941, to aid the Dutch in defence of their islands surrounding the Banda Sea if Japan went to war. As the situation to the north deteriorated during the year, the RAAF had the foresight to commence flights for aircrews to familiarise themselves with their forward operating bases in the Netherlands East Indies. This also allowed operations personnel to observe the conditions for controlling missions at Laha and Halong (on Amboina, or Ambon, Island) and Namlea (on Boeroe Island), in the Molucca Group, and at Penfoei, near Koepang (on Dutch Timor). The composition for this first detachment from 2 Squadron was:

Aircraft	**Pilots**	**Crews**	**Ground Staff**
A16-6	*W/C Headlam*	*F/L McGilvray*	*Cpl Hendrickson*
		S/L Ryland	*FSgt Myers*
		F/L Anderson	*FSgt Wiburd*
		F/L Hepburn	*Sgt Oldfield*
A16-12	*F/O Law-Smith*	*Sgt Warburton*	*Sgt Kempt*
	F/O Bonython	*LAC Roberts*	*Sgt Fox*
			LAC Houston

42 Wiburd flew 574 hours as a wireless operator, and his devotion to duty in manning wireless watches at the advanced bases led to his receiving the AFM from AVM Wrigley on 4 Jun 1942.

43 Brig Gen M MacCloskey, *The United States Air Force*, Praeger, New York, 1967, p.31.

44 PRO AIR 10/1658, "Handbook of the Air Service of the NEI", SD.136, Air Ministry, London, 1939, *Treaties and Commitments*, para.6.

45 PRO AIR 10/1658, para.8.

A16-18	*F/L Hemsworth* *P/O Gorrie*	*Sgt Franklin* *LAC Williams*	*Cpl Humphris* *Cpl Prance* *Cpl Thornton*
A16-80	*F/L Cuming* *P/O Taylor*	*Sgt Blake* *Sgt Sutherland*	*P/O Hood* *LAC Gibbs*[46]

Leaving Laverton on 24 September, the Hudson crews surveyed airfields en route and made thorough studies of their deployment bases, as well as carrying out patrols in what would become their operating area. On return from his inspection of the forward bases on 6 October, Wing Commander Headlam noted the following:

> LAHA
> 1. W/T installations will be completed by 31/10/41.
> 2. Accommodation provided for officers and airmen was good.
> 3. Accommodation for 100 men will be available by 31/10/41.
> 4. The operations room building is completed.
> 5. One receiver was used satisfactorily on the second operation.
> 6. The only channel of communication between LAHA and AMBON is one telephone approximately one mile from the operations room and other buildings.
>
> It is considered that efficient operational work could be carried out from this base when the installation of W/T and telephone communications is completed, subject to the limitations of the aerodrome.
>
> HALONG
> 1. The ACH[47] building is nearing completion.
> 2. Accommodation for officers and airmen is excellent.
>
> NAMLEA
> 1. W/T installation will be completed by 31/10/41.
> 2. Accommodation for 100 men will be available by 31/10/41.
> 3. Operations room building is completed.
>
> From the point of view of weather, aerodrome and approaches, this base is much better than LAHA.
>
> KOEPANG
> W/T installations and buildings are not nearly completed. Work at this base is much behind that at NAMLEA and LAHA. No date of completion was given.[48]

Never before had an Australian flying unit been given the opportunity to study its area of operation in the detail that this familiarisation visit permitted. Also the facilities that the Dutch were providing at the bases, with the exception of Pentoei aerodrome at Koepang, would allow for efficient forward operations. On this trip Kym Bonython made another movie "epic"—*Tropical Paradise*—which he filmed in Koepang and Ambon, with some clips of Melbourne's Fitzroy Gardens among the equatorial scenes.[49] When filming on the island of Amboina he remembered:

---

46 AA CRS A1196/2 60/501/77, "Flight of No 2 Squadron Hudsons to NEI 25/9/1941—6/10/1941", of 29 Oct 1941.

47 Area Combined Headquarters.

48 AA CRS A1196/2 60/501/77.

49 The earlier movie *Guardians of the Trade Routes* was copied for retention at the AWM Canberra, but unfortunately the sole copy of *Tropical Paradise* was lost in the 1983 Ash Wednesday bushfires in Adelaide. Kym Bonython, letter to author, of 13 Oct 1993.

> I shot one sequence on a beautiful promontory. I could not know that within a few months the Japanese would massacre a number of Australian servicemen on that very spot, including some of my squadron mates.[50]

This reconnaissance operation was repeated during November by another four crews, again reconnoitring Koepang, Namlea and Laha.[51] The severity of the situation in the Pacific was now reaching a crisis, and on 5 December two major events occurred. Flight Lieutenant John Ryland, the "B" Flight Commander, departed on posting to command the Advanced Operational Base (AOB) at Namlea.[52] The same morning, the four Hudsons of "A" Flight, under Flight Lieutenant Rob Cuming, moved to Darwin. Squadron pilot, Flying Officer Bob Law-Smith, related:

> When we arrived in Darwin from Laverton on the 6th of December, the aircrew briefing informed us we were to move to action stations at our designated base at Koepang. When we asked why, the answer was that Japan was about to attack Pearl Harbor and war with Japan was imminent. It is now clear in retrospect, and especially in view of declassification of much formerly secret material, that from whatever sources our briefing statements were derived, the Australian Government was in no doubt that we should be in a state of war readiness after arriving at Koepang.[53]

Cuming had led "A" Flight to its new base, Penfoei Aerodrome, near Koepang, on Dutch Timor, on 7 December (6 December Hawaii time), and in the early hours of that morning, the remaining seven aircraft of No 2 Squadron, under the CO, set out for the north.[54] When en route through Oodnadatta, the crews learnt of the Pearl Harbor attack,[55] but Rob Cuming's advance party were advised in more tense conditions. Bob Law-Smith continued:

> Before daybreak on the morning of Monday December 8th, Australian EST, our Hudsons were despatched on diverging sea searches north of Timor. At our pre-flight briefing we were told to expect a signal while in the air, that the anticipated Pearl Harbor attack had taken place, and if so, any Japanese shipping found would be a legitimate target.[56] Sure enough, while on patrol, the W/T Op Air Gunner brought me a message which, when decoded, broke the news that we were at war with Japan.

---

50 Bonython, p.48.

51 This second operation was over the period 6–15 Nov. No 2 Sqn A50, of Nov 1941.

52 He travelled by DC-2 (A30-4) to Townsville, then on to Darwin the next day. From here, he flew by Empire flying-boat via Ambon to Namlea. Ryland's Flying Logbook.

53 S R Grantham, *The 13 Squadron Story*, self published, Sydney, 1991, p.34.

54 No 2 Sqn A50, of 7 Dec 1941.

55 Kym Bonython, letter to author, of 13 Oct 1993.

56 Sir Robert Law-Smith later repeated this interesting insight: "While we were out on patrol we would be sent a signal that the Japanese were going to bomb Pearl Harbor—when the signal came through, if we found a Japanese ship we may bomb it. This is an interesting bit of history as this was several days before the Japanese did bomb Pearl Harbor—it is all in my log book. Now, I was the lowest form of life in the Air Force—any lower and I'd be out the bottom—so it wasn't a matter of very senior people being privileged to this information." Extract from his speech made in Victor Harbor, SA, on 26 Oct 1991, at the book launch of *Trapped on Timor*.

At about 7am Hawaiian time on Sunday December 7th, Japan had attacked Pearl Harbor. That would correspond to about 2.30am on Monday morning in Timor. Given that it would take an hour or so for the news to be flashed to a sleeping Australia, we would have received our confirmation signal probably about dawn.

Shortly afterwards, I found a Japanese vessel of about 300 tons north of the eastern end of Timor. I attacked it, starting a fire, but scoring mostly near misses with my first bombs ever dropped in anger. The Jap seamen quickly took to their lifeboats and rowed away. Apparently realising after several minutes that I had expended my bombs, they started to row back towards their vessel. A few bursts from my forward guns persuaded them that the distant coastline was more inviting.

2 Squadron had the honour of achieving the first strike in our particular theatre of war.[57]

57 Grantham, pp.34–5.

CHAPTER 12

Baptism of Fire

Bill White won his DFC for his gallant attempts to fight against impossible odds. His aircraft, by that time known as "The Flying Colander", was badly damaged but he insisted on "one more lash at the Japs". It was the last, for after that his aircraft was too shot-up to fly any more and a few days later he was captured and beheaded by the Japanese.[1]

The islands of Timor and Ambon, on the edges of the Banda Sea in the Netherlands East Indies (NEI), had been selected as suitable locations for future forward airbases as early as 1936 or 1937 by a secret RAAF mission. Led by Squadron Leader Charles Eaton,[2] the small group had travelled incognito through the islands in civilian clothing.[3] This intelligence gathering activity led to No 13 Squadron's reconnaissance visit in May 1941. Five Hudson crews had inspected airfields and facilities and, by wearing clothes without badges, were to appear as civilian ferry crews![4] In August, parties of five RAAF men each had been detached to Koepang, Ambon and Namlea to provide a communications link back to Australia. Again, all members wore civilian clothes to avoid military appearance, and were joined by an intelligence officer and additional airmen the following month.[5]

These pre-war reconnaissance flights had provided great value for the crews. The Penfoei AOB, an airfield 13 kilometres east of Koepang, consisted of five runways varying in length from 900 metres to 1,300 metres, and up to 200 metres wide. The surfaces, being crushed coral, were serviceable in all seasons. Approaches were generally good, with those from the south and south-east over coconut trees 15 metres high.[6] However, as a target:

> ... it was unmistakable and unmissable, especially since the Japanese learned that our "defences" consisted only of a few army machine-guns. They were sited in pill-boxes with such narrow slits that the guns had only a ten degree elevation.[7]

Aircraft servicing facilities were very limited, and the Hudsons had to be flown back to Darwin for maintenance. Half of the groundcrew had deployed to the islands by Hudson, the other half had followed from Laverton to Darwin by road and rail

---

1 Bonython, p.53.

2 Gp Capt C Eaton OBE AFC, 24. RAF 1918–19; comd RAAF Darwin 1940–41, 2SFTS 1941–42, 72 Wg 1943, 79 Wg 1943–44; AOC SA 1945. Consul, Dili 1946–47; Acting Consul-Gen Indonesia 1947–49. Regular air force offr of Melbourne; b London 12 Dec 1895.

3 Coulthard-Clark, p.450.

4 Grantham, p.27.

5 Gillison, p.191.

6 AA CRS A1196/2 60/501/77, of 29 Oct 1941.

7 Bonython, p.50.

about a week later. They were able to provide the rear servicing at Darwin for both Hudson squadrons, as 13 Squadron had also advanced to its forward NEI base at Laha, on Ambon. Operations staff were established at the ACH, on the opposite side of Ambon Bay at Halong, to control the forward bases and task the Hudson units. All Hudson operations were under the command of Headquarters North-Western Area, which controlled the land masses of the NEI, western (Dutch) New Guinea, and the Northern Territory, and the seas in between.

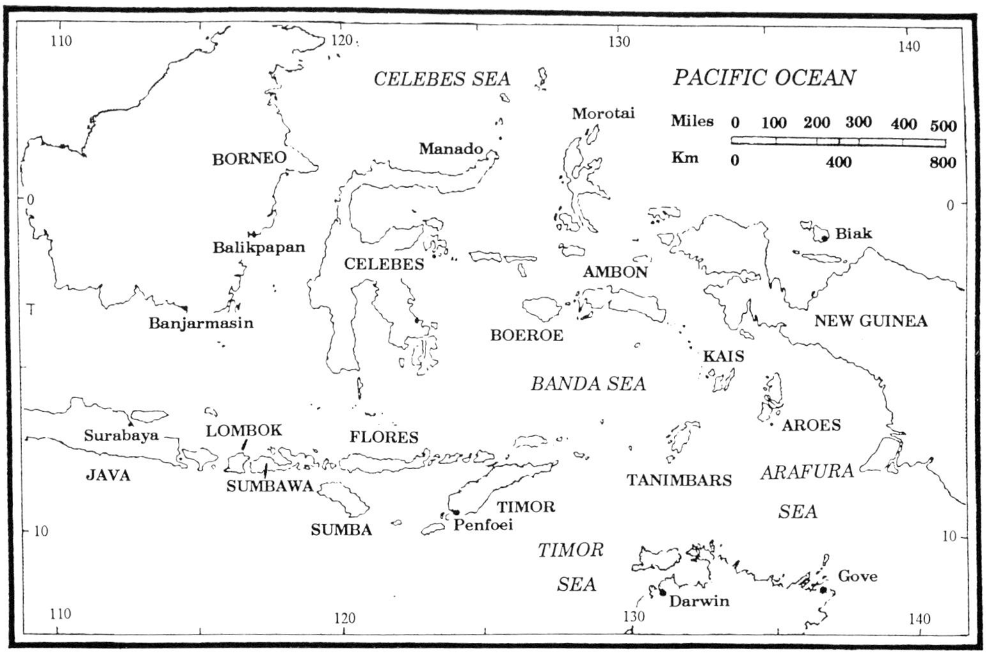

No 2 Squadron area of operations 1941–1945 RAAF

On 8 December, the coastal/aeradio operator at Dili had intercepted a message instructing a vessel to anchor in Dili Harbour as a guide for Japanese aircraft. This, the Japanese radio ship *Nanyo Maru*, was the vessel located and attacked by Law-Smith's formation, which forced it to run aground off Timor.[8] The crew was captured.

LAC Richard Taylor,[9] an engine fitter on the Squadron who had arrived with the first flight at Penfoei, described the first preparations for combat:

> Long hours of reconnaissance flying were now taking place while the Japanese were pressing towards Singapore. The long flights meant that the Hudsons needed more and more maintenance and repair work. At the time the air crews began to look more strained and tired each day. The pace was accelerating.
>
> Shipments of high octane fuel (44-gal drums) were arriving together with bombs and stores—these items were transported to our base by the army where every mustering helped

---

8 Grantham, p.34.

9 Cpl A R Taylor, 11035. No 2 Sqn, RAAF Cd. Engineering and purchasing officer of Melbourne; b Melbourne 23 Feb 1918.

to unload and store them in the various dumps. Natives were completing the extension of the runway with crushed coral (all by the means of hand-held baskets), while at the same time aircraft anti-bomb blast earth revetments were being constructed alongside the runway. All in all, a very busy time with little rest—day or night.

The operations section and wireless hut were set up some distance from the runway in a dried up creek bed for protection against future air raids.[10]

Bad news was to follow for the Hudson squadrons on 10 December with the loss of the CO of 13 Squadron. Wing Commander Josh McDonald[11] crashed into the sea soon after take-off, while leading his formation on a long-range raid on the Japanese flying boat base at Tobi, in the Caroline Islands. The Australians were cheered, however, by the news that reinforcements were on the way. Not air reinforcements, as nothing more was available, but the 2/21 Battalion, the main combatant unit of "Gull Force", was on its way to add to the defence of Ambon.[12] Arriving on 17 December, its fighting capability was, from the beginning, limited; there was no sea protection or fighter cover. This was never to be rectified, despite the repeated requests of Gull Force's commander, as no fighters were available. It was, of course, the same problem that was experienced elsewhere across the fragile forward defensive arc over Australia's north.

On 28 December, the Allied Combined Headquarter at Halong received a sighting from an American Catalina of a destroyer about 500 kilometres north of Namlea. As no friendly shipping was reported in the area, ACH despatched three 2 Squadron Hudsons to attack. Unbeknown to the Hudson crews and the ACH operations staff who had ordered the mission, the destroyer was the *USS Peary*. After an inconclusive exchange of identification codes, the *Peary's* gunners opened up in self-defence; naturally, the Hudsons pressed home their attack. Their accuracy disabled the destroyer's steering gear, and she limped into port in the Halmaheras for repairs. Having been damaged in attacks in the Philippines, at Cavite and off Corregidor, the Americans had survived more attacks as they had headed towards Java. This latest blow prompted the American naval historian to record "there was slight satisfaction in knowing that our allies were better marksmen than our enemies!"[13]

Two days later, as the Japanese struck Babo in Dutch New Guinea, three 13 Squadron Hudsons under John Ryland (A16-98) were despatched there from Laha, in desperation, to act as "fighters".[14] This was, in retrospect, carrying air power's attribute of flexibility too far, but these were desperate days—Australia was facing

---

10 Richard Taylor, Record of 2 Sqn service, Jun 1993, No 2 Squadron Association, Melbourne, pp.7–8.

11 Wg Cdr J R G McDonald, 86. Nos 1 & 2 Sqns, 1FTS, RAF Coll, 3SFTS RAF, RAAF HQ; comd No 13 Sqn 1941. Regular air force offr of Melbourne; b Melbourne 24 May 1915. Killed in action 10 Dec 1941. Also lost in crash of A16-69 were 13 Sqn crew members Flg Off R H Pope, Sgt W R Foreman and Sgt W J Nagel.

12 "Gull Force" was assigned to Ambon, "Sparrow Force" to Timor, and "Lark Force" to Rabaul.

13 S E Morison, *History of United States Naval Operations in World War II*, Vol III, Little & Brown, Boston, 1948, pp.196–7.

14 Gillison, p.274, and John Ryland's Flying Logbook.

invasion and the RAAF possessed no fighters.[15] To fill the gap of their departure, a flight of four Hudsons from No 2 Squadron, under Flight Lieutenant Rob Cuming, was deployed from Penfoei to Namlea.

The New Year saw the Japanese relentlessly thrust southwards. The first air raid on Laha came without warning in the early hours of 6 January, when seven Navy Type 97 flying boats bombed and strafed the aerodrome.[16] However, five days later Hudsons from Laha and Namlea struck back, attacking the Japanese invasion force off the north-east Celebes near Menado and Kema. The Hudsons scored a direct hit on one cruiser and hits on smaller ships. Another five Hudsons from Laha scored hits on a destroyer, and reported that another destroyer appeared damaged from the previous attack. A further four Hudsons from Namlea bombed a large transport and were engaged by Japanese float-plane fighters. The Hudson gunners shot two down in flames, a third was sent down in a spin (counted as a "probable"), and a fourth was damaged. Credit for this action went to 2 Squadron crews piloted by Flight Lieutenant Parker Hodge[17] and Flying Officer Peter Gorrie.[18] These missions were over six hours flight time, and many crews flew twice that day. The only casualty was one of the air gunners, who was slightly wounded.

Success was short-lived as the Hudsons were destined not to maintain their brief moment of air superiority. The next morning, 12 January, five Hudsons from Namlea departed to strike at the enemy invasion forces at Menado. They were intercepted by three floatplanes and five Zeros, the first time this Japanese fighter had been seen in the area. Both the Hudsons of Hodge (A16-46)[19] and Gorrie (A16-12)[20] were lost, in addition to two 13 Squadron aircraft.[21] The only survivor from these four aircraft was Parker Hodge's navigator, Pilot Officer Ted Howard,[22] who parachuted into the sea, and took 16 hours to swim ashore.[23] He was captured and survived his captivity at a POW camp in Batavia (Jakarta) on Java, being repatriated to Australia on 30 September 1945. The fifth Hudson, flown by Flight Lieutenant Rob Cuming, was the only aircraft to return.

15 Indeed, the defence of Rabaul was to lay in the hands of Wirraway "fighters"—on 20 Jan 1942, six Wirraways of No 24 Squadron were lost or damaged engaging a superior attacking force.

16 No 13 Sqn Unit History Record of Operations, Form A51, of 6 Jan 1942. RAAF Historical Section, Canberra.

17 Flt Lt P H R Hodge, 451. No 2 Sqn. Regular air force offr of Melbourne; b Melbourne 15 Jun 1913. Killed in action 12 Jan 1942.

18 Flg Off P C Gorrie, 407168. No 2 Sqn. Clerk of Adelaide; b Peterborough, SA, 6 Jun 1918. Killed in action 12 Jan 1942.

19 Crew: Sgts W L Morgan (408112), J E Farrington (17633) and K R T McCrackan (406219).

20 Crew: Sgts H C Shore (3898) and J Mawdsley (407367).

21 The two 13 Sqn crews lost were led by Flt Lts G Sattler (A16-7) and A R Barton (A16-67).

22 Flt Lt E D G Howard, 407715. Nos 14 & 2 Sqns. Commercial traveller of Adelaide; b Adelaide 20 Oct 1916. POW 12 Jan 1942.

23 RAAF Historical C.D.102/45 A16-46, of 1 May 1946.

Flight Lieutenant Ron Cornfoot[24] was also airborne during the time of this attack, with three 13 Squadron Hudsons from Laha. He recalled:

> The Hudsons would have been sitting ducks to the fighter tactics that the Japanese pilots were using against them. It was clear to us that they knew the full characteristics of the Hudson and the limitations to its range of fire.[25]

The capabilities of the RAAF Hudsons, fighting in Malaya since early the previous month, had evidently been disseminated through the enemy fighter units. The standard Hudson armament had been two fixed Browning .303-inch guns in the nose and two in the powered dorsal turret. A field modification, of two side guns protruding through openings cut in the perspex windows, proved useful, but the field of fire was still restricted. The Hudson was no match for the manoeuvrable Zero.

The side gun modification to the Hudson, added to provide as much firepower as possible to counter the Zero RAAF

By 13 January, the enemy held the northern Celebes, bringing Ambon under constant air attack, and two days later 28 Japanese bombers and 15 fighters attacked the base.[26] The attack, lasting two hours, set one Hudson on fire, destroyed two United States Catalinas on the water nearby, and rendered the runway temporarily unserviceable. In addition two Dutch Buffalo fighters attempting to intercept were shot down.[27] To reinforce Namlea, one Hudson (A16-79) had returned to Penfoei on 20 January. Flown by Rob Cuming, the aircraft was to collect extra fitters, and load supplies of food, spare parts and ammunition for the AOB.

The crash of A16-79 on 20 January 1942 cost the lives of 11 members of No 2 Squadron. F JEFFERIES

> We were shocked with the changed appearance of F/L Cuming and his crew. They were looking tired and worn out and had all lost weight. Their Hudson was refuelled and loaded with the extra fitters, and boxes, sacks, etc.
>
> F/L Cuming taxied to a position for maximum runway length. The aircraft was then held by the brakes until the engines had been taken up to maximum revs (full power). With the brakes released the aircraft began to accelerate rapidly but, as it passed our position, it was

24 Sqn Ldr R G Cornfoot AFC, 250630. Nos 13 & 14 Sqns, 1 OTU, 36 Sqn; comd 38 Sqn 1944–46. Clerk of Melbourne; b Melbourne 22 Nov 1918.

25 J D Balfe, *War Without Glory*, MacMillan, Melbourne, 1984, p.183.

26 John Ryland's Flying Logbook.

27 No 13 Sqn A51, of 15 Jan 1942.

> obvious that the tail would not lift. I vividly remember a terrible feeling of apprehension. Suddenly the tail began to rise only to fall back with a jolt which must have sent the unsecured load sliding towards the rear. The next moment the aircraft with screaming engines climbed at an almost vertical angle, turned on its back at a height of approximately 100 feet, and then crashed nose first into the end of the runway in a ball of flame with ammunition exploding in all directions.
>
> A shocking tragedy.[28]

Up until this stage, the enemy had seemed unaware of the RAAF's use of Namlea as a forward base. That was to change on 24 January, as Bob Law-Smith encountered an enemy Kawanishi Navy Type 97 flying boat (later nicknamed the "Mavis") just as his Hudson was airborne:

> It was early morning, and I had just taken off—undercarriage up, easing back on takeoff power and about to arm the guns, when I exclaimed to my co-pilot, "Look ahead." There about 200 yards ahead was a Japanese flying boat. I banked towards him as quickly as I could and fired a couple of bursts from the forward guns. As the "Mavis" broke away, my rear gunner Llew Allen[29] fired from his turret.
>
> Moments later he ceased shooting, and shouted something quite unintelligible through the intercom. I sent the W/T Op Air Gunner back to see what the matter was. He found that Allen was wounded in the head and bleeding profusely. I had no option but to land and replace my gunner if I were to continue with my recce of the Celebes. Fortunately, it was only a grazing wound.[30]

By this stage the situation facing the Australians was indeed grim. The enemy had reached the southern Celebes to the west, and a Japanese invasion force from Guam had taken Rabaul to the east. The enemy was approaching Ambon and attacks could be expected on Timor. Because of this, a flat open area at Mina River, about 70 kilometres east of Koepang, had been designated as "available for emergencies",[31] and a satellite airstrip was being constructed about 12 kilometres north-west of Penfoei.[32] In addition, living and working conditions at Penfoei were extremely difficult with little fresh food, and the men suffered so much from dengue, malaria, dysentery and tropical sores that up to 30 per cent were on sick parade daily.

The two Hudson squadrons, by 25 January, had lost 13 aircraft, and the enemy, through persistent reconnaissance, was aware of the disposition of the Australian aircraft. The following day, Australia Day Monday 26 January, enemy Zeros attacked the airfields at Laha and Namlea, and with no warning and inadequate air defence, the Hudsons were caught on the ground and a further four were destroyed.[33] The first

---

28 Richard Taylor, pp.16–17. Rob Cuming's heavily laden Hudson had claimed 11 Squadron members. Also killed were: Plt Offs R H Martin (401464), P H Richards (407592), Sgts J Bessell-Browne (406205), A E Blake (4560), H J Dean (3934), K H Kelaher (403287), LACs O F E Crawford (19485), E G Mears (19266), V H Smith (30308), AC F Morris (16698).

29 Flt Lt W L Allen DFM, 407633. No 2 Sqn, 1 OTU, 13, 1, 2 & 32 Sqns, 1 OTU. Clerk of Adelaide; b Kadina, SA, 14 Feb 1917.

30 Grantham, p.44.

31 Kym Bonython, letter to author, of 13 Oct 1993.

32 Gillison, p.417.

33 Gillison, pp.374–5.

attack on Penfoei was also that morning at 9 a.m., and from then enemy strafing raids became a daily event. On that first attack seven enemy fighters swept in on a low-level attack, shooting down a Dutch Dragon Rapide and then intercepted an RAAF DC-2 (A30-8) over Soemba Strait on its return from flying a radio party to Soerabaya. The aircraft was flown by Flying Officer Noel Webster[34] and Sergeant Lionel Van Praag.[35] The crew spent 30 hours in the water, beating off shark attacks, until reaching Soemba Island, 50 kilometres away. After being cared for by natives, they were able to reach safety and were evacuated by Hudson to Darwin on 12 February.

From left, Flight Lieutenants Neville Hemsworth and Bryan Rofe, with Pilot Officer Peter Thompson, at Penfoei in January 1942 K BONYTHON

At the end of January, Flying Officer Kym Bonython was sent with his crew and a ground servicing party to the Mina River emergency landing ground (ELG), in the south-west corner of the island.[36] Here it was hoped that not only could overdue maintenance be performed away from the constant threat of air attack, but also fuel, which was to be delivered by the American destroyer *USS Peary*, could be stored for deployed operations. Bonython described this episode:

> We had flown over this area and it looked to be a nice big open space, but no one had ever landed there. As I approached I saw what looked like soft green grass, and made what I expected to be a perfect landing. Unfortunately the grass was six feet high and I was misled

34 Sqn Ldr N W Webster GM, 270847. Nos 30 & 36 Sqns; comd 35 Sqn 1945. Store manager of Brisbane; b Brisbane 17 Aug 1919.

35 Flt Lt L M Van Praag GM, 60431. Nos 36, 35 & 38 Sqns; comd Para Trg Unit 1944. Professional motor cyclist of Sydney; b Sydney 17 Dec 1908.

36 Gillison, pp.417–8.

into trying to land on top of it. It felt as though the bottom had dropped out of the Hudson when we actually hit the ground.[37]

Flying Officer Bonython turned the aircraft (A16-9), so that the crew could push it tail first into the edge of the jungle and cover it with a large camouflage net. One of the group, LAC Richard Taylor, stated how their attempts were soon thwarted:

> Within 30 minutes of our camouflage effort someone in the group (always on alert) spotted a bright glint very high in the sky—although no sound was audible we eventually made out the silver reflection from a Jap recco plane circling and probably taking photographs. Then we realised that the Hudson had left deep wheel marks in the grass during the landing and taxying.
>
> In the morning ... the time now was approximately 8.45am, we were to walk to the Hudson to begin the maintenance work, when without any warning 5 low-flying Jap Zero planes seemed to come from nowhere with guns blazing at the Hudson, which immediately burst into flames. The five Zeros now slowly circled the ELG in line astern each giving a burst of fire at the burning Hudson.[38]

Flying Officer Kym Bonython on the trek back to base from Mina River, early February 1942.
K BONYTHON

It was also on this day, 30 January, that the British Imperial Airways Empire flying boat "Corio" (G-AEUH) was shot down by enemy fighters at the mouth of the Mina River, while flying from Darwin to Singapore. The attackers were most probably this same flight of Zeros. After desperate evasive action just above the water, with two engines on fire, the huge flying boat broke its back striking the sea and sank, killing 13 people on board.[39] The pilot, Captain Koch, and five others were thrown clear and after several hours all managed to swim the 10 kilometres to shore.[40] Also during an attack on Penfoei a Hudson was destroyed and its pilot wounded as he attempted to taxy to a revetment.

Having cut a track through the swamps and dense jungle from the strip to the beach for the transport of the petrol drums, the party, now stranded at the Mina ELG, were joined after an abortive attempt to land the fuel from the destroyer by a group of United States

37 Bonython, p.51.

38 Richard Taylor, pp.11–12.

39 P W Moss, "BOAC At War", *Aeroplane Monthly*, London, Aug 1975, p.412.

40 C H Barnes, *Shorts Aircraft Since 1900*, Putnam, London, 1989, p.331.

Marines. The RAAF men were forced to undertake a three-day trek back to base, where the Squadron Equipment Officer, Flying Officer Arthur Cole,[41] collected the squadron tractor and drove back through the jungle tracks to ensure that the petrol storage was completed.[42]

Meanwhile, to the north, on the afternoon of 26 January, a Hudson reconnaissance sortie located an enemy convoy of 22 ships—transports with a naval escort—moving on a south-easterly course, to the east of Menado.[43] Since early in January, the RAAF forward bases had been commanded by the newly created ABDA Command,[44] formed in desperation as the Japanese advanced down the Malay peninsula. The ABDA area included Burma, Malaya, the NEI, the Philippines and northern Australia, with headquarters in Bandoeng, on Java. On 24 January, ABDA had instructed that if Laha and Namlea became untenable, reconnaissance would continue from Koepang and Darwin.[45] The only personnel to remain were a skeleton staff, to continue radio communication with Darwin for as long as possible, and a small servicing team to maintain, refuel and rearm Hudsons on reconnaissance patrols. Faint hopes had been held for reinforcements from Malaya or the United States that may have arrived in time to repulse the Japanese. There were no reinforcements and now, as there was no doubt that Ambon was to be invaded, NW Area Headquarters ordered an evacuation of Laha.

While some Hudsons continued to shadow the approaching force on its progress through the Molucca Sea, others prepared for the evacuation of personnel from Ambon. Aircraft were stripped of all non-essential equipment, and one 13 Squadron machine, flown by Ron Cornfoot, staggered off the runway crowded with 23 men, standing for the flight to Darwin. One of Ron Cornfoot's passengers was photographer LAC Aubrey McEgan,[46] a veteran at Laha since the original deployment seven long weeks previously:

> It took the whole length of the runway to get off. There was no flare path and only a pale moon to guide the pilot. We climbed to 12,000 feet for safety. All passengers stood for the whole of the journey, parked between the pilot and the second main spar of the aircraft—only one person being aft of the main spar (air gunner Sgt A Ducat[47] who was afterwards awarded the DFM). The journey took 4 ½ hours and we were glad to be relieved from our cramped positions.[48]

---

41 Sqn Ldr A F Cole MBE, 261733. No 2 Sqn, 4AD, HQ WA, 1AD. Company manager of Sydney; b Glastonbury, Somerset, Eng, 26 Aug 1906.

42 No 2 Sqn A50, of 1–20 Jan 1942.

43 Gillison, p.375.

44 The abbreviation of American, British, Dutch, Australian Command.

45 Gillison, p.373. (Scott and Ryland had flown A16-57 to Bandoeng for instructions on 23 Jan 1942, Ryland's Flying Logbook.)

46 WOff A H McEgan, 2404/A3893. Nos 2, 4, 6, 13, 22 & 24 Sqns, HQ NWA, ARDU, CPE. Photographer of Adelaide; b Adelaide 24 Jun 1914.

47 Flt Lt A W Ducat DFM, 406327. No 13 Sqn, 45th Sqn USAAF, RAAF Adv HQ. Salesman of Perth; b Melbourne 22 Apr 1913.

48 "Interview with WOff A H McEgan (2404), Evacuation of Laha and Initial Raid on Darwin", undated, courtesy of "Strawb" McEgan, No 2 Squadron Association, Melbourne.

The RAAF's operations from Laha and Namlea were later summarised by Cornfoot, and this probably expressed the views of all the Hudson crews:

> Whatever might be said, I shall never believe that the Ambon-Namlea affair was anything but a debacle, though certainly not attributable in any way to those who were its victims. Try as we did, I doubt we delayed the Japanese drive south for more than moments. Our losses were terrible in proportion to the numbers of Army and Air Force engaged. Seventy per cent of our precious Hudsons were lost with their crews and their irreplaceable captains ...
>
> Again the Allies at Ambon were without effective fighter protection. The Japanese had complete command of the Moluccas and Celebes air. Most of our losses occurred in daylight while we were trying to bomb the Japanese invasion fleets that assembled up around Menado in the north of the Celebes. Over 11, 12 and 13 January, the three days of those operations, we lost five crews of our eight, all shot down by Zeros. The survivors were Squadron Leader John Ryland, who replaced Josh McDonald as commander of 13 Squadron and after the war was chief pilot of Trans Australia Airlines, Flight Lieutenant Archie Dunne[49] and myself.
>
> Empty bunks were conspicuous in our barracks.[50]

As the enemy approached Ambon on the 30th, demolitions around Laha were carried out by the men of Gull Force.[51] The final two Hudsons, which had been retained by the ACH commander, Wing Commander Dallas Scott, for "last minute" reconnaissance reports of the enemy invasion force, were then made ready. As the aircraft were being refuelled that evening for the evacuation to Darwin, Flying Officer Bill White's Hudson (A16-121) was found to have a massive fuel leak. Scott then ordered 17 personnel out on the serviceable aircraft, flown by Flying Officer Jack Haythorn,[52] leaving 11 for White's stranded aircraft. Bill White's crew were his two wireless operator-air gunners (WAGs) Sergeants James Baker[53] and Isaac Read,[54] and a copilot from 13 Squadron, Flying Officer Francis Meyer.[55] Scott's last signal sent on 31 January read:

> All cyphers burned. Demolition completed. Known Japanese landing at Leahari and Hukurila. Japanese have reached Laha from overland and engagement proceeding. Will call whenever possible.[56]

---

49 Wg Cdr R A Dunne DFC, 260420. Nos 2 & 13 Sqns, 1 OTU, 528th Sqn USAAF; comd 23 Sqn 1944–45. Commercial pilot of Sydney; b Warracknabeal, Vic, 10 Oct 1913.

50 Balfe, p.169.

51 The men of Gull Force on Ambon fought until their ammunition was exhausted and they were overwhelmed, capitulating on 3 Feb 1942.

52 Flt Lt J Haythorn, 260722. Nos 13 & 6 Sqns, 1 OTU. Draftsman of Sydney; b Young, NSW, 23 Jan 1920. Died of injuries 4 Apr 1943.

53 Sgt J Baker, 402700. No 2 Sqn. Of Lithgow, NSW; b Lithgow 6 Sep 1919. Executed by Japanese 20 Feb 1942.

54 Sgt I W Read, 406222. No 2 Sqn. Of Perth; b Perth 3 Mar 1911. Executed by Japanese 20 Feb 1942.

55 Flg Off F N Meyer, 290745. Nos 21 & 13 Sqns. Ass't works manager of Perth; b Perth 2 Oct 1921. Executed by Japanese 20 Feb 1942.

56 Gillison, pp.376–7.

The RAAF party had been communicating by a portable transceiver, concealed under camouflage, operated by LAC Laurence Walker[57] for 24 hours after the last Hudson had left. Until batteries were exhausted, Walker had been able to pass the landing places of the invading forces and their approximate strength to Darwin. Bill White, who had earned No 2 Squadron's first DFC of the war, with his crew and the rest of Wing Commander Scott's party, set off across the hills to the north of Ambon with the intention of crossing by boat to Ceram Island. In doing so they were intercepted in Piru Bay by a Japanese patrol boat and taken prisoner. Most were executed on 20 February.[58] The Japanese killed 234 men in mass executions at Laha between 17 and 20 February; the Australians died by decapitation, shooting, or some by bayoneting.[59] The barbarism of this enemy was still unknown in Australia when Wing Commander Headlam later wrote:

John Ryland and Dallas Scott, who carried out a mission to Java in January 1942 for the final appraisal of the Japanese southern advance
NATIONAL LIBRARY OF AUSTRALIA

> During the whole of the operations in this area F/O W V D White had been an outstanding Captain. He had shown exceptional endurance, devotion to duty, ability and personal courage. He was later awarded the DFC.
>
> F/Lt R B Cuming as a Flight Commander was also outstanding. Both on the ground and in the air he showed great ability, devotion to duty, excellent judgment, dash and bravery and was an inspiration to every man under his command. F/Lt Cuming was recommended for the award of the DFC[60] but was later killed in an aircraft accident at Koepang.[61]

Over on Boeroe at Namlea, the RAAF force, under the command of Squadron Leader "Tich" McFarlane,[62] had been signalled by Scott on 30 January advising that three Hudsons were being dispatched for evacuation of base staff. From his group of 60–70 men, McFarlane had to retain a small demolition and maintenance party:

57 LAC L D Walker, 27521. SHQ Darwin, AOB Halong. Electrical mechanic of Adelaide; b Cooringa, SA, 26 Mar 1918. Executed by Japanese 20 Feb 1942.

58 Gillison, p.377.

59 Balfe, p.193. These men are remembered at the Ambon War Cemetery at Galala.

60 The DFC cannot be recommended posthumously.

61 No 2 Sqn A50, of Jan 1942, summarised by Wg Cdr Headlam later in 1942.

62 Gp Capt A B McFarlane DFC, 250207. Nos 21 & 6 Sqns; comd AOBs Ambon & Namlea 1941–42, Rear Party Namlea 1942, 2 Sqn 1942; OTLO O'seas HQ 1943–44; DD Op Trg 1944–45, DDPS 1947. Secretary Dept Air 1956–68. University student of Melbourne; b Melbourne 4 Jun 1916.

> I had the unenviable task then of sorting out who would stay behind ... we needed about nine people, a couple of fitters, a couple of sigs people, a couple of armourers ...[63]

With all his force volunteering to stay, McFarlane picked his group, comprising the required musterings. He later recorded:

> I can only say how proud I was of the behaviour of the men of Namlea base and how privileged I was to have been in command. Every man of the 60 or 70 airmen there responded to the call for volunteers. I shall never forget the instant raising of arms and their complaints at my decision to sort out evacuees on the basis of their responsibilities back home.[64]

That evening the three evacuation aircraft arrived, flown by Flying Officers "Mickey" Finlayson,[65] Bob Law-Smith and Norm Lamb.[66] On landing, Finlayson's aircraft was damaged, necessitating a change of tailplane. Fortunately, a wrecked Hudson from an earlier Japanese raid had an undamaged tailplane, which could be hurriedly changed under torchlight. Most of the group were retrieving the undamaged tailplane, when McFarlane walked up to the Hudson which needed repairs:

> One fellow was working on the tailplane that was damaged ... I didn't recognise him. I said "Who are you?" He said "I'm the cook, sir." "What the hell are you doing out here?" He said, "I'm taking the tailplane off!" I walked away thinking, oh my God! Anyway, they finally made the change, but without the trim tab connected, so Finlayson took the aircraft off without the trim tab.[67]

During the course of these repairs, McFarlane had planned an escape route, for when the time came for the rear party to evacuate. The only conceivable rescue would be by flying boat, and McFarlane considered the obvious place for this would be at Tifoe, on the south of the island. The direct route was some 100 kilometres, and being unsure of how long his group would continue to man Namlea, McFarlane told Law-Smith to advise Darwin that the party would be at Tifoe in 14 days. Law-Smith, with tears in his eyes, then begged McFarlane and his small group to return on the already overloaded Hudsons. McFarlane had to refuse; he had his orders to keep Namlea open for forward operations.[68]

The three heavily laden Hudsons staggered off in the half-light, just before dawn—15 in Finlayson's damaged aircraft, 22 in Lamb's, and 23 in Law-Smith's.[69]

> How they ever rose from the water-logged 800-yard-long runway, climbed over the hills, and flew to Darwin safely is one of the squadron's mysteries. But as the demolition party

---

63 Tich McFarlane, interview with author, of 1 Nov 1993.

64 Grantham, p.55.

65 Sqn Ldr J W Finlayson, 260523. No 14 Sqn, HQ NWA, 74 Wg, 21 & 23 Sqns. Cost clerk; b 4 Nov 1916.

66 AirCdre N F Lamb, 290732/O5788. Nos 21, 14 & 2 Sqns, 8 Ftr Sector; CI Armt Instrl Sqn Nhill 1944–45; RAF Empire Air Armt Sch, MOS; comd Armt Sch 1953–55; HQ TC, RAAFSC; comd Wagga 1964–66; HQSC. Regular air force offr of Perth; b Perth 14 Feb 1916.

67 Tich McFarlane, interview with author, of 1 Nov 1993.

68 Tich McFarlane, interview with author, of 1 Nov 1993.

69 Gillison, p.378.

watched with bated breath, they swayed drunkenly down the strip, lifted sluggishly, and disappeared into the murk. Eight minutes after the take-off, the second Hudson had laboured up to four hundred feet. Five hours later it landed at Darwin, to find the others already arrived.[70]

From dawn until midday on 30 January, Japanese fighters, bombers and dive bombers attacked Namlea, and "did the place over".[71] The defenders counted 78 enemy aircraft in continuous waves attacking the base.[72] The runways, which had previously been mined, were destroyed when detonated by the enemy raids. As Allied aircraft would no longer be able to use the AOB, there was now little point in remaining. McFarlane sought the advice of the friendly Mayor of Namlea, on the route to Tifoe, only to learn that the direct route was impossible, being marshy and impenetrable at that time of year. The best option was to the north, around the coast to the top of the island, then south to Tifoe across the 8,000-foot mountain range. This would be a trek of nearly 250 kilometres.

McFarlane's party then set about to demolish anything that would be of value to the enemy. The fuel dump, weapons storage and operations room were all destroyed. The party then departed by road to the north, in a truck and a car:

> ... it was a practically brand-new Studebaker, of about 1939–40 vintage ... which was owned by the Mayor. And as the Mayor had decided it wasn't going to be of any use to him, and he didn't want to have the Japanese use the car, it was better that we took it. Now we had nowhere to take it but to the end of the road, which was about 20 miles north of the base, along the coast, and probably it's still there, in the jungle, with the truck too, both of which were immobilised and pushed down a steep hill into the jungle.[73]

McFarlane's party was now forced to live off the land on the arduous trek across the mountains, rivers and swamps towards Tifoe, on the south coast.[74] Their daily routine became rising at 5 a.m., eating whatever could be found, walking some 15 to 25 kilometres per day, and sleeping each night in the open. On the eighth day, 6 February, as they arrived at the village of Wiametea, their native guide refused to go further. They continued south across the mountain ranges, despite the lack of visible tracks, and by the twelfth day, 10 February, they reached Tifoe. Posting lookouts around the bay, they waited, realising any rescue attempt would be risky, as there was constant traffic of enemy aircraft.[75] At 6.30 p.m., on the fifteenth day, they were picked

70 "Timor Treks", in *RAAF Log*, AWM, Canberra, 1954, p.89.

71 "Timor Treks", in *RAAF Log*, p.89.

72 *Day By Day Across Namlea, NEI, 1942*, submitted by Reg Taylor, courtesy of Tich McFarlane, p.15.

73 Tich McFarlane, interview with author, of 1 Nov 1993.

74 In addition to McFarlane, the party comprised Flt Lt John Hanbury, Cpl C P Lohrey, LAC Clarrie Chambers, LAC Mal Michie, AC Tony Dignam, AC Les Meadows, AC F G Russell, and AC Reg Taylor. "Casualties in RAAF", *The Sun*, Sydney, Saturday 16 Feb 1942; Reg Taylor, telephone interview, of 29 Nov 1993 and letter to author, of 2 Dec 1993; Tony Dignam, telephone interview with author of 13 Dec 1993.

75 *Day by Day*, p.17.

up by Flight Lieutenant John Hampshire[76] in an Empire flying boat and flown to Darwin. It was their *lucky* day—Friday 13 February—and the rescue aircraft was the Empire numbered A18-1356.

However, the next day, 14 February, bad news continued on Timor. Sergeant Keith Rayner,[77] a WAG of 2 Squadron flying as crew with No 13 Squadron under Flight Lieutenant Geoff Mitchell[78] (A16-61), was lost soon after take-off from Penfoei:

> A fully loaded Hudson roared off the runway and while still within our sight exploded in a ball of flame over Koepang Bay. (Possibly the trailing radio aerial had been struck by lightning which ignited the fuel vapours from the breather holes in the full wing tanks.) Nothing could be done to help and we experienced a terrible feeling of sadness and frustration.
>
> Again, one can only state the great admiration for aircrews.
>
> This crash made us realise the horrific responsibility which the CO, Wg Cdr Frank Headlam, had in having to allocate aircrews for operations which were extremely dangerous.[79]

With Ambon and Namlea lost, it was now only possible to use Penfoei under the cover of darkness as the base was under constant attack. Long-range reconnaissance patrols operating from Darwin and Penfoei were flown daily beyond the Banda Sea, as far as Kendari in the southern Celebes and around the Flores Islands.

On the 18th, Headlam received orders to evacuate all RAAF personnel except small maintenance and communications parties.[80] All the men of No 2 Squadron volunteered to remain behind, and six officers and 23 other ranks were selected.[81] At dusk, a flight of Hudsons, led by 13 Squadron's CO, Squadron Leader John Ryland (A16-142), arrived to take the Squadron to Darwin. On the transit to Penfoei they had strafed an enemy submarine, but being stripped for the evacuation had not been carrying bombs[82]. Now, with a feeling of hopelessness, the last RAAF Hudson to operate from the NEI for the next three years departed into the darkness bound for Darwin. It was the early hours of Thursday, 19 February 1942.

---

76 Wg Cdr J MacL Hampshire DFC, 256. Nos 6, 11 & 33 Sqns; comd 41 Sqn 1942–43, 461 Sqn 1944; HQ Cstl Cd RAF 1944–45. Regular air force offr of Perth; b Port Macquarie, NSW, 27 Feb 1916.

77 Sgt N K Rayner, 407649. No 2 Sqn. Of Yarcowie, SA; b Terowie, SA, 28 Jul 1920. Killed in action 14 Feb 1942.

78 13 Sqn crew: Flt Lt G G Mitchell, Plt Off J N Goodfellow, Sgt C L Motteram.

79 Richard Taylor, p.15.

80 Gillison, p.421.

81 See Chapter 13.

82 In addition to Ryland, the other pilots were Arch Dunne, "Mickey" Finlayson and Nev Hemsworth. Grantham, p.59.

TO THE HUDSON MEN

by Evelyn Pitman

I saw you march on Anzac Day
I watched you passing by,
Brave heroes of a bygone war
Measured tread and steadfast eye.

I saw memories in your eyes
There was both pain and pride,
I knew you were remembering
All your friends who fought and died.

I saw those long-departed mates
Striding side by side with you,
Young men with straight sturdy limbs
Did you feel their presence too?

I saw the jets fly overhead
Great monsters of the sky,
How different the Hudson 'planes
Which you flew in days gone by!

And I looked heavenward again
The jets had roared away,
And I saw a flight of Hudsons
Saluting Anzac Day.

One of the 2 Squadron Wireless Air Gunners killed on the Menado raid, on 12 January 1942, was Sergeant Jack Mawdsley. This poem, written by his sister, Mrs Evelyn Pitman, is reproduced with her kind permission.

CHAPTER 13

PENFOEI PEDESTRIANS

We entertained no thoughts of surrender

Colin Humphris[1]

As the last Hudson departed Penfoei airstrip at 3.30 on the morning of Thursday 19 February, the small band of 2 Squadron volunteers left behind felt a sense of isolation and foreboding. The CO had placed the Meteorological Officer, Flight Lieutenant Bryan Rofe,[2] in command. Only one aircrew officer was left behind, Pilot Officer Peter Thompson.[3] That morning, with the AIF, they began the pre-arranged demolition and mining of the airfield,[4] and at 5 p.m., news of the Japanese invasion fleet was received. Radio contact could not be established with NW Area Headquarters; Darwin had just been subjected to the devastation of its first day of heavy air attacks.

Now with no immediate hope of evacuation, Rofe planned the group's withdrawal from the airfield. However, there was still a small group separated from him, under the Squadron Equipment Officer, Flying Officer Arthur Cole. They had been despatched three days earlier to salvage what they could from American Kittyhawk fighters which had crash landed on the north coast, near Atamboea. Cole's party arrived back that night, as news was received of enemy landings on the coast near Koepang and at Dili.

Early the next morning, Japanese warships began shelling the aerodrome. As the members of "Sparrow Force", the army defenders on Timor, set about destroying their communications centre and other facilities around the airfield, the RAAF demolished the bomb dump and fuel supply. Sparrow Force comprised the 2/40th Battalion from the Australian 8th Division, commandos of the 2/2nd Independent Company, support troops from 23 Brigade headquarters, together with elements of Dutch and British forces.[5] "C" Company of the 2/40th had been responsible for the airfield defence at Penfoei, and were now denying its use to enemy forces. As instrument fitter Corporal

---

1 C B Humphris, *Trapped on Timor*, self-published, Victor Harbor, 1991, p.32. This chapter is primarily based on Colin Humphris' experiences, as related in his book.

2 Sqn Ldr B Rofe MBE, 282539. Directorate of Met Services RAAF 1941–46. Of Adelaide; b 14 Sep 1917.

3 Plt Off P H A Thompson, 400845. No 2 Sqn. Of Albury, NSW; b Albury 14 Sep 1917. Died of illness 31 Mar 1942.

4 Gillison, p.421.

5 This included the 2/1st Heavy Battery, B Troop/18th Anti-Tank Bty, 2/1st Fortress Engs, 2/11th Field Engs, 2/12th Field Ambulance, and 75th Light Air Det, amounting in all to 70 officers and 1,330 men. In addition, 371 Dutch troops were under the command of Sparrow Force, as was the recently arrived British 79th Light Anti-Aircraft Bty. Meanwhile the 2/21st Bn had been sent to Ambon, where it was known as "Gull Force", while a third force was sent to Rabaul, this being code-named "Lark Force". T Paley, *The Sparrows*, Self-Publishing Assoc, Worcester, 1992, pp.81–7; and C C H Wray, *Timor 1942*, Mandarin, Melbourne, 1990, pp.20–1.

Colin Humphris[6] recorded, Rofe divided his party, now comprising 29 members, into two groups:

> "Nothing else you can do here, chaps" Rofe said. "Take the Chev ute and get away while you can. Make for Babaoe. The rest of us will follow soon in the truck and catch up with you there. If we are not there by 0700, drive on to army HQ at Tjamplong."[7]

Three officers and 13 men went ahead with the radio equipment, planning for the demolition group to join them later at the village of Babaoe, the main Sparrow Force base, to the east. Wireless mechanic LAC Clyde Pappin's[8] diary recorded:

> 1am—News that Japs landed at Dili and other points on coast.
> 5am—Drome being blown up by AIF.
> 6am—Petrol dump at Tanau and bomb dumps as close as 200 yards to us being blown up. Dashed to end of transport lines—100 yards—and shell hits 25 yards from transmitter. More dumps go up while we three reached op room road and met Sgt Sumner with truck. Great relief.[9]

Meanwhile a Hudson, flown by Flight Lieutenant Archie Dunne, was approaching Timor to evacuate the RAAF rear party[10] when they encountered the Japanese invasion of Koepang, where 16 warships were sighted. Dili, the capital of Portuguese Timor, was similarly being attacked. The rescue of the rear party had been forestalled by the enemy, by only a matter of hours, and the Hudson was forced to return empty handed to Darwin.[11]

The two groups did not rejoin at Babaoe, so the first party pressed on to army headquarters at Tjamplong to meet with the Sparrow Force commander, Brigadier Veale. From Tjamplong, the RAAF men saw 300 enemy paratroopers drop from 20 to 25 aircraft over Oesaoe Ridge, north-east of Koepang.[12] It was here the two RAAF groups finally rendezvoused, and Rofe conferred with Brigadier Veale. It was agreed the RAAF party should move to the hills overlooking Koepang Harbour to set up its radio equipment and establish contact with Darwin.[13]

As the RAAF evaders took to the hills, the main body of Sparrow Force was confronted by an invasion force of 23,000 men. The Australians, always intended as a

---

6 FSgt C B Humphris, 5170. 2AD, No 2 Sqn, 5AD. Company manager of Mt Barker, SA; b Gladstone, SA, 23 Sep 1918.

7 Humphris, p.13.

8 FSgt C W Pappin, 27530. SHQ Darwin, GR Sch, 1AD. Electrical mechanic of Stirling, SA; b Renmark, SA, 26 Nov 1916.

9 Clyde Pappin's diary, of Friday 20 Feb 1942, cited in Humphris, p.14.

10 A second Hudson also later departed for Timor, but turned back. Telephone interview by author with Arch Dunne, of 5 Nov 1993.

11 No 2 Sqn A50, of 20 Feb 1942.

12 This force of 308 paratroopers, from the 3rd Yokosuka Special Naval Landing Force, was in a coordinated invasion with the 228th Regiment. A further 323 paratroopers were dropped the next morning. Gillison, p.423.

13 Gillison, p.422.

limited defensive force to delay any enemy advance, were overwhelmingly outnumbered, and surrendered on 23 February. The 2/40th suffered 84 officers and men killed and 132 wounded, exacting a heavy toll on the Japanese invaders. Over 1,000 of the enemy were killed in the four days of fighting, and of the two drops of 600 Japanese paratroopers involved in fighting the Australians, only 78 survived.[14] Sparrow Force had shown the Japanese were not invincible, and the commandos of the Independent Company, who had been detached to Dili, were able to carry on guerilla harassment of the Japanese on Timor for the remainder of 1942.

The RAAF group now totalled 31 members as they escaped to the mountains, being joined by a RAN intelligence officer, Lieutenant Alan Bridge, and an army signaller from the RAAF cipher office at Penfoei, Private Clem Clements. Driving north-east towards Dili, the party was soon forced to take to foot and trek north, lugging the cumbersome radio equipment through the rough mountainous terrain. After slow progress, on Sunday 22 February they reached the friendly village of Naifalo where, on setting up the radio equipment, they made contact with Darwin to advise of their escape from the invading forces. Of critical importance was the advice to NW Area Headquarters that the secret ciphers, used for secure signalling, had not fallen into enemy hands.

They reached the village of Noeataoes on the 25th, but members were now suffering physical debility from malaria, tropical ulcers, and dysentery. The discovery that there was no quinine in the medical pack, so labelled, was a cruel blow. There was now no defence against the fever of malaria.

At the next village of Tarkow, when radio contact was again established with Darwin, they were advised of a further rescue attempt: a rendezvous by flying boat was planned for 1 March, at Kapsali. The radio operator, Corporal Ron Bell,[15] maintained contact with Darwin and the evaders were able to report on the disposition of enemy warships in Koepang Harbour, and provide a weather report. At dusk, they were rewarded with the sounds of a Hudson attack on the enemy in Koepang, and felt they were still effective members of No 2 Squadron.

They were now descending towards the coast, reaching Mosoe on the 27th, then down the Kapsali River to their rendezvous on the north coast. The plan had been for the four-engined Empire flying boat to pick them up near the mouth of the river at dusk, on Sunday 1 March. Clyde Pappin's diary recorded:

> Mr Rofe holds impressive church service—most sincere I have ever witnessed—even the hardest were touched I reckon. Tried unsuccessfully to make raft. Darwin now tells us that flying boat will pick us up tomorrow—2nd. Scrub on coast very thick—unable to get through.[16]

14 Paley, pp.158–63.

15 Flg Off R J Bell, 18349. SHQ Darwin, HQ SA, 12 OBU, 1st TAF, No 2 Sqn. Clerk of Melbourne; b Eaglehawk, Vic, 14 Oct 1915.

16 Clyde Pappin's diary, of 1 Mar 1942, in Humphris, p.27.

The tired party on reaching the coast after the strenuous trek across the mountains of West Timor: Standing: (from left) Clyde Pappin, Roy Andrews, Jock Burchall. Seated: Ted Mitchell, Jim Sumner, Bob Donald, George Ettridge. Standing, in background on right: Phil Kean C HUMPHRIS

Electrical fitter LAC Roy Andrews[17] and fitter-armourer LAC Les Borgelt[18] persisted with finding a way through to the coast; the others made ropes from the rushes and reeds, and attempted to fabricate a raft from bamboo and palm timber. The ever-present danger of discovery by the enemy was highlighted by wireless operator AC Len Bourke's diary entry: "Five enemy ships in convoy passed along the coast in the direction of Dili".[19]

Their rescue had been delayed a day, without explanation from Darwin, as the Empire flying boat required extra fuel for the long return journey from Darwin to Timor. Fuel tanks had been lashed together in the cabin, but had broken loose on take-off, requiring the pilot to put back down on the water. By the time this was rectified, it was too late to make a daylight rendezvous in Timor, and the pickup had been put back a further 24 hours. As there was now concern over more air raids at Darwin, the

17 Cpl R E Andrews, 18533. No 2 Sqn. Of Portland, Vic; b Portland 25 Mar 1916. Died of illness 13 Apr 1942.

18 Sgt L R Borgelt BEM, 12801. No 2 Sqn, 3BAGS, 79 & 32 Sqns, 2 OTU. Farm labourer of Nhill, SA; b Nhill 10 Dec 1917.

19 Len Bourke's diary, of 1 Mar 1942, cited in Humphris, p.30.

flying boat was directed to Broome on the afternoon of 2 March, to depart from there the next day for the rescue. The party moved to the beach expectantly, but the flying boat failed to appear. The next afternoon Darwin advised: "Broome raided unlikely rescue attempt tonight".[20] It had been another cruel blow.

All hands at raft making C HUMPHRIS

20 Humphris, p.30.

Unbeknown to the group of evaders, Broome had just been subjected to an air raid, which has since been termed Broome's "one day war". Ahead of the Japanese advance through the East Indies, an aerial shuttle service had begun between Tjilatjap, on Java, and Broome in Western Australia. In the second half of February some 8,000 refugees passed through Broome, and on some days up to nearly 60 aircraft passed through on the southerly journey to Perth or Sydney.[21] Japanese reconnaissance flights on the afternoon of 2 March and the early morning of the next day, confirmed the disposition of a large number of Allied flying boats and land-based aircraft. At 7.05 on the morning of Tuesday 3 March, nine Mitsubishi A6M2 "Zeke" fighters (the Navy Type 0 fighter, and hence the nickname "Zero") and a C5M2 "Babs" (Navy Type 98) reconnaissance aircraft departed Koepang for the 900 kilometre flight to raid Broome.

The Penfoei Pedestrians awaiting rescue by flying boat:
Back row: (from left) G Ettridge, P Kean, C Clements (AIF), R Lloyd, C Gibbs, W Smith, R Donald, E Mitchell, J Burchall, F Rusconi, W Arthur, C Humphris. Middle row: R Holloway, V Henrickson, R Oliver, J Sumner, J Graham, L Bourke, F Meeking, R Bell, K Greaves, W Grigg. Front row: P Thompson, L Borgelt, R Andrews, A Cole, B Rofe, A Ackerman, C Pappin, E Park
C HUMPHRIS

The raiders, from the Japanese Navy's 3rd Air Group,[22] arrived over Broome at 9.30 a.m., finding 15 flying boats in Broome's Roebuck Bay. In the horrific attacks that followed, some aircrew and refugees were able to jump from the crammed flying boats to safety. It has been estimated that over 70 people perished in the raid—the

21 A Prime, "Australia's Pearl Harbor", in *Australian Aviation*, Aerospace Publications, Canberra, December 1991, p.24.

22 This unit had been formed in April 1941 for land attack operations: I Hata & Y Izawa, *Japanese Naval Aces and Fighter Units in World War II*, Airlife, Shrewsbury, 1989, p.123. The Japanese Naval Air Force designation for this unit was the 3rd *Kokutai* of the 23rd *Koku Sentai* (Air Flotilla).

exact number will never be known, as the panic of the chaotic evacuation had overridden the necessity for passenger manifests. As the enemy raiders returned to Timor, the capability of Japanese air power in their relentless push south had again been demonstrated by the audacity of this long-range strike by single-engined fighters. Broome, as an escape route to the south, had been devastated.[23] This was not only the case for refugees, which included the planned evacuation by flying boat of No 1 Squadron personnel from Java,[24] but also for the 2 Squadron group trapped on Timor. The news from Darwin of "no rescue possible this end" did not help morale.

Darwin maintained the hopelessness of any rescue attempt over the next days, reiterating that there were no flying boats available. It would now be a battle of survival—foraging for food and fishing, erecting makeshift huts and maintaining a radio watch with Darwin. On the few occasions when a deer was shot, the cook, Corporal Bob Lloyd,[25] was able to satisfy appetites. While camped here, the party was joined by four soldiers of the 2/40th Battalion, who had evaded capture and made for the hills after the surrender of the main force. The soldiers then maintained their own camp nearby the RAAF group.

Spirits were lifted on 12 March, when NW Area Headquarters advised of a supply drop the next day, and instructed Rofe to pass details on any suitable emergency landing ground (ELG). Two groups then set out to the north for Naiklioe, a village 40 kilometres up the coast, to seek food and search for an ELG. No suitable landing site could be found to the north, so Arthur Cole led his team back to reconnoitre south of Kapsali. But malaria was taking its toll, as Col Humphris succumbed to the shivering and shakes of the fever:

> Col Humphris down with fever. Mr Rofe and Ralph[26] go on while I remained to look after Col. I felt a little better. Col sweating badly—ill all day. Natives gave us super meal.[27]

Rofe and Holloway set off the next morning, with the assistance of a porter and ponies. By mid afternoon, they reached the village of Manuello, two kilometres from Kapsali, which had been deserted by the Timorese, and where a new base camp had been set up in their absence. The main group had six of their members down with malaria, and had taken over a large hut to care for the sick. Arthur Cole's group had passed through on its ELG expedition to the south, leaving LAC Roy Andrews, who was down with malaria, and replacing him with AC Bob Oliver.[28]

---

23 In addition to this great loss of life, the following Allied air losses were incurred: eight Dutch, USN and RAF Catalinas, five Dutch Do-24K flying boats, two USAAF B-17 and two B-24 bombers, one RAAF Hudson (A16-119), one RAAF and one QANTAS Empire flying boat, and a Dutch DC-3 and a Lodestar transport. In addition, a further DC-3 was encountered by the returning force, and shot down to the north at Carnot Bay. Prime, p.28.

24 Gillison, p.444.

25 FSgt R Lloyd, 14576. SHQ Darwin, SHQ Richmond, 1EFTS, 5WU. Cook of Sydney; b Manchester, Eng, Feb 1902.

26 FSgt R G Holloway, 11093. SHQ Darwin, 1STT, 1AD. Grazier of Mt Gambier, SA; b Mt Gambier 15 Jul 1917.

27 Clyde Pappin's diary, of 15 Mar 1942, cited in Humphris, p.39.

28 LAC R W Oliver, 27996. No 2 Sqn. Mining engineering student of Adelaide; b Adelaide 9 Feb 1920.

Unsuccessful supply drops had been attempted by Hudsons on the unlucky Friday the 13th, when the stores parachute had become entangled with the bomb-bay doors, and again on 15 March when Len Bourke recorded:

> Received message that supplies would be dropped at 7.15am. Just about everyone strung out along beach for miles. We waited until 10am without an aircraft appearing and then returned to camp.[29]

The sad story continued the next day. They were advised supplies would be dropped at dawn, but by the time they reached the beach, the Hudson had gone. Any parachutes were easily lost in the dense scrub.

Peter Thompson's poor condition was now causing concern, and Bob Donald did all he could to care for the sick men with his limited medical supplies. Morale was not helped by receipt of a message from Darwin on the 17th, advising: "No repeat no further help possible from this end". By the next day, the fever had 10 men out of action, including the leader, Bryan Rofe.

Arthur Cole's party returned on Thursday the 19th, with the heartening news they had located an ELG south of the Keroes headland, suitable for a Hudson, on a stretch of hard beach at Toeakaoe, 20 kilometres away. This was advised to Darwin and, in spite of its close proximity to Koepang (approximately 25 kilometres), the group considered that this would provide the only chance of escape.

The party now dispersed to various huts along the beach at Kapsali from where, acting as coastwatchers, enemy shipping movements could be relayed to Darwin. On Saturday 21 March, after supplying a weather report to NW Area, the evaders had the satisfaction of hearing a dusk raid by Hudsons on Koepang. They were also advised of another attempt to drop supplies at dawn.

Arthur Cole and Col Humphris were accompanied by Sergeant Jim Sumner[30] and LAC Wally Grigg[31] to position for the drop the next morning. But the Hudson did not appear, and they were forced to take cover as three enemy aircraft swooped over their position. Advice came through that the drop had been postponed, and the next day the enemy aircraft returned. A Hudson drop was successful that evening, but all that could be located was a small parachute with a box containing three dry radio batteries. The other box dropped, probably containing the badly needed quinine and other medical items, could not be located.

Peter Thompson's condition was becoming alarmingly worse and one of the radio operators, AC Jim Graham,[32] was also now seriously ill. On the 31st, after a message from Darwin to move to the ELG, Thompson succumbed to his illness. He was buried

29 Len Bourke's diary, of 16 Mar 1942, cited in Humphris, p.41.

30 Flg Off J P Sumner, 3525/O210039. No 2 Sqn, 1 OTU, 2CRD, 44 OBU. Aircraft fitter of Sydney; b Birmingham, Eng, 11 Dec 1910.

31 Cpl F W Grigg, 20701. No 2 Sqn, 2AD, 5 OTU, 1FBRD, 41 Sqn, 114 ASR Flt. Cabinet maker of Sydney; b Sydney 10 Dec 1911.

32 AC J A Graham, 35933. SHQ Darwin. Of Sydney; b Geelong, Vic, 15 Jun 1921. Died of illness 2 Apr 1942.

beside the Kapsali River, and Flying Officer Bill Arthur[33] conducted a short burial service.

Bryan Rofe had decided, on Darwin's advice, to immediately move south to the ELG, and the group, in three separate parties, set off on Thursday 2 April. They eventually climbed a rough track over the Keroes promontory and descended to Arthur Cole's ELG, with the advance party arriving in the evening, and the radio party late that night. The following morning—Good Friday—the third party arrived, with the news of Jim Graham's death the night before en route, as they had negotiated the dangerous tides around the headland.

A large native hut was taken over at Toeakaoe to shelter the sick, now sapped of their strength from the strenuous trek from Manuello. On Easter Saturday, the message was received that a supply drop would be attempted the next morning at 3.30, on Keroes promontory, five kilometres to the north-west. Rofe, Cole, Humphris, Borgelt and Grigg set out, and in the morning darkness the Hudson appeared. Only one drop from the Hudson's two runs was observed, and finally a parachute was located. But there was nothing attached to it.

Ron Bell, one of the wireless operators
C Humphris

Now the radio receiver malfunctioned, and Ralph Holloway's skill and improvisation came to the fore. With the crudest of tools he was able to effect repairs. This then enabled the passing of a message on 7 April to Darwin of the presence of 25 ships and many flying boats in Koepang Harbour.

Members of the group were now suffering a lack of saliva in their mouths, in addition to stomach pains from dysentery and malaria attacks. In the early hours of Wednesday 8 April, the AIF member of the team, Private Clem Clements, was found in a coma. His groanings and convulsions baffled the others, who could find no explanation for his condition. He died later that morning, and as he was buried, the evidence of a snake bite became apparent.

Later in the day, 8 April, islanders brought news that 300 Japanese had disembarked from boats at Naiklioe and were searching the area. They were now only 15 kilometres away at Kapsali, heading towards the RAAF party. The campsite was abandoned early the next morning, for a jungle hide-out two kilometres back from the beach. Darwin was advised of this move and the necessity to cancel a supply drop for

33 Sqn Ldr W S Arthur, 262122. SHQ Darwin, 2 MW Sqn; comd 2 MW Sqn 1944, 62 AC Wg 1945. Civil engineer of Murrurundi, NSW; b 16 May 1903.

that night, but the small force was asked if there was the possibility of shining a light out to sea. NW Area Headquarters had considered the risk of overloaded aircraft, attempting a rescue from the ELG, as being too high, and was evidently planning another form of evacuation.

On Sunday 12 April, a villager reported that the Japanese had turned back before reaching Toeakaoe, and embarked on a vessel, presumably back to Koepang. That night, as they turned in at dusk they were thrilled to hear the sound of exploding bombs as the Hudsons again struck Koepang. Corporal Roy Andrews lapsed into unconsciousness the next day, and radio men Corporal Ron Bell, and ACs Phil Kean[34] and George Ettridge[35] were all very ill, with Clyde Pappin and Ralph Holloway getting progressively worse. Roy Andrews died peacefully during the night; now four of their number had been lost.

The next morning, the 14th, a move was made back to the coastal site at Toeakaoe, with the assistance of the villagers. There was no doubt their help was required as there were now only five reasonably fit men. Also, there was no doubt that any rescue would need to be soon and fortunately, on Wednesday 15 April, Darwin advised of arrangements for rescue by a United States submarine. However, there was no submarine that night—only, once again, bitter disappointment.

The radio batteries were by now almost spent, and a friendly villager made the journey from Toeakaoe to Manuello to recover a used battery left there before the move south. He returned the same night with the heavy battery, which still held enough power for Ron Bell to transmit the final messages for rescue arrangements.

Now awaiting another rescue attempt on the 17th, the group was joined by two exhausted and bedraggled RAAF pilots from 2 Squadron. Flight Lieutenant Harold Cook[36] and Pilot Officer Viv Leithhead[37] had been shot down by Zeros five days earlier, and had ditched at the edge of Koepang Bay. Sadly, both their WAGs were badly wounded, and were taken ashore and assisted by villagers. As the gunners could not travel, the pilots were confronted with the frightful decision: escape across to the north coast, aware of the rescue attempt for the stranded party, or remain with their gunners and certain capture. They were forced to leave the two wounded airmen in the care of the villagers.[38] The village chieftain of Pariti, Jermias Koanfora, aided them to escape. For this assistance, he was later imprisoned by the Japanese, and never fully recovered from his ill-treatment. However, his role in the successful rescue of the pilots was later acknowledged, in 1961, by the RAAF Air Attache in Indonesia.[39]

34 Cpl P J Kean, 24508. SHQ Darwin, 13 Sigs Unit, 40 OBU. School teacher of Brisbane; b Brisbane 16 Apr 1921.

35 Cpl G W Ettridge, 29688. SHQ Darwin, SHQ Pearce, 72 OBU, 4AD, 87 OBU. Shop assistant of Quairading, WA; b Perth 9 Oct 1922.

36 Sqn Ldr H O Cook, 280229. Nos 14 & 2 Sqns, 1 OTU, 38 Sqn; comd 33 Sqn 1943–44. Commercial pilot of Melbourne; b 18 May 1901.

37 Flt Lt V C Leithhead, 406858. Nos 2 & 31 Sqns. Of Perth; b Perth 23 Jan 1921. Killed in action 21 Aug 1943.

38 The two gunners, Sgts H F Hearle (406758) and W D Witham (404761), were captured, nursed back to health, and executed in May 1942. RAAF Historical Section C.D.102/45 for Hudson A16-159, of 24 May 1948 refers. RAAF Casualty records give date of deaths as 8 May 1942.

39 "Bravery Rewarded After 19 Years", *RAAF News*, Vol 2, No 4, Canberra, May 1961, p.5.

Word of an impending rescue attempt was also passed to the four AIF members who were camped nearby. With the remaining 27 Penfoei Pedestrians, and the other six, the party now totalled 33. In the twilight, they made their way down to the beach, carrying the very weak members by improvised stretchers. Large waves were breaking 50 metres out, which was to make for a treacherous passage to the calmer water beyond. They lit a fire, to signal the submarine, and waited. The rescue attempt would be risky, with the constant threat of Japanese patrol vessels and the difficulties of negotiating the dangerous surf.

USS Searaven US NAVY, VIA C HUMPHRIS

Then a light was spotted out to sea—a morse signal of "SR"—from the crew of the United States Navy submarine *Searaven*. In the dim light, an officer waded ashore, and after greetings, the problem of getting the party to the submarine's small boat was resolved. Two journeys would be necessary to carry everyone out to the waiting submarine. A villager then arrived with the alarming news that an enemy patrol was at the other end of the bay, on its way towards the Australians. A quick decision was necessary, and 16 of the RAAF men were selected to make their way out through the shark-infested waters, and the strong current, towards the boat.

The boat returned with some difficulty to the huge dark shape of the submarine. The time taken to get this first group on board necessitated the submarine commander, Lieutenant Commander Hiram Cassedy, delaying the rescue of the second group until the following evening.

On the night of the 18th, the remaining 17 waited for rescue. The three very ill members were carried beyond the breakers to the boat on the backs of three of the sailors. Huge waves then crashed over the small craft, tossing it up to the beach. It was now necessary to bail it out and, with a miraculous lull in the breakers, they were able to push it back through the surf with superhuman strength, to the calm water. Now back on the submarine, the rescue attempt had succeeded.

The submarine, with the extra 33 men on board, commenced the return journey to Australia. Darwin signalled its congratulations: "Well done Searaven stop proceed to Fremantle at full speed". However, the trip south to Perth was not without concern. Phil Kean was very close to death, his body wasted to half its normal weight, with tropical ulcers deep into his emaciated flesh. Bob Lloyd and George Ettridge were seriously ill, and Ron Bell was suffering badly with dysentery. Then, on the evening of Thursday 23 April, a fire broke out which disabled the submarine's power. The fire was adjacent to the aft torpedo room, which housed 15 tonnes of anti-aircraft shells, originally intended for the Philippines.[40] The *Searaven* survived this disaster, and the following day was taken in tow by another submarine, the *USS Snapper*, and finally by

40 Col Humphris, letter to author, of 21 Oct 1993.

the RAN corvette, *HMAS Maryborough*. The Penfoei Pedestrians reached Fremantle on 26 April.

Over the following months, they were nursed back to health, recovering from the ravages of their tropical diseases. All were able to return to active RAAF service. The MBE was awarded to Flight Lieutenant Bryan Rofe and Flying Officer Arthur Cole, and the BEM to LAC Les Borgelt. The captain of the *USS Searaven* and his junior officer, who had rescued the Australians from the beach, were both awarded the Navy Cross. Later submissions for acknowledgment of the leading part played in the rescue by the radio operator, Corporal Ron Bell, were, however, to prove fruitless.[41]

Farewell to Hollywood Hospital, Perth, after a month's nursing of the Penfoei Pedestrians: Back row: (from left) Flg Off Arthur, LAC Ettridge, Flt Lt Rofe, LAC Borgelt, Cpl Lloyd, Flg Off Cole, LAC Pappin, a Perth movements officer, LAC Grigg, Sgt Donald, LAC Holloway, LAC Smith, Sgt Humphris, Cpl Gibbs, Flt Sgt Sumner, Plt Off Greaves, Flg Off Burchall. Middle row: LAC Bourke, LAC Ackerman, Cpl Bell, Dr Robinson (hospital medico), LAC Park, Sgt Henrickson, LAC Meeking, LAC Oliver, Cpl Mitchell, Lt Bridge RAN. Front Row: Sgt Shields, Cpl Bell, Pte Jones, Pte Davies (all AIF) C HUMPHRIS

The success of this remarkable saga was a result of the efforts of these men and others in the team, but was in no small way due to Bryan Rofe. His authority had never been in question, and his leadership encouraged them through this epic of survival.

41 Letter from Col Humphris to Alexander Downer MP, of 21 May 1985, stated: "Bell ... displayed wisdom and judgement in assessing the need for the radio equipment he prepared ... a high degree of technical skill in its preparation and in operating it ... tenacity, dogged persistence and resolute determination ... in a condition when most men would have given up."

THE PENFOEI PEDESTRIANS

MA.2539	Flt Lt Bryan Rofe
A.2122	Flg Off Bill Arthur
A.2031	Flg Off Jock Burchall
A.1733	Flg Off Arthur Cole
A.2236	Plt Off Ken Greaves
400845	Plt Off Peter Thompson
10397	Sgt Bob Donald
3525	Sgt Jim Sumner
18349	Cpl Ron Bell
6448	Cpl Vernon Henrickson
5170	Cpl Colin Humphris
13576	Cpl Bob Lloyd
9169	Cpl Ted Mitchell
18533	LAC Roy Andrews
12801	LAC Les Borgelt
19259	LAC Charles Gibbs
20701	LAC Wal Grigg
11093	LAC Ralph Holloway
27530	LAC Clyde Pappin
10440	LAC Eddie Park
16269	LAC Bill Smith
10440	AC Arthur Ackerman
35131	AC Len Bourke
29688	AC George Ettridge
35933	AC Jim Graham
24508	AC Phil Kean
13374	AC Frank Meeking
27996	AC Bob Oliver
35453	AC Frank Rusconi
	Lieut Alan Bridge RAN
	Pvt Clem Clements AIF[42]

42 This is the list of personnel that appears in the No 2 Sqn Operations Record Book (Form A50), of 18 Feb 1942. Col Humphris notes that his and Snow Henrickson's promotions to Sergeant had been promulgated to be effective from 1 Feb 1942, but they had not as yet been advised. This appears the case for several other members of the group, and the ranks are those as listed in the A50 at that time. Similarly, some service numbers changed later.

CHAPTER 14

THE BOMBING OF DARWIN

The raid of 19th February 1942 ... was the first attack by any power on Australian soil and as the bombs rained down on Darwin, and on the ships in Darwin harbour, no one knew that it was not the beginning of the invasion of the continent of Australia.

John Gorton[1]

The raid on Darwin on Thursday 19 February 1942 was opened by 27 Nakajima B5N Navy Type 97 "Kate" carrier-based bombers flying at 14,000 feet, followed by 27 Aichi D3A Navy Type 99 "Val" dive bombers. These were escorted by three fighter groups of nine Mitsubishi A6M2 Navy Type 0 "Zekes"—known as the Zero. Operating from four carriers, this was part of the same naval force that had devastated the United States Navy in Hawaii. They were led by Commander Mitsuo Fuchida, the leader of the attack on Pearl Harbor, only two months earlier, who "was responsible for the infamous message, 'Tora! Tora! Tora!'".[2]

> In the post office a telegraphist got out three words, "Raid in progress", before the post office was blown to pieces. Alongside it, the postmaster Harry Bald, his wife and daughter had just jumped into a slit trench. They were killed, too. These four people were the first Australians killed on Australian soil by the Japanese.[3]

After an interval of two hours, two formations each of 27 land-based heavy bombers of the Japanese Navy's 23rd *Koku Sentai* (Air Flotilla), with no fighter cover, attacked the RAAF base at Darwin.[4] The total strength of aircraft in North-Western Area was just 17 Hudsons of Nos 2 and 13 Squadrons (of which nine were at Darwin, including those that had just evacuated personnel from Koepang), 14 Wirraways of No 12 Squadron, and 10 USAAF[5] Kittyhawks then in transit to Java.[6] Four No 2 Squadron Hudsons were destroyed in this raid.[7] The timing had been planned to neutralise Darwin prior to the invasion of Timor the following day. To achieve this, the total force launched for this first day's assault on Darwin amounted to 188 aircraft,

1 Introduction by ex Prime Minister John Gorton, in R N Alford, *Darwin's Air War*, Aviation Historical Society of NT, Darwin, 1991, p.2.

2 G Wallace, *Up in Darwin With the Dutch*, Melbourne, 1983, p.63.

3 "The Day The War Came To Us", *The Sun*, Thursday 19 Feb 1959, cited in Wallace, p.67.

4 27 were G4M1 Betty bombers from Kendari, in the Celebes, with 27 G3M1 Nell bombers from Ambon. Alford, p.19.

5 The US Army Air Corps (USAAC) had become US Army Air Forces (USAAF) on 20 Jun 1941. The USAAF was subsequently renamed the US Air Force (USAF) in 1947.

6 Gillison, p.425.

7 A16-6, A16-57, A16-78 and A16-135. In addition, 13 Sqn's A16-72 was destroyed.

comprising 36 Zero fighters, 71 "Val" dive-bombers, and 81 "Kate" high-level bombers.[8] The death toll was assessed at 238.[9]

Engine fitter LAC Eddie Perkins,[10] who had been at Darwin with 2 Squadron since 9 December, was on the aerodrome during the raids. Dive bombers in the first raid hit both ends of the Hudson hangar, which then exploded stores of oxygen bottles and ammunition. He recalled:

> We had a fair display of fireworks for a while ... later a high level raid [was] from two lots of 27 aircraft, which dropped all their bombs on the RAAF 'drome.[11]

The RAAF base at Darwin in 1941, with the Hudson hangar at the top of the photograph
E PERKINS

Bombs fell near Perkins' slit trench, which blew in on top of him, and has affected his hearing ever since. Of these first bombing raids on Australian territory, photographer LAC Aubrey McEgan remembered:

> ... medium ack-ack opened up on the leading formation which was flying at approximately 5,000 feet. The aircraft continued flying on through the ack-ack and when directly over the town ... peeled off to dive bomb shipping in the harbour.

---

8 Alford, p.13–14.

9 This total partly comprised at least 157 who died on ships, 14 civilians killed in the town and 21 on the wharf, seven RAAF personnel, seven USAAF personnel, and two army personnel. Gillison, p.430. Another assessment is 243 dead, and around 350 injured or missing. Alford, p.17.

10 Sgt E B Perkins, 11874. Nos 2 & 31 Sqns, 18 & 120 NEI Sqns. Farmer of Cowangie, Vic; b Cowangie 28 Apr 1920.

11 Eddie Perkins, taped recollections, of 5 Jul 1993. No 2 Squadron Association, Melbourne.

Our communication centres also came in for attack. Immediately following the dive bombing, fighter aircraft indulged in strafing gun installations and aircraft parked on RAAF Station Darwin. Our aircraft were lined at intervals on the edge of the runway and it was possible for an enemy fighter to have three aircraft in line on one strafing run. The Zeros were so low that it was possible to distinguish the pilots leaning from their cockpits observing details of buildings and choosing targets at will. Following the strafing, there was a lull for a few minutes and then a dive bomber group of nine aircraft which had been sitting up high above RAAF Darwin swooped down singly and dive bombed alternate targets giving the impression that the attack had been well rehearsed, so accurate was the bombing. There was no screaming roar from engines and the silent motors were a feature.

Dive bombing finished, we considered that the raid was over. Hangars were left burning, all the aircraft set on fire were now complete wrecks.

We then received notice of another raid and ... saw a formation of 27 bombers in line abreast proceeding in a northerly direction and bombs could be heard falling ... Turning around and looking to the north directly above, I noticed another formation converging and flying in a head-on direction to the first formation, proceeding south. The two formations passed one another directly over the hangar area. It was at this stage that a bomb dropped less than 8 feet from the trench I was in. The crater made came to the edge of the slit trench which I shared with the Orderly Officer. Our tin hats were knocked from our heads and the trench vibrated for approximately three seconds after the hit.[12]

A RAAF Hudson burns beside the wreckage of a USAAF P-40E Kittyhawk beside the Darwin strip
A McEgan

For the personnel on the ground, this attack of pattern bombing on the aerodrome was absolutely terrifying:

From the time of the release of the bombs, from about 16,000 feet until the explosions began, was a nightmare. There was a constant swishing noise, forever increasing in volume, when eventually our L-shaped trench seemed to rock from side to side. I remember the noise, dust and the smell of cordite which was followed by a vacuum making it impossible to breathe.[13]

12 McEgan interview, pp.2–4.

13 Richard Taylor, p.20.

Although, naturally enough, morale was not high, radio mechanic LAC John Dorsett[14] remembered that the major problem was the great deal of confusion:

> ... and because the airdrome was targeted, the safest spots were well away from the area ... so many people became excess to establishment, the best idea was to order them away from the drome.[15]

The wreckage of Hudsons at Darwin A McEgan

The station commander, Group Captain Scherger, had indeed given this order to deploy away from the aerodrome, and the next day the forward echelons on Nos 2 and 13 Squadrons had 100 per cent attendance at parades.[16] Airframe fitter LAC Reg

14 Flt Lt R J Dorsett, 27653. SHQ Darwin, No 2 Sqn, SHQ Pearce, 73 OBU, 1st TAF. Radio mechanic of Murray Bridge, SA; b Adelaide 4 Apr 1917.

15 "Raid on Darwin, 19 Feb 1942", notes by R J Dorsett, RAAF Hudson Squadrons Assoc, Adelaide.

16 Gillison, pp.429–30.

Daymon[17] recalls two Hudsons then being despatched to strike the Japanese carrier force,[18] but it was most probably fortunate that they couldn't find it.

The devastation caused by the Japanese air raids on Darwin resulted in dispersal of these meagre air assets, separating operational and administrative control.[19] As an outcome of the first raids on Darwin, steps were taken to disperse the area and station headquarters from the aerodrome. Fortunately there had been some foresight with dispersing the operations centre before the raid. Signals equipment had been installed at the new site, with a communications centre set up in the bush, and lines installed to the "11 Mile" transmitting station.[20] Telephone lines were also laid to the Hudson dispersals at the base.

The two war-worn Hudson squadrons were also dispersed. Nos 2 and 13 Squadrons moved to Daly Waters, 500 kilometres to the south-east. One of the 2 Squadron groundcrew, engine fitter LAC Murry Lawson,[21] remembered the move:

> On the day after the first air raid on Darwin, that is February 20th 1942, three Hudsons were available to transport ground staff to Daly Waters which was being prepared as a servicing facility for the Hudsons of 2 and 13 Squadrons.
>
> One aircraft made the trip without incident,[22] another (A16-142)[23] overflew Daly Waters and crash landed in Lake Woods. The third aircraft (possibly A16-18), flown by Ivan Black[24] ... forced landed at Katherine at midnight.
>
> We were about to board the aircraft at about dusk when an air raid red alarm sounded. Needless to say we disembarked very rapidly and headed for the scrub, which in those days closely bordered the runway. Happily it was a false alarm but by the time the plane was reloaded, with 24 of us on board, it was quite dark.
>
> The pilot had given instructions about pre-arranged signals, one of which was for a designated person to move aft and the other signal was for someone to move forward to trim the aircraft as needed.
>
> We took off and the wheels were retracted before we were fully airborne so we had a momentary slight sinking feeling. No sooner had we become airborne than we were picked up by searchlight. This was a very eerie feeling for most of us groundies who had not previously experienced it. We hoped that we gave the correct recognition signal and that the AA gunners weren't trigger-happy—everyone at that time was well keyed-up.

---

17 Sgt R Daymon, 10879. No 2 Sqn, 55 OBU, 1 OTU, GR Sch. Auto electrician of Melbourne; b Melbourne 21 Jan 1920.

18 Reg Daymon, taped recollections, of Jun 1993. No 2 Squadron Association, Melbourne.

19 Gillison, p.463.

20 Dorsett notes.

21 Sgt M A Lawson, 26325. No 2 Sqn, 1 OTU, 1FBRD. Motor mechanic of Adelaide; b Adelaide 7 Oct 1920.

22 This Hudson was flown by Sqn Ldr Ryland. After an electrical storm, they circled a military convoy and were attempting to land when "some dim lights which could be the kerosene flare lights at Daly Waters had been sighted. This turned out to be the case and we landed safely with only a few minutes fuel left". Richard Taylor, p.22.

23 A16-142 was flown by Flg Off Norm Lamb. AirCdre N F Lamb, telephone interview with author, of 13 Dec 1993.

24 Flt Lt I O Black DFC, 248. Nos 4, 6, 1, 13 & 6 Sqns, 1 OTU, 9 OG, 21 Sqn. Chemist of Melbourne; b Melbourne 12 Jun 1917.

Shortly after heading south we ran into a tropical storm which gave us a hectic ride and the frightening experience of seeing lightning run along the mainplane and flick off the wing tip. If this wasn't enough, we were lost. I saw the moon first on one side then the other so I realised we were flying in circles.

Some ground lights appeared and Jim used the Aldis signal lamp to let someone know that we were in trouble. Shortly afterwards a vehicle was seen to move away from the stationary lights, drive some distance then stop with the headlights full on. The pilot took this to indicate a landing area (which it was) and made four cautious approaches before touching down. It was one of the locals at Katherine who had the nous to use his light truck to indicate the landing strip. I have read elsewhere that it was the Americans, but this is not so.

We slept under the mainplane that night and in the morning we realised what a wonderful job the pilot had done. The unmade strip was bordered on one side by the Katherine River and on the other by rocky outcrops, and the strip itself was not all that long.

As the fuel supply was dubious, the crew and a couple of others left in the morning, again heading for Daly Waters, promising that we would be picked up later.[25]

The Sergeants' Mess, RAAF Darwin A McEGAN

The following day, the stranded party was picked up by two DH-84 Dragons and finally made it to Daly Waters. At Daly Waters, under uncomfortable and unsuitable conditions, ground staff carried out major servicing and rectification, maintaining only a small operations and servicing team with the aircrew in Darwin. Conditions at Daly Waters were primitive and unhygienic, and spares and equipment scarce. LAC Reg Daymon summed up Daly Waters: "No water to drink, wear gaiters for

25 "Two Flights From Darwin", Murry Lawson, *RAAF Hudson Squadrons Association Newsletter*, No 20, Feb 1993, Adelaide, p.4.

protection against death adders, fight the flies for your food".[26] These conditions endured until roads, adequate accommodation and sanitation improved, but the site remained unsatisfactory.[27]

An attempt had also been made on 20 February to rescue the stranded rear party at Penfoei. Flight Lieutenant Arch Dunne set out at dawn to arrive in Timor mid morning.[28] He recalled:

> As I neared the coast I saw a destroyer and evidence of a landing in process. There was little cloud and I couldn't see how I could land at Koepang. I moved to skirt the low hills about 800 feet high on Roti Island at the tip of Timor. We two pilots were instinctively counting the vessels engaged in the landing, when suddenly out of the corner of my eye I saw we were right over an aircraft carrier with aircraft lined up on its deck.[29]

With all the forward Allied bases around the Banda Sea now lost, Darwin had become the front line. To reinforce the Darwin units No 8 Squadron, forced back from Malaya and Singapore to Java, had been instructed on 20 February to hand over its remaining Hudsons and some of its crews to No 1 Squadron and to disband. Ten crews were subsequently flown to Darwin to join Nos 2 and 13 Squadrons.[30]

Intelligence reports were now pointing towards an imminent invasion. With the few Hudsons available, it was decided to continue armed reconnaissance of Koepang and Dili searching for enemy convoys. Aircrew were briefed to use their discretion as to how far to press their surveillance towards the fighter-protected port of Koepang. On clear days high flying was ordered, the pilots approaching only close enough to detect the presence of ships; close approach would be made only if cloud cover was available.[31]

On 28 February, NW Area Headquarters was moved to a site in the bush about three kilometres south of the base, and on 3 March the administrative staffs moved to Birdum, between Katherine and Daly Waters, to establish a rear headquarters. The next day, Area operations headquarters moved again, further south to a camp at "22 Mile",[32] just as Darwin was raided again, this time by six Zeros. A No 2 Squadron Hudson (A16-65), about to take off, was destroyed, but with no casualties. Raids continued on Darwin throughout March, in addition to attacks on Katherine, Broome, Derby and Wyndham.[33] Indeed, bombing raids would continue on Australia's north until the end of the following year.[34]

---

26 Daymon tape. No 2 Squadron Association, Melbourne.

27 Gillison, p.469.

28 Arch Dunne, telephone interview with author, of 1 Nov 1993.

29 Grantham, p.61.

30 Gillison, p.437.

31 Bladin's Report on NW Area, numbered "734", RAAF Historical Section, of Jan 1946, para.5.

32 Gillison, pp.468–9.

33 No 13 Sqn A51, of Mar 1942.

34 The last raid (number 64), on 12 Nov 1943, attacked Parap, Adelaide River and Batchelor. Alford, p.79.

CHAPTER 15

BANDA SEA

No 2 Squadron, Royal Australian Air Force, is cited for outstanding performance of duty in action during the period from April 18, 1942, to August 25, 1942. Operating from bases in northwestern Australia, this squadron, equipped with airplanes highly vulnerable to enemy fighter airplanes, made repeated attacks on enemy shipping, airdromes, troops, and installations on and near Timor, Amboina, and other islands in the Banda Sea, inflicting heavy damage to enemy material and causing numerous casualties. It successfully maintained continuous long-range reconnaissance over the waters to the northwest of Australia. The courage of its members in combat and the high morale of the unit under hazardous combat situations contributed greatly to the success of operations in the area.[1]

With the 2 Squadron party still stranded on Timor during March 1942, a Hudson captained by Flying Officer Ian McCombe[2] made the first, and unsuccessful, attempt to drop supplies. The crew then reconnoitred Penfoei, reporting 36 enemy aircraft lined up on the aerodrome; it had not taken the Japanese long to repair the damage from the AIF demolition work. On the morning of 18 March, as none of the expected heavy bombers had arrived in northern Australia, six Hudsons bombed Penfoei, starting fires and encountering heavy anti-aircraft defences and fighter interception.[3] The aircraft flown by Flying Officer John Venn[4] (A16-100) was damaged, and with only one good engine and fuel exhausted, he managed to ditch off Cape Fourcroy on Bathurst Island.

Over 19 and 20 March, Hudsons reconnoitred Saumlaki, on the south of Jamdena, the main island in the Tanimbar Group (500 kilometres directly north of Darwin), bombing there on the second mission. The following night seven aircraft flew against Penfoei airfield, with one 13 Squadron crew, captained by Arch Dunne, seeing its bombs hit the runway and starting fires in the barracks area. Two days later, 23 March, on a sortie over Koepang, a Hudson crew claimed the first Zero to be shot down over Timor. Flight Lieutenant Jack MacAlister[5] (A16-109) had just successfully bombed a flying boat at Koepang, as 15 enemy aircraft were seen taking off to intercept him.[6] He was attacked 10 kilometres off the coast and as he attempted to

1 US Presidential Citation, made on the recommendation of General MacArthur, 1942.

2 Sqn Ldr I A McCombe DFC, 250813. Nos 23 & 2 Sqns, 380 Gp USAAF, 24 Sqn. Farmer of Casterton, Vic; b Casterton 7 Nov 1912.

3 Gillison, p.557.

4 Flg Off J H Venn, 406063. Nos 7 & 2 Sqns. Insurance clerk of Perth; b Dardanup, WA, 29 Sep 1921. Killed in action 13 May 1942.

5 Flt Lt J L MacAlister MBE DFC, 390. Nos 25 & 6 Sqns, HQ SA, 2 Sqn. Commercial pilot of Melbourne; b Melbourne 28 Dec 1917. POW 24 Mar 1942.

6 Citation for the award of MacAlister's DFC, RAAF Personnel Records, Canberra.

fight his way to safety, his crew succeeded in downing a Zero. However A16-109 had been hit in the starboard engine, which burst into flames and with one gunner dead in the rear turret, the others attempted to bale out.[7] MacAlister was the only one to successfully escape; the other crew members were killed as their parachutes streamed. The aircraft crashed near Oepoela,[8] and MacAlister landed two kilometres away in a tree, and was strafed by the Zeros. He was found by natives and taken to the nearby village of Tjamplong, where he was captured.

During his imprisonment, MacAlister tried several daring escape attempts by trying to capture enemy aircraft. In the first, an enemy DC-3 transport[9] had landed at an airfield near to the prison camp on Timor. MacAlister had flown this type before the war so, having reconnoitered outside the wire that evening, the following night he led out seven other fellow prisoners for their flight to freedom. As his army colleagues removed engine covers and dechocked the wheels, MacAlister broke a cabin window to open the overwing escape hatch and gain entry. The attempt was thwarted—MacAlister could not get any power as the battery had been removed. The group returned to camp before daybreak, in time for a surprise roll call. MacAlister was soon moved to Changi prison in Singapore, where he began to hatch another plan. This involved the cooperation of local Chinese, but the courier of messages between the conspirators betrayed them to the enemy. MacAlister and his fellow plotters were court martialled, and he received the most rigorous torture at the hands of the Japanese Military Police. After nine months of solitary confinement, under the most degrading conditions, he was finally removed to hospital "more dead than alive".[10] MacAlister miraculously survived captivity, being released on 19 August 1945 and disembarking in Melbourne a month later. For his exemplary behaviour as a POW he was awarded the MBE, to add to the DFC he had won on his last mission.

With the fall of the Philippines, General Douglas MacArthur, of the United States Army, arrived in Australia and was appointed the Supreme Commander of all Allied forces in the South-West Pacific Area. SWPA encompassed the Philippines, the South China Sea bordering French Indo-China and Malaya, the NEI, New Guinea and Australia. Within SWPA, the Allied Air Forces Headquarters were soon established in Melbourne under USAAF Lieutenant-General George Brett, with the RAAF's Air Vice-Marshal Bostock[11] as Chief of Staff. All Allied air power in SWPA—American, Australian and Dutch—was centralised under Brett's command, with operational control exercised through the existing RAAF Area organisation.

---

7 Jack MacAlister, telephone interview with author, of 11 Nov 1993.

8 Crew: Sgts J G Wright (405530), D G J Horsburgh (11640), J R Maddern (401824).

9 20 DC-3s had been delivered to Japan between 1937 and 1939. In addition, the Japanese copied the type, and produced nearly 500 as the L2D, nicknamed *Tabby*.

10 Citation for the award of MacAlister's MBE, RAAF Personnel Records.

11 AVM W D Bostock CB DSO OBE. 1st AIF; 48 Sqn RFC; D Ops & Int RAAF HQ; DCAS 1939–41; CoS AAF SWPA 1942; AOC RAAF Cd 1942–46. MHR 1949–58. Regular air force offr of Melbourne; b Sydney 5 Feb 1892.

No 2 Squadron aircrew, March 1942:
Front, on ground: Flg Offs Pinn, Dolphin, Sutherland, Robertson, Erikson, McKenzie. Second row, seated: Plt Offs Davidson, Coglan, Flt Lts Gilletic, Overheu, McCombe, Wg Cdr McFarlane (CO), Flg Offs Watson, James, Cambridge, Morgan (Defence Officer), Rigg. Third row: "Doc" Hall, Intell Officer, W Off Grossman, Sgt Firth, Flt Sgt Stevenson, Sgt Breckenridge, Flt Sgt McDonald, Sgts Smyth, Kealey, Francis, Frazer, Lang, Strambini, Flt Sgt Hawkesford, Ops Officer, Plt Off Dunning. Fourth row: Sgt Keen, Barnes, Bundy, Adamson, Ryan, Duff, Kenihan, O'Shannesy, McLean, Manin, Cooper. Back row: Flt Sgt Allen, Sgt Laity, Flt Sgt Oldfield, Sgts Lindsey, Smith, Keech, Childs, Rhodes, Thame, Gunn, Connor
A McEGAN

As the Japanese were now preparing to thrust through New Guinea, the limited strike aircraft available in NW Area were directed to harass the occupying forces on the islands around the Banda Sea. This was to become the primary role of No 2 Squadron in the air campaign throughout 1942, and to some extent for the remainder of the war, as NW Area protected the flank of MacArthur's main concentrations in New Guinea. In addition to its patrolling, targets selected for 2 Squadron's attention were airfields, military installations and shipping. On one such shipping attack on 27 March, three Hudsons attacked Tenau Harbour, five kilometres south-west of Koepang, scoring a direct hit on a 7,000 ton ship.

On 30 March, four Hudsons, led by Flight Lieutenant Bob Dalkin of 13 Squadron, destroyed at least six grounded aircraft at Penfoei, damaged four flying boats moored in Koepang Harbour, scored two near misses on ships and damaged shore installations. Bob Dalkin, who was awarded the DFC for his leadership in this raid, later recorded:

> Coming in low and fast ... we found the whole of Koepang airfield lit up ... with taxiway flares, runway flarepath, and aircraft with navigation lights switched on.
>
> There were many aircraft on the ground ... with some flying in the circuit area ... selected some parked and taxying aircraft as the aiming point, got the bombs away, realised that I was too low ... and with ... the explosions felt that awful thump ...
>
> Not quite knowing whether we were still in one piece, one quick turn to give the crew a chance with their .303s, and then it was low level ... to the nearby flying boat base at Hansisi. There were some aircraft on the water, and these were attacked with gunnery ... the rear gunner shouted that the [anti-aircraft gun] fire was passing between the forward stub aerial and the ... fins ... I came very close to digging a wing into the sea. It was quite a night![12]

The attack on the aerodrome continued on the first day of April with two Hudsons, one captained by Flying Officer Arthur Sharp[13] and the other from 13 Squadron, claiming six aircraft destroyed and another six damaged. The mission was repeated on the night of 3–4 April, with 13 Squadron scoring four aircraft destroyed and two damaged.[14] For these sorties, an advanced operational base (AOB) was being used near Drysdale Mission, on the Western Australian coast, halfway between Broome and Darwin. Crews operating from there would maintain radio silence, and attack at first or last light, or in moonlight to reduce the risk of encountering enemy fighters.[15]

These sorties were not without loss as they continued through April. On the evening of the 13th, Flight Lieutenant Harold Cook led two Hudsons to attack shipping off Koepang. His wingman, Flying Officer Roger Blanchard[16] of 13 Squadron, managed to shoot down a Zero, and reported a large ship of 7,000 tons damaged

---

12 Alford, p.32.

13 Sqn Ldr A J Sharp DFC, 402140. Nos 7, 2 & 15 Sqns. Farmer and grazier of Forbes, NSW; b Sydney 21 Feb 1916. Killed in flying accident 11 Mar 1945.

14 Attack Report A.108(B) No 14, of 3 Apr 1942, RAAF Historical Section, Canberra.

15 Gillison, p.558.

16 Flt Lt R J Blanchard, 260776. No 13 Sqn. Of Sydney; b Sydney 22 Mar 1916. Killed in action 23 Apr 1942.

when he returned alone to Darwin.[17] Cook's aircraft (A16-159) had been shot down and crashed in the sea near the northern shore of Koepang Bay. As previously related, Cook and Leithhead made the trek to the northern side of the island, leaving the wounded gunners in the care of natives. After his rescue Cook was able to report his attack at Koepang:

> I did a low power glide approaching from the protection of a cloud bank towards an 8 to 10,000 ton vessel anchored there. At that moment I saw enemy Zero fighters in the air and retracted the bomb doors and started to turn away.
>
> I then realised that I had not time to get to cloud cover before the enemy reached me so I decided to do my approach run over the target but just as I opened the bomb doors again three Zeros tried to head me off. I continued the run however and dropped a stick from 800 feet two of which definitely hit the target. The last I saw of F/O Blanchard's aircraft was about 3000 feet away with a Zero attacking it. I went down to almost water level and tried avoiding action. The three Zeros followed and we soon began to be badly shot up despite very heavy fire by my two gunners and my second pilot P/O Leithhead. A cannon shell burst inside the cockpit wounding me in the right arm and putting it temporarily out of action.
>
> The enemy pilots adopted a clever system of attack. Two were astern and kept up an alternate attack on my blind side. They would get in a burst then turn away and repeat the attack. The third a/c was trying to get me by attacks from the starboard and port bow alternately.
>
> One enemy aircraft astern of us went down in flames. P/O Leithhead told me that he distinctly saw his tracer bullets enter the belly of the plane. Sgt Hearle also stated that bullets from his turret guns also entered the enemy a/c.
>
> As soon as the third pilot saw his compatriot out of the fight, he swung in to take his place in the attack astern of us. He used his cannon well and one burst put the turret guns out of action, severely wounding Sgt Hearle in the right hand and right leg. Sgt Witham was also badly wounded. A bullet went into his cheek and lodged between the muscle of the chin, rendering him unconscious.
>
> I sneaked along the beach and then turned in across the land and finally came out again over Koepang Bay. All this time—about 30 minutes in all—we were being shot at by the enemy a/c.
>
> The aircraft became more and more difficult to handle and the controls had been badly shot up. Then the starboard engine started to lose power. I did my utmost to coax her and began to feel her respond when another burst from the Zeros put my port engine out of action. An instant later it burst into flames.
>
> We were then 80 to 100 feet off the water. There was nothing else to do but pull both throttles back and flop onto the water. We crash alighted about 1 mile west of Pariti in about 3 feet of water and practically low tide.
>
> The Zeros circled around us a couple of times but did not molest us again. I believe they thought we were all dead.[18]

Ten days later, on 23 April, a No 2 Squadron Hudson A16-182, flown by 13 Squadron crew,[19] departed from Darwin on a night reconnaissance and bombing

17 Attack Report No 37, of 13 Apr 1942.

18 HQ NWA Interviews Book 1, Item 80, undated. RAAF Historical Section, Canberra.

19 13 Sqn Crew: Flg Off R J Blanchard, Sgts D Jacobs, D V Woods, O B P Gray.

mission to Koepang and was never heard of again. This was not the last time such a fate would befall crews on these lonely solo patrols. This would prompt the Squadron to raise with higher headquarters the need for formation patrols to be flown.[20]

Just as No 2 Squadron's harassment was constantly disrupting the enemy's activities from the air, so too were Australian commandos active on the ground. The Independent Company, the sole active survivors of Sparrow Force, had now resumed the role as envisaged for this type of commando group—that of staying behind, living off the land or provisioned by air to "be a thorn in the flesh of the occupying enemy, emerging in true guerilla style to attack vital points and then disappear again into the jungle".[21] For the remainder of the year, until the commandos were eventually taken off Timor by the RAN, the Japanese were unable to prevent supplies being brought in by air and sea. The first Hudson drops to Sparrow Force were on 24 April, and the first Catalina flying boat resupply on 24 May.[22] Support for Sparrow Force would remain a task of 2 Squadron for the duration of the guerilla campaign.

In late April, Squadron Leader Tich McFarlane assumed command of the Squadron. This was intended as a temporary appointment as Squadron Leader James McGilvray[23] had been posted to take over from Headlam. However, he was killed at No 1 OTU during his refresher training at Sale, and McFarlane remained in command until October. One of the first changes McFarlane implemented was instilling more discipline in the operations of the unit, devoting more training to formation flying and emphasising the importance of accurate gunnery. He demanded very tightly knit formations, capable of penetrating any weather, day or night, and at very low level, to offer the maximum in mutual defence and firepower. McFarlane was also to exercise operational control over both 2 and 13 Squadrons, as the CO of 13, Squadron Leader John Ryland, was supervising the completion of the new base for his squadron at Hughes.[24]

By this stage, the Squadron had again moved forward to Darwin, and was camped outside the RAAF base beside the main north–south road. This was clearly unsatisfactory, as personnel could not get sleep between their work shifts. McFarlane had delegated all administrative and domestic matters to his adjutant, Flight Lieutenant Hughie Hunt,[25] so that he could concentrate on the operational aspects of running the unit—the aircrew's operations and aircraft servicing. A week later, Hunt organised for Squadron personnel to move onto the base and occupy the empty

20 After a later sortie, the Squadron reported: "This operation, it is suggested, is further justification for the contention that these patrols should be undertaken by at least two aircraft in company." No 2 Sqn Air Reconnaissance Report No 60, of 12 Jun 1942.

21 F Spencer Chapman, *The Jungle is Neutral*, Chatto & Windus, London, 1949, p.8, cited in Wray, p.14.

22 Wray, pp.99, 106.

23 Sqn Ldr J W McGilvray, 232. Nos 2 & 7 Sqns, 1 OTU. Regular air force offr of Brisbane; b Brisbane 11 Jun 1910. Killed in flying accident 9 Jun 1942.

24 Tich McFarlane, interview with author, of 1 Nov 1993.

25 Flt Lt H Hunt, 272784. Nos 6 & 2 Sqns, SHQ Rathmines, 3WAGS. Insurance broker of Sydney; b 7 Aug 1901.

married quarter accommodation. Even though Darwin was still subjected to constant enemy air raids, this was considered preferable to the unsatisfactory conditions which had been endured at Daly Waters. However, there was concern with now having so many personnel quartered on the enemy's prime target of the RAAF base, so ultimately the men were dispersed, a few kilometres away, to a new site at Rapid Creek. Again, at Rapid Creek, there was to be a high incidence of illness, mostly malaria, dengue and tropical ulcers. Furthermore, the new camp was to prove even more unsuitable, as it had been built *in line* with the decoy "dummy" strip on the aerodrome,[26] certainly inviting any stray bombs!

The Air-Officer-Commanding (AOC) NW Area, Air Commodore Francis "Dad" Bladin,[27] directed his Hudson crews to continue reconnoitring enemy ports for signs of invasion fleets. As the Japanese were receiving the setback in the Coral Sea battle, it became evident that invasion was no longer imminent. The new task was to prevent the enemy from consolidating newly won bases around the Banda Sea on Timor and Ambon. The harbour and aerodrome at Koepang were primary targets, raids being made also on Ambon and Dili, with resupply drops to the 2/2nd Independent Company continuing the fight in Portuguese Timor. These missions were conceived, planned and executed wholly within the Hudson squadrons. Long-range Hudson reconnaissance sorties were also maintained—one crew airborne for nine and a half hours reached Soembawa Island, west of Flores. For attack missions, the standard bomb load was maintained at 1,000 pounds, despite being overloaded with auxiliary fuel tanks in the cabin. On 11 May, Squadron Leader McFarlane carried out a daring daylight reconnaissance of the major enemy Banda Sea bases. He reported a destroyer, a possible light cruiser and heavy merchant vessels at Ambon. At the airfields he observed:

LAHA	Two new runways in course of construction and a group of new white huts adjacent to new N–S runway being built.
NAMLEA	Aerodrome in exactly the same condition as when abandoned. Bomb craters still in runway. Buildings in the same condition. No sign of any activity.[28]

Following this reconnaissance, two days later on 13 May, McFarlane led eight aircraft off in the early afternoon for a dusk shipping attack in Ambon Harbour.[29] Approaching the target, McFarlane split his formation into two flights; he led four other aircraft flown by Flight Lieutenant Simon Fraser,[30] Flying Officers McCombe and Sharp, and Pilot Officer Venn, and achieved surprise by flying low through the

26 Dorsett notes.

27 AVM F M Bladin CB CBE. D Ops & Int; AOC SA 1941–42; AOC NWA 1942–43; SASO 38 Gp RAF 1943–44; DCAS 1945; CoS BCOF Japan 1946–47; AOC EA 1948–49. Regular air force offr of Melbourne; b Korumburra, Vic, 26 Aug 1898.

28 Aircraft Reconnaissance Report No 40, of 11 May 1942. RAAF Historical Section.

29 Gillison, p.560.

30 Sqn Ldr S J Fraser DFC, 524. Nos 7 & 2 Sqns. Stockbroker of Melbourne; b Melbourne 10 Dec 1914. Killed in ground accident 11 Oct 1942.

hills towards the town. Attacking from the south-east, the five Hudsons were in line abreast and each chose a target in Ambon Harbour.[31] Pilot Officer John Venn (A16-196) and his crew made a mast height attack on a 3,000 ton vessel in the face of heavy anti-aircraft fire. As his bombs impacted, his aircraft was seen to explode and fall in two pieces into the sea.[32] One crewman, Sergeant Leonard Omsby,[33] miraculously escaped from the crash, only to be executed by the Japanese.[34]

The second flight—flown by Flight Lieutenant Bob Dalkin, Flying Officer John Alcock[35] and Pilot Officer Gordon Jaques[36]—then swept in from the east at 1,000 feet.[37] They dive-bombed their targets, and Dalkin's aircraft was hit by groundfire, which holed his fuel tank. He barely made it back to base. Despite the loss of Venn's crew, the strike was considered a success claiming a 3,000 ton vessel sunk, a 2,000 ton vessel hit, and a near miss on a 5,000-tonner.

An attack on Japanese shipping at Ambon by nine Hudsons, 13 May 1942 RAAF

McFarlane led a second shipping attack back to Halong Bay on 22 May, with eight aircraft from both the Hudson squadrons. They attacked a destroyer and four merchant vessels in the face of a heavy anti-aircraft barrage from all the ships.[38] Five Zeros attacked immediately, but two 2 Squadron Hudsons, which had hesitated in their attacks, had lost formation. McFarlane reduced speed markedly to allow his scattered force to move into close formation. Pursued by two Zeros and with one engine in flames, Flying Officer Graham Allchin[39] and Sergeant Laurence Montgomerie[40] were able to bale out as their stricken aircraft (A16-187) headed towards Ambon, but the other two crewmen

31 Load for each Hudson was four 250 lb General Purpose (GP) bombs, fused with 11 seconds delay.

32 In addition to Venn, crew also killed were: Sgts W J James (411496), B A Kilpatrick (408552). Venn was remembered by the naming of Venn airstrip near Katherine.

33 Sgt L A Omsby, 407871. No 2 Sqn. Compositor of Adelaide; b Adelaide 23 Sep 1915. Executed by Japanese 13 May 1942.

34 RAAF Historical C.D.102/45 A16-196, of 21 Jan 1948 and undated.

35 Flg Off J E Alcock, 408056. Nos 7, 2 & 13 Sqns. Bank clerk of Launceston, Tas; b Deloraine, Tas, 30 Jan 1918. Killed in action 1 Oct 1942.

36 Flt Lt G A F Jaques, 260806. Nos 7, 6 & 13 Sqns, 65th Sqn USAAF. Articled clerk of Sydney; b Sydney 7 Aug 1921.

37 These aircraft each carried four 250 lb semi-armour piercing (SAP) bombs, to ensure penetration, with one-second delay fuses.

38 Aircraft Reconnaissance Report No 49, of 22 May 1942.

39 Flg Off G W C Allchin, 401321. No 2 Sqn. Of Melbourne; b Ballarat, Vic, 28 Aug 1916. Executed by Japanese 22 Nov 1942.

40 Sgt L J Montgomerie, 406865. No 2 Sqn. Of Perth; b Marble Bar, WA, 7 Mar 1915. Executed by Japanese 22 Nov 1942.

perished in the crash.[41] (The two survivors were captured by the Japanese, and subsequently executed six months later.) A second aircraft (A16-174), flown by a 13 Squadron crew under Flying Officer Philip Brooks,[42] was shot down in flames, crashing into the bay with no survivors.

CO, Wing Commander "Tich" McFarlane
A MCFARLANE

The sudden change in speed initially confused the attacking Zeros, and the defensive ring of fire from the "V" formation prevented further losses. The formation was by then flying very low, with the lead aircraft at 50 feet, thus protecting their bellies and adding to the difficulties of the enemy. Even at this extremely low level, the Zeros still tried to keep below and behind the Hudsons, avoiding fire from the turrets, and only daring to shoot from long range. Flying Officer Dick Overheu[43] reported that enemy fighters had learnt to:

> ... keep away from arc of fire of turret guns and out of range. Enemy closed to about 1000 yards, fired a burst and then broke away.[44]

Flying Officer Arthur Sharp's aircraft was attacked from below and the tunnel gunner, Sergeant Martin McDonald,[45] returned fire. The Zero abruptly disengaged, pulling up to expose its belly, allowing the turret gunner, Sergeant Ernie Keith,[46] to give it a burst, sending it down in flames. The Zeros also attempted some head-on passes, but after 25 minutes of combat, another two of them had been hit by the formation's defensive fire, so they broke off combat, trailing black smoke.[47]

The enemy fighter pilots were reluctant to close on the Hudsons because the field of fire was now formidable. McFarlane maintained close formation at very low height, and the combined firepower of the turret and tunnel guns achieved the desired

41 Crew: Plt Offs K I Kersten (406857), W C McLay (408865).

42 Flg Off P G Brooks, 407219. Nos 7, 6 & 13 Sqns. Bank clerk of Cootamundra, NSW; b Adelaide 26 Aug 1917. Killed in action 22 May 1942. Crew: Plt Off A W Moss, Sgts P C Roll, C A Vercoe.

43 Sqn Ldr R F Overheu DFC*, 406115. Nos 6, 13 & 2 Sqns, 529th Sqn USAAF, 24 Sqn. Farmer of Perth; b Pingelly, WA, 10 Jul 1920.

44 Combat (Bomber) Report A.108(D) No 14, of 22 May 1942.

45 Flt Lt M A McDonald, 401702. Nos 7 & 2 Sqns, 1 OTU, 67 Sqn, 9 OG, HQ NWA, 24 Sqn. Mechanic of Rockbank, Vic; b Melton, Vic, 10 Dec 1912.

46 Flt Lt E A Keith, 406808. Nos 7 & 2 Sqns, 1 OTU, 2 & 18 Sqns. Carpenter of Perth; b Black Range, WA, 8 Jun 1915.

47 Combat (Bomber) Report A.108(D) Nos 9 to 13, of 22 May 1942.

The Adjutant, Flight Lieutenant Hugh Hunt
A McFarlane

defence. The newer Hudsons, Mark III and Mark IV aircraft delivered from the beginning of the year, were fitted with a retractable aft-firing tunnel gun, lowered by a handle and a bicycle type chain, under the belly. In May 1942, the policy for Hudson aircrew had been changed; air observers replaced second pilots, who were to be given conversion and navigation courses to become captains.[48] As a fifth crew member was required to man the ventral position, the crew composition became pilot, observer, and three air gunners. This simple modification of the tunnel gun in the belly had eliminated the dangerous blind spot, reducing the Hudson's vulnerability, and had raised the morale among air-gunners.[49] One who wore the "AG" wing was Sergeant Murray Childs,[50] who remembered the role of the air gunner was:

> ... mainly to deter enemy fighters rather than to shoot them down. I think the most intelligent gunners all realised that the protection of a bomber was far more important than to shoot down a fighter of much less value to the war effort ... One tactic used by a crew I knew was to suddenly reduce speed, as arranged by signals between pilot and gunner. This would cause the fighter to over-shoot, thus disrupting the fighter pilot's planned mode of attack while presenting himself to unexpected danger from the air-gunners. I suspect that cool nerve was needed also by our boys![51]

Early June brought victory for the Allies at Midway, and this was to transform the balance of power in the Pacific in favour of the United States Navy. However, in NW Area, as Intelligence reports suggested that the Japanese intended to occupy Dobo, in the Aroe Islands (towards Dutch New Guinea), the Squadron maintained its reconnaissance of the Banda and Arafura Seas, in addition to carrying the war to the enemy with attacks on Timor. But aircraft were still scarce, and Bladin expressed his frustration to Allied Air Force Headquarters concerning the lack of resources NW Area was receiving. Bladin had been advised in late May that his squadrons were low priority to receive new Hudsons, so he made the point that the Hudson was not a good bomber, carrying only a small load, and his few aircraft were limiting his strike capability. His frustration showed with his closing statement:

> When suitable bombardment aircraft become available, Hudson aircraft will be employed in their correct reconnaissance role.[52]

---

48 No 13 Sqn A51, of 11 May 1942.

49 Murray Childs, "Hudson Gunners and their Equipment", *RAAF Hudson Squadrons Association Newsletter,* No 19, 1992, Adelaide, p.3.

50 Flt Lt M C Childs, 407386. Nos 8, 2, 13 & 2 Sqns, 1 OTU, Pac Ferry Unit, 24 Sqn. Clerk of Adelaide; b Penola, SA, 9 May 1922.

51 Childs, p.3.

52 AOC HQ NWA letter to HQAAF, "Operational Employment of No 2 and 13 Squadrons" 7/70 AIR, of 3 Jun 1942.

The wreckage of Hudson A16-176, which crashed on take-off on 7 May 1942 A McEGAN

The remains of A16-175, photographed in March 1989 RAAF

This aircraft situation was apparent on 16 June, when Squadron Leader McFarlane led seven Hudsons against Penfoei. It had been intended to make this a maximum effort raid with four flights of four, but unserviceabilities rendered this impossible. Approaching the area it was difficult to distinguish the target clearly, but the position was soon identified by anti-aircraft fire.[53] The runway was successfully bombed, but near the target Arthur Sharp's aircraft (A16-175) developed engine trouble, and became separated from the formation. He eventually limped home and reached friendly shores, ditching in Hyland Bay, about seven kilometres from Port Keats. Years later, the wreckage of A16-175 was rediscovered by an overflying RAAF F-111.[54]

Bombing also continued to be concentrated on Dili in support of Sparrow Force, by attacking buildings, barracks and shipping at Dili and Atamua. Five Hudsons, led by Flight Lieutenant Simon Fraser, attacked barracks at Dili on the night of 1 July. The

53 Aircraft Reconnaissance Report No 61, of 16 Jun 1942.

54 "Mystery of Hyland Bay Hudson Solved", *RAAF News*, Vol 20, No 5, Canberra, Jun 1979, p.5.

problem, often encountered, of trying to locate the target at night was accentuated when the briefing was indefinite on the exact target location. Fraser commented:

> It is disconcerting to crews to have to make attacks of this nature where the location and description of the target is so extremely vague. Effective attacks upon any secondary target suffer accordingly.[55]

Better targeting information was available at dusk on 9 July, when McFarlane led six Hudsons to attack barracks at Tibesse, south-east of Dili. McFarlane flew Group Captain William Hely,[56] Senior Air Staff Officer for NW Area, on this mission as the turret gunner. Although two aircraft were hit by groundfire, the CO was able to report:

> Numbers of buildings and huts were seen to be hit. No fires were observed. Our aircraft returned to the target area, diving to 200 feet and machine gunned the buildings in the area, including the barracks. 3300 rounds were fired.[57]

The dangers of these sorties were being exacerbated by the high weights at which the aircraft were now operating. The normal all up weight (AUW) of the Hudson, with four 250 lb bombs was 20,614 pounds (9,203 kg). However, with the third gunner now being carried, two 100 gallon auxiliary fuel tanks, and overloading with six 250 lb bombs, strike sorties to Ambon were departing at 22,877 pounds (10,213 kg) AUW. The hazard of operating at these weights was that the Mark IV Hudson had no single-engined performance beyond 20,600 pounds (9,196 kg), but this was accepted as a risk of war.[58]

Several missions were also flown as "demonstrations of force", the first over Saumlaki led by Flight Lieutenant Fraser (A16-200) on 12 July, another led by the CO (A16-200) over Toeal the following day. These flights over the newly acquired Japanese strongholds were attempting to show the Allied resolve in carrying the war to the enemy. Reconnaissance patrols were also maintained, and on an afternoon sortie over Koepang on 17 July, Flying Officer Ian Hay[59] (A16-234) was intercepted, just short of the coast, by two Zeros approaching from astern. The navigator, Pilot Officer Henry Carter[60] manning the starboard side gun, directed the belly gunner, Sergeant Frank Webster,[61] onto the first Zero:

55 Air Reconnaissance Report No 65, of 2 Jul 1942.

56 AVM W L Hely CB CBE AFC, 53/O324. Nos 1 & 3 Sqns, RAAF HQ, SASO NWA 1942–43; comd 72 Wg 1944, 84 Wg 1944–45; AOC WA 1951–53, DCAS 1953–56, AMP 1960–66. Regular air force offr of Sydney; b Wellington, NSW, 24 Aug 1909.

57 Air Reconnaissance Report No 66, of 10 Jul 1942.

58 AOC HQ NWA letter to HQAAF, of 4 Jul 1942.

59 Wg Cdr I H A Hay DFC, 250720. Nos 7 & 2 Sqns, 1 OTU; comd 15 Sqn 1944, 467 Sqn 1945–46. Timber merchant of Melbourne; b Melbourne 10 Jan 1917.

60 Flt Lt H J Carter, 405190. Nos 100 & 2 Sqns, 1 OTU, 71 Wg. Stock agent of Charleville, Qld; b Charleville 14 Jul 1907.

61 Flg Off F V Webster, 416470. 48 MG Bn, Nos 7, 2, 15 & 8 Sqns. Clerk of Adelaide; b Low Fell, County Durham, Eng, 27 Dec 1921.

> Webster brought gun to bear and directed two bursts at enemy aircraft. Second long burst entered enemy aircraft which began to smoke. Its pilot banked sharply and baled out, the aircraft breaking into flame and crashing. The other Zero which had first been seen approaching on port side from astern was then observed more than once following but was evaded in cloud as our aircraft proceeded east along coast.[62]

Hudson refuelling at Hughes AWM 27607

The damage inflicted by No 2 Squadron's repeated attacks on Penfoei had forced the enemy to increase its air defence around the airfield. This became obvious to the Australians on the night of 26 July, a night with perfect visibility and described as "clear as day". Leading seven Hudsons, Flight Lieutenant Fraser (A16-200) split his force into two flights for a coordinated attack from the east and from the north. They dropped their bombs successfully on the runway, hangars and barracks, and intense heavy anti-aircraft fire lit the sky with tracer. Red tracer was reaching their height at 14,000 feet, with smaller calibre green and white tracer at lower levels. Although it was assessed that there were now up to 10 heavy guns at the airfield, on this night no aircraft were hit.[63]

Long-range reconnaissance over the Banda Sea by Flying Officer Hay's crew (A16-234) on 29 July was again a fruitful sortie (they had destroyed the Zero twelve days previously). Just off Ambon, near Laha, a small 150 ton vessel was attacked and:

> … was observed to rise high by the bows and almost "stand up on end" but further results were not seen … Immediately bombs were released a Zero was observed on starboard quarter on same level at about 400 yards turning towards port and lower. Cloud intervened. Shortly afterwards enemy aircraft seen again lower approaching underneath at about same distance and it is thought that it gave one short burst. Belly gunner fired about 60 rounds observing tracer entering starboard side of fuselage. A little smoke was observed over Zero starboard wing as our aircraft again entered cloud.[64]

The following day, 30 July, Flying Officer Robert Muecke's[65] crew flew the same aircraft, tasked to patrol Saumlaki, in the Tanimbar Islands. The crew's track was

62 Air Reconnaissance Report No 71, of 17 Jul 1942.

63 Air Reconnaissance Report No 72, of 27 Jul 1942.

64 Air Reconnaissance Report No 73, of 29 Jul 1942.

65 Flg Off R C Muecke, 283858. No 2 Sqn. Of Adelaide; b Adelaide 11 Apr 1918. Killed in action 30 Jul 1942. Crew: Flg Off F S Moss (401991), Plt Off L M Miners (405797), Sgt J E Laman (405794).

south-west of the Kai Islands, and a Japanese broadcast later reported that a Hudson crashed into the sea off the Kai Islands after colliding with a Japanese seaplane. This was not the only loss to the Squadron that day. During an air attack on Darwin by 27 enemy bombers, which damaged the bomb and fuel dumps, Flying Officer Ron Tregonning[66] was killed instantly in a slit trench as bombs also struck the domestic site at Rapid Creek.

The last day of July saw the "A" Flight Commander, Flight Lieutenant Simon Fraser (A16-200), on reconnaissance around Banda Island, where he located an enemy *Kako*-class cruiser, of 6,000 to 8,000 tons, and two destroyers. The post-mission report summarised:

> The Cruiser immediately got under way. The destroyers remained at anchor. F/Lt Fraser made two runs to bomb but cloud interfered. However, as the cruiser got out to sea it became clearer with cloud at 6000 feet. He then made a third run just under cloud from astern. The Cruiser appeared to be travelling at at least 20 Knots and at the beginning of run, turned completely round and the attack was made from the bow 30 degrees to Port. The first bomb was seen to fall just short of ship about quarter of length of ship from bow. The other was not seen to fall in water but smoke and flame observed on deck where guns had been firing. It is considered bomb must have hit. Immediately after bombs dropped, the Float Plane ... a biplane with twin floats and a large radial engine[67] ... appeared out of clouds dead astern. Our aircraft had rated power and its speed 200 Knots. The Float Plane, however, closed rapidly to 800 yards when turret gunner fired three bursts and saw his trace entering aircraft. Evading action was taken in cloud and this aircraft was not again seen.[68]

Major-General George Kenney assumed command of Allied Air Forces SWPA in early August, with the Allied situation in New Guinea growing desperate. The enemy was establishing the Gona-Buna beachhead and advancing on Milne Bay, so Nos 2 and 13 Squadrons were ordered to be ready to move to New Guinea from NW Area if needed.[69] However, their presence in the North-Eastern Area was not ultimately required, after the decisive victory by Milne Force.

The two hard-worked Hudson squadrons continued to harass the enemy's bases in the islands around the Banda Sea and supporting the guerilla force on Timor. The need for this support of Sparrow Force was now more urgent as the enemy opened a determined offensive aimed at enveloping and destroying the Australian–Dutch force.[70] Sparrow Force continued to provide accurate information of the enemy's positions, and the Hudsons were able to air drop requested supplies and lend support by bombing Japanese-occupied towns; the Independent Company was indeed cheered by the familiar Hudsons as they overflew to strike the enemy.[71] In addition to this support, reconnaissance and attacks around the Banda Sea continued into August.

---

66 Flg Off R D J Tregonning, 401839. No 2 Sqn. Of Melbourne; b Melbourne 11 May 1914. Killed in action 30 Jul 1942.

67 This was most probably a Mitsubishi F1M2 Navy type 0 seaplane, nicknamed the "Pete".

68 Air Reconnaissance Report No 74, of 31 Jul 1942.

69 Gillison, p.609.

70 Gillison, p.643.

71 Wray, p.125.

Early in the morning of the 7th, Wing Commander McFarlane led nine Hudsons of No 2 Squadron against two transport ships escorted by an *Akikazi*-class destroyer off Suai, on Timor's south coast. The attack came from the north at 6,000 feet, and one 2,000 ton transport was left burning.

On 21 August, Flight Lieutenant Simon Fraser (A16-178) led five Hudsons out again to support Sparrow Force by attacking Maubisse, near Dili, for the second successive day. Bombs were dropped on the town and the Hudsons reconnoitred the area for enemy activity. Two Zeros attacked, and the Hudson crews soon became aware of "the ability of the enemy pilots and their obvious knowledge of the Hudson defences".[72] The Zeros set on fire the aircraft flown by Flying Officer Sid Wadey[73] (A16-209). He was able to evacuate the aircraft, but his crew were unable to escape.[74] He described the engagement and his escape from his stricken Hudson:

> When the Zero attacked from ahead, several bursts went through the instrument panel. These I observed, as in slow motion; individual holes appearing, and the panel disintegrating, with a splintered (star effect) look around the holes pointing towards me. Simultaneously, I was aware of my navigator passing me, and heading towards the body of the aircraft, when "whoosh"—flames surrounded me as the incendiaries and cannon hit the inside fuel tank.
>
> Behind the pilot's seat there is armour plating, but the tank extended a couple of inches past the vertical side of the plating, and that was where some of the projectiles went. I saw some of the bullets hit Stan Faull, the navigator, in his back as he was passing through the entrance from the cockpit into the body of the aircraft, also he would have been directly alongside the exploding tank. The other members of the crew were similarly in impossible predicaments.
>
> In order to escape from the plane it was necessary for the crew to move forward in the body of the plane to one side or the other, grab the parachute, and clip it on the harness. For the crew it was literally impossible in the intense heat and flames to find their respective (or any) parachute pack, grab it, clip it on, dash to the exit door in the back of the cabin and jettison the door, before they could jump out. For the tail gunner, his position was even more desperate. He had to swivel the turret, align it with an opening into the body of the aircraft, his only means of escape, then leap into what was a fiery furnace in order to obtain his pack. I had been protected from the direct blast of the explosion of the petrol tank by the armour plating. The sound was (Whoosh) muffled, and not at all similar to the sound of a bomb; and the actual pressure wave did not subsequently affect my hearing abilities, so the body impact was not great.
>
> As we were flying in formation, my right hand was on the throttles, and I instinctively reacted very quickly, flicked the seat belt undone, and jumped at the correct angle, toward the escape hatch in the top of the aircraft. In the process, I knocked back the throttles, and as I jumped vertically head first through the escape hatch, I was aware of being hit in the lower back by the top of the fuselage, as the slipstream forced me backward. I fell clear of the aircraft on the right side, facing forward and could see A16-209 dropping back out of the formation with flames streaming back behind like a comet tail. I looked around hoping to see other parachutes, but realised that there would not be any.

72 Air Reconnaissance Report No 83, of 22 Aug 1942.

73 Flt Lt S G Wadey, 407068. Nos 6, 14 & 2 Sqns. Accountant of Adelaide; b Adelaide 2 Apr 1918.

74 Crew: Plt Off S W Faull (401779), Sgts W R Edeson (406716), F M O'Reilly (406730), W H Gould (414224).

An SE-5A fighter, previously registered A2-4 in RAAF service, is displayed in the Australian War Memorial, Canberra, to represent Captain Henry Forrest's 2 Squadron machine, C5939. RAAF

Upon reforming at Laverton, Victoria, in 1937, No 2 Squadron was equipped with a flight of Hawker Demons. This aircraft, A1-8, is on display at the RAAF Museum, Point Cook. RAAF

A beautiful shot of Australia's only remaining airworthy Hudson. This aircraft, owned by Malcolm Long of Wangaratta, has been restored to represent KO-Y of 2 Squadron.
R Fox

The flight line at RAAF Butterworth, looking west towards the island of Penang, during 2 Squadron's time at the Malaysian base in the 1960s L HALVORSON

Canberra formation—No 2 Squadron and RNZAF aircraft (with underwing rocket pods) over the Malay peninsula, October 1966 L HALVORSON

No 2 Squadron's home in Vietnam, the Phan Rang air base, in 1967. The white buildings, right of the hill (in the centre), were 2 Squadron's domestic area. L HALVORSON

View from the cockpit—2 Squadron Canberra over South Vietnam R HOWE

Checking the bomb load for a night mission L HALVORSON

Night bombing R ALLCHIN

Bombs away J WILKINSON

A84-230 returns to base. G WALTON

Vietnam farewell—the traditional watering-down for aircrew after their last sortie was normally provided by fire truck hoses, but in this case buckets were the next best thing. R HOWE

A low-flying Canberra overflies the airfield at Biak, during survey operations in the 1970s. R OWEN

2 Squadron Canberra A84-236 over Perth RAAF

Laying up the Squadron colours, with past members of 2 Squadron, at Point Cook, November 1993
RAAF

> The formation continued along a straight flight path away from me, and they were still in perfect formation. All the other aircraft were OK. I scanned the sky for Zeros—none in sight. Decided I was now at about 1000 feet above the mountain—so pulled the ripcord—felt a jerk—looked up and saw the parachute open fully. I watched A16-209 continue its rate one turn and disappear into the valley between the mountain for which I was aiming and the adjoining mountain. The aircraft still had its comet tail of flames streaming behind it. As I saw the plane disappear, simultaneously I observed a flight of 3 Zeros, in formation in the valley below, flying low above the trees, as they emerged from behind the opposite side of the mountain below. To my surprise I landed legs together in the middle of the clearing at which I had aimed, slipped, then slid on to my behind a few yards. Looking around I found myself in the clearing, which was a very small and a fairly steep rocky slope, the open space roughly circular and about fifteen yards in diameter, and to my amazement the trees surrounding me were, of all things, Gum Trees, growing densely amid dry grass which was about three to five feet tall. I had expected jungle, not eucalypti.[75]

Sid Wadey was found, severely injured, by natives and with their assistance found his way to the neutral Portuguese administration, and finally through to Sparrow Force. His evacuation was then arranged. Meanwhile, another aircraft in the formation, flown by Flying Officer Neil Badger[76] (A16-241), became separated and was pursued by a Zero low along the valleys to the sea. The enemy fighter was making varied attacks, and then tried to head the Hudson off as it coasted out, and each time the Australians were able to evade.

> Finally the Zero made two quarter attacks from astern, level to slightly lower than our aircraft, which was then 80 feet above the water. On the second of these attacks when pulling away to starboard at 50 yards it was raked by our belly gun and crashed into the sea.[77]

Ten crews from No 2 Squadron had been lost in the campaign against the enemy in the islands surrounding the Banda Sea. Crews flew missions repeatedly back to harass the enemy and support Sparrow Force, into hostile groundfire and the nightmare of combat with Zeros. Typical of the aircrew in the campaign was WAG Flight Sergeant "Bill" Oldfield,[78] who was awarded the DFM for his service over the Banda Sea. His citation, in part, read:

> His willingness to participate in operations at all times has been outstanding and his conduct in the course of operations has been exemplary. When stood down from operational flying at the end of an operational tour he continued to fly at his own request with his Commanding Officer on every occasion.
>
> Flight Sergeant Oldfield's character and leadership have been outstanding and the example set by him at all times has materially aided the other wireless air gunners who have operated with him.[79]

---

75 Sid Wadey, The Operation Order for the Day Read, unpublished manuscript, courtesy of his widow Mrs M Wadey. RAAF Hudson Squadrons Association, Adelaide.

76 Flt Lt N T Badger DFC, 407161. Nos 6 & 2 Sqns, 528th Sqn USAAF. Horticulturist of Willunga, SA; b Adelaide 27 May 1916. Killed in action 8 May 1944.

77 No 2 Sqn A50, of 21 Aug 1942.

78 Flt Lt E R Oldfield DFM, 406221. No 2 Sqn, 1 OTU, 2 Sqn, HQNWA, HQWA. Clerk & commercial traveller of Perth; b Perth 11 Jun 1918. POW 26 Apr 1945.

79 Citation for Bill Oldfield's DFM, RAAF Personnel Records.

After returning to No 2 Squadron for a second tour in 1943, Bill Oldfield became the Area Gunnery Officer at HQ North-West Area. When serving in a similar capacity in Western Area and flying on a Liberator mission with No 25 Squadron, Bill Oldfield was posted as missing. After a night attack on the airfield at Den Pasar, his Liberator (A72-133) caught fire, after possibly being hit by ground fire.[80] Staying airborne for a further two hours, the crew were eventually forced to abandon the aircraft south-east of Soemba Island. Oldfield was captured, serving the rest of the war in Cycle Camp at Batavia, and returned to Australia in October 1945.

In general, the fate of crews fortunate enough to escape their burning wrecks was capture by a barbaric enemy and execution. Only two POWs from No 2 Squadron survived the war. The remainder were executed. It was in recognition of this courage, of flying mission after mission in the face of these hazards, that General MacArthur recommended the award of the United States Presidential Citation to the Squadron.

80 Bill Oldfield, telephone interview with author, of 23 Feb 1994.

CHAPTER 16

STRIKE FROM THE NORTH-WEST

I was told that there were about 600 aircraft out there and that should be enough to fight a pretty good war ...

General George C Kenney[1]

The strategy of the Allies in the conduct of the war meant a concentration on the European theatre, with only limited resources being made available for the Pacific. This resulted in only 15 per cent of the total Allied resources being employed against Japan in 1942—the remaining 85 per cent against Germany.[2] Within these constraints, the United States 5th Air Force was formed in early September 1942. At the same time a similar RAAF organisation was being established to control Australian units through the Area staffs. This was to become RAAF Command under AVM Bostock which, with the 5th Air Force, comprised Kenney's Allied Air Forces, now headquartered in Brisbane. Kenney's primary mission was to achieve air superiority in the South-West Pacific Area.

In New Guinea, while the Allies were gaining air superiority over the areas for the planned ground offensives, the squadrons in NW Area conducted a concurrent campaign to strike at the more distant Japanese strongholds. This air campaign was to be the pattern for the remainder of No 2 Squadron's war. Missions were now being flown north-east of Darwin to hinder consolidation of the enemy's new bases in the Arafura Sea. Targets were the Japanese-held Dobo (on the Aroe Islands) and Toeal (Kai Islands), with patrols to the coast of New Guinea. Patrols were also maintained to the west of the Banda Sea, with attacks on Timor, some missions of almost nine hours duration, fighting through the elements of the foul weather in the north-west. Even though the Australian and British POWs on Timor were always at risk from Allied air raids until their movement to Java in September 1942, they had always been heartened by the sound of the RAAF Hudsons.[3]

An early morning raid, led by Wing Commander Tich McFarlane (A16-160), was flown on 15 September 1942 against enemy vessels at Saumlaki. Three Hudsons dive-bombed a ship, which they hit, but not without loss. Flying Officer Ken McDonnell[4] (A16-172) was seen diving on the target when his aircraft was hit by intense anti-aircraft fire from all along the shore. The Hudson's auxiliary fuel tanks were

---

1 Gillison, p.570.

2 Morison, Vol VI, 1950, pp.4–5.

3 Paley, p.169.

4 Flg Off K L McDonnell, 401656. No 2 Sqn. Driver of Melbourne; b Dubbo, NSW, 20 Dec 1916. Killed in action 15 Sep 1942. Crew: Flg Offs V Treloar (416171) and K J Ahern (401754), Sgts G S Dobbs (414292) and G Ward-Smith (413294).

apparently hit, and it exploded and crashed in flames into the sea.[5] Once again the vulnerability of the aircraft with the auxiliary tanks in the cabin was tragically shown.

On 23 September, members of the newly formed 2/4th Independent Company were unloaded from *HMAS Voyager* to reinforce the men on Timor. Unfortunately, *Voyager* became beached and had to be scuttled, and its crew was evacuated onto the *Warrnambool* and *Kalgoorlie*, along with Sid Wadey,[6] who had been shot down the previous month. Four days later, Wadey was picked up in Darwin by the CO, Wing Commander McFarlane, and returned to the Squadron at Batchelor. Although high in spirits, he was very weak from dysentery and injured by the extensive burns sustained from his exploding auxiliary tank.

Two days later, 25 September, three 2 Squadron aircraft bombed Dili. The crew of Pilot Officer Ralph James[7] (A16-160), flying their first mission, were delayed with a faulty engine and arrived over the target four minutes after the others. They were attacked by two Zeros. James tried to evade his attackers in cloud; one Zero attempted a belly attack and was shot down, smoking in a spin, by a sustained burst by Sergeant Hugh Reilly.[8] Five minutes later, over the sea north of Dili, a second Zero was shot down by the turret gunner, Sergeant Pat Reen.[9] The other crew members, Sergeants Ron Ryan[10] and Ken Keech,[11] continued to call the approaches of another two Zeros which persisted in their attacks as the Hudson fought its way home. From the Squadron Operations Record Book:

> Thirty minutes after first attack main wing tanks were practically empty, belly gun out of ammunition, also side guns and a few minutes later turret reported to be unserviceable. Plt Off James then dived ... levelling out several feet above water, side gunner meanwhile taking 100 rounds from front guns. Zigzagged across Wetar Strait, side gunner using short bursts to ward off attacks from stern quarter. Turret gunner reported port gun serviceable again and fired short burst into Zero attacking from starboard quarter. Zero broke off attack and headed for Dili. Our aircraft on reaching Timor coast flew through valleys close to ground ... Zero could only attack from dead astern. The turret gunner was compelled to load port gun by hand, with the result that he tore most of the skin off his hands ... He was only able to fire single shots and odd bursts up to five rounds when the Zero came close. The bottoms of the valleys were obscured by fog and Zero was shaken off, but midway across island showed up again 50 yards astern and turret gunner immediately fired short burst of five rounds into motor. Zero pulled away sharply and broke off engagement. Proceeded at 0 feet towards Darwin, no further attack, the whole engagement taking 45 minutes.[12]

5 RAAF Historical C.D.102/45 A16-172, of 23 Sep 1946.

6 Wray, pp.137–141.

7 Flg Off R R James, 408811. Nos 4, 6 & 2 Sqns. Store manager of Shepparton, Vic; b Shepparton 13 Apr 1912. Killed in action 28 Apr 1943.

8 Flt Lt H Reilly, 414081. Nos 2, 73 & 7 Sqns, 1 OTU. Clerk of Mackay, Qld; b Ipswich, Qld, 19 Jul 1920.

9 F Sgt P S Reen, 412185. Nos 7, 6 & 2 Sqns. Student of Casino, NSW; b Toowoomba, Qld, 15 Mar 1918. Killed in action 28 Apr 1943.

10 Plt Off R D Ryan, 404989. Nos 7, 6 & 2 Sqns. Clerk of Brisbane; b Brisbane 31 Dec 1916. Killed in action 28 Apr 1943.

11 F Sgt K G Keech, 8966. Nos 24, 6 & 2 Sqns. Insurance representative of Lismore, NSW; b Orange, NSW, 10 May 1917. Killed in action 28 Apr 1943.

12 No 2 Sqn A50, of 25 Sep 1942.

Squadron Leader Ralph Moran congratulates Flying Officer Ralph James, Sergeant Pat Reen and Sergeant Hugh Reilly after shooting down two Zeros on their first mission. RAAF

This, being the crew's first strike, was a remarkable performance in professionalism and crew coordination. The value of having an extra gunner in the belly position had again been proven.

"A" Flight Commander, Squadron Leader Simon Fraser, killed in a jeep accident, 11 October 1942 A McFARLANE

Tragically, on 11 October, Squadron Leader Simon Fraser, the popular "A" Flight Commander, was killed in a jeep accident. He had been travelling with Flying Officer Ian Hay on the main road to Darwin, south of Batchelor. It was a shock to the Squadron; as the CO remembered "you expect to lose them in the air, not in a vehicle".[13]

During October, the enemy had continued to press the Australian commando force on Timor, and No 2

13 Tich McFarlane, interview with author, of 1 Nov 1993.

Squadron's efforts were directed against Aileu and Maubisse, on the main Japanese line of advance southward from Dili, and the airfield at Koepang. On 24 October, Wing Commander McFarlane handed command of the Squadron over to Squadron Leader Ralph Moran.[14] There is no doubt that McFarlane's inspiration during the dark days of 1942, and his ability to "lead from the front", were reflected in his Squadron's performance, and the unit's recognition by the Presidential Citation.

On return from a Dili strike during October 1942: Sgt Winzar, Plt Off Rigg, Wg Cdr McFarlane, Flt Lt Johnston, Flg Off Overheu, Plt Off Mills, unknown, Sgt Childs AWM 27606

After a similar raid that night, three more Hudsons were sent against Penfoei the following morning, 25 October. Flight Sergeant Alan Hawkesford[15] (A16-181) led Flying Officer Keith Mills[16] (A16-160) and Flight Sergeant Raydon Dunning[17] (A16

14 Gp Capt R H Moran, 83. Comd No 2 Sqn 1942–43, 13 Sqn & RAAF Stn Canberra 1943, 76 Wg, 71 Wg & Aitape Air Task Force 1944; SASO HQ WA 1944–46. Regular air force offr of Narre Warren, Vic; b Melbourne 13 Sep 1914.

15 Flt Lt A W Hawkesford DFC, 406484. Nos 7 & 2 Sqns, 1 OTU, 9 OG, 24 Sqn, 7 OTU. Commercial traveller of Perth; b Perth 13 Nov 1915.

16 Flg Off K R Mills, 405637. No 2 Sqn. Of Brisbane; b Brisbane 15 Jan 1918. Killed in action 20 Apr 1943.

17 Plt Off R K R Dunning, 407317. No 2 Sqn. Bank clerk of Kimba, SA; b Ramco, SA, 15 Oct 1916. Killed in action 25 Oct 1942. Crew: Sgts M L Kenihan (407437), L H S Fraser (5103), D W Francis (416665).

170), but just before making landfall on the south coast of Timor at Cape Mali, Dunning's crew lost formation and his Hudson was not seen again.[18] At this critical time the leader's crew were having intercom problems and were unable to talk to each other. Missing his attack run, Hawkesford had to turn back from Koepang. In the turn, Mills became separated and, in an intense heavy anti-aircraft barrage, the two Hudsons attacked the airfield individually. The considerably strengthened ground defences were putting up fire to 17,000 feet, as Mills' crew bombed the south end of the runway. Having survived this, they had more problems on returning to base. With an air raid in progress at Batchelor, the crew could not obtain a radio bearing to fix their position. As his fuel ran low flying along the Australian coastline, Mills saw a clear stretch of beach and forced landed above the high water mark. The aircraft was flown off the beach at Cape Ford and back to base two days later.

The RAAF's Beaufighters were over Timor for the first time on 17 November, while Hudsons from Nos 2 and 13 Squadrons bombed Maubisse, Dili, Bobonaro and Baucau.[19] For the remainder of the month 2 Squadron's Hudsons continued their attacks over Portuguese Timor, attacking sometimes in company with the Beaufighters of 31 Squadron.

Flying Officer Frank Morcombe[20] (A16-197) led another Timor strike on 4 December, in company with Flying Officers Harold McDouall[21] (A16-161) and Arthur Cambridge[22] (A16-232), tasked with a dawn offensive sweep to the northern tip of Timor, Kisar Island, the Wetar Strait and Alor Island. Entering Wetar Strait and turning towards the Laga anchorages, the formation was attacked by a Zero. The attack was concentrated on the third Hudson, flown by Flying Officer Cambridge. The Zero's close-in attacks severed the hydraulic lines of A16-232, making the hydraulic system unserviceable, including the turret. The enemy pilot apparently noticed that the turret could not function and continued his attacks on Cambridge's aircraft.

> In the third attack the starboard engine of this aircraft was hit and the aircraft immediately fell behind and lost height. In the next attack it received the full burst of machine gun fire ... and the tunnel gunner, Sgt Thame,[23] was killed.[24]

With the assistance of the other two Hudsons, Cambridge was able to regain formation, and after a seventh attack the Zero shadowed the formation as far as Dili,

18 Air Reconnaissance Report No 103, of 28 Oct 1942; and RAAF Historical C.D.102/45 A16-170, of 27 Aug 1946.

19 Gillison, p.647.

20 Flt Lt F K Morcombe DFC, 406572. No 2 Sqn, 1 OTU. Of Waddi Forest, WA; b Ceduna, SA, 18 Aug 1913. Killed in flying accident 5 Oct 1943.

21 Flt Lt H S McDouall, 403014. Nos 6, 2, 34, 37 & 21 Sqns. Station hand of Barraba, NSW; b Barraba 3 Mar 1920.

22 Flt Lt A A Cambridge, 406750. Nos 7, 6, 2 & 24 Sqns. Audit clerk of Perth; b Perth 23 Dec 1915. Crew: Flg Off Barnes and Sgts Cooper, Thame, Gunn.

23 Sgt G J Thame, 403609. No 2 Sqn. Of Sydney; b Barraba, NSW, 6 Apr 1912. Killed in action 4 Dec 1942.

24 No 2 Sqn A50, of 4 Dec 1942.

when it was lost in cloud. On return, the damaged aircraft had its undercarriage hanging half down and, having jettisoned bombs, could not close the bomb doors. Cambridge carried out a skilful belly landing at Batchelor, without further injury to the crew. All the Hudsons had been holed by the Zero's persistent attacks, which had lasted for 35 minutes, and A16-232 had to be written off. The following comments finalised the official attack report for this mission:

> Whilst the actions of all members of the three crews were in the highest traditions of the RAAF it is felt that the conduct of F/O Morcombe (the leader of the formation) in
>
> His meeting of the attacks made against the formation,
>
> The nursing of the third aircraft so badly hit,
>
> The determination exhibited in continuing the search through an even more dangerous zone,
>
> is especially deserving of commendation and at the same time that of F/O Cambridge in successfully bringing his aircraft back to base under the circumstances described deserving of the highest praise.[25]

After being in action for nine months, the 2/2nd Independent Company of Sparrow Force was evacuated from Timor at the end of 1942. In January 1943, the 2/4th was also withdrawn, leaving small Intelligence parties—S Force. The next month, S Force departed too and the Australian ground campaign on Timor came to an end. There was now no useful purpose being served by the small force, nor was there any prospect to mount a major campaign in Timor while engaged in New Guinea.[26] For almost 12 months the Australians had harassed the Japanese, killing an estimated 1,500 of the enemy for a loss of 40 of their own, successfully tying down enemy troops, who could have been used to a possibly decisive effect in the New Guinea campaign.

Christmas Eve 1942 saw both the Hudson squadrons sending out flights of four Hudsons to attack a destroyer and four transports in the Wetar Strait off Lavai, near the north-eastern tip of Timor. Flight Lieutenant Sid Austin[27] (A16-217) led the 2 Squadron flight and, in spite of scoring some near misses, none of the vessels were hit. No 13 Squadron did suffer a loss, however, when A16-166 was shot down by enemy fighters, with the loss of the crew.[28]

Three days later, 27 December, Wing Commander Moran (A16-197) led six 2 Squadron Hudsons back to Lavai to attack storage at the port, and a motor transport concentration at the Lavai River crossing. Making landfall on the south coast of Timor, Moran split his formation, taking his flight of three to the stores, with Flight Lieutenant Austin (A16-217) leading the other flight against the Motor Transport concentration. It was at this stage his formation was attacked by three Zeros, with the fighters concentrating on Austin's flight and disabling his rudders. The last man in

---

25 Report on Operation BAT 23 No 3, of 4 Dec 1942, RAAF Historical Section.

26 D McCarthy, *South-West Pacific Area—First Year*, AWM, Canberra, 1959, p.620.

27 Wg Cdr S Austin DFC, 260359. Nos 7 & 2 Sqns, HQ EA, 1 SFTS. Grazier of Trangie, NSW; b 20 Nov 1912.

28 13 Sqn crew: Flg Off G P Thomson, Sgts K G Chote, J McA Dunbar, R M Clark, R S West.

the flight, Flying Officer Merv Johns[29] (A16-210), was straggling continuously, and the rest of the formation gave him as much protection as possible. Flight Lieutenant John Robertson's[30] belly gunner (A16-160) was able to put a long burst into a Zero at 50 metres range, and the fighter fell away in flames. However, with two Zeros concentrating their attacks on A16-210, Johns' aircraft was raked with fire:

> ... and burst into flames from half way under the fuselage and shot up in the air out of control and finally spiralled into the sea.[31]

Now 10 kilometres from the Timor coast, the Zeros attacked for a further 15 minutes, but the remaining five Hudsons, all holed bar one, returned safely. The report on the operation concluded:

> It is considered that special mention be made of the splendid formation flying shown by Flight 1, which kept close formation at high speed during the violent evasive action taken during the whole conflict.[32]

The best defence for the Hudsons was shown, yet again, to lie in keeping close formation and providing mutual support. To add to this effective tactic, a gunner was now tasked as a "fire control officer", to act as gunnery leader for the formation, by controlling the engagement of the Hudsons' defensive fire over the radio.

Flying Officer Morcombe (A16-178) led a formation on 6 January 1943 to carry out photographic reconnaissance of the Kai Islands and to attack shipping at Saumlaki. Turning south towards their target from Toeal, they spotted an *Otori*-Class torpedo boat steaming northwards. The Hudsons, diving from 6,000 feet and releasing their bombs from 1,500 feet, scored hits on the ship's smoke stack and aft gun turret. The ship stopped dead in the water, belching black smoke, and exploded.[33] On further reconnaissance around the Kai Islands on 28 January, Sergeant Syd McDonald's[34] crew (A16-197) experienced an unusual interception. After spotting an enemy convoy, the Hudson was attacked by a Mitsubishi Army Type 97 Bomber, known as the "Sally". The superior speed of the Sally enabled it to make repeated intercepts on the Hudson from astern below the tail, firing with its front guns and heavy calibre turret, and persisting with attacks for an hour. During this inconclusive engagement, the Hudson fired nearly a thousand rounds, and returned to base undamaged.[35]

---

29 Flg Off M W C Johns, 407242. No 2 Sqn. Clerk of Deniliquin, NSW; b Kadina, SA, 12 Aug 1919. Killed in action 27 Dec 1942. Crew Plt Off J F Nixon (401142), Sgts J D Horseman (416339), D L Tyler (416160), K Ross (416192).

30 Flt Lt J W Robertson, 250658. Nos 8, 14 & 2 Sqns, 1 OTU, GR Sch, 1st TAF. Builder of Melbourne; b Melbourne 16 Aug 1917.

31 Report on Operation BAT 17 No 1, of 28 Dec 1942; and RAAF Historical C.D.102/45 A16-210, of 2 Jun 1947.

32 Report on Operation BAT 17 No 1, of 28 Dec 1942, RAAF Historical Section.

33 Report on Operation BAT 33 No 1, of 6 Jan 1943.

34 Flt Lt S L McDonald, 411524. Nos 7, 6, 2 & 24 Sqns. Agricultural inspector of Leeton, NSW; b Sydney 30 May 1916. Killed in action 6 Apr 1945.

35 Report on BAT 32, of 28 Jan 1943.

Attacks on Timor continued throughout February, and reconnaissance patrols occupied a large part of the Squadron effort as the enemy built up its strength on the Aroe Islands. One reconnaissance sortie to Timor on 24 February was unusual, when Flying Officer Norm Ralston[36] (A16-240), carrying a naval specialist in his crew, flew off Koepang with detection equipment to indicate any enemy radar activity. It had been suspected that Radio Direction Finding (RDF) stations were operational around Koepang because of the repeated accuracy of fighter interception. On this sortie there were no positive indications of any enemy radar (i.e. RDF) transmissions.[37] These attempted intercepts of radar activity were among the first electronic warfare operations of the RAAF.

A Hudson bombload of two 250 lb bombs and six 100 lb bombs, with armourers about to load A16-242 at Batchelor, October 1942 AWM 27611

Although operations still ranged as far as Timor during March, with the rout of the Japanese fleet in the decisive aerial engagement of the Battle of the Bismark Sea, NW Area's main effort was directed against the islands to Darwin's north in the Arafura Sea. Although MacArthur considered northern Australia was still threatened, there were now signs that the enemy was experiencing a growing shortage of naval and

36 Flt Lt N G Ralston, 411185. Nos 7 & 2 Sqns, 7SFTS, 1 OTU. Salesman of Sydney; b Sydney 6 Jun 1914.

37 No 2 Sqn A50, of 24 Feb 1943.

merchant shipping.[38] Flying Officer Ralph James (A16-207) led three Hudsons out on the 15th to attack a convoy of an escort and two cargo vessels of 1,000 tons, off the Aroe Islands. Individual mast height attacks were made, each Hudson releasing five 100 lb bombs, all of which failed to explode. All vessels were then raked by the Hudsons' guns. The escort was successful in putting James' starboard engine out of commission, but he was able to return safely.

This was not the only mission to have had bomb detonation problems, and clearly sorties could not be wasted releasing dud bombs. The Hudson's normal war load had been four of the standard British 250 lb General Purpose (GP) bombs. In late 1942, the United States 100 lb demolition bombs became available, and typical loads became two 250 lb bombs and five or six 100 lb bombs. For anti-shipping use, a semi-armour-piercing (SAP) bomb, with a strong case and delayed fuse, to allow penetration before detonation, was preferred. Fusing problems were overcome with the 100 lb bombs and by mid-year they had completely replaced the 250 lb bomb on the Hudson.[39]

Six aircraft were tasked on 3 April with a low-level attack on the Japanese-held airfield at Timoeka, on the south-west coast of Dutch New Guinea. Operating from Millingimbi, Flying Officer Harold McDouall (A16-238) led the first flight, with "A" Flight Commander Squadron Leader Malcolm Burns[40] (A16-169) leading the second. Having dropped his bombs, Burns was half way along the runway on a strafing run at about 200 feet when he was hit by groundfire, and a red flash was seen on the port side of his aircraft. Burns pulled the aircraft into a climbing turn to starboard, and appeared to have full control, but after another 10 seconds the aircraft dived, crashing into the ground with no survivors.[41] Several other aircraft were holed by enemy fire, and when Flying Officer Gordon O'Neil[42] (A16-247) returned there two days later, the accurate anti-aircraft fire severely damaged his aircraft. He was, however, able to crash-land back at Millingimbi with no casualties, but A16-247 was another write-off for the Squadron.

This was the day, too, when the sister squadrons—Nos 2 and 13—swapped COs. Wing Commander Moran took over 13 Squadron as he was tour-expired, and 13 was leaving operations to move south and re-equip with the Beaufort. No 13's CO, Wing Commander "Jock" Whyte,[43] took command of No 2 Squadron and the following

---

38 G Odgers, *Air War Against Japan*, AWM, Canberra, 1957, pp.38–9.

39 Ten 100 lb load was standard by May 1943, and the last 250 lb bombs were used by Jul 1943. Other stores included packs of 15 4 lb incendiaries to mark targets and start fires, a reconnaissance flare to illuminate the target area, and the M46 photo flash to allow photography for damage assessment. Loads could be mixed to match target requirements.

40 Sqn Ldr M C Burns, 271331. No 2 Sqn. Of Mossman, Qld; b Brisbane 29 Oct 1912. Killed in action 3 Apr 1943. Crew: Plt Off L R Jope (400434), FSgt A A Dean (404328), Sgts K Jones (410240), R J Johnstone (410073).

41 RAAF Historical C.D.102/45 A16-169, of 25 Oct 1946.

42 Flt Lt G L C O'Neil, 406949. Nos 13 & 2 Sqns, 1 OTU. Civil servant of Northam, WA; b Geraldton, WA, 1 May 1918.

43 Gp Capt J M Whyte DFC, 154. Nos 4 & 6 Sqns; comd 1ANS 1941–42, 13 Sqn 1942–43, 2 Sqn 1943, 2ANS 1943–44, 86 Wg 1945; SASO Darwin 1945–46, SASO HQ SA 1946–48. Regular air force offr; b 1 Sep 1916.

week, on 12 April, the Squadron moved from Batchelor to Hughes, which had been vacated by No 13. Hughes Strip, named after Mr D D Hughes, the Director of Mines in Darwin, was located 46 kilometres south of Darwin, immediately to the north of Livingstone airstrip. It was one of the first of the many strips prepared along the main highway, the initial work having been carried out by United States engineers and finished by the RAAF. It was a strip cut out of the timber which, like the neighbouring airfields along the main highway, certainly presented the crews with a challenge to find on a dark night.

This move took place as the Pacific war entered a new phase. Sufficient resources were now becoming available and, coinciding with this growth in Allied strength, enemy strength was weakening. A limited offensive was now possible by the Allies, and this was to take place north of New Guinea—the breaking of the "Bismark Barrier", the obstacle between MacArthur and the Philippines. The Allies had almost gained air superiority in the Pacific as new aircraft arrived; but there was yet to be new equipment made available for 2 Squadron. The Hudson was to equip No 2 Squadron longer than any other unit in the RAAF.

A maximum effort attack was planned for the night of 18 April against the well-known target of Penfoei airfield. Thirteen Hudsons of No 2 Squadron, led by Squadron Leader Keith Kemp[44] (A16-236) and staging through Drysdale, were coordinated to strike with nine B-25 Mitchells of No 18 (NEI) Squadron. The Dutch Mitchells led the way, attacking from 3,500 feet just after midnight. Despite heavy anti-aircraft fire over the target, the Mitchells escaped unscathed. Kemp had been forced to return to Drysdale with engine problems, caused by water in the fuel, so Flying Officer John "Bunny" Austin,[45] a veteran from the war in Europe,[46] led the Squadron. In the face of very heavy defensive fire, Austin remained over the target area for over an hour at 20,000 feet, flying back and forth between Penfoei and Koepang to confuse enemy defences. He finally made his attack and returned safely to Drysdale.[47] The intense anti-aircraft fire had been so accurate that the crews could smell the acrid fumes of the bursts, but no aircraft were hit.

The following night, soon after their two Hudsons set off from Millingimbi to bomb Timoeka in Dutch New Guinea, Flight Lieutenant Keith Daniel[48] (A16-197) and Flying Officer Keith Mills[49] (A16-183) collided over Rapuma Island. Both crews were

44 Wg Cdr K M Kemp DFC, 250728. Nos 13 & 2 Sqns; comd 463 Sqn 1945. Clerk of Melbourne; b Melbourne 9 Feb 1918.

45 Flt Lt J S Austin DFC, 400363. No 608 Sqn RAF, Nos 32, 13 & 2 Sqns. Stock agent of Lara, Vic; b Melbourne 15 May 1918. Died of illness 9 Nov 1943.

46 Austin flew 270 operational hours on Hudsons in No 608 Sqn (Coastal Command) in the UK, was mentioned in dispatches and his crew was credited with shooting down a German Ju88. "Timor Treks", p.87.

47 No 2 Sqn Report No 72, of 19 Apr 1943, RAAF Historical Section.

48 Flt Lt K E Daniel, 402852. Nos 13 & 2 Sqns. Of Moree, NSW; b Menangle, NSW, 17 Oct 1916. Killed in action 20 Apr 1943. Crew: Flg Off J Laws (412979), Plt Offs W J H Gove (401636) and J G Butler (408820), FSgt T W Wilson (408899).

49 Crew: Flg Off N E John (409551), Plt Offs J L Barnes (404946) and T W Bassan (411985), FSgt H H K Hadley (411904).

lost, and Daniel's navigator, Sergeant Bill Gove,[50] was remembered by the naming of the Gove airfield, near Melville Bay, in the Northern Territory. Postwar, the entire peninsula became known as Gove, even though the base ceased operations in August 1945.

Four days later, on 24 April, Pilot Officer Aloysius Kilgariff[51] (A16-217) was carrying out an armed reconnaissance of Dobo, in the Aroe Islands, and found four floatplanes beached south of Taberfane village. Although some were hidden under branches, they were identified as three Aichi E13A "Jake" reconnaissance seaplanes and a Nakajima A6M2-N "Rufe" fighter, a seaplane variant of the Zero. Kilgariff's crew reported:

> A strafing run was made in the dive from 1000 feet to 30 feet. About 200 rounds were expended. As the Hudson was about to make a second attack a biplane with single float (probably a "Dave"[52]) was sighted. Hudson was at 400 feet, the enemy aircraft 3000 feet higher. "Dave" attacked in a dive from 2 o'clock position, approaching to within 150 yards. Trace was seen close to the Hudson. The Hudson dived under "Dave" and headed west for cloud.[53]

On a similar sortie to Dobo the following day, Anzac Day, Flying Officer Austin's crew (A16-204) was also intercepted by an enemy floatplane. Coming in to reconnoitre Taberfane at tree-top height, Austin spotted a Rufe on patrol. After a "cat-and-mouse" standoff with the Rufe feinting attacks, and not engaging at close range, the enemy then:

> ... made an attack from 5 o'clock from under 400 yards, firing cannon with trace. The Hudson held its fire till Rufe was 400 yards off at which point the turret got in some good shots. Rufe broke off making steep turn to port, exposing its undersurfaces. Hudson's belly gun and starboard gun fired. Trace was seen all around the Rufe from which a good deal of black smoke came as it increased speed.[54]

The Squadron now concentrated its activity around the islands north of the Arafura Sea, and the Dutch New Guinea coast. On a reconnaissance to Dobo on 21 April, Flying Officer James (A16-230) had received groundfire, which wounded his tunnel gunner Flying Officer Hugh McCullough,[55] and three days later his aircraft was again hit near the Kai Islands. The following day, Flying Officer Harry Warlow-Davies'[56] aircraft (A16-219) was hit while bombing Timoeka airfield and the turret gunner, Flight Sergeant Hart Stockley, was wounded.[57]

---

50 Plt Off W J H Gove, 401636. No 2 Sqn. Of Melbourne; b Melbourne 22 Feb 1921. Killed in action 20 Apr 1943.

51 Flt Lt A K Kilgariff, 407324. No 2 Sqn, 1 OTU, 35 & 24 Sqns. Clerk of Adelaide; b 16 Feb 1917.

52 "Dave" was the Nakajima E8N Navy Type 95 Reconnaissance Seaplane.

53 No 2 Sqn Report No 82, of 24 Apr 1943.

54 No 2 Sqn Report No 87, of 25 Apr 1943.

55 Flt Lt H C McCullough, 406810. Nos 2 & 14 Sqns, 7 OTU. Clerk of Perth; b Perth 17 Sep 1920.

56 Flt Lt H Warlow-Davies, 408221. Nos 13 & 2 Sqns, 1 OTU, AA & G Sch. University student of Hobart; b Hobart 2 Jan 1921.

57 Flt Lt H R Stockley, 407394. 18 MG Regt, Nos 8, 13, 2 Sqns, 1 OTU, 11 & 20 Sqns. Butcher of McLaren Vale, SA; b Norwood, SA, 23 Feb 1919.

Three days later, 28 April, seven Hudsons carried out an armed reconnaissance for shipping around the Aroe Islands. One aircraft failed to take off, another returned to Hughes after engine problems, while the remaining five refuelled at Millingimbi. After departure, half an hour north of the staging base, Flying Officer Ralph James' crew[58] (A16-241) reported losing an engine, but the aircraft disappeared when attempting to return to Millingimbi. Despite an intensive search, carried out by four Hudsons every day for a week, there was no sign of the aircraft or the crew.[59] It had been Ralph James' crew that had shot down two Zeros on its first mission.

Flying Officer Francis Kavanagh[60] (A16-219) led five Hudsons out before dawn on 7 May, to attack floatplanes seen the previous day parked at Maikoor and Taberfane, in the Aroe Islands. Approaching Maikoor, flying in line abreast, the formation was intercepted by two Rufes. This seaplane development of the Zero, with a large single float beneath the fuselage which caused considerable drag, was still manoeuvrable and could still reach speeds of 235 knots,[61] enough to overhaul the Hudson by 20 knots. Warrant Officer Viv Jackman[62] (A16-171) was straggling behind the formation and was attacked inside 20 metres by one Rufe, which just avoided a collision. Crippled, A16-171 climbed away to starboard, turned over on its back then dived vertically from about 1,000 feet, streaming flames. It exploded on hitting the ground in mangrove swamp just south of the Maikoor River.[63] As a result of damage in this engagement, Flying Officer Ray Hornby's[64] aircraft (A16-211) nicknamed "The Tojo Busters", crash-landed on return to Millingimbi, and was written off without casualties to the crew.

Casualties continued when on 12 June the "B" Flight Commander, Flight Lieutenant John Mason,[65] was tasked with test flying A16-186. Being airborne for almost half an hour, he made a single-engined approach to Hughes, then retracted the undercarriage to initiate an overshoot. The Hudson then descended, disappearing and impacting with the ground two kilometres north-east of the strip, killing all on board, including three Hudson captains.[66]

---

58 Crew: Plt Off R D Ryan (406810), WOff H E Cutten (3403), F Sgts K G Keech (8966), P S Reen (412185), Sgt R E Norris (414157).

59 RAAF Historical C.D.102/45 A16-241.

60 Flt Lt F J Kavanagh, 264906. Nos 13, 2 & 25 Sqns. Clerk of Sydney; b Sydney 6 Nov 1919.

61 R J Francillon, *Japanese Aircraft of the Pacific War*, Putnam, London, 1990, pp.426–8.

62 WOff V L Jackman, 400715. No 2 Sqn. Of Culgoa, Vic; b Melbourne 14 Sep 1917. Killed in action 7 May 1943. Crew: Flg Off D R Hicks (405161), F Sgts E Quinn (403203), I W MacKenzie (404782), Sgt D A Emery (416080).

63 No 2 Sqn Report No 95, of 7 May 1943; and RAAF Historical C.D.102/45 A16-171, of 3 Mar 1947.

64 Flt Lt R H Hornby AFC, 403053. No 48 Sqn RAF, Nos 32, 2, 37 & 38 Sqns. Clerk of Sydney; b Dorrigo, NSW, 17 Aug 1912.

65 Flt Lt J F Mason AFM, 252763. Nos 13 & 2 Sqns. Of Koroit, Vic; b Mt Helena, WA, 28 Jul 1909. Killed in flying accident 12 Jun 1943.

66 No 2 Sqn A50, of 12 Jun 1943. Crew: Flt Lt J R Wood (280849), Flg Offs L G Knauer (405180), T W Swann (411403), Sgt G R Rowland (409745). Also on board two members of No 13 Sqn: FSgts R J Flaherty, R D Woods.

Searching for enemy shipping and anti-submarine escort patrols in the Arafura Sea now became the major task. These searches were in pre-designated areas, with code names such as "Peanut" and "Zebra". The "Zebra", flown north towards the Aroe Islands, necessitated the continued use of Millingimbi (known as "Gimbi") as a forward operating base. Carrying out a "Zebra" search on the morning of 20 June, Flying Officer John Cameron[67] (A16-227) was at 4,000 feet when he was attacked out of the sun by a Rufe. As the fighter pressed home its attack inside 30 metres, its cannon fire could be heard as the Hudson dived for the water. Defensive fire from the Hudson's turret gunner possibly damaged the fighter, and the Hudson was able to escape unscathed at water level.[68] That night, in retaliation for a Japanese raid on Darwin earlier in the morning (raid number 55 on the Darwin area), No 2 Squadron returned to Koepang to bomb Penfoei. Following a Catalina raid, Squadron Leader Kemp (A16-236) led four Hudsons, which bombed from 14,000 feet, untroubled by inaccurate and disorganised anti-aircraft defences. Referring to both the hostile fire and the enemy's searchlights, one of Kemp's crew commented: "They knew there were aircraft above, but they didn't know where."[69]

Flying Officer Don Thomas[70] and his crew (A16-230) were tasked with carrying out an anti-submarine patrol on the afternoon of 2 July for a small convoy being escorted by *HMAS Latrobe*. As the Hudson arrived two enemy Jake floatplanes were attacking the convoy. Thomas engaged one of the single-engined Jakes, and after a 50 kilometre chase shot it down with his forward firing guns. His navigator, Flying Officer Steve Grey,[71] later related:

> In spite of accurate return fire Thomas pressed his attack to close range to shoot the bomber [sic] down in flames using fixed forward firing machine-guns. Another Jap bomber withdrew from the action.[72]

The successful crew of A16-230 on 2 July 1943 with their Jake "kill": Geoff Morgan, Len Garret, Bob Nicholls, Steve Grey, Don Thomas S GREY

67 Flt Lt J O'G Cameron, 406827. Nos 7, 2, 13 & 2 Sqns, 1 OTU. Farmer of Perth; b Latrobe, Tas, 20 Jan 1921.

68 No 2 Sqn Report No 114, of 21 Jun 1943.

69 No 2 Sqn Report No 115, of 21 Jun 1943.

70 Flt Lt A D Thomas, 415560. No 2 Sqn, 1 OTU, 2 Sqn. Horse trainer of Perth; b Perth 21 Dec 1919.

71 Flt Lt S W Grey, 408182. No 2 Sqn, 1 OTU, 2 Sqn. School teacher of Launceston, Tas; b Smithton, Tas, 18 Jun 1916.

72 S W Grey, letter to the editor, *Wings*, Jun 1991.

The second Jake was not seen again. Steve Grey had also emptied the magazine of his side gun during the encounter, and still has vivid recollections of the persistence of the Japanese rear gunner during the close engagement.[73] Thomas brought his aircraft, which had sustained slight damage, back to Hughes with the distinction of being the only Hudson pilot to shoot down a Japanese aircraft with the front guns.

During May, General Kenney, as Allied Air Forces Commander in SWPA, had directed RAAF Command to concentrate offensive action against Koepang. The importance of Penfoei to the enemy in launching attacks on northern Australia was realised, and Bostock repeated this directive to NW Area: "with maximum effort during the first part of July".[74] Mastery of the skies over the Japanese bases in 2 Squadron's areas of operation had still to be wrested from the enemy's 23rd Air Flotilla and the 7th Air Division.

Timor was the target for the night of 6 July, when Flight Lieutenant Bunny Austin (A16-207) led five Hudsons to bomb Koepang town. The next night he led back seven aircraft against the airfield at Penfoei, attacking before dawn on the 8th prior to United States Liberators bombing the runways and barracks.[75] Austin's aircraft was caught by Penfoei's master defensive searchlight, and Austin used violent evasion and switching on and off of his IFF[76] equipment, a tactic used in Europe. The radar-controlled searchlights went out.[77] Flying Officer Mick Helsham[78] (A16-160) attacked first to act as a pathfinder and mark the target for the other Hudsons and Liberators:

> The Liberators were preceded over the target by a Hudson, which dropped high explosives and incendiary bombs on the runways and dispersal pens during the night. Flying Officer M Helsham, who made sure of hitting the target by diving to 2000 feet before releasing his bombs, saw aircraft burning in the dispersal pen.[79]

The new CO, Squadron Leader Cyril Williams,[80] took over on 7 July from the outgoing commander, Wing Commander Jock Whyte, who had paid a tribute to his aircrew:

> These boys are good. They are full of fighting spirit and fly with enthusiasm and daring against the best that the Japanese can send against them throughout their long operational tour ...[81]

---

73 Interview with Steve Grey's son, Wg Cdr Bob Grey, of 18 Nov 1993.

74 Odgers, pp.56–7.

75 Odgers, p.66.

76 Identification Friend or Foe radar transponder equipment to discriminate between friendly and hostile aircraft.

77 No 2 Sqn Report No 124, of 8 Jul 1943. During 1942, RAF bomber crews had believed that pulses transmitted by their IFF sets could jam German searchlight control radars, affording them a measure of protection. This belief was erroneous, but was encouraged by Bomber Command as it aided crew morale, when life expectancy was short! R V Jones, *Most Secret War*, Hamish Hamilton, London, 1978, p.492.

78 AVM M M Helsham AO DFC, 411780/O215065. No 2 Sqn, 1 OTU, 2 Sqn; Judge Advocate Gen 1968–1982. Law student of Sydney; b Sydney 30 Nov 1921.

79 N Ottaway, "Head-On Clashes in Timor", *The Sun*, Sydney, Saturday 10 Jul 1943.

80 Sqn Ldr C C Williams, 377. Nos 8, 6 & 7 Sqns; comd 2 Sqn 1943. Regular air force offr of Glen Innes, NSW; b Glen Innes 10 Oct 1919. Killed in action 9 Jul 1943.

81 G P Jones, *Attacker—The Hudson and its Flyers*, William Kimber, London, 1980, p.98.

Williams' first operation, on 9 July, was a four-aircraft raid back to Penfoei aerodrome, led by Flight Lieutenant Austin (A16-244) with the new CO (A16-195) flying at the rear. The flight encountered abysmal weather approaching Timor, and although the other aircraft turned back, nothing more was heard from Williams and his crew.[82] One of the other pilots, Pilot Officer Ted Stacy[83] (A16-189) recalled:

> Of the four aircraft which took off, two turned back when they hit a very heavy front about 100 miles from the target. I went through a towering cumulus at 10,000 ft but failed to see the target owing to cloud layers beneath us, and experienced the roughest possible conditions on turning for home through the front. Throughout the operation nothing was heard of S/Ldr Williams, as formation was not maintained at any stage.[84]

Flight Lieutenant Bruce Watson (Adjutant), Wing Commander Jock Whyte (CO) and Squadron Leader Keith Kemp (Flight Commander) F RIGGS

Immediately before taking over No 2 Squadron, Squadron Leader Cyril Williams married Joy MacLellan at St Hillary's, Kew, in Melbourne. VIA D PHILLIPS

Natives later found wreckage near the summit of Mount Kekneno, 100 kilometres north-east of Koepang, and this was presumed to be A16-195.[85] Apparently the new CO had pressed on in unsuitable weather, when his more experienced colleagues in the area had aborted any attempt to reach the target; he undoubtedly felt the pressure of taking over a unit with such a hard-won reputation. Squadron Leader Keith Kemp was appointed temporary Commanding Officer until the arrival of Wing Commander Keith Hennock.[86]

The new AOC for NW Area, former 2 Squadron AFC pilot AVM King Cole, was aware of Kenney's determination that the priority for air assets would continue to NE

---

82 Crew: Flt Lt A McP Fowler (635), Flg Offs W A Hepworth (406851), R C Lindsey DFM (406218), FSgt D C Farmer (416251).

83 Flt Lt E B Stacy AFC, 403962. 18th Bn, Nos 6, 13 & 2 Sqns, 1 OTU, 1CU. Salesman of Sydney; b Coolah, NSW, 19 May 1914.

84 HQ NWA Brief, Interview with Flg Off E B Stacy, of 17 Nov 1944, p.2. RAAF Historical Section.

85 RAAF Historical C.D.102/45 A16-195, of 23 Jan 1946.

86 AVM K S Hennock CBE DFC, 211/O364. Nos 4, 6, 23 Sqns, 206 Sqn RAF; comd 459 Sqn 1942, 2 Sqn 1943–44; RAAFHQ; comd 391 Sqn 1951–52; RAAF Wash 1958–60; comd 82 Wg 1960–62, Richmond 1967, AOC HQOC 1967–69, COS FEAF 1969–70, AOC HQSC 1970–73, AMP 1973–75. Regular air force offr of Kingsvale, NSW; b Young, NSW, 8 Feb 1918.

Area. The duties of NW Area were seaward reconnaissance (to provide 36 hours warning of major raids or invasion), protection of Allied shipping, and the disruption of enemy airfields and shipping.[87] Cole, like his predecessor, recognised that new equipment was required to continue long-range reconnaissance and carry the war in an effective air strike campaign. However, MacArthur's strategy for the recapture of Rabaul had changed during August. All plans and efforts had been directed towards this since 1942, but now this fortress was to be bypassed in an island-hopping campaign. This change of strategy was not to alter priorities, and did not alleviate NW Area's shortages. In desperation at the Allied advances, the Japanese air regiments of the 7th Air Division at Ambon were moved forward to Wewak, to bolster their defences on the northern New Guinea coast.

No 2 Squadron "B" Flight crews, July 1943 D PHILLIPS

The focus of attention for 2 Squadron remained the Arafura Sea, as far north as the New Guinea coast. Strikes were maintained against the 23rd Air Flotilla's bases in the Kai and Aroe Islands, from where enemy floatplanes had continued to harry Allied shipping. On 15 August, a single Hudson was tasked with seaward reconnaissance followed by an attack at Keaukwa village, on the south coast of Dutch New Guinea. After its departure from Millingimbi, there was no further communication with

87 Odgers, p.111.

Sergeant John Lamb's[88] aircraft (A16-181), which disappeared without trace. The pilot of another Hudson on a shipping search reported two enemy aircraft from Timoeka over Keaukwa,[89] and it was presumed A16-181 was most probably shot down by Zeros.

2 Squadron Hudsons AWM NWA 0244

The United States massed attack over 17–18 August on Wewak, which destroyed some 200 enemy aircraft, finally turned the balance of air power to the Allies in New Guinea. As MacArthur pushed towards Lae and Nadzab, offensive action by the NW Area squadrons attempted to bluff the enemy into believing there was another front, and continued to wear down his strength by constant harassment and destruction. In addition, by forcing Japanese fighters to be retained in these flank areas, enemy air power was diverted from opposing the main Allied advance.[90] As part of this strategy, No 2 Squadron bombarded installations at Taberfane and the airfield at Langgoer.

88 FSgt J B Lamb, 416770. No 2 Sqn. Of Melbourne; b Melbourne 7 Jul 1916. Killed in action 15 Aug 1943. Crew: F Sgts C A Ikin (408151), M H Jackson (416776), N D McLean (416771), T W Wise (418033).

89 RAAF Historical C.D.30/46 A16-181, of 25 Oct 1946.

90 Odgers, p.112.

Aircrew on the way to the airstrip: Sgts R D Nicholls, F T Riggs and D G Phillips, Flt Sgt R D Jordan VIA D PHILLIPS/F RIGGS

Shipping reconnaissance patrols of the Arafura Sea continued via new routes, designated "Giraffe" (around Timor), "Horse" (around the Tanimbars), "Jackass" (between Timor and the Tanimbars) and "Koala" (around the Aroes). On 8 September, Flying Officer Lynn Martin[91] (A16-233) departed with his crew on a Horse patrol. No communication was received from the aircraft after take-off, but an enemy broadcast on the Tokyo Home Services News in Japanese on 15 September gave a clue to the crew's fate. It stated that a small merchant ship had been attacked north of Tanimbar Island by a twin-engined bomber on 8 September, and that the aircraft had been shot down. Two of the crew had baled out and had been captured.[92] Martin and two of his crew were killed immediately.[93] The two crewmen who were captured, Flight Sergeants Maurice Graham[94] and Jack Boanas,[95] were taken to Toeal in the Kai Islands, and later to Ambon, where they were executed the following month.[96]

Attacks on Langgoer airstrip continued through September and into October. On the night of 10 October, Flying Officer Mick Helsham (A16-213) was hit by the accurate groundfire over the airfield:

> The aircraft went into a right hand spiral dive, dropping towards the sea, with the controls shot away. The radio had been shot out ... the crew hastily prepared to abandon ship, and the emergency exit above the pilot was jettisoned.
>
> Meantime intense ack-ack fire was streaming at the falling aircraft, and the slipstream racing past the open hatchway was making an awful din. After the aircraft had spun down 3000 feet, Helsham ... decided to make a last second attempt to save the bomber and crew ... there was some port aileron control sufficient to keep the left wing down, and so prevent the

91 Flg Off L A Martin, 408816. No 2 Sqn. Of Mornington, Vic; b Kaniera, Vic, 21 Sep 1913. Killed in action 8 Sep 1943.

92 RAAF Historical C.D.102/45 A16-233, of 23 Sep 1946.

93 These crew members were: Plt Offs M C Wettenhall (408818), R H Williams (416391).

94 F Sgt M A Graham, 410337. No 2 Sqn. Of Thornbury, Vic; b Geelong, Vic, 6 Jan 1923. Executed by Japanese 31 Oct 1943.

95 F Sgt J W R Boanas, 417331. No 2 Sqn. Of Lochiel, SA; b Muston, SA, 7 Dec 1921. Executed by Japanese 31 Oct 1943.

96 RAAF Historical C.D.102/45 A16-233, of 21 Jan 1948.

> aircraft continuing spiralling to the right. There was also sufficient elevator control to keep the nose up. But the rudders were useless and the starboard aileron was out. All trimming tabs were uncontrollable.
>
> He distributed his crew about the plane to keep the aircraft straight and level ... using humans to balance a ship that should be balanced mechanically. He climbed to 5000 feet, with the Hudson still inclining to spiral to starboard. With no radio the crew was unable to flash an SOS to base, and with no rudder control it was a tossup whether Helsham could keep the aircraft on course.
>
> But by skilfully manoeuvring the motors, and with lots of luck, Helsham made the coastline. The roar of the slipstream continued to make the trip a nightmare.
>
> ... the aircraft had to be landed flaps up. The aircraft's speed was so great that only a belly-landing could be made ... the crew was amazed to see Helsham bump the bomber into land and finish up on one side of the strip. Beyond a few bruises and scratches no one was hurt.[97]

Helsham's A16-213 was so badly damaged it had to be written-off. All too often, though, crews did not have Helsham's luck. Four days later, Flying Officer John Oldridge's[98] crew (A16-178) departed from Millingimbi to the north on a Koala shipping reconnaissance. Their plan was to fly about 50 kilometres off the Aroe Islands, then a similar distance off the coast of Dutch New Guinea.[99] As had happened before on these lonely patrols, nothing more was ever seen or heard of the crew.

Flight Lieutenant Bunny Austin had been one of the Squadron's Hudson pilots during 1943. In November he received an insect bite and died a few days later from blood poisoning. Acknowledging his award of the DFC, to add to his previous Mentioned in Dispatches, a contemporary magazine article described his exploits:

> They recalled the time Bunny stooged over Penfoei for an hour after leading the Hudsons in. Bunny just defied the screaming ack-ack fire for a grandstand view to observe the bombing results—and then took his own aircraft in to bomb. But Bunny had been in plenty of other hot shows. I went along with his crew one night to Langgoer, Kai Islands, 300 miles over water. The flak pelted at us like red hot tomatoes, but Bunny never turned a hair. *Yet Bunny was just a typical member of this Hudson unit.*[100]

97 "The Tale of the RAAF Hudsons", *Wings*, Melbourne, 12 Dec 1944, p.4.

98 Flg Off J P Oldridge, 417509. No 2 Sqn. Of Melbourne; b Melbourne 30 Nov 1921. Killed in action 15 Oct 1943. Crew: F Sgts D N Rumble (413260), J F G Cullen (423078), Sgts F R Drake-Brockman (427451), D W Petch (437040).

99 RAAF Historical C.D.102/45 A16-178, of 3 Mar 1947.

100 "The Tale of the RAAF Hudsons", p.2–3.

CHAPTER 17
Beaufort Interlude

... those who flew the Beaufort operationally—those whose lives depended on its qualities—swore by it and its inherent strength; whereas those who flew it only as a trainer and as a stepping stone to later designs frequently swore at it ... [1]

By October 1943 the Japanese had become alarmed at their inability to hold the Allies in New Guinea, and had considered withdrawing from Rabaul and abandoning the Bismark Sea area, to a new defensive perimeter to the west, from the Marianas, through Truk to Wewak.[2] NW Area continued supporting the New Guinea offensive by destroying as much as possible of the enemy strength, on the flank to Darwin's north, and to the rear in Timor.

Armed reconnaissance patrols around the Arafura Sea continued for No 79 Wing—the Hudsons of No 2 Squadron and B-25 Mitchells of No 18 (Dutch) Squadron—in addition to nightly attacks on Koepang, Lautem, Fuiloro and Langgoer. The long and monotonous shipping patrols were the standard introduction for new crews joining the Squadron. One crewman was Sergeant Bruce Wallace,[3] a WAG with Flight Lieutenant John Scott's[4] crew, whose vivid description, flying in A16-199 on 12 November, recounted such a sortie:

> We took off at 9AM on our first Horse. It is an eleven hundred mile flip around the Tanimbar Islands. These islands are volcanic in origin and are different from the ones to the east. North-west of the Tanimbars lies Seoea, an extremely fertile, small precipitous island rising abruptly out of the sea to the height of two thousand odd feet. It is an active volcano, a line of white smoke drifting lazily towards the sky. Clinging to the sides of the mountain are several native villages. The huts rise tier upon tier towards the summit. Terraced fields slope towards the sea. The vegetation is thick and luxurious. A couple of minute rocky coves give access to the sea. It presents a beautiful sight, an emerald set in a sapphire sea with fleecy white clouds like a horseman's cloak floating from the shoulder of the smoking volcano.[5]

Two days later, 14 November, Bruce Wallace's crew, flying A16-177, was sent with three other Hudsons to attack Penfoei. Wallace recorded in his diary:

> We saw the target, the dispersal area at the eastern end of the strip at about 11.15PM. A kite or two had been in before us and left a couple of small fires. We came in at 9000 feet from the

---

1 R Hayward, *The Beaufort File*, Air-Britain, Tonbridge, Kent, 1990, p.5.

2 Odgers, p.89.

3 FSgt B E Wallace, 410763. No 2 Sqn. Farmer of Bendigo, Vic; b Bendigo 30 Jul 1923. Executed by Japanese 16 Aug 1944.

4 Sqn Ldr J L Scott, 290602. No 2 Sqn. Of Mt Lawley, WA; b Adelaide 8 Apr 1914. Executed by Japanese 16 Aug 1944.

5 Bruce Wallace's diary, of 12 Nov 1943, courtesy of his brother, Mr John Wallace, Melbourne.

NW over Koepang towards the strip. Over the town I was very busy throwing pamphlets out the belly gun.[6] The navigator thought he saw another kite so we dived down 2000 feet to port. There were pamphlets from one end of the kite to the other. As we made our run over the target the Nips came to life. On "Bombs gone" I began pushing out bottles, pamphlets and the incendiaries which were carried inside instead of on the bomb racks underneath. A searchlight came sweeping towards us and a red line of bofors shells laced upwards on our starboard quarter. One shell left the line and passed right under us. It looked a moral from the belly gun. As it raced towards us I made a dive for the camera bracket and my chute. Then we were out of the target area and making for the sea. Our bombs and incendiaries landed on the target and left four fires in their wake.

November 21st. Millingimbi again today. Took off again at 11.30AM on the Koala. Reached the Aroes at 2PM. We took another 100 odd photographs of the coastline. On our return we heard that 1 B-25 was shot down over Taberfane. Five B-25s and 6 Beaufighters did the joint over. They got 2 Rufes and 4 ships.

January 4th. Left Millingimbi at about 10AM on the old Koala now named Kiwi. We dodged several fronts on the way up to the New Guinea coastline ... turned into a bay to strafe some Jap soldiers ... on our right we suddenly saw another village with Japs dashing about. Beside the village lay Timoeka aerodrome running east west from the water's edge. Our fire straddled the canoe and the Japs crouched down in the bottom ... others sat up and stopped fire from the turret. Suddenly there was a crash and a couple of machine gun bullets went through the kite fired from the edge of the strip. One hit the magazine on the starboard beam gun—and wrecked it. Had anyone been on it they would have been hit in the face. Another one passed right through both sides of the fuselage leaving a hole the size of a dessert spoon beside the door. Quite an interesting trip. One of the engines coughed on the way home and gave us all a fright.

January 13th. Went up to do some formation flying with a couple of Beauforts this morning. The rest of the crew went up again with them this afternoon to do some Spitfire Co-op. During the action two Spits collided and crashed near Humpty Doo Station and gave everyone quite a shock.[7]

Plans had finally come to fruition in late 1943 for the replacement of 2 Squadron's ageing Hudsons. No 12 Squadron had already exchanged its Hudsons for Vengeances, and No 13 Squadron had been out of the front-line for six months re-equipping with the Beaufort. This coincided with the change in character of the war, as enemy air strength was weakening and not more than a few of their aircraft could be maintained at forward bases. Being no longer vulnerable to air attack, logistics and repair facilities were moved forward to Darwin from the back areas. From the end of 1943 the Japanese discontinued raids altogether, and withdrew their Army's weakened 7th Air Division from Wewak to Ambon, to assist in the protection of the Celebes area.[8]

6 Often on these raids the Hudsons dropped propaganda pamphlets to the Timorese and the enemy. Also bottles were dropped, to present the enemy with the terrifying whistling sound simulating a bomb.

7 Bruce Wallace's diary.

8 Odgers, pp.120–1.

During November, the introduction of the Lockheed B-34 Ventura to the Squadron was in progress.[9] The "A" Flight Commander, Squadron Leader Keith Kemp, with a ferry crew from 1 Aircraft Depot at Laverton, delivered 2 Squadron's first Ventura (A59-9) to Hughes on 22 November. A further nine Venturas were allotted to No 2 Squadron as the AOC, AVM King Cole, visited Hughes to inspect the first of the new aircraft in the Area. This USAAF B-34 variant of the Ventura was soon found by Wing Commander Hennock to be unsuitable for operations. He reported:

> I tested it with S/Ldr Kemp, but of the B-34 type, we found that it was useless for our job, having neither range nor being capable of carrying a decent bomb load. The Navy Ventura, as distinct from the Army [variant], is a far better machine for the type of work which we required.[10]

A9-380 was one of the 13 Squadron Beauforts due for transfer to 2 Squadron, but was damaged at Canberra before being delivered. RAAF

The planned re-equipment with the Ventura was, however, to be cancelled on 16 December,[11] as there were only enough of the improved naval variant (the PV-1 Ventura) to equip one RAAF unit (No 13 Squadron). The next day, the following memorandum was issued detailing 2 Squadron's re-equipment with Beauforts of No 13 Squadron:

---

9 No 2 Sqn A50, of 8 Nov 1943.

10 Interview with Wg Cdr K S Hennock DFC, DDOR RAAFHQ, of 12 Feb 1945, p.4. RAAF Historical Section.

11 HQQO 902, of 16 Dec 1943, cited on A59-9 Aircraft Status Card, RAAF Historical Section.

ROYAL AUSTRALIAN AIR FORCE
ORGANISATION MEMORANDUM No 404

Nos 2 and 13 SQUADRONS—RE-ARMING
(151/2/268) (151/2/99)

1. No 2 Squadron will be re-armed with Beaufort aircraft to replace the Hudsons presently held on strength. The re-arming will be effected as follows.
2. The Beaufort aircraft of No 13 Squadron at Canberra will move as early as possible into North-Western Area and will arm a flight of 2 Squadron. Beauforts crews of 13 Squadron will be transferred with their aircraft. As the Hudsons of 2 Squadron become due for repair or overhaul they will be withdrawn and replaced by further Beauforts, until the squadron is re-armed to a strength of 18 Beauforts, plus one dual Beaufort.
3. Following the transfer of the Beaufort aircraft, 13 Squadron will be re-armed by the allotment of 9 Venturas (type B-34) as they become available from 1 AD.
4. No 13 Squadron will then consist of one flight of 9 type PV-1 Venturas and one flight of 9 type B-34 Venturas.

17th December 1943[12]

However, this schedule for re-equipment was not to progress exactly according to plan either. Nine Beauforts were expected to arm a flight of No 2 Squadron,[13] but when Beaufort A9-380 of 13 Squadron crashed at Canberra, only eight aircraft were to reach 2 Squadron by the end of the year. The maximum number of Beauforts ever operated at one time by 2 Squadron was 13 aircraft, with the last of the Hudsons soldiering on well into 1944. The Beaufort was a four-place torpedo and general reconnaissance bomber which had served with the RAAF since 1941. The crew comprised pilot, navigator and two wireless-operator air gunners (WAGs). The performance was better than the Hudson, with a maximum speed of 241 mph at sea level, and a maximum load of 1,000 kg (2,200 lb). Being an aircraft of British design, the standard weapon load became six 250 lb bombs, reverting to the older British weapons.

The Beaufort program had, in fact, been one of the miracles of Australian wartime production, paving the way for a viable aircraft industry. Unable to obtain bomber aircraft in quantity from England in 1940, the Beaufort Division of the Department of Aircraft Production was established at Fishermen's Bend, Melbourne, and Mascot, Sydney, as final aircraft assembly factories. Local manufacture of Beaufort components extended from the railway workshops in three states to General Motors-Holdens in South Australia. Hundreds of private engineering workshops were also organised for production of parts, the policy being to subcontract the manufacture of components to the maximum extent. Planned production at the rate of 20 aircraft per month had been attained by June 1942, reaching its peak in September 1943 with an output of 37 aircraft.[14]

12 Grantham, p.78.

13 No 2 Sqn A50, of 20 Dec 1943.

14 B Graham & F Smyth, *A Nation Grew Wings*, Winterset House, Melbourne, 1946, pp.275–6. The last Beaufort, A9-700, was delivered in Aug 1944.

The navigator's position in the nose of the Beaufort RAAF

The flight of eight Beauforts arrived with their crews on Christmas Eve, led by Flight Lieutenant Joseph Lee,[15] who became "B" Flight Commander. Some of the other pilots transferred from 13 Squadron with the aircraft were Flying Officer Keith Coughlan,[16] Flight Sergeant Jim Henderson[17] and Sergeant "Peter" Hocking.[18] Remembering the short Beaufort period on 2 Squadron, Jim Henderson recalled:

> The Beaufort aircraft gave a good deal of trouble; mostly small things but they were being flown from 4 1/2 to 8 hours, there was no room for malfunction. The Beauforts did not perform well in the tropical air of Darwin. Being an NCO pilot I always flew in the "box".[19] This was a very tiresome position to fly as a loaded Beaufort wallowed like a pig in the slipstream of the leader. Having no co-pilot and no auto pilot, a long trip was an exhausting business. It is no wonder I vividly recall those times.[20]

The Beaufort crews began working up in their new environment and converting the older 2 Squadron crews to the new type. Meanwhile Hudsons maintained the over water searches into 1944—the routes were now known as the "Gull", "Heron", "Jackass" and "Kiwi". The first Beaufort convoy protection patrols were flown on Wednesday 12 January by Keith Coughlan's (A9-386) and Jim Henderson's (A9-375) crews, and these fairly mundane sorties became the pattern for the first three months of 1944 as Beaufort experience was gained.

February saw the remnants of Japanese air power leave Rabaul and their fleet withdraw from Truk. Japanese hopes of expansion were now in retreat, as No 2 Squadron returned to the skies over Timor again on the night of 18 February, when three Hudsons, led by "A" Flight Commander Squadron Leader Les Ingram[21] (A16

15 Sqn Ldr J S Lee DFC, 4211/734/O211122. Nos 23 & 22 Sqns, 1 OTU, 13 & 2 Sqns, 7 OTU, 78 Wg. Flight fitter of Sydney; b Sydney 15 Nov 1916.

16 Flt Lt K E Coughlan, 427298. Nos 13, 2 & 35 Sqns. Indent agent of Perth; b Perth 21 Feb 1913.

17 WOff J I Henderson, 410332. Nos 13 & 2 Sqns. Farmer of Wairewa, Vic; b Orbost, Vic, 9 Sep 1922.

18 Plt Off B M Hocking DFM, 418839. Nos 13 & 2 Sqns. Orchardist of Healesville, Vic; b Melbourne 11 Jul 1922.

19 The number 4 position at the rear of the formation.

20 Jim Henderson, letter to the author, of 20 Mar 1993.

21 Wg Cdr L A Ingram, 379. Nos 14, 13 & 100 Sqns; comd 2 Sqn 1944. Regular air force offr of Geelong, Vic; b Melbourne 1 Apr 1917.

203), bombed Manatoeto. No 79 Wing now comprised 1 Squadron at Gould and 2 Squadron at Hughes, both with Beauforts, 18 (Dutch) Squadron at Batchelor with Mitchells, and 31 Squadron at Coomalie with Beaufighters. Wing Commander Les Ingram took command of 2 Squadron on 23 March, by which stage only four Hudsons remained, with most Hudson crews either posted, or converted to the Beaufort.

NW Area was now to support MacArthur's landings at Hollandia and Aitape by an air strike campaign against the airfields to the north and in western New Guinea, with emphasis on the Kai Islands. Further tasks included harassing airfields bordering the Timor Sea, to prevent any staging of enemy aircraft, which could be used against the beachheads. On 27 March, Squadron Leader John Scott (A16-230) searched islands to the north, in the Wetar Straits between the Banda and Timor Seas. Scott's aircraft was shot down near Loeang Island, but the subsequent search by Beauforts failed to locate any trace of the crew. All of the crew, in fact, had been able to reach the island, and were cared for by natives before being captured. It is believed the crew were taken to Babar Island, and then to Saumlaki on Jamdena Island. When the crew were being transferred by barge to Toeal on 24 May, they were attacked by an 18 Squadron Mitchell near Tenaman Island. Pilot Officer Don Beddoe[22] was wounded in this attack and then shot dead by one of his captors. Scott and the rest of his crew (Flight Sergeants Robert King,[23] Bruce Wallace and Keith Wright[24]) were then moved to Ambon, and were all executed by the Japanese at Galala, on 16 August 1944.[25]

In the early morning darkness of Monday 3 April, Pilot Officer Colin Brockhurst's crew[26] (A16-203) departed from Hughes on a convoy protection patrol over Allied shipping in the Arafura Sea. The aircraft failed to return. (The wreckage, sunk in mud in the Adelaide River, was not discovered until 1946.[27]) These two losses over the last weeks of Hudson operations marked virtually the end of No 2 Squadron's association with these war-weary bombers.

The first Beaufort strike on Penfoei was the night of Thursday 30 March, when Squadron Leader Lee (A9-578) led three aircraft, staging through Drysdale, to bomb the airfield. The exercise was repeated the following night with Squadron Leader Alan Hayes[28] (A9-578) leading six aircraft. On this occasion, heavy anti-aircraft fire was encountered, but caused no problems because of its inaccuracy. Raids such as this in

---

22 Plt Off D C Beddoe, 401731. No 2 Sqn. Of Melbourne; b Melbourne 29 Jun 1917. Killed in captivity 24 May 1944.

23 FSgt R I M King, 427122. No 2 Sqn. Of Perth; b Kalamunda, WA, 16 Apr 1923. Executed by Japanese 16 Aug 1944.

24 FSgt K C Wright, 434064. No 2 Sqn. Of Kangaroo Point, Qld; b Brisbane 10 Jul 1924. Executed by Japanese 16 Aug 1944.

25 RAAF Historical C.D.102/45 A16-230, of 3 Mar 1947 and 31 May 1948.

26 Plt Off C D Brockhurst, 425593. No 2 Sqn. Of Nambour, Qld; b Nambour 23 Aug 1921. Killed in action 3 Apr 1944. Crew: Plt Offs R J Philpot (403877), R L Black (411275), H J Gillies (422042), FSgt R G Frew (417771).

27 CD 102/45 for A16-203, A/L 9, of 23 Sep 1946.

28 Wg Cdr A S Hayes, 260512. 2SFTS, CFS, No 2 Sqn; comd GR Sch 1944–45, Darwin 1945. School teacher of Ariah Park, NSW; b Goulburn, NSW, 10 Dec 1912.

the NW Area strike campaign against Ambon and Timor had led the enemy to suspect an Allied offensive would be mounted in the NEI. For this contingency, the Japanese had kept some 300,000 troops waiting for the expected invasion from Darwin.[29]

Night formation strikes against Timor continued into April for 2 Squadron Beauforts, striking with 18 Squadron Mitchells against Koepang, Dili and Lautem. These aircraft were joined in operations by the Beauforts of No 1 Squadron. For these night operations, a modification had been carried out on the Beauforts: the engines were made less conspicuous by fitting flame arresters, which produced a subsequent loss of 5 knots of airspeed. These strikes continued as diversions from MacArthur's push along the northern New Guinea coast, and for the first time NW Area's Spitfires were used in an offensive role. Staging through Bathurst Island on 18 April, the Spitfires struck Babar Island, 600 kilometres to the north of Darwin.

The last Hudson sortie was flown by Flying Officer Roger Kuring's crew: (from left) Leo Zarnke (WAG), Bob Hebbard (nav), Roger Kuring, Ray Hilton (WAG) and Jack Carolin (gunner).
R KURING

Searches and convoy protection sorties were flown during May. Taking off for one such Heron search of the Arafura Sea and Aroe Islands on the evening of 18 May, Flying Officer Roger Kuring[30] (A9-560) crashed, but with no injuries to the crew. Kuring was one of the few pilots to fly three types of aircraft (Hudsons, Beauforts, Mitchells) on operations with No 2 Squadron, and had the honour of flying the Squadron's last Hudson operation, in A16-199, on 8 April 1944. During this period of

29 Odgers, p.227.

30 Flt Lt R N Kuring, 401516. Nos 2 & 38 Sqns. Clerk of Melbourne; b Melbourne 20 Mar 1917.

Beaufort operations one crew had the experience of flying a strange formation on return from a shipping sweep:

> This occurred at night and the crew thought it was one of our planes until they got home and found out none of our planes was flying that night. It is believed that the extra aircraft was Japanese.[31]

Meanwhile the decision had been taken to replace the Beaufort on No 2 Squadron with the B-25 Mitchell. These aircraft had been procured by the Dutch and were now excess to their requirements. As new Beaufort crews arrived on the Squadron, so did extra co-pilots and gunners to build up the crew of four of the Beaufort to the six required for the Mitchell. Joining the Squadron with his Beaufort crew over this period was a new navigator, Flight Sergeant Brian Hawthorne,[32] whose vivid recollections of Hughes and life on the Squadron have not faded over the half century that has passed.

LIFE AT HUGHES—1944

First Impressions

> The long flight from Parafield to Gorrie (Birdum)—via Oodnadatta and a freezing overnight on the airfield at Alice Springs—was an eye-opener to our crew after training in the cold green South. The long stages over featureless desert, with the vast sandhills extending from horizon to horizon like ocean rollers, and the incessant turbulence, impressed us with the emptiness of our continent, and the sturdiness of the Dakota, which was, in those days, a huge aircraft.
>
> Two or three days in the transit camp at Gorrie (55 Operational Base Unit) was our introduction to life in the Top End, a life which was to be ours for the next nine months (the length of an operational tour for aircrew in North West Area—groundcrew did two years). It was to be the same open air life at Hughes, at Millingimbi, and at Truscott—tents under scrubby eucalypts the smell of new canvas, musty mosquito nets, citronella oil, and burnt kerosene doused latrines. The rough and ready mess kitchens, opensided, built of saplings and corrugated iron, and, underfoot, hot grey sand. The mosquitoes and sandflies were torture at Gorrie—but not quite so bad at Hughes.
>
> Hughes Field lay on the eastern side of the North–South Highway, about seventy kilometres south of Darwin. Unlike the fighter strips, at Livingstone, Sattler and Coomalie Creek, which were beside and parallel to the highway, Hughes ran off at an angle, and was not visible from the road. The living quarters and administrative area were on the western side, a few hundred yards from the highway, being screened by the same light scrubby bush that characterises the area.
>
> I have a hazy recollection of the administrative HQ and orderly room, the three messes all grouped close together, and, I think, a hospital—all built of iron on sapling frames, with oiled hessian screens. There was also a large marquee, the Operations Tent, and I suppose there must have been an Equipment Store as well.
>
> Then there was the haphazard scatter of tents through the bush, with open air latrines and shower blocks. It was impossible to shower before 9pm as the water pipes lay along the surface, and were heated by the sun during the day.

31 Jim Henderson, letter to author, of 20 Jul 1993.

32 Sqn Ldr B D Hawthorne, 410662/O33129. Nos 2, 34 & 77 Sqns, 82 Wg; comd Momote 1952–54; 82 Wg, 1 OCU. Clerk of Melbourne; b Melbourne 10 Dec 1922.

"B" Flight, No 2 Squadron, May 1944:
Back Row: (from left) Plt Off Polgreen, WOff Brooks, WOff Campbell, Flt Sgt Davies, WOff Clarke, WOff Griesbach, Plt Off Kaiser, WOff Trevor, Flg Off Lutze, Plt Off Kemp, Flg Off Carter, Flg Off Hebbard. Middle Row: Flg Off Ingram, Flt Sgt Guthrie, Flt Sgt Malley, Flt Sgt Dinsdale, Flt Sgt Weston, Flt Sgt Larkin, Flt Sgt Wood, Flt Sgt Elliott, unknown, Flg Off Fileman. Front Row: Flt Sgt Field, Flg Off Zarnke, Flt Sgt Hughes, Plt Off Crossthwaite, Flt Lt Ditchburn, Sqn Ldr Lee, Flg Off Kuring, Plt Off Moriarty, WOff Brown, Flg Off Smith, Flg Off Davies
J HENDERSON/R KURING

Our tent was furnished with canvas stretchers, a couple of tin lockers, and little else. Someone had tried to outline a short entrance path with half buried, up-ended beer bottles, but nearby tents had more extensive improvements. Later we were to find that our tent had housed a Hudson crew who failed to return from a mission a week or two before.

We unpacked our duffle bags and sat and looked at one another. This was going to be a rough tour. Then a diffident voice at the entrance said "Excuse me! Would you chaps care to join us in a cup of tea? We're just brewing up."

So we joined the crew next door in a very formal "getting to know you" afternoon tea. They made us comfortable on butter box chairs, gave us the cleanest enamel mugs, and fussed over us like old ladies at a church social. I have never forgotten that simple neighbourly act of courtesy and civilised conduct, carried out in such an unlikely setting. It seems to have symbolised the comradeship of life in an operational squadron, and the values we believed we were fighting for.

* * * *

Beaufort KO-Z RAAF

Two days later we made our first familiarisation flight over the local area in a Beaufort. Two days after that we made our first operational reconnaissance flight over the north coast of Arnhem Land—a five hour search of islands and beaches for any signs of enemy activity.

The Squadron had just become operational with Beauforts, and still had the last Hudsons, although these were now phased out. In the revetments were a couple of rakish looking B-25s. There were rumours that 2 Squadron was to operate them, but nobody knew when. Just a rumour. Probably crews were training elsewhere.

Then, after a month of patrols in Beauforts, operations ceased, and we found ourselves allowed to enter the B-25s, where groundcrew explained the equipment and controls and systems. After a few days, Dutch pilots from 18 Squadron arrived and we started flying, under instruction. After Beauforts, B-25s were impressive. I never entered a Beaufort without losing a piece of skin somewhere, but the Mitchell was spacious, and almost luxurious. The pilots had sunblinds and ash-trays, heavy armour plate, and padded armrests. The B-25 was armed with more than a dozen .5 machine guns, as against the Beaufort's miserable four .303 peashooters. And the invariably fatal manoeuvre (in a Beaufort) of turning into a "dead" engine could be carried out with ease in the Mitchell.

The conversion took 3 weeks and about 21 hours in 15 sorties. Pilots usually went solo on their second flight. After that there was instrument flying, formation practice, and high and low level bombing, with gunnery to master.

Then we went looking for trouble.

* * * *

Field dentist, Territory-style, October 1943 F RIGGS

Operations

Most sorties were planned, and many sweeps of known enemy haunts were almost routine. Very few sorties were the result of an immediate scramble.

Usually crews were summoned to the Ops Tent for briefing by a call on the Tannoy. Speakers were mounted in trees at strategic places around the camp. All too often they were employed to broadcast music from cracked 78rpm records from an amenities record player. "Carry Me Back to Ol' Virginnie" crackled through the bush again and again, hour after hour. It became almost relief to hear the music stop and a voice say "Flying Officer Avery and Crew, Flying Officer Slater and Crew, Flt Lt Carter and Crew, report to Operations immediately!".

Crews would assemble in the big marquee and the Operations Officer would outline the operation on a large map. Intell, Nav, Sigs and Met briefings would follow. Sometimes the CO might be there to inspire us.

A late afternoon briefing was normal for an early take-off next day. If take-off was the same day, we would usually have plenty of time for flight planning, collection of safety equipment, inflight rations, etc. In either case, one of the non-flying crews would have preflighted your plane. We did this for one another. It was a good system and it worked.

Oh, those dawn take-offs! Early call was usually at about 4am. I would wake just as the heavy boots of Corporal steward "Fitz" approached our tent. "Wakey wakey! All right youse! Git yer 'ands off it and rise and shine! Up youse all git. All youse aircrew bastards, breakfast in the Airmens' Mess at 4.30."

Rubber eggs and sausage and bacon in a cheerless tin shed. A rough jolting ride in the back of a battle buggy or 5-ton truck across to the dispersal area. Stifling dust, cool air and silent thoughts to match the darkness under the shadowy trees.

This was a time to wonder what you were doing in this place, at this time. It was not a time for jokes, or excited chatter. You tended to envy the groundcrew, who could be pretty sure of being here at the end of the day.

But, after take-off, with the crew settling in to the routine of the flight, the joining up of the formation, the testing of the guns, the sun coming up in the east to warm through the perspex, and the prospect of hot coffee coming up soon, doubts and fears receded, and it was possible to actually enjoy the adventure of approaching strange coastlines, and perhaps seeing a little action.

The islands around Timor are remarkably scenic, being volcanic cones rising abruptly from the waters, or large plateaux with sheer jagged cliffs that extend for miles, or, as in the Aroe and Tanimbar and Kai groups, low and flat and swampy, with huge mangrove coastlines. We often marvelled at the beauty as we dived past picturesque native villages, coconut palms, fish traps, lakatois and lap-lap clad natives who would either wave at us or flee into the jungle. This was the sort of trip a millionaire could not buy.

So now we pitied the groundcrews, stranded in the monotony of the bush around Darwin, for two years at a stretch. At least we were seeing something different, having a little excitement, and seeing the fruition of their hours of painstaking labour. We were grateful to them, confident in the serviceability of our planes, <u>their</u> planes, and we knew how much they must have envied us, when we returned after a successful mission, laughing and chatting and exchanging notes on our way to debriefing in the Ops Tent.

But there were also days, when the AA had been a little too close, the weather a little too frightening, or the fuel a little too low on return. There was no laughter, no chatting on these days. Just a group of silent men, their expressions far away, their voices abrupt, and their eyes very, very thoughtful. That was when we all learned that there were no heroes in battle—just very frightened men.

* * * *

Recreation

When not flying, crews were on standby, or rostered for preflight duties. As far as possible, one day per week, usually Sunday, was a standdown. A truck would be provided to take personnel on standdown to Darwin, if they so desired. After one or two trips to Darwin, where there were no theatres, no cafes, no shops and no women, only Army, most of the chaps preferred to remain at Hughes, doing exactly what they had been doing all week when not flying.

These activities were:

Sleeping in their tents.
Writing letters home.
Reading books or magazines.
Playing cards.
Watching ant lions building their nests in the ground.
Pushing ants into the ant lions' nests.
Walking over to the Operations Tent to see if any new Intelligence Bulletins had arrived.

There were no sporting facilities, and no gymnasium. There was no social life in the messes. Beer was rationed to two bottles per man per week perhaps. On Thursday nights every tent saw the ritual of the Pouring of the Beer. Non-drinkers collected their ration, of course, and sold, gave or exchanged their bottles with other crew members. The flickering generator-driven electric light bulbs shone on the small circles of eager drinkers, who carefully uncapped each bottle, and with dreadful concentration poured the contents in exactly equal portions, to the last agonised, bottle-squeezing drop.

The alternative to beer was "lollywater", a passable soft drink brewed by the Army, using empty beer bottles, which abounded around Darwin. The North–South road to Adelaide River was called the "Crystal Highway", because of the glitter of smashed and whole bottles along the edges.

Once a week, clutching our bottles of lollywater, we sat on rough sapling benches, or the ground, in the open air theatre carved out of the bush, to watch the latest Hollywood or British epic.

In the wet season, we wore raincapes and hats fur felt, because not even a tropical downpour was allowed to deprive us a glimpse of Rita Hayworth, or Jimmy Stewart, or Looney Tunes. For an hour or two we lived in a world of cities and houses, trains and automobiles, like the world we used to know.

We used to complain about the absence of women in the area. Strangely enough, when some of us visited Charleville, I realised that what I had missed, more than female company, was the sight of curtains in windows, and the sound of children at play.

* * * *

The CO

As a lowly flight sergeant, I had little contact with our commanding officers. So, when I walked over to the revetment area one day, to carry out some chore, I was disconcerted to see the tall figure of Wing Commander Dave Campbell DFC pacing the taxiway in my vicinity. Automatically, I flung him a triple quiver salute, the first I had given in a long time. Too late I remembered that one does not salute on the tarmac. But the CO returned my salute, with an amused smile, and resumed his lonely pacing to and fro. He seemed deep in thought, lost in contemplation.

I wondered if he had heard bad news, and wanted to be on his own. Or was this how he planned operations? Perhaps he just wanted to get away from the pressures of office.

Years after the war, I read a fascinating book by David Campbell, who, I learned, was now one of Australia's finest poets. I intended to write to him, and ask him if he had been composing a masterpiece on that day, and to thank him for having recommended me for my commission. But he died before I could do so.[33]

33 These notes were supplied by Brian Hawthorne to the author on 3 Sep 1993. Brian's eloquent description is included, in full, to evoke the memories of those who served in Australia's north, and provides an insight into life on No 2 Squadron in 1944.

CHAPTER 18

ROGERS

The Commander of the Japanese forces in Timor later stated that after April 1944 Allied air attacks in Dutch Timor were so skilful and heavy that vehicles could not move by day, and coastal shipping could only travel at night.[1]

The last Beaufort sorties were flown on 22 May, and then No 2 Squadron was withdrawn from operations to enable re-equipment with the Mitchell.[2] The North American B-25 Mitchell, which was code-named the *Roger* by Operations staff,[3] was to provide a very capable mount for the 2 Squadron crews. The B-25 bristled with guns, the pilot had four fixed forward-firing .50 cals, with nose guns for the navigator, waist guns, an upper turret, and a tail position. The earlier B-25D version had a single gun in the tail, the later B-25J had the sting of twin 0.5s. The B-25D had a manually operated dorsal turret in the rear fuselage with twin 0.5s, the later variant had a hydraulically operated turret mounted further forward, just aft of the cockpit. This modification had required the navigator's position to be moved to the glasshouse nose.

The Mitchell was a delight to fly, and arguably the finest medium bomber of the war. In six years, No 2 Squadron had progressed through the Anson, Hudson and Beaufort, each at the time an innovative type, and an improvement on its predecessor. However, these were all now woefully obsolete. The Mitchell was another example of how American ingenuity was the leading force in the development of air power.

It was not only necessary to convert to the new aircraft, but the additional crew members carried on the B-25 meant devising new procedures. The Hudson had a crew of five, the Beaufort four. New crew members for the Mitchell were the co-pilot and tailgunner who, with the captain, navigator and two WAGs, comprised the crew of six.

An Air Board directive, in April 1944, described to Headquarters NW Area how the transfer of B-25s from the Netherlands Army Air Force to No 2 Squadron was to proceed.[4] Eighteen B-25D and two B-25C were to be taken over by the RAAF from the Dutch at Canberra on 19 April. The B-25D aircraft were new, having been flown to Canberra from the United States, whereas the B-25Cs had completed between 300 and

1 Wray, p.180.

2 Odgers, p.228.

3 Other Operations code-names for RAAF aircraft using the phonetic alphabet were *Easy* for the Catalina, *Fox* Mosquito, *George* Beaufort, *Hugh* Ventura, *Mike* Liberator, *William* Beaufighter, and *X-Ray* Spitfire.

4 Air Board letter from AMSE 9/41/195, of 17 Apr 1944.

400 hours, and were not considered suitable for operations.[5] After inspection and minor modification, the aircraft were to be ferried to Hughes, and conversions conducted on the Squadron. Three Dutch pilots, Lieutenants Renaud, Radenaker and Jansen, together with four B-25s, were to be made available over May to begin conversions to the new type.[6] Pilot Flight Sergeant Jim Henderson recalled:

> The only drama I had was while being converted by a Dutch veteran. He told me to land on the Darwin strip but he did not see the huge concrete blocks in the middle of the runway (the runway was being strengthened to take Flying Fortresses and the concrete was to test the surface for strength). I made the approach to land on the far side of the concrete block but he grabbed the controls from my hands and proceeded to land in front of the block. He then saw the block and managed to haul the B-25 over it. The men on top of the block leaped for safety—about 20 feet.[7]

No 2 Squadron's B-25 operations commenced in June, on searches and patrols. MacArthur's campaign was now focused on the amphibious assault to the north of Dutch New Guinea on Noemfoor Island and, as with the earlier landings, NW Area was again to play the supporting role by harassing enemy installations on Timor and Ceram. With no staging airfields to the north for refuelling, there were now few worthwhile targets that the medium bombers of 79 Wing could reach.[8] As the war passed on to the north, this would continue to be the case, causing concern at the futility of some operations.

A strike on 27 June, at Lautem West airfield on Timor, opened this distraction from the Noemfoor campaign. Nine Mitchells were led by "A" Flight Commander Flight Lieutenant Charles Hill[9] (A47-7), each dropping seven 500 lb bombs from 10,000 feet. Moderate anti-aircraft fire was encountered over the target. Two days later, Flying Officer John ("Joe") Simpson[10] (A47-16) led Flight Sergeant Peter Hocking (A47-6) and Pilot Officer Bob Avery[11] (A47-9) to bomb and strafe Doka Barat airfield, on Trangan Island, the largest in the Aroe Group. Bob Avery's navigator, Flight Sergeant Brian Hawthorne, recorded:

> We were Number Three, and took off last. The Mitchell had 4 five hundred pound bombs aboard, but one hardly noticed that. We were quickly alongside the other planes, and we set

5 The AMSE letter cites five Dutch B-25C aircraft available, with the RAAF having the pick of the best two: N5-154 (ex 41-30584), N5-155 (41-30586), N5-157 (41-30588), N5-158 (41-30589) and N5-160 (41-30713). In fact, these aircraft were early model B-25Ds. J M Andrade, *US Military Aircraft Designations and Serials*, Midland Counties, Earl Shilton, Leics, 1979, p.48. The only difference between the B-25C and B-25D, in any case, was that the former was manufactured at the North American plant at Inglewood, California, the latter at Kansas City.

6 Signal RAAF HQ A630, of 25 Apr 1944. The first B-25s made available from this date were A47-16, A47-17, A47-19 and A47-20.

7 Jim Henderson, letter to author, of 20 Mar 1993.

8 Odgers, p.231.

9 Flt Lt C N Hill, 406528. 44th Bn, No 2 Sqn, 1AD, 7 Sqn. Audit clerk of Perth, WA; b Perth 25 Sep 1918.

10 Flt Lt J C Simpson DFC, 400558. Nos 32 & 2 Sqns, CFS. Grazier of Oaklands, NSW; b Urana, NSW, 13 Nov 1914.

11 Flt Lt R L Avery, 408360. No 2 Sqn, 2AD. Student of Hobart; b Adelaide 27 Nov 1922.

off in a NNE direction, only three minutes behind time. By the time we were over the coast of Van Diemens Gulf, the formation was holding nicely.

The crossing over the water was quite uneventful ... By now we were a couple of hundred miles out, halfway across the Arafura Sea. So I took my papers and maps down to the front compartment. It is rather a struggle to crawl through the tunnel leading from the navigator's compartment to the nose[12] ... Sitting on the ammunition bins for the two fixed guns which are mounted on the starboard side of the compartment is not the most comfortable position in the world, but it affords excellent all round visibility. By now we were fairly low, 100 feet or so, as we were approaching enemy territory, and wanted to avoid being detected by possible radar units.

The other two planes were in loose formation beside us. I could see the navigators sitting in the nose, the pilots at their controls, the turret gunners swinging their guns in what I hoped was an efficient search, and I could make faces at the tail gunner of the next plane. So far, so good.

A brief crackling on the very high frequency intercom, and we were diving gently, testing our guns. In the nose, the poor observer is surrounded by guns. He has two at his right side, poking out through the front perspex. Outside, in blisters on each side of the fuselage, are four more pilot-controlled guns, two on each side. And he has one free gun mounted in the nose, which is fired by himself. When all seven guns are firing, the vibration and noise is rather terrific.

I estimated another ten minutes before our arrival at Enoe Island. I checked the bombing gear, and made what settings I could. I knew that my intervalometer, or automatic bomb distributor, was unserviceable, and that I would have to "toggle" by hand, judging the spacing.

... and then over the mainland. I was concentrating on the terrain ahead, watching for the target. Yes! The leader saw the clearing over to port, and we wheeled round towards it.

Almost before I knew it we were over the strip. It opened out below us, suddenly. I was surprised at its comparatively small size, and, its emptiness. For a moment I thought "Why, its hardly worth bombing!"

... we wheeled round in a broad circuit to the left. I opened the bomb doors as I saw the leaders opening. So far we seemed to have the place to ourselves.

We straightened up on the run in. Though I didn't think of it at the time, this was the vital 20 seconds in the whole 6-hour job, the peak of the effort which organised production, maintenance, training, supplies, and the activities of thousands necessary to put crews over a target.

I wasn't thinking of this, nor did I notice the trace from machine guns leaping at us from the opposite side of the strip. The leading crew only saw that. We were concentrating on formating on the leader, and my eyes were glued on his bomb bay.

Our approach was a little too direct. We were tending to cross the strip at about 40 degrees, instead of 20 degrees. I saw the first bomb fall from the leading plane. I toggled my switch, remembering in time to space the four releases over about 3½ seconds.

"Bombs gone!" I called.

With a yank of the lever I closed the bomb doors. We had an 11 second delay on those bombs. I hoped it was no less. Almost immediately Les [Pilot Officer Les Smith], from his position in the tail turret, yelled *"Bee-yoo-tiful!"* He had seen our bombs go off. They had skipped a little but had exploded on and near the strip.

12 A47-9 was a B-25D model.

2 Squadron "A" Flight, Hughes Field, 1944 A HIGGINS

2 Squadron "B" Flight, Hughes Field, 1944, Squadron Leader Lee (centre front) Flight Commander *RAAF*

By now we were swinging round beside the Barat River. The planes ahead were diving at spots along the bank, strafing. Then the first plane went in along the strip. Trees and bushes along the edge crumpled, and clouds of smoke marked the strafed positions. The second plane did its work, also among the trees. I heard and felt our guns as Bob dived at promising clumps of trees.

We went back and strafed the strip again, putting long bursts among the trees. So far I was feeling a little disappointed. I had seen no Japs, nor any buildings or vehicles, and not even my own bombs bursts.

We worked back down the river ... every now and then a devastating stream of red trace would be seen crashing into the trees, smoking and ricocheting, and behind that destructive fire would be a B-25, trailing twin streams of smoke and spent shells. No wonder the Nips were conspicuous by their absence. They have always shown a marked fear and hatred of the Mitchell, beyond their inevitable respect for its dozen odd heavy calibre guns.[13]

During July, there had been signs of increased shipping activity between the islands of Flores and Timor, and Liberators, Beaufighters and Dutch Mitchells had already conducted sweeps against them. On the morning of Wednesday 19 July, Wing Commander Dave Campbell (A47-10), a Hudson pilot in the Squadron in 1940 who was now about to take command, led six Mitchells against a concentration of barges in Kalabahi Bay, near Dili. Number six in the formation, Pilot Officer Arthur Pierce's[14] aircraft (A47-15), was hit by anti-aircraft fire, killing the turret gunner, Warrant Officer Ian McCallum,[15] seriously wounding the side gunner, Warrant Officer Vic Muir,[16] and also wounding the navigator, Warrant Officer "Paddy" Ryan.[17] The enemy fire had damaged the bombardier's compartment and crawlway, the bomb-bay and the rear fuselage. In retaliation, that evening another six B-25s, led by Flying Officer Joe Simpson (A47-10), conducted a shipping sweep north of Dili. He then led the first flight of three in for a mast height attack on shipping in Dili Harbour. Flight Sergeant Peter Hocking (A47-16) was Simpson's wingman, flying as number two, with Pilot Officer Bob Avery (A47-7) as number three. Brian Hawthorne, as Avery's navigator, was in the nose of A47-7, when they found a 500 ton vessel in the harbour:

I made lightning settings on the bombing controls as we wheeled round in a steep turn, into the harbour. I just caught a few scattered phrases over the RT: "Leader to formation—I will bomb first! Numbers two and three—go in ahead—strafe ahead of me!"

Then it all began!

With our bomb doors down, we ran in along the edge of the harbour, towards the ship. At the same time as our front guns began to blaze away, the shore became alive with gun flashes and smoke. We were heading almost straight for the white buildings of the town itself, with the church steeple looking strangely peaceful and out of place.

---

13 Brian Hawthorne's diary, of 29 Jun 1944.

14 Flt Lt A J A Pierce, 414595. Nos 32 & 2 Sqns. Clerk of Brisbane; b Brisbane 29 May 1920.

15 WOff I S McCallum, 413226. No 2 Sqn. Of Goulburn, NSW; b Goulburn 23 Sep 1922. Killed in action 19 Jul 1944.

16 WOff V A C Muir, 413224. Nos 32 & 2 Sqns, 1 OTU, 35 Sqn. Student of Uralla, NSW; b Belfast, Nthn Ireland, 8 Aug 1922.

17 WOff S P A Ryan, 416233. Nos 22, 32 & 2 Sqns, 1 OTU. Shop assistant of Port Lincoln, SA; b Port Lincoln 7 Oct 1918.

I felt the plane vibrating and shuddering under the recoil of the guns and the speed of our shallow strafing dive. Of the ship, I have but the haziest recollection of a smallish rusty brown freighter, riding high and pointing inshore, with red bullets spattering its sides. Then we were over the ship. We screamed around hard to the left, in a 90 degree steep turn, a few feet above the water, and almost under the muzzles of those Naval heavies. At least, a couple of hundred yards seemed that close.

How they missed us I will never know!

Weaving and skidding we sped out of Dili Harbour as fast as two Wright Cyclones would take us. The ack ack was wicked! Black puffs of smoke were bursting all around us, uncomfortably close, and the lighter shells were exploding just above the surface or else plopping into the water.

I was scared stiff!

Everybody else later confessed to having been scared but I plumbed the depths of sheer stark fear when we made our second run. Joe Simpson's bombs had fallen short of the target. It wasn't exactly through bad judgement on his part. Just before he reached the ship an anti-aircraft shell burst in the water in front of him, throwing water over his windscreen, and he couldn't see out of his cabin. It must have been a pretty grim moment.

Now he was outside the harbour, circling with the second formation. He says he tried to call us back, out of the harbour. But we didn't hear him, and Pete Hocking went in to bomb, with us going ahead to strafe.

I was torn between a desire to scream with pure rage and fright at Joe for having led us into such a hot spot, and an even more irresistible desire to laugh hysterically at his pure unadulterated impudence in doing this thing to the Japs. But what I really did was clutch my gun more tightly than ever as we went in for the repeat performance!

Almost as soon as we opened fire, our guns jammed!

And there, helpless in the frail transparent nose, I endured the horrible experience of being rushed at over 200 mph towards a confused landscape of dilapidated looking ship, belching, flaming, flashing, shoreline, spouting water, peaceful church and town buildings, and a background of towering mountainsides.

It was horrible, really horrible, wondering just which flash presaged the almost instantaneous end of everything. The air must have been thick with machine gun fire and Bofors shells from the gun pits along the shore. The noise must have been deafening. But to us it was just the rushing roar of engines and air, and a series of blinding flashes.

This time we turned away from the ship, wheeling steeply to the right. I had a glimpse down into ruined white buildings and deserted gardens as we passed over the end of the town. Then we were over the air strip itself, still turning, getting back to the formation.

At last we got out of range, and looked back into the harbour. Pete Hocking had put his bombs right on that ship. From it rose a column of dark smoke, billowing at the top like a mushroom.

We still had four bombs. But never was I so happy in all my life as when Joe Simpson called us up: "I think that will do her! We will reform, and go home!"[18]

Pete Hocking's A47-16 had been holed between the turret and tail, shooting away his elevator trim cable, and damaging the port tail fin and rudder. With one gunner, Flight Sergeant Jack Purcell,[19] suffering through gun shot wounds of the back, and radio and intercom out of action, Hocking had sunk the ship. The formation believed

18 Brian Hawthorne's diary, of 19 Jul 1944.

19 WOff J Purcell, 424215. Nos 13 & 2 Sqns, GR Sch. Assembler of Sydney; b Sydney 23 Oct 1923.

this to be the ship that had killed their comrade that morning. Hocking was awarded the DFM for this mission, and it was cited in Simpson's award of the DFC.

Domestically, the members of No 2 Squadron, like others serving in the fighting forces, were fortunate in the support provided by the Salvation Army. The efforts of the "Salvos" added small comforts, which were always appreciated. One who remembers, instrument fitter LAC Rex Freeman,[20] recalled:

> We were fortunate to have with us Major Bill Parkinson, our Salvation Army Welfare Officer, who was attached to No 2 Squadron at Hughes Field in 1944 and later at Balikpapan in 1945. He was a very popular "officer", especially at Hughes Field when he drove his mobile canteen van on to the strip each day at morning and afternoon tea time to dispense ice cold "lolly water" to all the airmen who emerged suddenly from huts, aircraft or behind trees, as soon as someone yelled that The Major (H.J.)[21] had arrived. With tin mugs in outstretched hands, everyone would swarm around the Major's van and with temperatures at "boiling point", whether or not you were in the shade or behind the control panel of a B-25 checking instruments or whatever, the "lolly water" was a real life-saver.[22]

Major Bill Parkinson, centre in shirt, with his mobile canteen, surrounded by 2 Squadron ground crew R FREEMAN

Wing Commander Dave Campbell DFC took command of No 2 Squadron on 21 July. Campbell, whose finger had been shot off over Rabaul while flying Hudsons, became a popular and well-respected CO. Four days later, he led a formation on a reconnaissance of the Tanimbar and Kai Islands. During a mast height bombing and strafing run against some powered barges off Koer Island, his aircraft (A47-23) was holed by shrapnel from a land mine detonated as the aircraft pulled out of the attack low over the trees. His tailgunner, Sergeant Bevan Gee,[23] was wounded in the hand,

---

20 LAC R C Freeman, 147730. No 2 Sqn. Auditor's clerk of Melbourne; b Melbourne 13 Jul 1925.

21 "Holy Joe", a term of endearment, not of disparagement. Major Bill Parkinson passed away in Melbourne on 29 Dec 1992 in his 91st year, after a lifetime of devoted service with the Salvation Army.

22 Rex Freeman, letter to author, of 18 Nov 1993.

23 FSgt B W Gee, 432693. No 2 Sqn. Audit clerk of Sydney; b Sydney 10 Oct 1922.

but despite bad damage to the fuselage, port wing and rear turret, the crew returned safely to Hughes. On the following evening, Wednesday 26th, Campbell led 12 Mitchells against the airfield at Langgoer, in the Kai Islands, 2 Squadron's first return there since the Hudson days of 1943. Flying in the number 10 position in this squadron-strength raid, on return to base Brian Hawthorne recorded:

> ... home again, under tranquil starry skies. I should imagine the Japs weren't feeling so tranquil. Some of our bombs had delays from 6 to 36 hours, and thus they got two raids for the price of one, and all bomb explosions seen had been well within the target area![24]

Three days later, at sunset on the 29th, the CO led nine Mitchells in behind 12 from No 18 Squadron to bomb Penfoei airfield. The Dutch Squadron was to silence the anti-aircraft defences, while 2 Squadron attacked the runways and revetments to neutralise the base. Each B-25 had a full load of eight 500 lb bombs, necessitating removal of the long-range bomb-bay fuel tank, thereby reducing endurance to seven hours. It took nearly three hours to reach Timor, and the cloud opened up to reveal the conspicuous white shape of Penfoei. In the face of intense target defences, which had been far from silenced, the Mitchells blasted the target area, to escape for the monotonous trip home over the Timor Sea. Another combined operation was flown on the night of 31 July, against the harbour installations at Maeomere in the Flores Islands. The extreme range of this target meant the 12 Mitchells of 2 Squadron, and the 15 of the Dutch Squadron, would return to the vast newly prepared airfield at Truscott, in Western Australia. The target was left in flames, with explosions visible from 80 kilometres, as the Mitchell force searched for night fighters on the return leg over Roti Island, off Koepang. Replenished, and with crews rested at Truscott, the 27 Mitchells struck Timor on 2 August, blasting enemy transport vehicles near Mina River.

Truscott, named after the RAAF fighter ace "Bluey", was a large airfield on the Anjo Peninsula, north-west of Wyndham, "about as far as you can get from Melbourne in that direction without getting wet".[25] It had been built by the RAAF's Nos 1 and 14 Airfield Construction Squadrons over April–June 1944, and was the only airstrip in Western Australia made of Marsden Matting steel mesh. This was convenient for the navigators—"if you landed at or on a steel mesh airstrip in NW Australia in 1944/45, you were at Truscott"![26] Truscott airstrip was planned, designed, constructed, operated and almost completely staffed by Australian airmen and soldiers. Unlike all the other bases and airstrips, this was an Australian show, and a good one. One of its successes was its secrecy, which accounts for the fact that so few Australians to this day know about it.[27]

No 2 Squadron, like other units of the Commonwealth air forces, had designed a unit badge and initiated the process of formal approval of the design by Chester Herald, in London. The badge, featuring an Australian magpie, or piping shrike, had

---

24 Brian Hawthorne's diary, of 26 Jul 1944.

25 Howard Young, Truscott Base Tribute Committee letter, of 12 Nov 1993, courtesy of Peter Helder.

26 Truscott Base Tribute Committee letter, of 26 May 1993.

27 Truscott Base Tribute Committee letter, of 26 May 1993.

been submitted in September 1943.[28] Sir John Heaton-Armstrong, Chester Herald, who acted as the Inspector of RAAF Badges (IRAAFB), advised the CO on 3 August 1944:

> College of Arms
> Queen Victoria Street
> 3.8.44
>
> You will, I am sure, be glad to know that the King has sanctioned the Badge put forward for your Squadron.
>
> The painting signed by His Majesty will be presented to No 2 Squadron in due course.
>
> Will you please let me have a cheque for £10-10-0 in respect of the fees due.
>
> Yours very truly
> J Heaton-Armstrong
> Chester Herald
> IRAAFB[29]

The approved badge also displayed a Fleur de Lys, which represented the Squadron's service in France during the First War. The original Tudor Crown (King's crown) was subsequently replaced by the St Edwards (Queen's) Crown in December 1952.

While not involved on operations, 2 Squadron conducted practice bombing at "J" bombing range at Adelaide River, or on the wreck of a ship off Bathurst Island. On one such practice sortie at Adelaide River on 6 August 1944, A47-13, flown by Flying Officer Stan Davies,[30] was seen to salvo a stick of bombs, before they exploded directly beneath the bomb-bay. The aircraft caught on fire, and Davies was able to crash-land in trees, about 20 kilometres east of the range. A47-13 was completely destroyed and three of the crew (Warrant Officers John Campbell[31] and Arthur Griesbach,[32] and Sergeant Fred Conaghan[33]) and an RAF Spitfire pilot[34] travelling as a passenger, were killed instantly. The navigator, Flying Officer David Lane,[35] succumbed to his injuries within two days, and the co-pilot, Pilot Office Arthur Buckland,[36] two weeks later.

Another maximum effort raid was flown against Langgoer airfield on 26 August. Led by Flight Lieutenant Philip Squires,[37] 11 Mitchells bombed the strip from medium

---

28 R J Cluley, *Unit Badges of the RAAF*, RAAF Association, Melbourne, 1989, p.17.

29 AA CRS 1969/100/39, 2Sqn/1/12/Air.

30 Flt Lt S Davies, 415512. Nos 32 & 2 Sqns. Laboratory assistant of Perth; b Perth 12 Feb 1922.

31 WOff J S McC Campbell, 415227. No 2 Sqn. Of Westonia, WA; b Perth 10 Feb 1923. Killed in accident 6 Aug 1944.

32 WOff A K Griesbach, 415321. No 2 Sqn. Of Perth; b Perth 2 Nov 1922. Killed in accident 6 Aug 1944.

33 Sgt F H Conaghan, 433289. No 2 Sqn. Of Sydney; b Sydney 4 Jun 1924. Killed in accident 6 Aug 1944.

34 Flg Off K J Hadley (131581) was from No 549 Sqn RAF.

35 Flg Off D G Lane, 416434. No 2 Sqn. Of Marryatville, SA; b Port Pirie, SA, 26 Jan 1917. Died of injuries 8 Aug 1944.

36 Flg Off A K Buckland, 7010. No 2 Sqn. Of Perth; b Roebourne, WA, 29 Mar 1919. Died of injuries 19 Aug 1944.

37 Flt Lt P J Squires, 411542. Nos 32, 100 & 2 Sqns. University student of Sydney; b Sydney 13 Apr 1921.

Faan Airstrip, on the Kai Islands, under attack by 2 Squadron's Mitchells A HENRY

level in the face of accurate anti-aircraft barrages. Two aircraft were hit, but were able to return safely to Hughes. Flight Lieutenant "Tige" Carter[38] (A47-27), heading out from the target area at incredible speed, later related:

> I just pushed the throttles against the dashboard, 42 inches, 30 hundred revs, and the old kite just gobbled up the juice by the plateful! I could hear it! Gobble! Gobble! Gobble![39]

In addition to such strikes, shipping sweeps—searching for enemy supply vessels around the Banda Sea and the islands to Darwin's north—had continued to occupy the Squadron effort during August, as AVM Cole handed over NW Area to another former 2 Squadron pilot, Air Commodore Charlesworth. The standard routes for these patrols had now been retitled the "Jig" and the "King". Any shipping targets attacked were normally classified as "Sugar Dogs", "Sugar Charlies" and "Sugar Bakers". A Sugar Dog was the code name for a small type of Japanese supply vessel (with one hatch) of 150 to 300 gross tons, used between major ports and outlying bases. Larger vessels with two hatches, of 300 to 1,000 tons, were known as Sugar Charlies.[40] A Sugar Baker was 1,000 to 3,000 tons, but enemy vessels of this size were now scarce.

The following transcript of an ABC broadcast, on Sunday 10 September 1944, describes a ship strike by three 2 Squadron Mitchells. Although the journalist stated it was flown "last Wednesday", this mission was actually flown on Monday 21 August 1944. The formation was airborne as a four-ship, led by the CO Wing Commander Dave Campbell (A47-12), who was forced to return with an unserviceability. The rest of the formation was flown by Flying Officer Keith Coughlan (A47-7), Flying Officer Les Ekert[41] (A47-3) and Flight Sergeant Peter Hocking (A47-26).[42]

THE END OF A JAP SUPPLY SHIP

> Recently, John Elliott, ABC War Correspondent in the South West Pacific Area, described the destruction of Penfoei Airfield, near Koepang, on Timor, by a RAAF Squadron, commanded by Wing Commander Dave Campbell, DFC, of Canberra flying American Mitchells, medium bombers, based in a northern operational zone. Last Wednesday, three planes from this Squadron, captained by F/O Keith Coughlan of Perth, WA, F/O Les Ekert, of Brisbane, and Flight-Sgt Peter Hocking, of Healesville, Victoria—Keith Coughlan being Flight

38 Flt Lt W A Carter, 415098. No 2 Sqn. School teacher of Claremont, WA; b Henley Beach, SA, 11 Nov 1915.

39 Brian Hawthorne's diary, of 26 Aug 1944.

40 Odgers, p.403n.

41 Flt Lt E L W Ekert, 404499. No 2 Sqn. Store mgr of Coolangatta, Qld; b Rockhampton Qld 12 Oct 1920.

42 No 2 Squadron A51, of 21 Aug 1944.

Commander—went out on a circumnavigating sweep of Timor, Alor, and Wetar Islands, and thereby bagged, with other minor prey a Jap supply ship.

John Elliott was aboard a bomber captained by F/O Les Ekert. Here is his story:

"I suppose I'd better start where we first sighted the Jap supply ship—which means any Japanese transport with a gross tonnage of between 500 to 1,000 tons—since nowadays in these parts, the preliminary ploughing through the skies above the seas comes out all much the same, what with the weather being stabilised at this time of the year, and good at that, and Japanese fighters apparently almost non-existent—which doesn't however, apply to Japanese anti-aircraft guns, rousing ack-ack receptions being as hot as ever over certain areas. Well, we first saw our Jap supply ship at 1.30 in the afternoon, just over four hours after setting out, and about half an hour after our observer-bombardier, Flt Lt Len MacDonnell[43] of Brisbane, had predicted that we might do so. At the time, this was to me either somewhat of a marvel in deductions, or a spot of very good luck.

It appeared that some time before setting out, a reported sighting had been received of a distant Jap 1500 tonner going in a certain direction at a certain speed. Len MacDonnell set to work. After calculations, he marked on the pilots' map, a certain area well worth searching at about 1pm.

Les Ekert's crew: Rear (from left), Earl Ralph, Doug Sandow, Les Ekert. Front, Curley Derrick, Darkie Hensman, Len McDonnell D SANDOW

I knew nothing of all this; and by noon, with Les Ekert, captain and No 1 pilot, and his co-pilot, Doug Sandow,[44] a young Flt Sgt from St Peter's Adelaide, beside him, with Warrant Officer Earl Ralph, a farmer from Mooroopna, Victoria, and Sergeant "Curley" Derrick, of Swan Hill, Victoria, both wireless air-gunners, at the waistguns; and Sergeant "Darkie" Hensman, an ex-clerk from East Coburg, Victoria, at the tailguns, we'd spent hours searching vainly every nook, cranny, inlet, and cove, in and around the coasts of the Islands—for nowadays Jap transports, like burglars, get around only in the dark, hiding during daylight, anywhere they can, up creeks, in coves, under cliffs, anywhere, camouflaged as trees, tiny islands, anything, to escape the notice of bombers and strafers.

I was somewhat restless—what with the monotony of the long uneventful flight, blue sky, blue sky everywhere, pounding of the motors, and nary a craft to be seen, excepting an occasional native boat, high on the beaches—and I touched Les Ekert on the shoulder, and when he glanced back, made signs without words—the RT (Radio Telephony) being silent in case Jap radio-detectors pick us up—and asked what time he expected to sight the target, if any.

The human voice is helpless with the engines of the Mitchell going for their lives. He passed over control to Doug Sandow, then leaning back, lifted his wristwatch and pointed confidently to 1 o'clock.

43 Sqn Ldr L W MacDonnell, 404439. No 12 Sqn RAF, 458 Sqn, 108 Sqn RAF, 2 Sqn. Clerk of Brisbane; b Brisbane 16 Oct 1913.

44 Flg Off D V Sandow, 417317. Nos 2 & 12 Sqns. Student of Clare, SA; b Balaklava, SA, 27 Jul 1923.

Well, came 1 o'clock; and still only mountain tops and deep ravines, paddyfields and grassless terrain, sea and sands, huddles of native villages, thatched huts with mushroom tops.

Sometimes these strafing missions miss out altogether. Came half-past one and we were nearing the northern tip of an Island, and I was just supposing that this, maybe, was another blank trip, when Les Ekert turned his head again, this time his eyes shining. Winking long and significantly, he nodded at something below on the starboard side.

Yes, there it was, a two masted, engines aft, demure as a violet, trying to hide around the corner of a pretty little bay, as near inshore against a cliff as it could get, apparently hoping that it wouldn't be seen.

Les Ekert banked towards it. A few moments later he turned his head again, and shaped his lips. "What a beaut" he said, soundlessly.

We were alone at the time, Keith Coughlan in No 1 and Peter Hocking in No 2 being elsewhere, and having, as we subsequently learned, a very enjoyable time. Sighting near the entrance to a bay, an 80-foot launch, going like a scalded hare, making for a hiding place somewhere up a creek, Peter Hocking promptly strafed it, setting it ablaze, the Japs aboard panicking. It was last seen heading for the mountains, leaping clean out of the sea and coming to its last, final resting place, halfway up a beach. Keith Coughlan set afire a 40-foot lugger, leaving it aground. Before attacking the Jap supply ship Les Ekert decided to find the others—it being considered bad policy tackling any armed ship alone, where necessary. And the Jap supply ship was not unarmed by any means, there being a pair of business-like anti-aircraft guns poking up from its bridge. Les Ekert opened up on his RT. Ten minutes later three Mitchells were in formation, encircling the Island, and bearing down on the ill-fated Jap supply ship. Ill-fated, perhaps; yet the Japs aboard didn't think so, since they kept a frantic barrage of ack-ack going any time a Mitchell came within range.

And as already mentioned, the ship was as near inshore against a cliff as it could get, in no position at all to be attacked recklessly.

Due to the terrain beyond the cliff continuing to rise inland to a height of possibly 5,000 feet, the Jap supply ship defied attack from off the sea. Owing to a quick twist at the bend of the small bay, an attack along the line of the coast was fraught with dangerous possibilities, also; not only from the risk of crashing into the face of the cliff, but of being used as a lame duck target by ack-ack gunners.

The only avenue of attack seemed to be from inland—which meant first encircling the lowest of the mountain ranges, then swooping down, ever vulnerable to those guns all the way.

Well, in we went. Tactics used were not novel, but probably the best to date, and adequate for this task.

Keith Coughlan led the way, accompanied by Les Ekert, strafing, descending from about 1500 feet to a few hundred, both planes spraying the ship with a mixture including tracers and incendiaries and concentrating on the guns on the bridge, and diverting attention from Peter Hocking who followed on a low-level bombing attack, scoring a near-miss, straddling the ship and surrounding it with swirling waters for a hundred yards or more.

The Jap gunners, for all their energy, had hit nothing. While reforming, we saw signs of fire—thin white wisps of smoke curling upwards in front of the bridge—making first blood to us, and all very encouraging.

Second run out from the mountains, Peter Hocking followed us in. Keith Coughlan bombing, only to miss due to a fault in his bombing mechanism. But our guns weren't missing much, splattering the bridge, deckhouse and cargo with fierce short bursts. Passing over this time, the smoke amidships was getting denser, waist and tailgunners, firing into the centre of the fire all the time the ship was under their sights.

Banking round the mountains, we could see the white wisps of smoke were developing into sturdy columns. The fires were taking. This time it was our turn to bomb.

The "Sugar Charlie", off Alor Island, bombed and strafed by the three B-25s on 21 August 1944 RAAF

Starting our run in, Keith Coughlan and Peter Hocking were already strafing. The red tongues of flame were rising from within the smoke. Doug Sandow was at the controls. Les Ekert turned from his guns and smiled gleefully, shook his hands over his head. That was the beginning of the end. The next instant he was trying to cut the Jap supply ship in half with his guns.

We passed over at about 200 feet, but didn't bomb. The Jap gunners, undismayed, despite the fire, despite our concentrated attack, kept the guns going. The cargo was aflame, the fires creeping towards the engine-room. That was it, now. The Japs had no chance of extinguishing that fire.

Well, that's about all. We made one more run, approaching along the line of the coast. Midships the Jap supply ship was like a blazing furnace as we passed overhead. The anti-aircraft guns were silent."[45]

DISTINGUISHED FLYING MEDAL
No 418839 FLIGHT SERGEANT BRYAN MITCHELL HOCKING

CITATION

Flight Sergeant Hocking, whilst a member of No 2 GR/B Squadron, completed 335 hours of operational flying, participating in 17 strikes, 4 shipping sweeps and 29 other sorties.

As a captain he showed good leadership, and all operations which he carried out, were characterised by commendable aggression and determination. On the 19th July, 1944, he was one of the formation which carried out two mast head attacks against a vessel in DILI HARBOUR in the face of intense anti-aircraft fire from the shore batteries and it was his bomb which finally sank a 500-ton enemy vessel from mast height off ALOR ISLAND.

Flight Sergeant HOCKING'S courage and determination have at all times been in the highest traditions of the Service.

This award was announced on 5 May 1945. Dept of Air 55/1/370, of 7 May 1945.

45 Transcript of ABC broadcast on radio 3AR, of 9.15 a.m. Sunday 10 Sep 1944. Courtesy of Peter Hocking, No 2 Squadron Association, Melbourne.

CHAPTER 19

SHIPPING STRIKES

The Mitchells have combed Timor, the Aroe Islands, Kai and Tanimbar Islands, and reduced the once formidable enemy strongholds to mere reduced outposts. In waters where destroyers and freighters once steamed unmolested, the Japs now sneak supplies in barges, which are sunk despite attempts at concealment.[1]

The "B" Flight Commander, Flight Lieutenant George Hodges[2] (A47-3), led a formation of four on an anti-shipping sweep around Alor and Solor Islands on the morning of Thursday 31 August 1944. Hodges found a "Sugar Dog" lugger off the north coast of Alor, near Limboer, and on his first attack was hit by hostile fire. The vessel was left burning fiercely, as Hodges flew back to Truscott. His co-pilot, Sergeant Ron Jones,[3] still remembers:

> … we were hit in the starboard engine which started to vibrate violently and necessitated the engine to be closed down and the propeller feathered.
>
> At this stage the aircraft seemed to be just maintaining height above the water until the bombs and ammunition were jettisoned, when the aircraft gained height immediately. I can distinctly remember the perspiration pouring from George's face whilst all this was going on and when we knew we were safe we just looked at one another and shook hands. From that point we continued to climb until we had sufficient height to clear Timor and then on to Truscott where we landed safely.
>
> There wasn't much talk during the return to Truscott as we were all listening anxiously to the sound of the engine which fortunately just did not miss a beat.
>
> It was initially the skill of George Hodges as a pilot, followed by the performance of the Wright Cyclone engine that got us home.[4]

Wing Commander Campbell (A47-6) led eight Mitchells of No 2 Squadron, coordinated with Dutch Mitchells, to attack the town of Toeal in the Kai Islands on the night of 1 September. Three aircraft carried 4 lb incendiaries to act as "pathfinders".[5] One Dutch Mitchell was shot down by night fighters in the target area. During his bombing run, the third B-25 in 2 Squadron's stream, A47-12 flown by Flight Lieutenant Tige Carter, was apparently damaged by flak which exploded under the starboard wing, rocking the aircraft.[6] Carter lost his flight instruments and developed a fuel leak; they soon became lost and, with an excessive fuel flow, it became evident

---

1 From the diary of FSgt B D Hawthorne.

2 Sqn Ldr G S Hodges DFC, 407429/O4624. No 211 Sqn RAF, 1SFTS, Nos 2, 37 & 34 Sqns. Salesman of Berri, SA; b Adelaide 15 Jan 1915.

3 WOff R D Jones, 431057. No 2 Sqn. Clerk of Melbourne; b Wonthaggi, Vic, 12 Mar 1924.

4 Ron Jones, letter to author, of 30 Sep 1993.

5 The load for each Mitchell in this role was eight 500 lb clusters of 4 lb incendiaries.

6 Alf Batten, telephone interview with author, of 26 Oct 1993.

they would not make base. After giving his crew the option to bale out, Carter was able to coax his aircraft back to Australia. He found the coast too rough to attempt a forced landing, so elected to ditch near Peron Island. His tailgunner, Sergeant Alf Batten,[7] who was injured in the ditching with a lacerated scalp, recalled: "When we hit the water, the aircraft filled immediately. There was no floating, it just sank."[8] Fortunately the water was fairly shallow. Batten was trapped below the water attempting to escape through the small photographic window, and Carter was able to pull him free.

The next night, a searching aircraft (A47-6), flown by "A" Flight Commander Flight Lieutenant Joe Simpson, crashed in the sea nearby while looking for the wreck. Two extra crew members were on board: the Squadron padre Flight Lieutenant Hilford Easton[9] and armourer Warrant Officer Gordon King,[10] who were friends of the missing crew and were acting as extra sets of eyes and assisting in the dropping of flares. The tailgunner, Sergeant Greg Jellis,[11] as part of Simpson's crew,[12] recorded the end of the flight:

The wreckage of A47-12, Tige Carter's aircraft, high and dry on the beach at Peron Island with the tide out RAAF

> After flying for about forty minutes the pilot, F/L Simpson, indicated that he had seen a light on the water and, although it was bright moonlight, he put on the landing lights to investigate. I (the tailgunner in the rear turret) immediately advised that we were flying too low and that we would hit the water ... we did ... I was caught on some projection in the turret and called for help. W/O Milligan swam round and got me loose.
>
> There was some confusion before we got the dinghy cleared, and loaded F/L Higgins and F/O Park (both being injured) in. P/O Peters was in the nose turret and managed to get free; he joined us in the dinghy. It was then that the discovery was made that our two passengers were missing. They had been located about four feet from the tail turret, just to the rear of the top turret, which was where the aircraft broke. A search was carried out and the body of F/L Easton was located; it was presumed that W/O King went down with A47-6.

7 FSgt A R Batten, 438308. No 2 Sqn. Assistant foreman of Melbourne; b King Island, Tas, 13 Aug 1923.

8 Alf Batten, taped recollections, of Jul 1993. No 2 Squadron Association, Melbourne.

9 Chaplain H C Easton. Padre No 2 Sqn. Of Henty, NSW; b Adelaide 8 Oct 1913. Killed in accident 2 Sep 1944.

10 WOff G S J King, 4795. No 2 Sqn. Of Melbourne; b Blandore, SA, 26 Mar 1910. Killed in accident 2 Sep 1944.

11 Flg Off G R Jellis, 431284. Nos 2 & 18 Sqns. Sales clerk of Melbourne; b Melbourne 15 Apr 1925.

12 Co-pilot WOff R K Milligan (406945), nav Flg Off V Peters (401016), WAGs Flt Lt A J Higgins (416335) and Flg Off R L Park (418871).

B-25J A47-31, KO-P, during September 1944 — Sgt Alf Batten carrying the Thermos, and Sgt Al Henry with radio calibration equipment
A BATTEN

Sometime later we were picked up by an Army water transport vessel. The events of the 2nd September 1944 came to an end with the pick-up of the crew and the body of F/L Easton by PBY-5 A24-100, piloted by F/L Roberts.[13]

By this stage, General Kenney had decided that the United States 5th Air Force would be the assault force, with the RAAF and other Allies assigned to supporting roles and garrison duties for the final offensives of the war. For those who had previously been unaware that the RAAF was being sidelined from the main thrust, it now became apparent. This, of course, led to discontent, as RAAF crews felt the war was passing them by and they were not being assigned to worthwhile targets. The demand for more involvement in operations led to Spitfires being used sparingly in an offensive role. Following their employment in April against Babar Island, Spitfires were now tasked on a joint operation on 5 September with 2 Squadron. Flying Officer Les Ekert (A47-21) escorted a formation of 14 Spitfires from No 1 Fighter Wing (comprising three RAF squadrons, Nos 54, 548 and 549)[14] to strike an enemy camp area at Lingat village, on Selaroe Island, south of Saumlaki in the Tanimbars. The Spitfires recorded:

> The navigation was good ... the target was approached as briefed. The CO and F/Lt Aiken, having sprayed the target, were following along a motor transport track to the north of the village. F/Lt Aiken called over the R/T to say that there appeared to be absolutely nothing on the road (that was not the exact expression he used, but what he said means "absolutely nothing"), when suddenly, woomff! a bullet came whizzing through his cockpit. There must have been something on the road, or very near it. Nearly famous last words.[15]

Having led the fighters successfully to Lingat, Ekert then dropped incendiary bombs in the target area. His co-pilot, Flight Sergeant Doug Sandow, remembered this first joint Mitchell/Spitfire strike:

> ... after watching and waiting for the Spits to do their strafing ... we were given "short leave" to drop our four 500-lb incendiary bombs which were like a firework display. However, the Japs soon gave us some of our own back and some nasty ack ack fire

13 Statement on "The Loss of Aircraft A47-6", by G R Jellis, No 2 Sqn Assoc, Melbourne, Jun 1993.

14 Four aircraft of 54 Sqn flew as "Red Section", four of 548 were "Blue", four of 549 were "Yellow". The B-25 was "Tom 1", and two Spits from 1 Fighter Wing HQ were "Toms 2 and 3".

15 No 548 Sqn RAF A50, of 5 Sep 1944.

> surrounded us, but as always, plenty of evasive action and "full bore" away from the scene of the action.[16]

Even though the Spitfires attacked without bombs, being restricted to strafing, later enemy interrogation showed that the strike on Lingat was highly successful. The barracks were severely damaged and the camp was subsequently moved.[17]

Enemy activity was at a low ebb during September, with shipping patrols conducted on the standard "Jig" route, up the Tanimbar Islands to north of the Kais, and the "King", around the Kais and east to the Aroes. A break from shipping patrols on the 13th saw the Squadron returning in force to neutralise the airstrip at Langgoer in daylight. This involved flying high above the airfield defences at 18,000 feet, necessitating the use of oxygen by crews for the first time. The following week, on the 21st, two Mitchells, flown by Flight Sergeant Peter Hocking (A47-9) and Flight Lieutenant Tige Carter (A47-31), carried out a "King" armed reconnaissance, striking targets in the Aroe Islands. Sergeant Alf Batten, in A47-31, remembered this as a fruitful sortie:

> We came upon this motor boat going up a river, and we tried to bomb it and strafe it. As we were down on this river ... right down on the deck ... we came up over the trees we noticed a clearing. In the clearing were some trucks coming towards a jetty. Tige decided he was going to strafe the trucks. We came right down to zero level just at the top of the trees, we got to the end of the clearing, down right onto the deck, and away he starts strafing. The next minute I hear the second pilot say "Trees!"—singing out at the top of his voice—and you could feel Tige pull back. It was squashing, you could feel instead of going up we were squashing forward and the next minute we were taking the top of the tree off. We still came back with some of it in our front wheel and in our bomb bay.
>
> Anyway we bombed Doka, Elat, the motor boat, the jetty, the trucks. We had a ball that day, and it's probably the largest amount of rounds I ever shot. I estimate about 1000 rounds I put in the trucks, the motor boat, the barges and anything else we saw. It was really a good day.[18]

The following day, 22 September, four Mitchells were dispatched on a shipping sweep along the north coast of Timor. During a strafing attack on a barge in the harbour at Manatuto, which was narrow with steep banks on both sides, Flying Officer Allen Slater's[19] aircraft (A47-3) crashed into the sea from 500 feet, about 400 metres from shore. All on board were killed. It was thought the enemy machine-gun emplacements were responsible, but A47-3 had been fouled by wires stretching between the headlands. Jim Henderson explained:

> Our strategy was to dive in low and bomb the ships but the Japanese had anticipated this and strung wire ropes from bank to bank. The first B-25 hit the wire ropes and dived into the

16 Doug Sandow, letter to author, of 27 Sep 1993.

17 No 548 Sqn A50, of 27 Nov 1944.

18 Alf Batten, taped recollections, of Jul 1993. No 2 Squadron Association, Melbourne.

19 Flg Off A W Slater, 414849. No 2 Sqn. Of Brisbane; b Nambour, Qld, 20 Feb 1916. Killed in action 22 Sep 1944. Crew: Flg Offs M S Millett (426641), J F Daggett (434341), B A Wisniewski (423966), FSgt K R Philipson (429669), Sgt D F Harberger (435448).

sea out of control. I was fortunate in that my turn to attack did not come until next day and we were told to bomb from 6,000 feet.[20]

Wing Commander Campbell (A47-1) led eight Mitchells out on 26 September, staging through Truscott, to strike shipping in Maoemere Harbour, on the north coast of Flores. They attacked at sunset but no shipping was present, so they bombed port installations. On the return, a 3,000 ton "Sugar Baker" was spotted along the northern coast of the island, and concentrated attacks were made from Truscott over the following days against the vessel. The ship sheltered in Wodong Bay, under the protection of shore defences, which thwarted attempts at accurate bombing until finally the weather closed to prevent further attack. The vessel was apparently damaged and immobile, however, as it was successfully attacked again the following month.

Down the nose — Dave McQuoid's navigator, Bernie Mahoney crouching over the bombsight
A HENRY

Replacement crews for No 2 Squadron were not arriving in NW Area in time to release crews who had completed their nine-month operational tours. Not only were replacements slow in arriving, but they were not trained on Mitchell aircraft. There was no B-25 Operational Training Unit and crews were still being trained on the Beaufort or Hudson at No 1 OTU, at East Sale. As a result, towards the end of September, No 79 Wing found it necessary to suspend nearly all operational flying in both Nos 2 and 18 Squadrons in order to train and convert seven Hudson replacement crews for No 2, and 10 replacement crews for No 18, who were not fully trained.[21] This increased training activity during October coincided with preparation for the planned movement of 79 Wing, comprising No 2 Squadron and two Dutch units (Nos 18 and 120 Squadrons), to New Britain under Northern Command, to support land operations there.[22] However, as the airfield was not completed, this plan was delayed, and No 79 Wing was reinforced by the return to the area of No 13 Squadron with Venturas, based at Gove.

20 Jim Henderson, letter to author, of 20 Mar 1993.

21 Odgers, p.247.

22 Odgers, pp.297–9.

Despite the intensive training commitment, the Mitchells were still able to maintain heavy pressure on enemy shipping, destroying or damaging 22 freighter barges during the month.[23] In support of MacArthur's assault on the Philippines, NW Area's anti-shipping campaign was directed at blockading the Banda Sea. On one shipping sweep of the Sawoe Sea, west of Timor, on 10 October, a formation successfully attacked a small ship towing a barge. The aircraft flown by Flying Officer Bob Ingram[24] (A47-22) was hit by fire from the ship, but managed to return safely to Truscott. The lack of substantial shipping targets indicated the enemy's desperation. Indeed, at this stage of the war, with the enemy almost broken, it must be queried whether it was worth the risk of losing a Mitchell against the negligible value of a barge. This was really a case of keeping the RAAF on the fringes, as MacArthur took the 5th Air Force on to glory in the main campaign.

The Mitchells maintained anti-shipping activity into November, with a formation on the 4th carrying out a shipping sweep around Timor, Wetar and Alor Islands. While the CO attacked the anti-aircraft position at Goerita Bay, in Dutch Timor, Flying Officers Dave McQuoid[25] (A47-14) and Jack Selway[26] (A47-8) strafed barges. Selway's aircraft was hit by groundfire as he approached from over the land, and crashed in flames into the sea. None of the crew was seen to bale out, nor were any of them seen during a search which lasted for 20 minutes.[27] One of gunners in A47-14, Sergeant Al Henry[28] recalled:

Goerita Bay, the target for 12 November 1944
A HENRY

> We were in a strafing dive and I was in the upper turret, with Selway in echelon starboard. He was hit in the starboard engine and attempted to pull up, but rolled right into a steep dive and knifed straight in. He just left a splash—there was no wreckage.[29]

It was losses like this that were not only felt by the aircrew, but also the groundcrew. Corporal Keith Dodd,[30] an engine fitter on "A" Flight, remembered:

---

23 Odgers, p.314.

24 Flt Lt R M Ingram, 427481. No 2 Sqn, 1 OTU. Clerk of Coogee, WA; b Fremantle, WA, 3 Jun 1924.

25 Flt Lt D A McQuoid, 415543/O57069. 2AP, No 2 Sqn, 4AD. Motor mechanic of Kalgoorlie, WA; b Kalgoorlie, 27 Dec 1919.

26 Flt Lt J H Selway, 403381. No 2 Sqn. Of Sydney; b Sydney 11 Feb 1914. Killed in action 4 Nov 1944. Crew: Flg Off A E Pott (405932), Plt Off H B Worman (417438), WOffs J F Stormon (415057) and A E Hawkins (418837), FSgt R C Palfreyman (436941).

27 RAAF Historical C.D.102/45 A47-8, of 23 Jan 1946.

28 WOff A C Henry, 431145. No 2 Sqn, 1 OTU. Trainee accountant of Melbourne; b Melbourne 16 Apr 1925.

29 Al Henry, interview with author, of 8 Oct 1993.

30 Cpl K R Dodd, 14596. No 133 (NEI) Sqn, 14 RSU, 2 Sqn. Motor engineer of Sydney; b Wollongong, NSW, 29 Dec 1916.

> It was named a "Bash" ... that made the airstrip just jump alive with crews moving to board the aircraft. Battery carts arriving to start the engines, personnel at the ready to see them on their way into radio silence.
>
> Then came the time of waiting, not a task but only waiting to count them in; sadly sometimes eleven not twelve aircraft would return. Later it was time for debriefing, also a service routine: the operations officer would request the daily inspection book (EE.77) together with the log books—blue and brown—for the aircraft which was missing.
>
> The greatest demand of all was to remove the crew names from the flight blackboard and replace them with new crews and aircraft—just to see it all again.[31]

Two days later, on the evening of Monday 6 November, four Mitchells from 2 Squadron with 12 from 18 Squadron carried out a coordinated attack on a vessel at Waingapoe, on the north coast of Soemba Island. Again it was necessary to stage through Truscott in Western Australia. The Dutch Mitchells were to attack the town and anti-aircraft positions from medium level, while the Australian Mitchells swept in low to attack the ships.

The four Mitchell captains were Flight Lieutenants Paddy Norriss[32] (A47-23), the "A" Flight Commander, and Tige Carter (A47-17), Flying Officer Elmo Thompson[33] (A47-31) and Warrant Officer John "Bluey" Leach[34] (A47-26). When south of Cape Melangoe the Dutch reported seeing "a 1000-ton motor vessel moving out from the wharf at Waingapoe".[35] With 18 Squadron coordinating the attack, the Dutch bombed the enemy defences from medium level to coincide with 2 Squadron's on-the-deck attack. The Australian Mitchells weaved through the valleys across the island to strike the port from the north. Tige Carter's tailgunner, Sergeant Alf Batten, recalled:

> We had to go in single file ... we flew up this valley, at zero feet, and you could look up at the fields. Our number 4 tucked himself into our starboard side ... and when we did get through to the other side, he stayed tucked in while we did the raid, down onto the deck to strafe and skip-bomb the shipping. He was so low ... and that close to us, it made it very difficult in the tail to be shooting back, in case the empty rounds flew out and hit his aircraft, or in case I shot him down![36]

This mast-height attack succeeded in straddling the vessel, which was estimated at 300 tons, and the last aircraft over the target saw it enveloped in smoke. Although only claimed by the Squadron as severely damaged, postwar Japanese records showed that their vessel, *Special Submarine Chaser No 118*, was indeed sunk.[37] Norriss' aircraft was damaged in the bomb-aimer's compartment.

---

31 Keith Dodd, letter to author, of 16 Sep 1993.

32 Sqn Ldr P J Norriss, 406551. No 2 Sqn. Mining plant operator of Perth; b Perth 27 May 1916.

33 Flt Lt W F E Thompson, 406741. Nos 35 & 2 Sqns. Chemist of Fremantle, WA; b Kalgoorlie, WA, 19 Apr 1915. Killed in action 20 Dec 1944.

34 Flg Off J B Leach, 410603. Nos 67 & 2 Sqns, 1AD. Process engraver of Ballarat, Vic; b Ballarat 13 Jun 1923.

35 Odgers, p.314.

36 Alf Batten, taped recollections, of Jul 1993. No 2 Squadron Association, Melbourne.

37 Odgers, p.315.

Simpson's Circus, by Flt Lt John Ditchburn A HIGGINS

Cherry Method A HIGGINS

Skip bombing used in attacks such as this were not only fraught with the danger of low flying, but also with the risk of "flying formation" with your own bombs! One 2 Squadron pilot, Flying Officer Len Power,[38] recalled:

38 Flt Lt L E Power, 427147. 1 N&WW, No 2 Sqn. Clerk of Perth; b Perth 29 Jul 1923.

> Sometimes we used low (50 feet) level bombing to try to destroy a barge on the beach. The bomb, if it missed the target, would bounce along the beach following the aircraft. Imagine the poor tail gunner seeing this bomb coming from underneath him, rising almost to his level, then down for another bounce, and him wondering when the 13 second delay set for detonation would end![39]

Twelve Mitchells were led back to Goerita Bay by Flight Lieutenant George Hodges (A47-14) during the afternoon of 12 November. Barges, port facilities and the anti-aircraft sites were bombed successfully. Hodges led the Squadron to Flores two days later, to attack the port at Ende, on the island's south coast. With half the formation carrying 500 lb bombs, and half carrying incendiaries, Flight Sergeant Brian Hawthorne (A47-34) recorded:

> These 4-lb sticks of incendiary magnesium are dropped in 500-lb clusters, which scatter over a wide area after release. Eight such clusters represents 1000 small fires, and six planes were carrying that load.
>
> After the photos had been interpreted, and further photos taken of the town a few days after (fires were still burning on the following day), all crews concerned were congratulated from Area on having destroyed 150 buildings, and having utterly razed 1000 feet of the water front. "Good show, chaps!"[40]

The Mitchells continued to obtain excellent results from shipping sweeps among the islands in the Flores-Timor group during November, sinking or damaging no fewer than 32 vessels, including four small freighters.[41] The Japanese were now so short of vessels, they had resorted to using small native craft. The anti-shipping operations that NW Area had mounted, while being on the sidelines of the main United States thrust, had indeed been successful in tying up, and cutting off, enemy resources in the NEI.

Sometimes, however, crews found that practising the anti-ship procedures, usually in Darwin Harbour, was probably more dangerous than actual operational flying. Flying Officer Len Power recalled the cooperation exercises the Squadron used to fly with a navy *Fairmile* motor torpedo boat:

> A flight of 4 would split into individuals with flight paths aimed to turn in towards the *Fairmile* from its four quarters and judged to arrive over it simultaneously. Descending from say 500 feet, each would weave to practise evading AA fire and yet would have to be at its pre-arranged 50, 100, 150 or 200-foot level as it passed over the ship at the end of a straight short bombing or strafing run. Of course pilots kept a very wary half-eye on just what level was actually being flown by his lower and higher neighbour. Nice to be assigned the highest! Even so, there was a probably unwarranted reliance on the accuracy of altimeters in this procedure. We never did have to use it in anger. The smoke arising from a ship fighting back would have made an interesting extra hazard.[42]

39 Len Power, letter to author, of 9 Nov 1993.

40 Brian Hawthorne's diary, of 14 Nov 1944.

41 Odgers, p.314.

42 Len Power, letter to author, of 9 Nov 1993.

Having carried out the Spitfire strike in the Tanimbars in September, another Spitfire escort was conducted on 27 November, this time to attack the Cape Lore radar station, on the northern tip of Timor. Being at the extreme range of the single-engined fighters necessitated their refuelling at the Snake Bay strip, known as "Austin", on Melville Island. Five Spitfires of No 549 Squadron RAF, and two from No 1 Fighter Wing, accompanied four Mitchells, led by Flight Lieutenant George Hodges (A47-25). Co-pilot to the fourth Mitchell, flown by Flying Officer William Coward[43] (A47-23), was Sergeant Keith Shattock,[44] who remembered this four and a half hour Timor sortie for the Spitfires:

> Because of the extreme range they were allowed only 10 minutes over the target and we had great difficulty getting them to formate for the return trip—"Just one more run" was the cry. The aircraft all returned safely with only a few minutes fuel in hand. The Spitfire pilots were not very impressed when they formated on our aircraft to see us sitting back scoffing sandwiches and coffee with "George" doing the flying while they were cooped up in their small cockpits having to fly "hands on".[45]

The Mitchells also bombed and strafed the target before escorting the Spits back to Darwin, while a Catalina was providing Air Sea Rescue (ASR) cover by orbiting south of Timor in case any of the fighters had to ditch. The press soon published reports of this long-range raid:

> Spitfires of the Royal Air Force based in Northern Australia have made the first attack by single-engined aircraft on Timor and completed the longest operational flight by Spitfires in any theatre of war.
>
> Piloted by men of the RAF and one Australian, Group Captain B R Walker DSO,[46] of Adelaide, the planes had to make the longest sea crossing yet attempted by Spitfires to reach the target, a Japanese radio-location station at Cape Lore, on the north-east corner of the island.
>
> The Spitfires were of a long-range, high-altitude type. The mission involved a round trip of nearly 800 miles.[47]

However, the value of such a long-range hazardous crossing in a single-engined aircraft, with its limited firepower, was dubious. It was a case of No 1 Fighter Wing's three Spitfire squadrons sitting idle, as there had been no enemy air raids during the year, and feeling their war was devoid of purpose. Again, these were assets which could have moved north to follow the main action, but were being left behind. This was a precursor of the futility the RAAF would feel in the final year of the war.

On 28 November, four Mitchells carried out a shipping sweep near Maoemere, on the north coast of Flores. An 80 ton lugger was strafed and sunk by the formation, and four barges were left burning. Flying Officer Dave McQuoid (A47-16) then bombed

---

43 Flt Lt W G Coward, 411005. 3AD, GR Sch, No 2 Sqn. Bank clerk of Sydney; b Sydney 23 Sep 1919.

44 FSgt D K Shattock, 434946. Previously Army; No 2 Sqn. Traveller of Melbourne; b Seymour, Vic, 18 Jul 1920.

45 Keith Shattock, letter to 2 Sqn Association, of 18 Jun 1993.

46 Gp Capt B R Walker DSO, 94. Comd No 12 Sqn 1941–42, 30 Sqn 1942–43, 5 OTU 1943–44, No 1 Ftr Wg 1944–45. Regular air force offr of Adelaide; b Lyndoch, SA, 27 Mar 1913.

47 "Record Spitfire Attack on Timor", *The Sun*, Melbourne, date uncertain, circa late Nov 1944.

and destroyed a further four barges, and another was set on fire on the way home. Wing Commander Thomas "Toss" Ingledew[48] took over command of the Squadron the following day, as the anti-shipping campaign continued. On one of these strikes, on 5 December, Flight Lieutenant Paddy Norriss (A47-11) was leading a flight of four on a sweep north of Timor, when he was hit by anti-aircraft fire at Laga. His aircraft was holed in the starboard engine and Norriss shut it down to prevent the possibility of fire with the escaping fuel. A subsequent failure of the port generator made transfer of fuel from the starboard tanks impossible, and the port engine eventually became starved of fuel. Norriss was able to safely crash-land on a clay pan on Bathurst Island. On 15 December, four B-25s from 2 Squadron destroyed a 400 ton vessel by mast-height bombing and strafing, followed three days later by a direct hit on a 150 ton vessel, near Roti Island off Timor. These contributed to No 79 Wing's tally for December of four freighters and 25 smaller vessels destroyed.[49]

Flying Officer Elmo Thompson's[50] crew (A47-33) departed Hughes on the morning of Wednesday 20 December for a shipping reconnaissance. This was a new route, the "Love", which took in the east coast of Jamdena Island, in the Tanimbars, and around the Kai Groups. No communication was received again from A47-33, but postwar searchers were informed the aircraft had been shot down with no survivors, and that wreckage marked "47-33" had been found 500 metres south of Saumlaki.[51]

During December 1944, both 2 and 18 Squadrons had been advised of their move to Jacquinot Bay, in New Britain. However, the supporting No 83 Operational Base Unit was not able to establish the base until February 1945, and 79 Wing's move north was then delayed beyond that. But No 80 Wing's Spitfires were able to move north and, to a limited extent, follow the war through the islands. No 2 Squadron assisted in the deployment of the first group, No 452 Squadron. Three Mitchells, led by Group Captain Clive Caldwell, deployed to Sattler to escort the fighters to Morotai, in the Halmahera Islands. The first B-25 was the weather "recce" ship, flown by Bob Avery (A47-7) 30 to 45 minutes ahead of the fighter formation, to send back weather reports. A second B-25, Les Ekert (A47-25), was the navigational escort, shepherding the 24 Spitfires, which trailed in three lines astern of eight planes each. Number three B-25, Dave McQuoid (A47-16), flew 15 minutes behind, acting in an ASR capacity, carrying inflatable dinghies for any unplanned letdowns, and to facilitate the arrival of the ASR Catalina.

The formation departed from Gove on 17 December on the two-hour overwater leg to Merauke, on the south coast of Dutch New Guinea. On landing at Merauke, the Spitfires experienced a mishap, which caused a deadly pile-up:

---

48 Wg Cdr T S Ingledew, 139. No 25 Sqn; comd 67 Sqn 1943–44, 1 Sqn 1944, 2 Sqn 1944–45. Regular air force offr of Sydney; b Sydney 11 Aug 1915.

49 Odgers, p.315.

50 The rest of Thompson's crew were: Flg Off L T Forsyth (418377), WOffs J E S Thompson (415921) and F H Mathews (418970), FSgts J A Rolfe (432434) and T H Rowlands (419918).

51 RAAF Historical C.D.102/45 A47-33, of 24 Sep 1946.

The eighth plane was just about to turn off when he suddenly slewed round. His undercarriage collapsed and one of the propeller blades went sailing in the air. The plane behind him had landed too close, had overtaken him and collided. The pilot was killed instantly by the four bladed airscrew of the second plane, which also lurched drunkenly to a stop, lying across the runway. A few seconds later, a third plane crashed into the wreckage of the two Spitfires. There was a sickening ball of flame, and the next minute two fighters were blazing furiously, while the first was knocked clear at the side of the strip. We learnt that the second pilot was unhurt, and the pilot of the third plane had been thrown clear, and sustained a fractured leg only. Three Spitfires and one pilot lost[52] in a single tragic accident![53]

One of 452 Squadron's Spitfires, A58-525 QY-X, wrecked at Merauke 17 December A HENRY

The following day, the formation took off for Noemfoor Island, planning to fly along the south coast then across the "Gooseneck", the narrowest part of New Guinea, but were forced to return due to bad weather. The aircraft did, however, reach Kamiri strip, on Noemfoor, on the 20th, and successfully arrived at Morotai the next day.

Morotai was a precarious beachhead, being the most northern island of the still Japanese-held Halmaheras group, and within easy range of enemy airbases. Consequently, Morotai was attacked almost nightly, but the base was vital for the Allied assault on the Philippines and the NEI. The Spitfires were to provide the air defence for the base and were welcome arrivals. Ultimately, Morotai would be the major base for RAAF fighter squadrons as they were relegated to "mopping up" operations in the closing stages of the war. Flying Officer Jack Bice, in Les Ekert's navigational Mitchell, remembered this deployment for the Mitchells as an interesting and successful exercise:

> The two strips at Morotai were very busy with vast numbers of USAAF kites. A Jap recce came over that night and was shot down by a Spit—a great start![54] The Spit is really not at all suitable for night fighting, but the specialised US "Black Widow" had been a washout.[55]

---

52 Flt Lt J R Sturm, 407133. No 452 Sqn. Of Adelaide; b Adelaide 15 Feb 1918. Killed in accident 17 Dec 1944.

53 Brian Hawthorne's diary, of 17 Dec 1944.

54 This first Spitfire victory was in fact three nights later, on Christmas Eve, when Flg Off Pretty shot down a "Nick" (Kawasaki Ki-45 two-seat night fighter) in flames into the sea. No 452 Sqn A50, of 24 Dec 1944.

55 The Spitfire was not equipped with radar, and had never been intended as a night fighter, as had the specialised Northrop P-61 Black Widow, which proved a disappointment in service.

22/12—back to Noemfoor and 23/12 back to Hughes direct, enjoying some barge strafing at Taberfane on the way.[56]

A Liberator had sighted two ships in the Banda Sea, so a flight of four Mitchells, led by Flight Lieutenant Les Ekert (A47-2), was sent on 26 December to conduct a shipping sweep. On an afternoon take-off, the number three, Flying Officer Bob Avery (A47-9), failed to get airborne, crashing in the overrun, fortunately without injury to the crew. The temperature had been hot, causing a high density altitude on the strip of 3,000 feet, making aircraft acceleration very sluggish. It appears that the brakes also were binding and Avery was unable to obtain more than 105 mph, when normally the aircraft showed 130 mph before leaving the strip.[57] The navigator, Brian Hawthorne, recorded:

As Bob opened the throttles, and the plane surged forward, it seemed to wander from its straight course a bit, and Bob had to kick the rudders pretty violently to correct this tendency. But passing the duty pilot's tower, everything was OK. I was standing in the navigator's cabin, looking over the shoulders of Bob and Peter.

As we approached the end of the strip, Bob looked at Pete, who grasped the undercarriage lever ready to pull the wheels up. But when Bob eased the stick back, the plane came unstuck for a second or two, then the wheels touched down again. Obviously we were not yet properly airborne. At last Bob lifted her off the runway, and Pete moved the undercarriage lever into the "Up" position.

Suddenly I saw Bob pull the throttles closed. I heard the engines die away, their harsh straining roar becoming a splutter. I felt the aircraft beginning to sink. It was rather a nasty sensation, feeling the power come off when we were only a few yards from the end of the strip, and a solid wall of trees a few hundred yards ahead at the other end of the flight gap. We were only 10 feet or so above the ground ... I leapt back onto the navigator's table, which folds over and becomes a padded seat for take-off and landing.

A second later I felt the plane tearing through the tops of a few light saplings and bushes. Then we hit the ground with a slight bump. We hit the ground again, harder this time, and for what seemed an age slithered and scraped and bumped through a blur of tall grass, bushes, stumps, and light scrub. The left hand side of the cabin opened up like a sardine can, and a shower of broken perspex flew into my face.

The plane came to a halt with a lurch, and the nose dropped. I heard Bob say to Pete "Let's get out before she catches fire", saw them fumble with the emergency hatch above their heads, and saw Pete disappear through his starboard window. Bob threw aside his parachute harness, and climbed out through the top. For a few seconds his boots dangled wildly in front of my face, then he was out, and I followed him.

I jumped to the ground. Out of the corner of my eye I saw that the plane looked fairly intact, but a small red flame was licking about the nacelle of the port motor. I remember a strange silence, broken only by a faint singing and hissing noise from the engines, and the crashing of the crew through the undergrowth.

Our first thought was for the bombs, and we ran off into the bush. When we had gone a few yards, I thought of the chaps in the back. Had they got out? Were they OK? I heard their voices ahead, with relief, and didn't even look back at the plane. When we were about a hundred yards away, I looked back, and saw an increasing spiral of black smoke.[58]

56 Jack Bice, letter to author, of 11 Jul 1993.

57 HQNWA letter 928/299/P1 (1110A), of 27 Dec 1944, A47 Crash Records, RAAF Historical Section.

58 Brian Hawthorne's diary, of 26 Dec 1944.

The plane did burn, and everybody seemed amazed to see the crew were alive. The Flight Commander congratulated them on their escape. The doctor was pale, but almost delirious with happiness. Everywhere groundcrew were coming to say how pleased they were that the crew had got out of it. The CO asked how they were feeling, then said he wasn't feeling too good himself! Then the fuel tanks blew, and a minute later the bombs exploded. Pieces of the plane fell around the maintenance section, some 400 metres from the scene of the crash, and that was the end of A47-9.

The wreckage of Bob Avery's A47-9 at Hughes RAAF

Meanwhile, the remainder of the formation was pressing on towards the Banda Sea. A 300–400 ton "Sugar Charlie" was located in the middle of the Banda Sea, near Schildpad Island, and attacked at mast height. The vessel was left heavily damaged and probably unseaworthy.[59] On return to Hughes, due to the weather, Les Ekert diverted his wingmen to land at Darwin, but attempted an approach. His gunner, Flying Officer Jack Bice,[60] recorded:

> ... we came back, night time by now, from a strafing job, into an intense electrical storm. Ekert ordered the other two to return to the main drome at Darwin, but, seeing a break, decided to press on and land at Hughes. On our landing leg, the storm closed in again. I was in my "landing position", sitting side on behind the pilot's seat.
>
> I was aware of the intense lightning and the fact that we momentarily dropped a wing, then levelled off but hit the tarmac deep and hard, shearing off our undercart. We slid sideways, generating lots of sparks ... My momentum broke the seat, and I received a good whack on the side of the face. I vacated pretty smartly, and ran to the rear of the aircraft to see if the three there needed any help in getting out. All were on the ground, surrounded by broken perspex, awaiting my arrival. We all had minor abrasions, and bruises ... I still bear a 2cm scar inside my left elbow. The kite never flew again, but was cannibalised for spares.[61]

Les Ekert's A47-2 after the Boxing Day mishaps RAAF

And that was the end of A47-2. From the flight of four, two Mitchells had been written-off, but fortunately without injury to the crews.

59 No 2 Sqn A51, of 26 Dec 1944.

60 Flt Lt J Bice, 403999. No 11 Sqn, 2BAGS, 3AOS, 2 Sqn, 1 OTU. Bank clerk of Sydney; b Nowra, NSW, 9 Mar 1919.

61 Jack Bice, letter to author, of 11 Jul 1993.

Flying Officer Len Power (A47-28) was sent on another Spitfire escort to Morotai at the end of the year. He left Darwin on 30 December, to stage through Gove, Merauke and Noemfoor. Power recollected:

> On this trip we acted as mother hen to the four chicks, mainly providing the navigation with the Spitties in formation. We arrived at Merauke with 4 but left with only 3. The 4th must have gone unserviceable as I have no recollection of it crashing there. On the flight from Noemfoor to Morotai we tuned into the control tower at Morotai and thereafter I listened to the continuously busy voice of "Bloomer" tower. On arrival, this was seen as the busiest airport I had yet encountered. All manner of American aircraft types, but mainly Liberators, were using the multiple parallel strips and the Yank twang made the control almost indecipherable. I allowed the Spitfires to get ahead of me in the circuit as they would have less fuel. A little bit dicey as on landing I was overtaking them fairly rapidly as they slowed and continued to taxy along the runway. I had allowed reasonable space but it was closing rapidly. My brakes got some fairly heavy use on that occasion. I'd have hated to chew up one of my nestlings!
>
> That night I slept heavily in a tent provided near the Spitfire squadron base and awoke in the morning to be told I should have been in a slit trench as a Jap patrol had penetrated the perimeter defence for a short time. A Spitfire pilot in a tent nearby had been wounded while he slept through the resulting gun fire. I really didn't hear a thing.[62]

No 79 Wing Mitchells carried out a series of widespread harassment raids on villages in Timor on New Years Day 1945, a time when the Japanese normally held festivities in Timor towns. However, the Wing's anti-shipping operations were seriously affected by weather during January. On one raid on Saturday 6th, one aircraft in a formation of four operating around Sawoe, west of Koepang, was hit by enemy groundfire. Flying Officer Coward's aircraft (A47-26) was holed in the port airscrew, fin, and had the nose wheel door shot away.

Further escorting duties for No 80 Wing were carried out in early February. Group Captain Caldwell led 24 Spitfires of No 457 Squadron along the same route, together with some replacement aircraft for No 452 and the first aircraft for the third unit, No 79 Squadron. Arriving at Morotai on 8 February, the Mitchells again staged back through Noemfoor. Operations for the rest of the month were all armed reconnaissance sorties, searching Timor and surrounding islands for signs of enemy shipping. The Japanese garrisons left in the Arafura and Banda Seas were now out on a limb, as the enemy tried to withdraw.[63] The few vessels that risked detection were successfully attacked.

Despite these losses, the enemy persisted in attempting to extract ground forces for redeployment to Asia. The seemingly sidelined activities of NW Area were, in fact, seriously impacting the enemy's war capability. The unrestricted activities of Allied aircraft had denied the enemy his planned withdrawal from the Arafura islands. As a result, the Japanese held important army units just north of Australia, until it was too late to return them for the defence of their threatened homeland.[64]

62 Len Power, letter to author, of 9 Nov 1993.

63 Odgers, p.403.

64 Odgers, p.404.

Operating from Truscott on 8 March, four Mitchells were tasked with a Timor shipping sweep and attacks on bridges near Ende. Led by Flight Lieutenant Neil Sharpe[65] (A47-14), the formation split into two flights with Sharpe and Flying Officer Ted Westbury[66] (A47-15) bombing the bridge at Nanga Roro River. Sharpe's navigator Warrant Officer Bill White[67] described their attack run:

> We approached through a misty shower of rain with F/L Sharpe the lead plane. I released the bombs at low-level. As we passed over the bridge ... my bombs fell alongside and destroyed the market garden. F/O Westbury in the other plane was more fortunate as they blew it sky high.[68]

Against this background of apparent Allied control of the air was a shipping strike carried out by Mitchells, renowned not for its success, but for the tragic losses of the accompanying B-24 Liberator heavy bombers. This mission was to become "the cruiser bash", of 6 April 1945. If this plan tried to emulate the precisely executed coordinated attacks of the successful Battle of the Bismark Sea in March 1943, then NW Area's attempt was a lamentable failure.

An enemy convoy was known to be leaving Koepang, after withdrawing troops, so an attacking force of 20 Mitchells of 79 Wing (10 from each squadron) and nine Liberators of 82 Wing planned to rendezvous off the south-east tip of Soemba for a coordinated strike. This was beyond the range of Allied fighter cover. Unfortunately, the Liberators had trouble in gaining formation in the early morning darkness, and arrived 20 minutes late. The Mitchells, at the limit of their range, could not wait and went in to attack. No 18 Squadron attacked at 10,000 feet, claiming two direct hits and many near misses with their 500 lb bombs. They were then attacked by fighters as 2 Squadron, splitting into three formations, concentrated its bombs on the Natori-class cruiser *Isuzu*.

The CO (A47-22) led four aircraft in at 10,000 feet from the starboard quarter, followed by three aircraft led by "A" Flight Commander Squadron Leader David Hannah[69] (A47-10) on the port beam at 11,000 feet, with Squadron Leader John McMillan[70] (A47-15) leading the last three aircraft at 10,000 feet from the port quarter. Attacked by fighters, and with the *Isuzu* weaving violently and throwing up a hostile barrage, Bill White, navigator to Neil Sharpe (A47-14) the "B" Flight Commander flying in the first formation, described his run:

---

65 Flt Lt E N Sharpe, 254899. Nos 100 & 2 Sqns. Fitter of Adelaide; b Lismore, NSW, 17 Oct 1917.

66 Flt Lt E J Westbury, 420318. 1AOS, No 2 Sqn, 2AD. Shop assistant of Sydney; b Uralla, NSW, 12 Aug 1922.

67 Plt Off W M White, 409011. Nos 22 & 2 Sqns. Schoolteacher; b Colac, Vic, 28 May 1920.

68 Bill White, recollections on service with No 2 Sqn, Nov 1944–Oct 1945. No 2 Sqn Assoc Melbourne.

69 Sqn Ldr D H Hannah DFC, 404551. No 55 Sqn RAF & 2 Sqn. Station overseer of Tanby, Qld; b Orange, NSW, 13 Jan 1913.

70 Sqn Ldr J A McMillan, 400726. No 60 Sqn RAF & 2 Sqn. Bank clerk of Inverleigh, Vic; b Geelong, Vic, 13 Nov 1916.

Our aircraft were in flights of three which attacked from different directions. The first flight made a hit on the stern after we had been attacked by 4 Oscar fighters. My bombs, unfortunately, hung up ... The result, of course, was a miss.

As we attacked the cruiser, the anti-aircraft guns put up a barrage, then the fighters attacked again.[71]

Also in the first flight was pilot Flying Officer Len Power (A47-19):

... having "bombs away" or a very short time thereafter the leader of my flight of four—the Commanding Officer—"Toss" Ingledew—took us into a steep diving left hand turn. I subsequently calculated the true airspeed reached in the manoeuvre was in excess of 400mph (our instruments were American—American gallons rather than Imperial and MPH rather than knots). Don't know exactly why this sticks in my mind except that 400 was about the maximum being flown by any service aircraft of those days. We got out of there quite quickly! I have a vague recollection that there was some cock-up with the timing of the attack, as between the three squadrons involved, and that this may have been an element in the losses of the two Liberators and consequentially the loss of a Catalina. We were on time of course![72]

Gunnery Leader Officer for the formation was Flying Officer Jack Bice:

Flying Officer Ted Westbury [A47-34] apparently lined up one of the E/A making a frontal, pressed the "tit", but had omitted to turn on the master switch. If only! F/Lt Rex Beck[73] flying turret in another aircraft [A47-14, F/Lt Neil Sharpe], had a shot at one of the E/A, and was heard by all, including base, to say: "If I didn't get that bastard, I'll give a f...... garden party!" When we got back to Hughes, someone, in the most elaborate script, had placed a card on the Mess notice board: "F/Lt Rex Beck invites the Members ... to a Garden Party."[74]

The Mitchells escaped unscathed, but the Liberators, attacking from 12,000 feet, 20 minutes later, were not so fortunate. The element of surprise had been lost, and the concentration of force had been denied. Enemy fighters were waiting for the "heavies". Two fighters attacked one Liberator flown by a former 2 Squadron Hudson pilot, Flight Lieutenant Syd McDonald (A72-81). Cannon shells burst around the flight deck and fire broke out, spreading immediately. McDonald ordered his crew to bale out, and two managed to escape successfully as:

... the aircraft immediately went up at a steep angle and appeared to reach the stall. Flames were seen coming from the nose wheel compartment and forward bomb bays. The aircraft turned over on the port wing and plunged vertically towards the sea, losing about 6000 feet. The aircraft recovered from the dive and went into a very steep climb and again stalled on the port wing and went into another dive of about 45 degrees heading towards the cruiser. A few seconds later the aircraft exploded.[75]

---

71 Bill White, recollections of 2 Sqn service, 2 Sqn Association Melbourne.

72 Len Power, letter to author, of 9 Nov 1993.

73 Flt Lt R B Beck, 407024. Nos 39, 69, 162 & 14 Sqns RAF, 454 Sqn, 2 BAGS, 2 Sqn. Interior decorator of Adelaide; b Adelaide 8 Oct 1918.

74 Jack Bice, letter to author, of 11 Jul 1993.

75 No 24 Sqn A50, of 6 Apr 1945.

One of those to jump from the aircraft was WAG Warrant Officer Shilling,[76] who had been crewed with McDonald on 2 Squadron in early 1943. During his parachute descent he witnessed another Liberator attack, and saw the aircraft flown by Flight Lieutenant Eric Ford[77] (A72-77) shot down, with six of the crew managing to escape. Shilling survived an attempted ramming by one of the Jap escorts, to be picked up by Catalina (A24-54) of No 112 ASR Flight. As the Catalina picked up other survivors it was attacked by a Zero, and caught fire. As it sank, and the crew and survivors scrambled out, Shilling was covered by blazing fuel. He was kept afloat by members of the Catalina crew until a further Catalina arrived and rescued the three remaining Liberator crewmen, and the six crew of the first Catalina. Shilling recalled, as the last survivor clambered aboard:

> … an Irving[78] was sighted making for us. The enemy made his first attack as we were taking off and scored hits. Our return fire did no visible damage. A running fight ensued for about twenty minutes. The attacks ceased and we headed for Darwin, reaching there about 2230 hours …[79]

That night one of the escorts was sunk by a submarine, the same fate befalling the *Isuzu* the next day. The failure of a coordinated air attack, beyond the range of escorting fighters, and the apparent lack of a back-up plan had spelled the failure of the strike. The execution of rushed plans is sometimes necessary in war, and rehearsals, such as conducted for the successful Bismark Sea operation, are not always possible. But the "cruiser bash" laid the Liberators open as "sitting ducks" over an area of the NEI where the enemy still controlled the air. Three aircraft, with most of their crews, were lost. The RAAF was fortunate there weren't more.

76 Plt Off K R Shilling, 406816. Nos 2 & 24 Sqns. Draper of West Midland, WA; b Narrogin, WA, 18 Jul 1923.

77 Flt Lt E V Ford, 255138. No 24 Sqn. Motor engineer of Melbourne; b London 14 Feb 1911. Killed in action 6 Apr 1945.

78 Nakajima J1N1-C naval fighter.

79 Odgers, pp.408–9.

CHAPTER 20

THE WAR'S END

The enemy was beaten through the forceful application and complete coordination of land sea and air power. No one service carried the war exclusively: all shared equally.

General Hap Arnold [1]

A warning order for No 79 Wing to provide air support for army operations in eastern New Britain had been issued in early December 1944, and indicated that the new airstrip would be ready for operations by February 1945.[2] Under the constant possibility of relocation, the Squadron's operations were consequently hampered through the early months of the new year. The Squadron had been advised on 10 February[3] that a Dutch ship, the *SS Bontekoe,* would arrive at Merauke later in the month to load No 120 (NEI) Squadron, the Dutch fighter unit attached to 79 Wing, and would then proceed to Darwin to load the other advance units for Jacquinot Bay. The rest of the Wing move would be conducted on the United States liberty ship, the *William H Seward*.[4] It soon became apparent that the capacity of these two vessels was completely inadequate to shift the entire Wing. A request was therefore made for vacant space on another ship,[5] but this was not forthcoming, so loading of personnel on the *Seward* was restricted to very small advance parties to make way for vehicles and ground equipment. As information coming through on the progress of the airstrip and facilities was meagre, both the Mitchell squadrons held back sufficient spares and maintenance personnel at Hughes to operate at a rate of 400 hours per month up until 21 April. At that stage, if the move had not been undertaken, a complete review would be necessary.[6]

On 25 February, the first advance party of groundcrew, under Flight Lieutenant Bill Stevens,[7] departed by Mariner flying boat for Jacquinot Bay. Flight Lieutenant Ron Edwards[8] then followed on the first ship lift with a party of 47 others,[9] departing in

---

1 General H E Arnold, Commanding General United States Army Air Forces, cited in MacCloskey, p.38.

2 HQ NWA postogram O503, of 2 Dec 1944. HQ NWA O613, of 28 Dec 1944, notified the following units would comprise 79 Wing: Nos 2, 18 (NEI) & 120 (NEI) Sqns, 18 RSU, 28 A/c Stores Park, 27 Med Clearing Stn, and 83 OBU. Cited in John Ryland's papers, National Library MS 5020 Box 1.

3 HQ NWA Signal O813, of 10 Feb 1945.

4 HQ NWA Signal O833, of 13 Feb 1945. The *Bontekoe* was identified as X13, the *Seward* as X2709.

5 79 Wing Signal O166, of 18 Feb 1945, requested the 5,000 tons that was vacant on the *P T Rowe*, X2991.

6 John Ryland's papers, National Library MS 5020 Box 1.

7 Flt Lt W W Stevens, 415281. Nos 13 & 2 Sqns, 1 OTU, 2 Sqn. Traveller of Perth; b Bonny Rigg, Scotland, 21 Aug 1910.

8 Flt Lt R C W Edwards, 265905. No 6 Sqn, 5EFTS, Armt Sch, 2 Sqn. Apprentice engineer of Sydney; b Sydney 15 Jan 1920.

9 No 2 Sqn Travel Instruction to No 8 T & MO, of 6 Mar 1945; and telephone interview with Ron Edwards, of 18 Oct 1993.

A47-5, KO-N, was a long-serving B-25D on 2 Squadron. It was one of the first Mitchells delivered in May 1944, remaining with the Squadron until the end of 1945 RAAF

2 Squadron Mitchells at Hughes in July 1945 AWM NWA 943

the *William H Seward* on 7 March.[10] The second ship lift of 90 personnel departed on 21 April on "Operation Budu"—the code-name for Jacquinot Bay—aboard the *Bontekoe*.[11] One of those who sailed was LAC Rex Freeman, who remembered:

> The *SS Bontekoe*, a troopship of approximately 8000 tons, called in to Bowen, North Queensland, to re-coal enroute to Jacquinot Bay. Whilst in Bowen, most of the troops were

10 AA CRS 1969/100/39, 2 SQN/2/54/Air 2 Sqn signal 032, of 7 Mar 1945.

11 AA CRS 1969/100/39, 2 SQN/2/54/Air 2 SQN signal A.11, of 22 Apr 1945; Keith Dodd, interview with author, of 16 Sep 1993; Rex Freeman, interview with author, of 10 Nov 1993.

> given a days shore leave, and having been starved of choice fresh food "up north", they descended upon the town like the locusts of Egypt, eating all the cream cakes and buns that laid in shops before them! The trades-people were delighted, but the locals no doubt were pleased to see us depart.[12]

However, the new airfield had not been ready for operations until late in March 1945, and the CO, with the commander of 79 Wing, Group Captain John Ryland (A47-22), flew there on the 27th to inspect the new facilities. By this stage, 40 per cent of the ground staff, and most of the Squadron's ground equipment, had left Hughes. In the meantime, with the rapid United States island-hopping advance, Dutch authorities had requested that 79 Wing's two NEI squadrons should be employed on Dutch territory, so the move to New Britain was subsequently cancelled:

> **HQ NW Area, Movt Inst No 40** **14 May 1945**
>
> **MOVE OF No 79 WING**
>
> No 79 Wing is now required to move to MOROTAI for subsequent employment in the BORNEO Area. The move to JACQUINOT BAY, now in progress, is cancelled.[13]

No 2 Squadron was now assigned to First Tactical Air Force operations in Borneo, and the vacancy left by not moving No 79 Wing to Jacquinot Bay was taken by the New Zealand Air Task Force. This period of reduced activity, without most of the groundcrew, had coincided with the arrival in NW Area of RAAF Liberators, with their increased range. The requirement for Mitchell sorties was, therefore, not so urgent. As the Squadron continued anti-shipping strikes, the operational opportunities were becoming less, and NW Area was indeed becoming a backwater.

On 28 April, a pair of B-25s, operating from Truscott, provided cover for an ASR Catalina attempting to locate the crew of a ditched Liberator, close to the coast of Cape Mangeh, on Soemba Island, to the west of Timor. While investigating the beach near the wreck, Flight Lieutenant James Legge's[14] aircraft (A47-39) was hit by machine-gun fire from at least four positions along the coast. The groundfire took out the port engine, and damaged the port side of the aircraft and the turret. Legge was able to make a safe single-engine return.

As destinations had been changed for 2 Squadron's forward deployment, the Squadron detachment at Jacquinot Bay embarked on the *William H Seward*, and sailed back to Darwin. Others boarded the *Andrew Carnegi* on 6 June for Morotai, followed two days later by others on the *Joe Fellowes*. Before arriving in Darwin, the *William H Seward* embarked No 13 Squadron groundcrew at Gove for their move to Morotai. With over 100 additional personnel from 2 Squadron joining in Darwin, this main party of some 600 personnel sailed north on 12 June to Port Moresby and Biak. Arriving at Morotai on 26 June, the ship was quarantined because of a meningitis scare, and personnel were unable to disembark to a transit camp until 2 July.[15]

---

12 Rex Freeman, letter to author, of 18 Nov 1993.

13 AA CRS 1969/100/39, File 2SQN/1/5 Pt.1, of 14 May 1945.

14 Flt Lt J L Legge, 408854. Nos 7, 67 & 2 Sqns. School master of Melbourne; b Murchison, Vic, 11 Jul 1916.

15 Grantham, pp.84–5.

The SS Bontekoe, *a Dutch troopship enroute to Jacquinot Bay with 2 Squadron ground staff, March 1945* R FREEMAN

The "Oboe" series of assaults by Australian forces was intended to take Borneo in three major landings at Tarakan, Labuan and Balikpapan. The Australian commander, General Blamey, had opposed the campaign, as the lack of sea transport had now denied the enemy use of Balikpapan's oil. MacArthur overruled the Australian command, with the justification that Borneo would provide the site for the overall objective of seizing the NEI. This appeared unsound as Java had been effectively isolated; however, at least the seizing of Borneo would provide quicker access to the enemy's main prison compounds.[16] Perhaps tying up the Australians in Borneo was really part of MacArthur's agenda for his involvement alone in the Philippines and then onto Japan's doorstep.

After the landing at Balikpapan on 1 July by the Australian 7th Division in the Oboe 2 operation, the airfield at Sepinggang was occupied the following day. Two days later, the larger airfield at Manggar, 10 kilometres further north, was taken, with Japanese resistance eventually being overcome on 8 July. This allowed for the immediate dispatch of the 2 Squadron party to Borneo in the three ships, arriving at Balikpapan on 17 July. Finally, land was prepared for a camp by United States Navy "Sea-Bee" (Construction Battalion) engineers, and the new base at Sepinggang airfield was established.

Japanese troops were still active around the perimeter of the field, and a captured diary detailed their movement and observations:

> 21 July. At 1000 hours departed towards destination, captured four suspicious natives on the way—shot three and kept one as a guide—he got away due to inattention …
>
> 22 July/0200. We are right in the midst of the enemy. Arrived at destination in the jungle at 1345 hours although a sentry has fired on us, we have suffered no casualties. I started to investigate the terrain in the vicinity of the airfield—enemy precaution is adequate—it is impossible to move in the day time, reconnaissance and infiltration will begin from now.
>
> /1940. It is painful to wait for my objective while looking at the enemy tent in front of me. Marched one kilo and came out to enemy tents and material dump—went through without notice of the enemy.

16 I Grant, *A Dictionary of Australian Military History*, Random House, Melbourne, 1992, p.25.

> 23 July. Slept overnight in the jungle ... Enemy position is in vicinity of Naisajima Force. It is easier to break through security positions by day than by night ...[17]

On the morning of 23 July, this group of 10 soldiers, commanded by a Lieutenant, infiltrated the 2 Squadron domestic area and attempted to destroy the camp water point. One of the 2 Squadron airmen on the scene was LAC Ray Green,[18] who later related:

> ... our 2 Squadron guards were on duty. We had a large steel tank of rain water for kitchen use, drinking and washing. The Japs decided to blow it up thus denying us fresh water. Our guards were stationed around the area which of course was jungle. They had rigged booby traps by stringing wire from tree to tree with milk and jam tins containing pieces of metal to rattle if the wires were disturbed. They were armed with Bren guns and Thompson sub-machine guns ... the Japs came into the area, the booby traps were activated, and our boys killed the lot of them ... the Japanese were buried nearby.[19]

LAC Rex Freeman noted:

> A week later, the No 2 Squadron Aussie Rules Football team played a team from the Army 7th Division at Sepinggang. During a break in the game, the captain of the Army team was heard to comment to our team captain: "We received a report from Headquarters saying that the 2 Squadron guards engaged the enemy at the Squadron water-hole and wiped them out—congratulations, well done!" No 2 Squadron claimed another victory—we went on to win the match![20]

While most of 2 Squadron's ground staff had been in transit between Jacquinot Bay and Balikpapan, back at Hughes on 31 May Wing Commander Lloyd "Smokey" Douglas DFC[21] had taken command of the Squadron. Little operational flying had been carried out over May, but on 3 June the "A" Flight Commander, Squadron Leader Hannah (A47-14), escorted Spitfires on their second strike against Timor. The target this time was parked aircraft at the Cape Chater airfield on the north-eastern tip of the island. Again the nine Spitfires, from Nos 54 and 548 Squadrons, staged through Snake Bay. Four Liberators from 82 Wing also participated on this strike, bombing the strip while the fighters strafed and provided top cover. The Spitfire pilots were still feeling frustrated with the lack of activity:

> No 548 Squadron does an operation!! the second since the Squadron was originally formed in February 1944.[22]

---

17 "An Extract from a Jap Officer's Diary", of 21–23 Jul 1945, courtesy of Ron Edwards, letter to author, of 19 Oct 1993.

18 LAC R H Green, 153390. 6SFTS, No 2 Sqn, 9 Tpt & Movt Off. Clerk of Adelaide; b Adelaide 9 Dec 1925.

19 Ray Green letter, to 2 Sqn Association, of 20 May 1993; and telephone interview with author, of 29 Oct 1993.

20 Rex Freeman, letter to author, of 18 Nov 1993.

21 Wg Cdr L A Douglas DFC, 403. Nos 1, 100 & 13 Sqns; comd 2 Sqn 1945–46. Regular air force offr; b 1 Jun 1919.

22 No 548 Sqn A50, of 3 Jun 1945.

With the Sepinggang airfield at Balikpapan ready, the first airlift move from Hughes finally went ahead on 14 August 1945, when a formation of six Mitchells departed for the Squadron's new location, via Biak and Morotai.[23] Flight Lieutenant Neil Sharpe led the formation, with the lead navigator Bill White. On the first leg approaching the New Guinea coast the formation encountered cloud. Sharpe descended to 3,000 feet and noticed the aircraft flown by Flight Lieutenant Edward White[24] (A47-37) swing away to the right and disappear behind a cloud. The leader then dropped to 1,000 feet and advised the formation that he was going to fly below clouds to the New Guinea coast, where he would orbit. This message was acknowledged by all aircraft except A47-37.[25] Bill White described the event:

> Having made a running line fix on an island cape, I knew we were on track ... so when we were about 50 miles off the narrow neck of the New Guinea coast and some cloud was forming I was confident when asked if it was safe to fly below it.
>
> Passing through the cloud, several of the aircraft broke formation so when we eventually arrived at the clear coast we had two aircraft not with us.
>
> Arriving at Mokmer [Biak] we were soon joined by F/O Ted Westbury's aircraft, which had flown higher, but the White/Morrell aircraft did not arrive.
>
> We spent the next day on a parallel search but there was no sighting.[26]

A47-37 was not found, and the remainder of the formation proceeded to Biak on 16 August. Years later, the wreckage of the lost Mitchell was found in Dutch New Guinea. This was a tragic loss for No 2 Squadron: the day after the accident, the war ended.

> **Washington, Wednesday 15 August.** The war is over. Japan has surrendered unconditionally to America, Britain, Russia and China. General Douglas MacArthur, appointed by the Big Four as Allied Commander-in-Chief, will lead the Supreme Command staff to accept the Japanese surrender, and will order the Japanese Emperor to carry out all provisions of the surrender terms as declared at Potsdam in July.[27]

It was ironic that after waiting for nearly a year to follow the war northwards, No 2 Squadron's aircraft did not finally arrive at its new base at Sepinggang until the war had finished. The second deployment of Mitchells left Hughes on 23 August for the 10 hour flight via Morotai, with the final flight of seven aircraft arriving at Balikpapan on the 30th. Peace brought relief from bombing missions, but there was still a lot of flying to be done. Reconnaissance of Japanese shipping became the first task, but flight authorisations were specific: "No attack to be made on any target without provocation."[28]

23 AA CRS 1969/100/39, 2 SQN/2/54/Air 2 SQN Movement Order No 1/45, of 11 Aug 1945. Six aircraft were A47-7, A47-14, A47-21, A47-35, A47-37, A47-41. A47-29 departed the next day.

24 Flt Lt E M White, 402097. No 2 Sqn. Of Sydney; b Melbourne 3 Mar 1912. Killed in accident 14 Aug 1945. Crew: Flg Off A K Morrell (7246), FSgts B M O'Brien (431510), R A MacGregor (433732), Cpl W J C Maxwell (34759), LACs B A Marshall (73231), V A Morgan (88726), I S Coleman (140811).

25 RAAF Historical C.D.102/45 A47-37, of 17 Feb 1947.

26 Bill White, recollections of 2 Sqn service, No 2 Sqn Association Melbourne.

27 "Unconditional Surrender by Japan: World at Peace", *Army News*, Darwin, Thursday 16 Aug 1945, p.1.

28 No 2 Sqn A51, of 19 Aug 1945.

SURRENDER OF JAPANESE FORCES

Address Delivered by

COMMANDER - in - CHIEF, AUSTRALIAN MILITARY FORCES
GENERAL SIR THOMAS BLAMEY

to

Lt - General Teshima, Commander Second Japanese Army

On the occasion of the signing of their Surrender

MOROTAI, 9th SEPTEMBER, 1945

LIEUT.-GENERAL TESHIMA, COMMANDER, SECOND JAPANESE ARMY :

"The Japanese Navy has been destroyed. The Japanese Merchant Fleet has been reduced to a mere fraction. The Japanese Air Force has been driven from the sky. The Japanese armies have been defeated everywhere and all that remained for them was to await their inevitable total destruction. Japanese cities lie in waste and Japanese industry has been destroyed. Never before in history has so numerous a nation been so completely defeated.

"To escape the complete destruction of the nation, the Emperor of Japan has yielded to the Allied Forces and an instrument of total surrender has been signed in his name. He has charged you to obey the orders which I shall give you.

"In carrying out these orders, the Japanese Army and Navy organisation will be retained for convenience. Instructions will be issued by the designated Australian Commanders to the Commanders of the respective Japanese Forces, placing upon you and your subordinate Commanders the responsibility for carrying out your Emperor's direction to obey all orders given to you by me.

"You will ensure that all Allied personnel, prisoners of war or internees in Japanese hands are safeguarded and nourished and delivered over to the Allied Commanders.

"You will collect, lay down and safeguard all arms, ammunition and instruments of war until such time as they are taken over by the designated Commanders. You will be given adequate time to carry this out. An official date will be named and any Japanese found in possession after that date of any arms, ammunition or instrument of war of any kind will be dealt with summarily by the Australian Commander on the spot.

"Orders will be given for these and other matters as I consider necessary and you will ensure the obedience to all such orders without delay.

"In receiving your surrender I do not recognise you as an honourable and gallant foe, but you will be treated with due but severe courtesy in all matters.

"I recall the treacherous attack upon our ally, China, in 1938. I recall the treacherous attack made upon the British Empire and upon the United States of America in December, 1941, at a time when your authorities were making the pretence of ensuring peace.

"I recall the atrocities inflicted upon the persons of our nationals as prisoners of war and internees, designed to reduce them by punishment and starvation to slavery.

"In the light of these evils, I will enforce most rigorously all orders issued to you, so let there be no delay or hesitation in their fulfilment at your peril."

Repatriated Dutch POWs at Maindai, in the Celebes, beside a Mitchell flown in by Group Captain Ryland (OC 79 Wing) and Wing Commander Douglas, 12 September 1945
NATIONAL LIBRARY OF AUSTRALIA

Wing Commander Douglas (A47-35) flew an eight hour reconnaissance on 19 August to Soerabaya Harbour, on Java, and encountered ack ack fire—peacetime flying still had its hazards. The Mitchells also flew to Bali, Lombok, Flores and the Celebes dropping leaflets advising that the war had ended. Another duty was to establish the location of the many POW camps in the region and determine what assistance was required. When the second airlift of four Mitchells left Australia on 22 August for Balikpapan, the 2 Squadron Headquarters closed at Hughes, and opened simultaneously at Balikpapan.[29]

Food drops to the POW camps then became the main role of the Mitchells. One typical drop was to the Pare Pare camp, in the southern Celebes, on 11 September flown by Flight Lieutenant Ian Ralfe[30] (A47-31). His navigator, Flying Officer Hugh Cope,[31] received a letter from the Dutch Camp Commander acknowledging this valued work by the Squadron:

---

29 AA CRS 1969/100/39, 2 SQN/2/54/Air 2 SQN Movement Order No 2/45, of 21 Aug 1945. Aircraft were A47-27, A47-31, A47-34, A47-36. Third airlift was 30 Aug: A47-5, A47-10, A47-15, A47-19, A47-23, A47-25, A47-26, A47-43. Fourth airlift of rear party was 12 Sep: A47-4, A47-28.

30 Flt Lt I D V Ralfe, 405501. 26th Bn, Nos 11 & 2 Sqns. Bank clerk of Longreach, Qld; b Longreach 3 Jun 1921.

31 Flg Off E H Cope, 433578. Nos 2 & 37 Sqns. Public servant of Sydney; b Sydney 11 Mar 1925.

Yesterday was a grand day not less than 12 parcels coming down from the air and nothing damaged.

Wish you could be here to see the pleasure that you gave us.

With joy we watch you every day and we hope soon to be able to shake hands.[32]

Flight Lieutenant Lawrence Kirk's[33] crew (A47-19) was sent on the morning of 15 September to escort a Catalina to Bandjermasin, near the southern tip of Borneo, in case the flying boat had trouble landing in the renowned treacherous waters. After the Catalina landed and a party went ashore for consultations with the natives, Kirk circled the area, then departed. A little while later, the Catalina captain was advised the Mitchell had crashed, and he proceeded to the scene. Here it was discovered only two of the crew had survived the crash, and had been removed to the local hospital. Only one, the tailgunner, survived.[34] Evidence showed that the B-25 was low flying and failed to pull out of a shallow dive, striking a coconut tree and bursting into flames on impact.[35]

The CO, Wing Commander Smokey Douglas, with the 2 Squadron rugby team at Balikpapan
R FREEMAN

32 Letter from Rev A Bikker, of 12 Sep 1945, enclosed in letter to author from Hugh Cope, RAAF Hudson Squadrons Association, Adelaide, of 29 Jun 1993.

33 Flt Lt L A Kirk, 407439. No 2 Sqn. Of Victor Harbor, SA; b Riverton, SA, 10 Aug 1914. Killed in flying accident 15 Sep 1945.

34 Crew killed: Flg Off L Bishop (433889), Flg Off P A Taylor (440106), WOff C R M Ricketts (429384), mess staff Cpl R O Byrne (33210) and LAC M S White (141075). FSgt F J Stolweather (439530) died soon afterwards. FSgt E A R Booth (433083), the tailgunner, was the sole survivor.

35 No 2 Sqn File 16/47/AIR, of 20 Sep 1945, in A47 Crash Records, RAAF Historical Section.

For the repatriation of members of the RAAF from the islands, the Liberators of 82 Wing and Mitchells of 79 Wing were stripped of armament and converted to transports. With guns and armour removed, the Mitchells were used as air ambulances, flying to Labuan, then to Singapore to repatriate the most urgent medical cases among the POWs. Troops and airmen were flown direct from Morotai, to the Australian capital cities, in one-day shuttles.[36]

Crew of Mitchell A47-10 about to depart Borneo on 11 November 1945 for demobilisation in Australia. Note how the fuselage guns have been removed for the Squadron's post-war transport role. From left: Don Thomas, Bob Nicholls, Geoff Morgan, Steve Grey and Jason Cruikshank S GREY

Escorting the many surplus RAAF fighters in the islands back to Australia became another task. Typical was one flight, during November, to lead 10 Kittyhawks back to storage at No 6 Aircraft Depot at Oakey, in Queensland. Flight Lieutenant Richard Seymour[37] (A47-4) departed on 6 November to rendezvous with the fighters at Tarakan, to lead them to Zamboanga, on Mindanao in the Philippines. Next day, the formation flew to Morotai, then on the 8th through Biak to Tadji, in New Guinea. Three legs were flown the following day to Port Moresby, Higgins and Townsville, with the final flight to Oakey on the 10th.[38]

36 Odgers, p.495.

37 Flt Lt R Seymour, 406879. Nos 14, 6, 100 & 2 Sqns. Clerk of Perth; b Perth 10 Aug 1920.

38 FSgt F T Riggs (418613), flying logbook, of Nov 1945. Courtesy of Fred Riggs, 2 Sqn Association.

Flying operations ceased at Sepinggang on 14 November, and No 2 Squadron commenced preparations for moving aircraft and personnel to the mainland. The first airlift of Squadron personnel left in A47-7 on 23 November, followed 10 days later by 62 personnel departing by boat. The following month, the formal notification of the future of the unit and its aircraft was issued:

14 DEC 45

No 2 SQN
MOVEMENT ORDER No 4/45

No 2 Squadron is to move in nucleus from Balikpapan, Borneo, to Laverton, Victoria, where it will come under Southern Area. All Squadron aircraft are to be ferried to Amberley, Queensland. Captains of aircraft will, on arrival at Amberley, ensure that their aircraft and equipment are handed over to No 3 AD.[39]

As the Squadron's activities were run down, eight Mitchells with 140 personnel of the Squadron departed Balikpapan on 15 December. The last Mitchell departed Borneo on 20 December, and returned to Laverton—from where No 2 Squadron had deployed four years earlier—with the CO and the administrative nucleus of the new unit. As the many operational aircraft types held in Australia were rationalised for the peacetime "interim" air force, the Mitchell was destined not to remain in the RAAF inventory. The aircraft were left at aircraft depots, eventually to be scrapped. Most of the aircrew and ground staff proceeded to personnel depots in their home states for discharge.

From Friday 21 December, No 2 Squadron operated on a cadre basis of five personnel, under Wing Commander Douglas. Although it was planned to reconstitute the unit as an attack squadron, and the CO was attached to No 5 OTU at Williamtown for a Mosquito conversion, this was not to be. Finally, No 2 Squadron was reduced to a "Number Only" basis and disbanded on 15 May 1946.

No 2 Squadron's last two war casualties were two groundcrew LAC Ernest Zosky[40] and LAC Esmond Kenny,[41] who both died of illnesses in Melbourne contracted during their duty in the tropics. As this era of 2 Squadron's operations ended, engine fitter Corporal Keith Dodd speaks for many as he recalls:

All those of 2 Squadron served us well,
those who survive remember them still.[42]

The revetments, taxiways, maintenance, operations and domestic sites at Hughes Field are now overgrown, and little remains to testify to No 2 Squadron's presence over those war years. Perhaps some still remember when:

---

39 AA CRS 1969/100/39, File 2SQN/1/5/AIR Pt.2, of 14 Dec 1945.

40 LAC E A Zosky, 53156. No 2 Sqn. Of Melbourne; b Melbourne 26 Jul 1901. Died of Illness 30 Jan 1946.

41 LAC E P Kenny, 10270. No 2 Sqn, 1AD. Engineer of Woodend, Vic; b Taradale, Vic, 12 Mar 1905. Died of illness 18 Feb 1946.

42 Keith Dodd, letter to author, of 16 Sep 1993.

... everybody would go down to see the take-off, from CO and Adjutant down to the cooks. Twelve planes would start up, twenty-four whirling airscrews would glint in the sunshine, and the noise was shattering. As the last plane raced down Hughes Field, the noise would die away in whispering echoes through the trees. And we would have to wait five or six hours, while those planes evaded the defences of Lautem West. Then, as we heard their engines overhead, how eagerly we rushed outside to count the navigation lights milling around in the starry sky. All back safely![43]

43 Brian Hawthorne, introduction to diary for 1944.

PART THREE

AIR SUPPORT

CHAPTER 21

THE HEAVIES

RAAF Lincolns were like all British-designed aircraft, an experiment in what the human airman could possibly endure physically. And with as many little hooks and jagged bits added afterward to make his existence in the beast even more trying. I think our Amberley Lincolns' decibel (db) rating in the cockpit and at the navigator's and w/op tables was about 100 … pain threshold.[1]

The first problem confronting a peacetime military organisation at the cessation of hostilities is the repatriation and demobilisation of large numbers of personnel. In October 1945 there were 152,000 men and women in the RAAF and WAAAF, being discharged at the rate of 5,500 per week.[2] During 1946, the Government fixed a figure of 12,000 men for the postwar "Interim Air Force"—the bridging period between the war to the re-establishment of a peacetime force—within an organisation of 16 squadrons of 144 aircraft.[3]

Together with mass demobilisation, another problem for the interim RAAF was planning the structure of the new force, the aircraft types to be kept for each role in the new organisation, and the squadron identities that were to be retained for a much reduced service. The bomber that had been selected to equip the restructured air force was the development of Bomber Command's Avro Lancaster—the Lincoln.

Originally, plans had been made to set up a Lancaster Mark III production line in Australia, and in November 1943 the local manufacture of 50 Lancasters had been approved, with a total run of 346 aircraft envisaged to re-arm all of the RAAF's Beaufort squadrons.[4] Because of the long lead-time, the RAAF had continued also to press for American bombers, and the USAAF approved the equipping of RAAF squadrons with the B-24 Liberator. By 1944, Australian production plans had changed to the Lancaster Mark IV which, by virtue of its greater range, was more suited to Pacific operations against Japan.[5] By March 1945 the RAAF was planning to supplement B-24s with 100 of the improved Lancasters,[6] now renamed the Lincoln, but delays in its introduction meant that Australian production was not underway by the war's end. Nevertheless, production continued for 85 aircraft.[7]

---

1 Alan Underwood, quoted in M Garbett & B Goulding, *Lincoln at War 1944–1966*, Ian Allan, London, 1979, p.96.

2 G Odgers, *The Royal Australian Air Force: An Illustrated History*, Ure Smith, Sydney, 1977, p.84.

3 Odgers, p.84 and Dept of Air, *The Golden Years*, AGPS, Canberra, 1971, p.35.

4 A Stephens, *Power Plus Attitude*, AGPS Press, Canberra, 1992, p.80.

5 Garbett & Goulding, p.92.

6 Stephens, p.82.

7 D Eyre & B White, "The Avro Lincoln in Australia", *AHSA Journal*, Vol IX, No 1, Jan–Feb 1968, p.4.

No 2 Squadron Lincoln, A73-28, in April 1948, with a line-up of 86 Wing Dakotas RAAF

The first five RAAF Lincolns had been assembled by the Department of Aircraft Production, Beaufort Division (which became the Government Aircraft Factories at Fishermen's Bend, in Melbourne) from components shipped out from the United Kingdom. The first aircraft, A73-1, flew on 12 March 1946. The Lincoln differed from the earlier Lancaster through more powerful Merlin engines, greater fuel capacity which led to a wider wingspan, heavier firepower from the gun turrets,[8] and more advanced radio and navigation equipment, including the H2S Mark 3 navigation and bombing radar and a radar-aided gun sight in the rear turret.

The Government Aircraft Factories ultimately produced 73 Lincolns for the RAAF,[9] which continued the progress made during wartime with indigenous production. The RAAF Lincoln B.30 had a crew of seven (two pilots, navigator, signaller, and three gunners) and had a bombload of 14,000 lb, typically 14 1,000 lb, or 500 lb bombs. These loads could be carried over a range of 3,000 miles, at a cruising speed of 215 mph (350 km/h), a performance superior to the B-24 Liberators they replaced in 82 Wing.

Tradition determines that the senior identities of units survive when forces cut back, and this was the case by the re-numbering of the squadrons of No 82 Bomber Wing, at Amberley. On 23 February 1948, No 12 Squadron was retitled No 1 Squadron, and No 23 Squadron became No 6 Squadron. No 21 Squadron was renumbered No 2 Squadron.[10] The three squadrons were not, however, autonomous,

8 The formidable firepower of the Lincoln had replaced the smaller .303 calibre guns of the Lancaster. The Lincoln was equipped with twin 12.7mm (0.5 in) guns in Boulton Paul nose and tail turrets, and twin 20 mm cannon in the Bristol dorsal turret.

9 These RAAF Lincolns comprised five built by Avro, and assembled in Australia by GAF, the bomber variant (B.30), and the long-nosed general recce (GR.31) and maritime (MR.31) variants.

10 No 2 (Communications) Squadron had been flying at Mallala, near Adelaide, since Jul 1947, with an assortment of transport aircraft in support of the Woomera Range area. No 2 (Comms) Sqn was retitled No 34 (Comms) Sqn on 1 Mar 1948.

relying on maintenance centralised in the Wing, and aircraft provided from the Wing "pool". The heavy bomber force came under Headquarters Eastern Area.

The first Commanding Officer of the re-formed No 2 Squadron was Wing Commander Arthur Cross,[11] whose task was to continue a program of conversion courses for 82 Wing and normal squadron bombing and gunnery training. The Lincoln was equipped with the Rebecca navigation system and H2S radar for ground mapping and bombing. The aged radar was nevertheless effective, and the squadron training program dedicated emphasis on radar techniques, which were new to the RAAF. Bombing was carried out at the Beaudesert Range and on the wreck of the *SS Maheno*, off Fraser Island, while gunnery training was conducted at Stradbroke Island. A site near Gatton was also investigated during 1948 for a bombing and gunnery range, but this was unsuitable, and the following year Evans Head, in northern New South Wales, was reopened for RAAF weapons training. This range remains today the primary weapons range for 82 Wing training. Emphasis was placed on training for the accurate delivery of weapons, as pinpoint delivery, and not the widespread area bombing of the RAF's Bomber Command:

> ... Pathfinder technique will not normally be used by RAAF bomber squadrons in peace or war unless acting jointly with British or allied forces. RAAF bomber aircraft will rely, under normal conditions, on the efficiency of the aircraft's own bombsight[12] to destroy small targets on land and sea.[13]

Squadron personnel settled into a routine common to peacetime units—deployments for mobility exercises in Darwin, naval cooperation support, bush fire patrols, and search and rescue duties, which included flood supply dropping and medical evacuation. For public relations, participation of RAAF squadrons in air pageants in the postwar years showed the public what this new peacetime air force could do.

Some of the Liberators were retained on the Squadron, however, primarily for the transport of VIPs. On one Liberator sortie on 20 August 1948, Wing Commander Cross (A72-172) was climbing out through cloud, 30 minutes to the south of Amberley, when a fire started in the inner port engine.[14]

> Despite the difficult situation this officer carried out his drill calmly and efficiently and after ten minutes eventually was able to douse the flames which had seriously damaged the mainplane. Notwithstanding this structural damage the aircraft was flown back to base in bumpy bad weather conditions.[15]

---

11 Gp Capt A E Cross AFC, 150/O352. Nos 5 & 9 Sqns, HQNEA, HQ 2TG; comd 99 Sqn 1945–46, 21 Sqn 1946–48, 2 Sqn 1948, 391 Sqn 1952–53, 2 Sqn 1955–57; HQ FEAF, HQSC; comd BSqn Pck 1962–65. Regular air force officer of Sydney; b Sydney 19 Apr 1915.

12 The Lincoln was fitted with the Mk.14 bombsight, designed for European "built up" targets, and had been fitted as an expediency.

13 AA CRS A1196/2, Dept of Air minute 60/501/186 "RAAF Operation of Heavy Bomber Aircraft", of 1 Mar 1948.

14 This transport sortie, known as "Operation Orient", was to Singapore, via Laverton. Departing the following day, the CO completed the task in Lincoln A73-27. No 2 Sqn A50, of 20–26 Aug 1948.

15 Citation to the award of the AFC to Wg Cdr Cross, RAAF Personnel Records, Canberra.

Commanding Officer of No 2 Squadron over 1949–1951 was Flight Lieutenant Jim Graney.
J GRANEY

A fuel leak had developed, which had dripped onto the turbo blower and ignited, the fire engulfing the mainplane until it was extinguished. The handling of this serious emergency resulted in the award of the Air Force Cross to the CO. The last Liberator was withdrawn from No 82 Wing soon after this incident, in January 1949. Over this period, Commanding Officers of the Squadron were to change fairly often; one significant appointment was Squadron Leader Bob Dalkin, a stalwart of the Hudson era, who commanded the unit in 1949, before handing over to Flight Lieutenant Jim Graney.[16]

The RAAF had been responsible for tropical flight trials of the Lincolns for the RAF, but another trial conducted in Australia was of equal importance to the RAF. During the second half of 1949 and early 1950, No 82 Wing conducted a series of long-range bombing and navigation trials, simulating the strategic profiles of RAF European missions. These trials, known as "Operation Cumulative", explored the limits of strategic navigation and bombing,[17] and were a maximum flying effort for both men and machines. Fourteen Lincolns were modified by GAF as Long Range Navigation (LRN) variants, with special radar, radio, oxygen, photographic and instrument fitments, and a position for an eighth crew member. In addition, a long-range fuel tank was fitted into the bomb-bay.[18]

Between October 1949 and February 1950, RAAF and RAF crews flew almost 1,700 hours on 12 hour routes from Amberley to Darwin or Kalgoorlie, at night, with aircraft on the same route, one minute apart.[19] This effort over summer not only involved 82 Wing crews, but also eight RAF crews attached from the United Kingdom, four of whom served with 2 Squadron for the trials.[20]

---

16 Gp Capt J P Graney AFC, 407944/O4409. 7EFTS, Nos 35 & 36 Sqns, CFS, ARDU, 1 Sqn; comd 2 Sqn 1949–1951; O'Seas HQ London; comd 1 Sqn 1958–62; AA Indonesia 1965–67; comd ARDU 1967–69; SASO HQSC; comd BSqn Wlm 1971–73; HQSC. Regular air force offr of Adelaide; b Sheffield, Eng, 9 Mar 1921.

17 Stephens, p.108.

18 S Wilson, *Lincoln, Canberra & F-111 in Australian Service*, Aerospace Publications, Canberra, 1989, p.30.

19 Training for "Cumulative" began in September, with the first wave of 10 Lincolns flown to Darwin on 30 Sep 1949. Nos 1, 2 and 6 Sqns A50s, of Sep 1949–Feb 1950.

20 The RAF crews with No 2 Sqn were captained by Flt Lt H J C Barnes (57648), Flg Off P C Wyche (59481), P.II R W Bradley (1817695) and P.III D B Ride (2360123). 1 Sqn RAF crews were under Flg Off R J Gaywood & P.II D A Taylor; No 6 Sqn Flt Lt N R L Bristow & Flg Off R L Tavanyar.

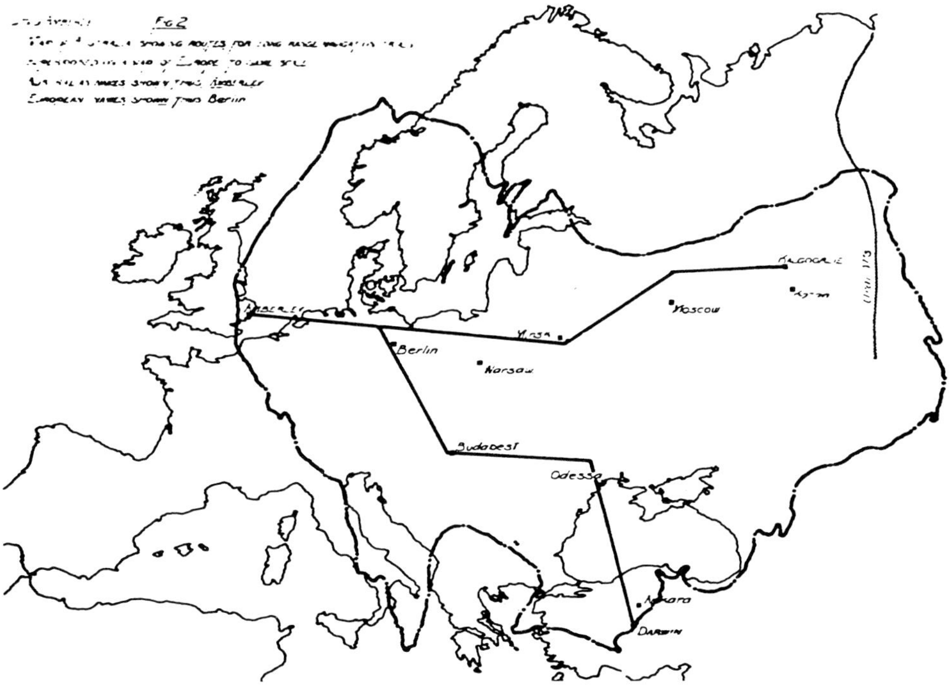

Routes for Operation "Cumulative" long-range trials 1949–1950 RAAF

Lincolns would take off at dusk each Thursday, with a time-on-target at dawn, terminating in alternating months at Darwin or Kalgoorlie,[21] dropping four 1,000 lb bombs from blind bombing runs, using the H2S radar, on targets before landing at their destination. Return profiles were flown to Amberley each Sunday evening. Each trip, flown at 20,000 feet, was the equivalent of flying from England to Moscow, and half-way back.[22] The eighth crewman was there to monitor the flight's progress and crew performance, and ensure safe navigation by use of the new Lucero navigation equipment. Information collected from the *Cumulative* flights was to provide the RAF with planning data for their strategic bomber force in its deterrent role during the Cold War.

In response to a British request in April 1950 for air reinforcements to join the RAF in the anti-communist campaign against Chinese terrorists in the Malayan Emergency, the Government announced on 27 June the immediate dispatch of a squadron of Lincoln bombers.[23] The unit selected for this campaign, which became known as "Operation Firedog", was No 1 Squadron, which deployed to Tengah, Singapore, that July.[24] This left Nos 2 and 6 Squadron forming 82 Wing at Amberley, with the latter being responsible for the conversion of crews to the Lincoln. In Malaya, No 1 Squadron's Lincolns were to become the mainstay of the Commonwealth offensive air

21 Sep/Oct and Dec 1949, and Feb 1950 destination was Darwin; Nov 1949 and Jan 1950 Kalgoorlie.

22 Garbett & Goulding, p.93.

23 P Edwards, *Crises and Commitments*, Allen & Unwin, Sydney, 1992, pp.87–97.

24 Initially six Lincolns were deployed, this being increased to eight during 1951.

operations against the communist terrorists (CTs), dropping 85 per cent (15,000 tons) of bombs used during the campaign.[25]

After dropping bombs in Malaya the standard procedure for 1 Squadron's Lincolns was then to descend and strafe the target area with all six guns, before returning to base on a low-level "flag wave" over the villages.[26] Such strafing was conducted in line astern by the large bombers, and this "Malayan Technique" was to be taught on No 2 Squadron for crews departing for their tour of duty in Singapore.

With the departure of No 1 Squadron from 82 Wing, the decision was made in November 1950 to make Nos 2 and 6 Squadrons independent units with their own groundcrew and aircraft, but remaining under the command of the Wing. No 2 Squadron had over the three years of Lincoln operations only comprised three or four RAAF crews. By early 1951, this strength was built up to five crews, and in May a formal flight structure was instituted. Under Squadron Leader James Sutherland[27] as the CO, "A" Flight was commanded by Flight Lieutenant Jimmy Wilson[28] and "B" Flight was led by Flight Lieutenant Cecil Miller.[29] This organisation provided a nucleus of two flights which could be rapidly expanded to bring the Squadron up to its full wartime establishment of eight crews by the addition of Reservists.

The peacetime RAAF rank structure had also undergone changes over this period. Following the RAF lead, non-commissioned aircrew had forgone the normal ranks of Sergeant, Flight Sergeant and Warrant Officer, and had inflicted upon them specialist ranks of airmen aircrew.[30] Although aircrew were entitled to continue using the Sergeants' Mess, they now did not have the authority of a senior NCO. As a result, having come off flying a long sortie, they could be rostered for the menial duties normally placed upon the lower enlisted ranks. This system was eventually recognised as being unsuitable, and the former ranks were re-introduced for aircrew in February 1951.[31]

---

25 Edwards, p.87; and Eyre & White, p.9

26 Garbett & Goulding, p.83.

27 Gp Capt J M Sutherland, 566/O398. Nos 21, 2 & 1 Sqns, POW 9.4.42. RAAFHQ, SAN; comd 2 Sqn 1951–52; RAAF O'seas HQ; comd 1ATU 1962–65; SEATO 1965–67; RAAF Acad 1970–73. Regular air force offr of Melbourne; b Melbourne 9 Oct 1920.

28 Gp Capt J S Wilson AFC, 412212/O22003. Nos 102 & 23 Sqns, 82 Wg, 2 & 1 Sqns, Canb Flt, Asst AA Wash, CFI CFS 1959–63; comd BSqn Edn 1964–67, CFS 1967–70, East Sale 1970–73; RAAF Acad 1973–75. Regular air force officer of Sydney; b Sydney 14 Aug 1921.

29 Sqn Ldr C G Miller DFC, 412451/O22004. Nos 620, 514, 7, 582 & 243 Sqns RAF, 37, 38, 2,1,6 & 34 Sqns; ATC Edn, Ric, Lav, HQSC. Regular air force officer of Yeoval, NSW; b Wellington, NSW, 21 May 1918.

30 This system, based on one introduced for a short period by the RAF, started in Feb 1949, and lasted for two years. The rank of Master Aircrew was equivalent to Warrant Officer, Aircrew I and II being equivalent to Flight Sergeant and Sergeant, Aircrew III being equivalent to Corporal, and Aircrew IV applied to trainee aircrew.

31 This also coincided with the introduction of new service numbers for personnel. The wartime Empire Air Training Scheme (EATS) numbers had been retained until this period, a six-digit number commencing with the year of enlistment. The new system comprised a prefix of O for officer or A for airman, followed by numbers, the first of which indicated the state of enlistment.

All of No 2 Squadron's crews were to be needed for the elaborate Air Force Week flying pageants in September 1951. For the previous year's displays at the main air bases, seven of 82 Wing's Lincolns had staged a massed flypast with 30 Mustangs. 1951, however, was the 30th Anniversary of the founding of the RAAF, and the air displays featured massed flypasts of over 100 aircraft. This included 18 Lincolns from the Wing and from the training units at East Sale, Mustangs, Vampires, Dakotas, and the first of the RAAF's new bombers, the Canberra.[32] This must surely have been a scene to behold.

Lincoln A73-10 at Amberley RAAF

Both of No 82 Wing's Lincoln squadrons were involved in the British atomic trials carried out in western and central Australia during the early 1950s. The first test was "Operation Hurricane" in October 1952, at Monte Bello Island, 80 kilometres from Exmouth Gulf in Western Australia.[33] The next was "Operation Totem", in October 1953 at Emu, in South Australia.

For the testing, the Lincolns were also supported by those of the newly formed general reconnaissance unit, No 10 Squadron, at Townsville. In the first trial, *Hurricane*, the bomb was exploded in a British frigate moored in the bay. Eight Lincolns from No 6 Squadron, and five from No 10 Squadron (two of which were flown by 2 Squadron crews) provided air communications, meteorological reports and atomic cloud sampling. No 6 Squadron operated from Broome, while the other Lincolns flew from Townsville. All of the aircraft from Townsville entered the atomic cloud for sampling, but no precautions were taken to protect the crews, nor were there any subsequent decontamination procedures for the aircraft.[34]

32 This first Canberra, A84-307, had arrived in Australia in July 1951 and, with a second aircraft, A84-125 which arrived in 1952, was a pattern aircraft for the Australian production line at GAF.

33 Wilson, p.55.

34 Wilson, pp.55–8.

One aircraft, A73-5, was flown by a 2 Squadron crew under Wing Commander "Garrie" Garrisson.[35] On the day of the sampling flight, 4 October 1952, Garrisson's crew flew for eight hours in the atomic cloud, collecting the radiation dust.[36] The navigator, Pilot Officer Trevor Owen,[37] recalled:

> We received a brief from the head British scientist in charge of the trials, covering where the bomb was to explode, and where the cloud was to drift. It was important for us to maintain our altitude and know our position for subsequent plotting of cloud movement. I later found out that Lincoln A73-5 was the most contaminated of all Lincolns. Surprisingly, groundcrew working on the engines were in the greatest danger of becoming contaminated. So much of the radiation was sucked in through the intakes, that with later washing and stripping down of the engines, several groundcrew were to suffer the ill-effects.[38]

No 2 Squadron continued training Lincoln crews during 1952, even though most conversions were conducted on No 6 Squadron. In September Flight Lieutenant Doug Harvey,[39] a Lancaster veteran of the past war, departed with his crew (A73-33) for No 1 Squadron at Tengah. Two months later, they were followed by Wing Commander Garrie Garrisson's crew (A73-36), which included as navigator Pilot Officer Trevor Owen. Garrisson returned from Tengah to take command of No 2 Squadron, and Owen was later to be Senior Navigation Officer of No 82 Wing in 1973 for the acceptance of the first F-111.

One interesting deployment the following year was in May 1953 when four Squadron aircraft were detached to Momote, on Manus Island, in the Admiralty Group to the north of New Guinea. The Lincolns were flown by the OC 82 Wing, Wing Commander Dave Colquhoun[40] (A73-19), the temporary CO, Flight Lieutenant Frank Montgomery[41] (A73-52), Squadron Leader Norm Geschke[42] (A73-54) and Flight

35 AirCdre A D J Garrisson OBE, 148/O350. Prev Army Militia. Nos 23 & 25 Sqns, HQSA, HQNEA, 9 OG; comd AAS 1944–45; BCOF, HQEA, 2 & 1 Sqns; comd 2 Sqn 1953, Fairbairn 1955–57; DepAir; AA Wash 1960–63; comd Amberley 1963–64, Fairbairn 1964–67; comd 1st TAF RMAF 1967–69. Regular air force offr of Sydney; b Brisbane 23 Jun 1915.

36 Of the 9 hours 55 minutes airborne on this sortie, 8 hours were spent in the cloud. Air Cdre Garrisson's flying logbook, of 4 Oct 1952.

37 AirCdre T C Owen, O32535. Nos 2 & 1 Sqns; 101 Sqn RAF, AAS, 82 Wg, 35th Nav Trg Wg USAF, DepAir, 82 Wg; Mil Adv Delhi 1976–77; comd East Sale 1977–79; DepDef DGR 1979–81, DPI 1981–83. Regular air force offr of Melbourne; b Bristol, Eng, 1 Dec 1930.

38 AirCdre Trevor Owen, telephone interview with author, of 28 Jan 1994.

39 Wg Cdr D C Harvey DFC* AFC, 413192/O22012. No 467 Sqn, ARDU, 2, 1, 2, 6, 2 Sqns; comd 2ATU 1961, RAAF Tpt Flt Viet 1965, 38 Sqn 1966–67; HQ Ric, HQ Dar. Regular air force offr of Gosford, NSW; b Hythe, Kent, Eng, 1 Dec 1922.

40 AirCdre D W Colquhoun DFC AFC, 250284/O33011. Nos 21, 1 & 6 Sqns, CFI 5 OTU, RAAF Cd, O'seas HQ, HQEA; comd 82 Wg 1952–53, Amberley 1953–54; HQHC; comd RAAF Coll 1956–57; DepDef, AA Saigon 1967–69; comd RAAFSC 1969–70, Point Cook 1970–71. Regular air force offr of Melbourne; b Warragul, Vic, 29 Jul 1916.

41 Wg Cdr F J Montgomery AFC, 2643/O3384. Prev Army. CGS, No 24 Sqn, 7 OTU, AAS; t/comd 2 Sqn 1953; 1 Sqn, 1AFTS; SATCO Pea 1958–61, Amb 1961–64. Regular air force offr of The Rock, NSW; b Wangaratta, Vic, 3 Oct 1912.

42 Wg Cdr C N Geschke OBE, 419448/O33159. Nos 36, 37, 38, 13 & 37 Sqns, AAS, Nos 2 & 1 Sqns, RAAF O'seas HQ 1956–59; comd 6 Sqn 1959; HQSC; comd 38 Sqn 1967–68; HQSC. Regular air force offr of Melbourne; b Windsor, Vic, 7 Mar 1924.

Lieutenant Frank Griggs[43] (A73-38). Staging through Townsville and Port Moresby, the Lincolns escorted eight Mustangs from No 23 Squadron at Amberley. Over a period of 10 days, navigation training and "flag-waving" sorties were conducted throughout New Guinea and New Britain, in addition to weapons training and fighter affiliation with the Mustangs. A visit was also made to the B-29s of the USAF 53rd Strategic Reconn Squadron, at Anderson Air Force Base in Guam.

Safety had not been improved for *Totem*, the second nuclear test series later in 1953, when No 6 Squadron operated from Woomera, and the No 2 Squadron detachment flew with a USAF B-29 contingent out of Richmond.[44] The right meteorological conditions had to exist to ensure the atomic cloud did not drift over south-east Australia, so the explosion was delayed for three weeks until the right conditions prevailed. Over this waiting period, the No 2 Squadron contingent lived rather primitively under canvas, the best accommodation, of course, being surrendered to the Americans.[45]

For the *Totem* trials, two bombs were exploded on 15 and 27 October.[46] No 6 Squadron carried out the initial tracking of the nuclear cloud after the first device, *Totem 1*, was detonated on 15 October 1953. The Americans then took over to ensure continuous tracking, and the following day No 2 Squadron took over from the Americans. Pilot Officer Andy Stapleton[47] (A73-25) tracked the thick, brownish layer at 10,000 feet, having to take geiger counter readings to pinpoint the centre of the cloud:

> Headings were flown back and forth through the cloud until readings reached their peaks, thus the centre and pattern of the cloud were established. Auxiliary fuel tanks were carried in the bomb bay, and the Lincoln remained airborne for just over 12 hours, landing back at Richmond very low on fuel and very dirty. The Americans gave the crew a wide berth. It was the early days for this sort of business and in their innocence of the effects of exposure to radiation the crew took their Lincoln back to Amberley, where it was banished to the farthest corner of the airfield. There it was to spend the rest of its days, its fuselage marked with large purple crosses.[48]

The lack of safety precautions for these initial atomic tests was deplorable. Both aircrew and groundcrew were put at risk from the radioactivity. The disregard for decontamination of aircraft and equipment ultimately led to confusion as to what was the fate of the four Lincolns that were finally classified as "highly contaminated".[49]

---

43 Gp Capt F M Griggs DFC* DFM, 400468/O35376. Prev Army. Nos 218, 214 & 109 Sqns RAF, 1 Sqn; t/comd 2 Sqn 1953; AA Wash 1954–56; comd 1 Sqn 1962, 2 Sqn 1963–65, Townsville 1972–76. Regular air force offr of Vinifera, Vic; b Swan Hill, Vic, 1 Dec 1921.

44 2 Squadron involvement was A73-21 (Sqn Ldr Geschke), A73-25 (Plt Off Stapleton), A73-26 (Sqn Ldr Fairbairn), A73-27 (Flt Lt Ross), A73-37 (Flt Lt Griggs) and A73-40 (Flt Lt Harvey). Wilson p.61.

45 Garbett & Goulding, p.96.

46 Wilson, p.58.

47 Flg Off A W Stapleton, 187264/O11577. Nos 1 & 2 Sqns, SAN. Photographer of Murwillumbah, NSW; b Murwillumbah 24 Aug 1931.

48 Garbett & Goulding, p.96.

49 These aircraft were A73-25, -47, -52 and -54. Wilson, pp.61–2.

With jet fighters now equipping air defence forces around the world, major improvements were seen to be necessary for the RAAF's strike capability, and the Air Board pressed for the acquisition of the new British-designed B2/N45 bomber.[50] This was to be the world's first operational high-speed, high-altitude jet bomber—the Canberra. With this "world-beater", the RAAF would supplement, and eventually replace, the Lincoln.

No 2 Squadron had been selected as the first operational unit to receive the new Canberra jet bomber and assume the training role of Canberra Flight, that had been operating in 82 Wing. Flying Officer Mike Ridgway[51] and Flight Lieutenant Bill Kerr[52] arrived from Laverton with the first aircraft, A84-307, on 8 December. Squadron Leader Peter Raw,[53] who had commanded No 2 Squadron since the beginning of the year, but had been attached to No 1 Long Range Flight for the England to New Zealand Air Race, resumed command as the last Lincoln was handed over to No 6 Squadron on 18 December 1953.

---

50 Stephens, p.107.

51 AVM M J Ridgway AFC, O33285. No 34 Sqn, ARDU, 2 Sqn, 101 Sqn RAF, 2ATU, 1AFTS, CFS; comd 1 Sqn 1969–72; RAAF London; AOC HQOC 1979–81; DefAir CAFP 1982. Regular air force offr of Adelaide; b Adelaide 5 Aug 1928.

52 Gp Capt W D Kerr, 12543/O34402. No 87 & 1 Sqns, SAN, 2 Sqn, RAAF London, RAAFSC, DepAir. Regular air force offr of Melbourne; b Ripponlea, Vic, 13 May 1922.

53 AirCdre P F Raw DSO DFC AFC, 409445/O33113. No 178 Sqn RAF, ANS, 78 Wg; t/comd 2 OTU 1952; comd 2 Sqn 1953–55; RAAFSC; comd 82 Wg 1965–66, RAAF Vung Tau 1966–67, RAAFSC 1970–72, HQ But 1972–76; HQSC. Regular air force offr of Melbourne; b Carnegie, Vic, 5 Jun 1922.

CHAPTER 22

Advent of the Jet Bomber

English Electric's policy ... was to create an aeroplane that was "the extreme in adventurous conventionalism".[1]

The 1953 England to New Zealand Air Race, from London to Christchurch, was to celebrate the 100th anniversary of the founding of the city of Christchurch, and coincided with the 50th anniversary of the first powered flight by the Wright Brothers. It was also the time that the first of the Australian-built Canberras came off the production line. It was an ideal opportunity not only to prove the long-range and high-speed characteristics of the RAAF's new bombers, but also demonstrate the high degree of technology that the Australian aircraft industry had achieved. In the speed section of the race, the RAAF entered two of its new Canberras:

Aircraft	Captain	Second Pilot	Navigator
A84-201	SqnLdr Peter Raw	FlgOff Noel Davis[2]	FltLt Bill Kerr
A84-202	WgCdr "Jel" Cuming[3]	FlgOff Bob Atkinson[4]	SqnLdr Col Harvey[5]

The crew members of A84-201 were destined to form the nucleus of No 2 Squadron. Aircrews were supported by groundcrew, led by Squadron Leader Tom "Mo" Walkley,[6] who deployed refuelling, air traffic and ground support equipment to cover the Cocos Island–Australia–Christchurch segments of the route. Early entrants for the speed section had also included three RAF Canberras and the new Valiant "V" bomber, a Danish air force F-84G Thunderjet, two Australian civil-registered Mosquitos, an Australian Mustang with a United States registered F-82 Twin Mustang, and a Spitfire Mark 24. As the event grew nearer, some entrants dropped out, and the eventual allocation of race numbers was:

---

1 R Jackson, *Canberra—The Operational Record*, Airlife, Shrewsbury, 1988, p.2.

2 Flt Lt F N Davis DFC, 420642/O23506. No 50 Sqn RAF, 1 Sqn, 82 Wg, 2 Sqn. Regular air force offr of Sydney; b Bathurst, NSW, 16 Mar 1922. Killed in flying accident 16 Jun 1954.

3 AirCdre D R Cuming OBE AFC*, 250553/O33012. Nos 2, 13 & 7 Sqns, 1APU/ARDU, 82 Wg; comd ARDU 1952–54, Edinburgh 1956–60; HQSC; comd 481 Sqn 1967–68; HQSC. Regular air force offr of Melbourne; b Melbourne 12 Feb 1917.

4 Flt Lt R J Atkinson, 435409/O11416. No 76 Sqn, 82 Wg, 1 & 2 Sqns. Regular air force offr of Gunalda, Qld; b Gympie, Qld, 2 Dec 1914.

5 Wg Cdr C G Harvey, 4285/O3492. Nos 12, 30, 24 & 34 Sqns, RAAF HQ, 86 Wg, SAN, ARDU, HQHC/HQOC, HQ FEAF NavO 1962–64, DepAir. Regular air force offr of Melbourne; b Armadale, Vic, 25 Sep 1920.

6 Sqn Ldr T L Walkley MBE DFM, O2110. Nos 461 & 460 Sqns, 7 OTU, 1 APU/ARDU, 2 Sqn, 2 OTU, 478 Sqn, DepAir. Regular air force offr of Mayfield, NSW; b Brisbane 2 May 1915.

Number 1	RAF Canberra PR.7 WE733 (Hodges)
Number 2	RAF Canberra PR.3 (Furze)
Number 3	RAF Canberra PR.3 WE139 (Burton)
Number 4	RAAF Canberra B.20 A84-202 (Cuming)
Number 5	RAAF Canberra B.20 A84-201 (Raw)
Number 6	Australian Mosquito Mk.41 VH-KLG (Oates)
Number 7	Australian Mustang Mk.20 VH-BVM (Whiteman)
Number 8	Australian Mosquito Mk.41 VH-WAD (Woods)
Number 9	RAF Valiant B.1 WB215 (Oakley)

Two of the Australian entrants, Woods and Whiteman, were forced to withdraw through lack of backing, and Oates ditched his Mosquito in the Indian Ocean, en route to England. The RAF Valiant, the second prototype of the untested bomber, was withdrawn at the last moment, perhaps so that it would not be shown up by the Canberras. This left five Canberras to fight for the speed prize in the race.

The RAAF crews trained initially on the two British built pattern aircraft[7] during mid-1953 with the Long Range Flight at Laverton, while awaiting the completion and modification of the first Australian aircraft. When the first GAF-produced aircraft became available in August, it was modified with long-range fuel tanks, in lieu of the bomb-bay, and additional navigation equipment.[8] Long-range training flights were also conducted to prepare the aircrews for the discomfort of flying long distances in their electrically heated flying suits and seated on the hard ejection seats in the cramped, cold cockpits. In addition, with the groundcrew, the pilots rehearsed fast refuelling and maintenance turnarounds, enabling a "pit-stop" of only 10 minutes on the ground at each point of landing. In fact, A84-201 recorded a time of only seven minutes between landing and take-off as it staged through Cocos Island during the race.[9]

The race commenced on the dismally wet London afternoon of 8 October 1953. Leaving at five minute intervals, the Canberras climbed through the cloud to 40,000 feet on their first leg to Bahrain. Taking off in race number order meant that the 2 Squadron crew in A84-201, under Raw, had been the last to depart. On this six-hour night leg, having passed Italy and over the Adriatic, Raw's navigator, Flight Lieutenant Bill Kerr recalled:

> ... in the midst of severe electrical storms, both of my radio compasses went haywire. An hour or so after passing Venice without a check on our position we locked on to Tunis, which showed our ground speed had increased by more than 100 knots. And there was joy as the other compass told us we were heading for Bucharest. The wind change had put us about 300 miles north of our intended track and we were well into forbidden Iron Curtain territory.[10]

---

7 A84-125 and A84-307.

8 Both GAF aircraft were fitted with a second radio compass to provide bearings from ground stations, and an Australian distance measuring equipment (DME) system.

9 Bill Kerr, telephone interview with author, of 10 Nov 1993.

10 Bill Kerr, "In the Last Great Air Race from London to Christchurch", *The Canberra Times*, Wednesday 9 Oct 1991, Canberra, p.25.

Crew of A84-201 at Takali Airfield in Malta
From left: Flt Lt Bill Kerr, Sqn Ldr Peter Raw, Flg Off Noel Davis B KERR

Despite this unintended deviation, Raw's crew arrived at Bahrain ahead of their predecessor out of London. The two pilots, Raw and Davis, swapped over, while a prepositioned navigator handed Kerr the weather forecast and flight plan for the next segment to Ceylon (Sri Lanka). They were off in 10 minutes. After a flight of four hours and 25 minutes, flown in half night and day, A84-201 arrived at Ratmalana aerodrome one minute in front of the favoured RAF Canberra PR.7, the latest model flown by their team leader Wing Commander Hodges. After another quick turnaround, a problem closing the crew entrance door delayed departure by some four minutes, enabling Cuming's A84-202 to beat them into the air. The most favourable tailwinds were at 25,000 feet for the next leg to Cocos Island, but the turbulence of this jetstream forced Raw to elect the clearer air at 40,000 feet. Cuming remained in the turbulence, which stopped one of his engines, and forced him to descend to get a restart. Nevertheless, Cuming reached Cocos first, by some 10 minutes, which put him as the race leader, but he blew a tyre on landing, forcing his retirement.

A jetstream across Australia's north-west now presented the opportunity of flying direct to Woomera, without the planned stop in Perth. Not only did this provide a faster speed and shorter distance to Woomera, it had the added advantage of giving extra fuel at Christchurch, by virtue of being a shorter final leg. The lead of the RAF competition, which had previously chosen to overfly Cocos for Perth, passed over the island as Raw leaped airborne. As the RAF Canberras had greater track distance via Perth, this placed A84-201 as the race leader. With the sun setting, Raw's crew crossed the Australian coast over the Monte Bello Islands—16 hours after leaving London—while the lead RAF aircraft was still 40 minutes short of Perth. At about 10 p.m. the lights of Woomera came into view; however, the nose wheel was iced up and would not extend and, as Davis landed, a large hole was gouged in the nose and the pitot head, which measures airspeed, was ground away. Staring at certain withdrawal from the race, Bill Kerr recollected:

> Our ground staff were not prepared to give in so easily. A Squadron Leader engineer, Carl Leopold, brushed us aside and said he'd get us on our way. He shouted to a group of airmen to clamber on top of the aircraft and sit on the tailplane to get the nose lifted so he could free the nose wheel. He then shouted for some more to go to the rear and hold the tail up so it would not hit the ground and be damaged. There weren't enough airmen there to do this and the many journalists and press photographers there pocketed their notebooks and dropped their cameras and came to our aid. At the same time, he sent an instrument fitter off to steal a pitot head from one of the RAF Canberras which was based at Woomera for the Atomic Trials Unit.[11]

11 Kerr, p.25.

As the emergency repairs were being carried out, the news came that the lead RAF Canberra had been forced to withdraw due to generator problems, and the second RAF Canberra had left Perth an hour before 201's arrival at Woomera, with the remaining aircraft 30 minutes behind. Carefully checking airspeed indications on take-off, Raw left Woomera at about midnight, having been delayed by an hour and a half. With favourable winds at 40,000 feet, A84-201 could still gain second place. However, the hole in the bottom of the nose, large enough for a man to fall through, forced cold air into the cockpit, and prevented a stable cruise at high level. The aircraft stalled, diving towards the ground until becoming controllable again at 30,000 feet. A climb was initiated back to the best winds at 40,000 feet, where the aircraft again dived. This porpoising continued across eastern Australia and the Tasman. A hole in a pressurised cabin has an even more sinister effect as Bill Kerr found as his microphone failed and he was forced to change his mask:

> Changing leads and tubes only takes about 10 seconds but, no doubt because of fatigue, it had not dawned on me that because of the hole in the nose we were no longer in a pressurised cabin. Around 40,000 feet, 10 seconds without oxygen causes one to pass out. Fortunately for me Noel Davis was watching and as he saw me fall forward was able to lean back and complete the oxygen connection and slap the supply regulator switch to the emergency position. I quickly recovered and we continued on our way.[12]

However, another problem had emerged since the nose landing: the loss of the radio transmitter. Although the crew could hear the radio calls from ground stations and other aircraft, nobody could hear the calls of A84-201. This was to cause problems for the arrival in the deteriorating weather at Christchurch. They passed over the final destination at 20,000 feet, and descended through cloud over the sea to the east, until the white caps of the water became visible at 500 feet. Fortunately, because of the shorter final leg, they had the reserve of extra fuel to allow a diversion to better weather of the North Island if necessary. Turning back towards Christchurch, they remained visual with the high ground around the coast, and on finding the local racecourse, turned in the direction of the airport. Bill Kerr remembered the relief of the ground radar shouting:

> "We have you, turn left two degrees, gear and flaps down, you are two miles out, descend and land straight ahead". The runway lights came into sight and we made a smooth touchdown. We were there at last, 24 hours and 32 minutes since leaving London.[13]

The winning aircraft, the RAF Canberra of Flight Lieutenant Monty Burton's crew, had beaten A84-201 by 41 minutes. The impromptu letdown at Christchurch had cost the Australians, and with the unscheduled delay at Woomera, this amounted to some two hours of lost time. As for actual flying times, the winner had spent 22 hours and 28 minutes in the air, and A84-201 exactly one minute longer. The third Canberra, flown by RAF Flight Lieutenant Furze, was only eight minutes slower. The performance of the Canberras in the race showed the superior characteristics of the aircraft that had been selected to equip the RAAF's bomber squadrons.

---

12 Kerr, p.25.

13 Kerr, p.25.

Preparations for the new aircraft's entry into service had already been in hand at Amberley since the middle of 1952 with the formation of Canberra Flight within 82 Wing, flying the two imported jet bombers.[14] The pilots and navigators selected as Canberra aircrew, which included Flight Lieutenant Jimmy Wilson and Flying Officer Charlie Reif,[15] had trained with the RAAF's original Canberra, A84-307. They then joined with the members of the Long Range Flight to form the new No 2 Squadron, as Squadron Leader Raw returned to assume command on 18 December. Two days later, he departed with Pilot Officer Bert Peut[16] (A84-307) on a three-hour flight to Auckland, and then established a record westward Tasman crossing to Sydney on 23 December. Meanwhile, the Squadron Navigation Officer, Flight Lieutenant Bill Kerr, was preparing the first Canberra conversion course for 2 Squadron navigators, which commenced in January 1954. The pilots who had been involved with the Long Range Flight for the air race were all allocated specialist roles for the introduction of the Canberra into operational service. Both Squadron Leader Peter Raw and Flight Lieutenant Noel Davis, as the flying instructors, worked on the conversion syllabus, while Flying Officer Bob Atkinson determined the best instrument flying letdown and approach procedures for the Canberra. A reserve pilot for the race, Flying Officer Mike Ridgway, became the weapons specialist, devising a test schedule and conducting bombing trials. This was to encompass dropping 25 lb practice bombs at 10,000 feet at the weapons range on Stradbroke Island, increasing heights by 5,000 foot increments as required accuracy was attained. Ridgway then continued trials at the Evans Head bombing range.

The conversion training in early 1954 was interrupted for the royal visit, and the flypast requirement for the Royal Salute at the Queen's opening of Parliament on 15 February. No 2 Squadron's three aircraft were to be involved, to overfly in a "vic" at the tail of the stream, and comprised:

Sqn Ldr Raw/Flt Lt Kerr	A84-202
Flt Lt Davis/Flg Off Reif	A84-125
Flt Lt Atkinson/Flg Off Bell[17]	A84-307

It was unfortunate that adverse weather prevented the flypast, and the Squadron resumed its training program. New pilots to the Canberra were first attached to No 2 OTU at Williamtown for jet familiarisation flying Vampires, as there then were no Canberras fitted with dual controls. One student pilot who joined the Squadron, having just completed a tour on Lincolns with No 1 Squadron in Malaya, was Flight

14 Charlie Reif, telephone interview with author, of 16 Nov 1993.

15 Sqn Ldr C W Reif, 171726/O1553. BCOF, No 1 Sqn, Canb Flt, 2 Sqn, AAS, 50 Sqn RAF, 2ATU, SAN, 1 OCU, 2 Sqn, TSF, 24 & 76 Sqns, DepAir. Regular air force offr of Mt Morgan, Qld; b Mt Morgan 2 May 1924.

16 Flg Off B N Peut, 187126/O11245. No 2 Sqn, 1 AD. Surveyor of Brisbane; b Charters Towers, Qld, 23 Feb 1928.

17 Flt Lt J Bell 169558/O21622. 82 Wg, HQEA, ARDU, No 2 Sqn, 101 Sqn RAF, SAN, 2 Sqn, 1AFTS, DepAir. Regular air force offr of Sydney ; b Sydney 12 Aug 1928.

Lieutenant David Nicholls.[18] Having finished his jet lead-in training at Williamtown, Nicholls was killed during his conversion on 16 June. Under the instruction of Noel Davis, the aircraft (A84-202) lost directional control while overshooting from an asymmetric approach, killing both pilots and the trainee navigator, Pilot Officer Fred Adler.[19] A week later, the award of Nicholls' DFC for his service in Malaya was announced.

One of the early Australian built Canberras, A84-203, displaying its sleek lines which made the world's first jet bomber such a technological advance AWM 127928

Deliveries of more aircraft permitted a Squadron deployment at the end of July to Pearce, in Western Australia. Exercise "Swan Lake" during August was an air defence operation with eight Vampires from Richmond's No 22 Squadron. The deployment involved:

A84-307	Sqn Ldr Raw/Flt Lt Kerr
A84-125	Flt Lt Kilsby[20]/Flt Lt Hughes[21]
A84-201	Flt Lt Goldner[22]/Flg Off Bell
A84-204	Flg Off Ridgway/Plt Off Pout

18 Flt Lt D C Nicholls DFC, 424445/O22075. No 461 Sqn, 82 Wg, 1 & 2 Sqns. Regular air force offr of Stockton, NSW; b Nabiac, NSW, 19 Jul 1924. Killed in flying accident 16 Jun 1954.

19 Plt Off F B Adler, O39704/O311305. Prev RAF. No 2 Sqn. Regular air force offr of Melbourne; b Vienna, Austria, 28 Mar 1923. Killed in flying accident 16 Jun 1954.

20 Wg Cdr C G Kilsby MBE AFC, 430211/O33178. No 356 Sqn RAF, SAN, 1 Sqn, ARDU, 2 Sqn, RAAF London; comd 6 Sqn 1960–61; 9 TRS USAF; comd Melb Uni Sqn 1965–66, BSupt Flt Viet 1966–67; DepAir. Regular air force offr of Strathdownie, Vic; b Casterton, Vic, 11 Aug 1924.

21 Sqn Ldr G Hughes DFC AFC, 402797/O21974. Nos 108, 148, 70 & 40 Sqns RAF, 2 & 38 Sqns, SAN, 2, 6 & 2 Sqns; t/comd 10 Sqn 1958–59, comd Qld Uni Sqn 1960–64. Regular air force offr of Newcastle; b Woodville, NSW, 18 Jul 1917.

22 Wg Cdr E B Goldner, 4365/O3567. Prev Army. 6 SFTS, 166 Conv U RAF, No 2 Sqn, ARDU, 2 & 6 Sqns; comd 1 Sqn 1956–57, Melb Uni Sqn 1958, SA Sqn ATC 1959–64. Regular air force offr of Adelaide; b Adelaide 28 Oct 1917.

The detachment enabled Squadron members to consolidate experience on the aircraft by flying navigation and photographic exercises, and weaponry training at the Flat Rock bombing range off the Western Australian coast. When the Squadron returned to Amberley, the prompt flow of aircraft from GAF allowed conversions to continue into 1955 for the formation of a second Canberra squadron. On one conversion sortie on 17 January 1955, a new pilot to Canberras, Flight Lieutenant Jim Stewart[23] (A84-204), with an experienced Canberra navigator on exchange from the RAF, Flying Officer Rod Martin,[24] were conducting instrument flying in the training area south-west of Amberley, near Stanthorpe. While doing steep turns on a limited instrument panel, the pilot became disorientated, which resulted in a steep dive from which he could not recover.[25] As it was apparent to the navigator that the aircraft was not about to recover, he ejected, but simultaneously the pilot was able to regain control. As he pulled out of the dive the pilot was "greyed-out" by the gravity forces, and on fully recovering consciousness he found himself flying low over the trees. He was able to land the aircraft safely back at Amberley, but the navigator had been killed. The high-speed, low-level ejection from the steeply diving Canberra was well outside the limits of the Martin Baker seat.[26] It was assessed that the aircraft was in a 40 degree dive when Martin ejected at 2,000 feet above the ground.[27] It is amazing that the aircraft could be recovered from such a steep attitude. A recommendation from the accident highlighted the need for dual Canberra trainers, and fully automatic ejection seats.[28]

Preparations were now underway for the Canberra's tropical trials, which were to be conducted at Darwin. These two weeks of testing, held from the end of March 1955, involved four Canberras in "Operation Comax". The CO (A84-214) led the detachment, which again was to include RAF members in the four crews involved,[29] flying aircraft A84-207, A84-208 and A84-211. The tropical trials, like those of the Lincoln, were indeed necessary. The initial testing by the RAF in England was under significantly different conditions to the equatorial climes where the RAAF was to operate its bomber force.

---

23 Flt Lt J G Stewart, 401544/O34438. Nos 23, 83 & 76 Sqns, 2 OTU, 77 & 2 Sqns, DepAir, 2 Sqn, 82 Wg, 1BFTS. Regular air force offr of Healesville, Vic; b Horsham, Vic, 2 Feb 1920.

24 Flg Off R U Martin, 4054424. RAF, No 2 Sqn. Killed in flying accident 17 Jan 1955.

25 RAAF *Crash Critique No 40*, for Canberra A84-204 on 17 Jan 1955, Directorate of Flying Safety, Dept of Air, Melbourne, para.3.

26 The Martin Baker Mark 1C ejection seat, fitted to the Canberra throughout its RAAF service, at that time was very restrictive in its use: the crew member had to manually release the harness to separate from the seat after ejection, the minimum altitude for ejection was 2,000 feet above ground level from an aircraft in *level* flight, and the maximum recommended airspeed was 400 knots.

27 Crash Critique No 40, paras 3,36.

28 Crash Critique No 40, para.56.

29 RAF crew members were Flt Lt H A Caillard (3054021), Flt Lt M D Fenner (607040) and Flg Off M J Taylor (607266). No 2 Sqn A50, of 31 Mar 1955.

During May 1955, the USAF exchange pilot on the Squadron, Captain Rex Stoner,[30] assumed temporary command—the only time an American has commanded No 2 Squadron. With a strength now of nine crews, which included pilots Flight Lieutenant George Turnnidge[31] and Flying Officer Russ Law,[32] the first full syllabus training course, No 1 Canberra Conversion Course, commenced. The students included the new OC 82 Wing, Group Captain Des Douglas,[33] and the incoming CO, Wing Commander Joe Fitzgerald.[34] No 2 Course, beginning in July, included among the students Pilot Officer Jack Boast,[35] who would later return to command the Squadron.

The effort of No 2 Squadron in converting new Canberra crews had now allowed a reorganisation of No 82 Wing into two jet bomber squadrons. On 11 July, Fitzgerald assumed command of the Squadron, and Stoner took command of No 6 Squadron. Eight of the Canberra trained crews were also posted to No 6, leaving five crews in No 2 to continue with the Canberra conversion role. All Lincolns passed to a Lincoln Conversion Flight, which was required to continue providing crews to No 1 Squadron in Malaya.

A highlight during 1956 was the departure in May of five 82 Wing Canberras on a deployment to the United States on "Operation Friendship". The 2 Squadron pilots on the detachment[36] were the "B" Flight Commander Flight Lieutenant "Bruce" Martin[37] (A84-225) and Flying Officer Jack Boast (A84-224), with Flying Officer John Jacobs[38] as a spare pilot. The 2 Squadron navigators were Flying Officer Bob Black[39] and Flight

30 Maj R K Stoner Jun, 16659A. USAF; t/comd No 2 Sqn 1955, 6 Sqn 1955. Regular air force offr of South Pines, North Carolina, USA.

31 Wg Cdr G C Turnnidge AFC, O33814. Prev RAF. Nos 77, 2 & 6 Sqns, 1AFTS, 1 Sqn; comd 1 OCU 1964–65; 2 & 34 Sqns; comd 22 Sqn 1971–72. Regular air force offr of Merewether, NSW; b Lambton, NSW, 4 Sep 1925.

32 AVM R N Law, 450963/O33833. Nos 76, 38, 36, 6 & 2 Sqns, A&AEE, ARDU, 3AD; comd 11 Sqn 1967–69; RAAFSC; comd ARDU 1974–75, Laverton 1976–77; HQSC; comd Edinburgh 1981–82; AOC HQOC 1982. Regular air force offr of Sydney; b Sydney 26 Jul 1928.

33 AVM D L G Douglas OBE DFC, 171/O359. Nos 5, 9 & 10 Sqns; comd 461 Sqn 1943–44; RAAF HQ; comd Fairbairn 1949–52, 82 Wg 1955–57, Townsville 1959, Laverton 1960–61; HQ FEAF; comd Richmond 1965–67, DepAir. Regular air force offr of Sydney; b Brisbane 5 May 1917.

34 Gp Capt J B Fitzgerald, 403649/O21986. Prev Army. No 500 Sqn RAF, 1415 Flt RAF, 13 Sqn, 81 Wg, 1 CU; comd 23 Sqn 1948, 38 Sqn 1949–50, 90 Wg; comd 6 Sqn 1954–55, 2 Sqn 1955–56; HQTC, HQ FEAF, RAAFSC. Regular air force offr of Moree, NSW; b Sydney 18 Nov 1916.

35 AirCdre J R Boast DSO AFC, O14165. Nos 2, 6 & 2 Sqns, 1AFTS, CFS, RAF CFS; comd 1 OCU 1968–69, 2 Sqn 1969–70; Asst AA Wash 1971–73; DefAir DPO 1973–76; comd Williamtown 1977–78; DefAir DGCD 1978–81. Regular air force officer of Brisbane; b Brisbane 8 Sep 1931.

36 Led by OC 82 Wing Gp Capt D L G Douglas (A84-221), with CO 6 Squadron Sqn Ldr H W Connolly (A84-226) and Plt Off W R Rayner (A84-227).

37 Gp Capt R H Martin CBE AFC, 443785/O22188. Nos 1, 6 & 2 Sqns, 50 Sqn RAF, 1BFTS; comd WA Sqn ATC 1961–63; 9 Sqn; comd 5 Sqn 1964–65, ASU 1967–70, Vung Tau 1970–71; DepAir; comd Fairbairn 1974. Regular air force offr of Canberra; b Queanbeyan, NSW, 2 Dec 1925.

38 AirCdre J A Jacobs, O34223. No 75 Sqn, 1ATU, 2 Sqn, 1AFTS, CFS, 3575th Trg Wg USAF, CFI CFS 1965–67, DepAir; comd BSqn Edn 1969–72; RAAFSC; comd Butterworth 1977–79; DefAir DGPS 1980. Regular air force offr of Brisbane; b Brisbane 11 Feb 1928.

39 Sqn Ldr R S Black, 450946/O33817. Prev RAN. Nos 2, 1, 6 & 2 Sqns, 1AFTS, 50 & 35 Sqns RAF, 1 OCU, SAN, HQOC, DepDef. Regular air force offr of Sydney; b Brisbane 18 Nov 1928.

Sergeant Tom Wright.[40] A giant USAF C-124 Globemaster transport had been provided to support the Canberras, which departed on 12 May and staged through Guam, Wake, Honolulu, San Francisco and Omaha, to Washington DC. While passing through Omaha, the detachment was entertained by the chief of USAF's Strategic Air Command, General Curtis Le May. In Washington, one Canberra provided a static display at Bolling Air Force Base for the United States Armed Forces Day.[41]

On the return leg through Hawaii, the detachment ran into the first serviceability problem of the journey. A84-227, flown by Pilot Officer Ward Rayner,[42] was taking off as last in the stream of five Canberras from Hickham Air Force Base. Rayner suffered severe vibrations, and on landing found the rudder stops were broken, which had caused the buffeting and had distorted the skin on the fin. Overnight, USAF technicians repaired the rudder stops, and reskinned the fin. On a test flight the following morning, 4 June, A84-227's starboard engine failed immediately after lift off. Rayner threw the aircraft back down onto the ground, retracting the undercarriage, and the aircraft skidded along the side of the runway.[43] The extensive damage necessitated its dismantling for shipping back to Australia, and it was received by No 1 Aircraft Depot at Laverton for repairs that November.

During July 1956, Home Command (as the main Headquarters of RAAF flying operations in Australia had now become[44]) instituted a Bombing Qualifying Scheme, requiring crews to meet certain accuracies at the various high levels. This emphasis on weapons training saw training conducted with live HE bombs at the Dutson range, near Sale in Victoria, and the 82 Wing involvement in an annual weapons competition. The competitive Bladin Trophy was first conducted with jet aircraft in November 1955, between Nos 2 and 6 Squadrons, and No 82 Wing Headquarters. No 2 Squadron won the competition that year, and again for the second time in October 1956.[45]

1956 had also seen Wing Commander Arthur Cross return to command the Squadron, Squadron Leader Hamilton Connolly[46] appointed as the "A" Flight Commander, and the new Officer Commanding Amberley, Group Captain Dixie Chapman,[47] an experienced pre-war pilot, commenced his Canberra conversion. The

40 Wg Cdr T D Wright, O27562. Nos 6, 1 & 2 Sqns, SAN, 1, 38 & 36 Sqns, HQOC, 2 Sqn; comd 21 Sqn 1970–73. Regular air force offr of Sydney; b Sydney 31 Dec 1932.

41 Bruce Martin, telephone interview with author, of 24 Dec 1993.

42 Sqn Ldr W R Rayner, 168105/O2921. 7CRD, BCOF, No 6 Sqn, 2ATU, 1AFTS, CFS, 5 & 9 Sqns, DepAir, 9 Sqn, SEATO Bangkok, DepAir. Regular air force offr of Yass, NSW; b Yass 20 Jul 1927.

43 Ward Rayner, telephone interview with author, of 23 Dec 1993.

44 The title of this Headquarters was again changed in 1960 to Operational Command.

45 No 2 Sqn also won the Bladin Trophy for the third time in Nov 1957, the last time it was to be competed for by the unit.

46 Wg Cdr H W Connolly DFC*, 402492/O23819. Nos 76 & 78 Sqns RAF, 466 Sqn, 86 Wg, 30 CU, 90 Wg; comd 6 Sqn 1955–56; 2 Sqn; comd Tas Sqn ATC 1957–59, t/comd 2 Sqn 1960–61, comd 36 Sqn 1961–64; DepAir. Regular air force offr of Sydney; b Sydney 31 Mar 1918.

47 AirCdre D R Chapman CBE, 80/O335. No 1 Sqn, DD Trg 1939–40; comd 23 Sqn 1940–41, 3 Sqn 1941–42, 451 Sqn 1942–43, 4 SFTS 1944–45, 84 Wg 1945; HQWA; comd 91 Wg 1952–54; AA Wash; comd Amberley 1956–59; HQOC; comd Darwin 1961–62; DepAir DG Org. Regular air force offr of Renmark, SA; b Tenterfield, NSW, 8 Sep 1912.

overseas deployment was to the Philippines to participate in the Manila air display during December, and comprised:

A84-225	WgCdr A E Cross/FltLt M M Alexander[48]
A84-224	SqnLdr B F N Rachinger[49]/PltOff N E Kilduff[50]
A84-226	FlgOff R N Law/FlgOff E R Cheney[51]

Another overseas task in early 1957 was transporting the CAS-elect, AVM Scherger, to South-East Asia for conferences with the Far East Air Force. Wing Commander Cross, with nav leader Miles Alexander, flew Scherger from Melbourne to Changi, via Darwin on 18 February, then on through Clark, in the Philippines, to Hong Kong. The return to Singapore on 24 February was through Labuan, and they arrived back in Melbourne on the 28th. The Canberra was a fast and efficient form of VIP transport. This was Cross' final journey with the Squadron; the following month he handed the reigns to Wing Commander Ron Hosking.[52]

The new CO led the Squadron detachment on the 1957 deployment to Malaya on 24 August for the Independence Celebrations on 1 September. Hosking led three Canberras from No 2 Squadron, and two from No 6. It was only three months later that it was announced that Malaya was to be the new home for the Squadron. No 2 Squadron was to deploy to the new base at Butterworth, and No 82 Wing was to be reorganised with 6 Squadron as the conversion unit.

This had coincided with the arrival of an upgraded Canberra for the RAAF. The RAF had switched production from the earlier Canberra B.2 to the more capable B.6. The Australians followed suit, modifying the Mark 20 with the more powerful Avon Mark 109 engines, which produced an additional 1,000 lb of thrust. In addition, the wings comprised integral fuel tanks, allowing carriage of an extra 3,000 lb of fuel. The first of the new aircraft for No 2 Squadron, A84-230, delivered on 19 August by the "B" Flight Commander Squadron Leader Basil Rachinger and Pilot Officer Terry Morton,[53] was also the first to arrive at Amberley with the "Green Satin" doppler and GPI mark 4 navigation system.[54]

48 Wg Cdr M M Alexander, O33289. Nos 1 & 2 Sqns, ARDU, HQOC, 2 Sqn, HQSC, SAN; comd 1 OSU 1969–70; DefAir, RAAFSC; comd BSqn Esl 1977–79. Regular air force offr of Adelaide; b Woodville, SA, 19 Aug 1929.

49 Gp Capt B F N Rachinger DFC AFC, 410091/O35055. No 268 Sqn RAF, 16 Flt, 4 Sqn, 1AFTS, 1FTS RNZAF, 2 & 6 Sqns; comd 1 OCU 1959, 6 Sqn 1964–65, BSqn Ubon 1965; DepAir; comd BSqn Lav 1968–70. Regular air force offr of Melbourne; b Daylesford, Vic, 26 Aug 1921.

50 Flt Lt N E Kilduff, O37602. Prev RAN. Nos 6, 1, 2 & 23 Sqns. Regular air force offr of Brisbane; b Southport, Qld, 26 Jan 1930.

51 Flg Off E R Cheney DFM, O21030. Nos 1, 6 & 2 Sqns. Regular air force offr of Parkes, NSW; b Parkes 19 May 1928.

52 Wg Cdr R A Hosking DFC AFC, 408721/O33102. Nos 94 & 87 Sqns; comd Surv Flt 1946, 87 Sqn 1948–50; 82 Wg, HQNWA; comd 2 Sqn 1957–58, BSqn WLM 1959–62; DepAir. Regular air force offr of Dromana, Vic; b Melbourne 4 Feb 1920.

53 Gp Capt T A Morton, O39437. No 2 Sqn, 2ATU, RAAF Acad, RAF CAW, DepAir, RAAFSC; comd OTS 1978; HQSC. Regular air force offr of Melbourne; b Melbourne 27 Jan 1935.

54 "Green Satin" not only drove the Ground Position Indicator (GPI) to provide the navigator with latitude/longitude position information, but also provided automatic groundspeed and drift inputs to the new T4 bombsight. The Squadron began bombing trials with the new equipment during October 1957.

The main reorganisation of the Wing commenced in January 1958, and Squadron Leader Doug Harvey returned to the Squadron to replace Rachinger as "A" Flight Commander. All the new "109" Canberras now being delivered were earmarked for No 2 Squadron's move to Malaya, with the older "Mark 1s" being transferred to 6 Squadron. Also in anticipation of the transfer north, Wing Commander Colin Steley[55] assumed command during April, and Flight Lieutenant Bill Collings[56] became "B" Flight Commander. (Collings, however, was posted again as a QFI on No 6 Squadron and did not move to Malaya with the Squadron.)

The first aircraft, flown by Wg Cdr Steley, touching down at Butterworth — a new home for No 2 Squadron AWM P448/212/120

In addition, a third flight was planned with transport aircraft to provide support to the sizeable RAAF presence being prepared for Malaya. The Dakota detachment that was to form "C" Flight arrived at Amberley on 2 June, under Flight Lieutenant Leigh Hindley.[57] "C" Flight was initially equipped with two Dakotas, A65-94 and A65-92, but eventually would be brought up to a strength of six aircraft during its term at Butterworth.[58]

The worldwide pressure of communist expansion, particularly from China, was still a concern in Asia, despite a peaceful settlement in Korea in July 1953. This uneasy peace, the victory of the Chinese-backed Viet Minh over the French garrison at Dien

55 Wg Cdr C W Steley DFC, 404554/O11332. Nos 10 & 461 Sqns, RAAF Cd; comd Surv Flt 1945–46; HQTC; comd Air Armt Sch 1957, 6 Sqn 1958, 2 Sqn 1958–60; HQOC. Regular air force offr of Ayr, Qld; b Maryborough, Qld, 7 Jun 1921.

56 AVM B H Collings AFC, O34495. No 77 Sqn, 1BFTS, 1AFTS, 6 & 2 Sqns, 1 OCU, 1 Sqn, ARDU, Asst AA Paris; comd 2FTS 1968–70; DepAir, HQ Amb, DefAir, HQOC; AOC HQSC 1981–83, DCAS 1985–87, CAFM 1987–89. Regular air force offr of Toowoomba, Qld; b Lismore, NSW, 21 Nov 1932.

57 Sqn Ldr L O Hindley, 421905/O23946. No 80 Sqn RAF, 86 Wg, 87 & 1 Sqns; comd 87 Sqn 1953; 36, 2, 38 & 9 Sqns; comd 5 Sqn 1965–66; 9 & 5 Sqns. Regular air force offr of Goulburn, NSW; b Goulburn 28 Feb 1923.

58 Ultimately as 2 Sqn was notified of its move to Vietnam in 1967, "C" Flt became Transport Support Flight (TSF).

Bien Phu in Vietnam during May 1954, and the continued terrorist activity in Malaya, had caused tension in the region. These threats of further communist efforts of dominance had led to the formation of the Manila Treaty on 8 September 1954, signed by the United States, Britain, France, Australia, New Zealand, Pakistan, Thailand and the Philippines.[59] This led to a collective security pact, the South-East Asia Treaty Organisation (SEATO), with the objective of deterring communist aggression in South-East Asia. The SEATO defence organisation, referred to as "an Asian NATO",[60] was headquartered in Bangkok, which had been a compromise between the British proposal of Singapore and the American preference for Manila.[61]

The security offered to Australia by SEATO's major powers led to Australia's policy of forward defence, which the Government felt should be a line drawn in Indochina.[62] Against a background that Britain may then wash its hands of South-East Asian commitments,[63] on 1 April 1955 the Government announced the provision of forces to the Commonwealth Strategic Reserve Force, based in Malaya. The significant RAAF contribution would comprise two fighter squadrons and a bomber squadron. This contingent would be based at an upgraded airbase on the west coast of Malaya near Penang, at the north of the Straits of Malacca.

The Reserve was to act primarily as a deterrent to Chinese aggression.[64] The significance of Australia's contribution was the acknowledgment that Australia's military involvement would in future be in South-East Asia. It would be to this region, and not the Middle East, that Australian forces would deploy and on which Australian foreign policy would focus.

Another quick VIP transport task by No 2 Squadron Canberras was for the Minister for Air during March 1958. The first No 2 Squadron Canberra to deploy to Butterworth was flown by Squadron Leader Tom Berry,[65] the Squadron's pilot-qualified Senior Technical Officer, with navigator Flight Lieutenant Eric Walker.[66] They arrived on 28 June,[67] just as Wing Commander Steley led another seven aircraft from Amberley:

A84-240	Wg Cdr C W Steley	Flt Lt H J Hurley
A84-243	Sqn Ldr D C Harvey	Flt Lt B W Overton
A84-236	Sqn Ldr W O Hewett	Flg Off D B Keenan
A84-242	Flt Lt A Turner	Sgt D G Little
A84-230	Flg Off R J Plummer	Plt Off Morton

59 Odgers, p.100.

60 Edwards, p.142.

61 Edwards, p.184.

62 Edwards, p.142.

63 Edwards, p.153.

64 Edwards, p.163.

65 Sqn Ldr T A Berry, 139856/O2662. 1APU/ARDU, No 2 Sqn, DepAir. Regular air force offr of Sydney; b Sydney 30 Dec 1922.

66 Wg Cdr E H Walker, O23606. Nos 11 & 43 Sqns, 82 Wg, 1 Sqn, AAS, ARDU, 2 Sqn, 3535th NTW USAF, 82 Wg, DepAir. Regular air force offr of Sydney; b Sydney 17 Apr 1924.

67 No 2 Squadron was officially farewelled by Prime Minister R G Menzies at Amberley on 12 Jun. The Squadron advance party arrived at Butterworth on 15 Jun 1958.

A84-246	Flg Off G A Collins	Plt Off A J Simmonds
A84-245	Flg Off R F Wyatt	Plt Off A T Hart

A84-233 Flt Lts J H Johnston/B A Johnson followed later, arriving at Butterworth on 13 July.

Staging through Darwin, the Squadron was met at Changi, Singapore, by Squadron Leader Berry (A84-244), and eight Canberras arrived at Butterworth on 1 July.

With the deployment of No 2 Squadron to Butterworth, No 82 Wing was restructured to include a dedicated training unit—No 1 Operational Conversion Unit. This policy decision in 1958 had impacted on the jet operators of both the fighter and bomber communities:

> The RAAF has decided to form the Operational Conversion Units because there has been a change in the system of pilot training this year.
>
> Since the war, pilots have gained their wings after a year's training course on Tiger Moths (which last year were replaced by Winjeels) and Wirraways. The Wirraway is being withdrawn and the Advanced Flying Training part of the pilot's course is now being done at Pearce, WA, on Vampire jet trainers.
>
> This means that all RAAF pilots, after the course now under training at Pearce, will be fully trained jet pilots. They will no longer need to go to an Operational Training Unit to be trained as jet pilots. They will only need to be converted to the type in which they will fly in the Squadron to which they will be posted.[68]

Accordingly in 1959, No 2 OTU at Williamtown became No 2 OCU, as part of 81 Wing, and No 1 OCU was formed at Amberley, as part of 82 Wing. No 1 OCU would supply aircrews to all Canberra units throughout the extended stay of No 2 Squadron in South-East Asia.

68 "RAAF Forms Two New Units", *Wings*, Dec 1958, p.27.

CHAPTER 23
BUTTERWORTH

You have a row of dominoes set up, you knock over the first one, and what will happen to the last one is the certainty that it will go over very quickly.

President Eisenhower, 1954[1]

The primary activity for No 2 Squadron at Butterworth was to provide a bombing capability for the British Commonwealth Strategic Reserve in South-East Asia, with a secondary role of carrying out attacks in support of the ground combat forces engaging the communist terrorists in the Malayan jungle.[2] The war against the CTs had been carried by No 1 Squadron's Lincolns from 1950 to 1958, until replacement by 2 Squadron. This campaign had nearly defeated the terrorists, but some were still at large.

The use of offensive air support had been a contentious issue, as the terrain was dominated by featureless jungle and the lack of strategic targets was a handicap. The enemy had no concentrations or strong points, so target information was rarely adequate to permit the proper use of air power.[3] The aims of the air support had been modified to drive the terrorists out of their areas into ambushes, moving the terrorists into country suitable for ground operations or containing them while ground forces swept an area. Probably the most effective use of offensive air power in the campaign had been the harassing of the terrorists to lower their morale.

The method of bombing by the Lincolns in Malaya since 1956 had been a technique known as Target Direction Post (TDP). This technique involved deployment of a self-supporting mobile ground radar unit, which provided a narrow beam passing over the target, down which attacking aircraft flew. A signal was then transmitted to bomb at a distance calculated on the basis of airspeed, bomb ballistics, height, stick length and slant ranges.[4] Beyond a range of about 25 kilometres from the radar site, a timed run could be flown from the limit of the beam, to attack more distant targets.[5] Aircraft could either fly individually or in formation, without being visual with the target. The technique, therefore, allowed for operation in cloud or at night, or when the target was nondescript jungle, which often was the case. It also preserved the element of complete surprise. During July the Canberras were briefed and practised TDP operation. In addition, the Canberras were now operating in the tropics, and the crews

1 Quoted from President Eisenhower's "falling domino" speech of 7 Apr 1954, in Edwards, p.127.

2 Odgers, p.102.

3 AP3410, *The Malayan Emergency 1948–1960*, Ministry of Defence, London, 1970, p.149.

4 AP3410, p.57.

5 AP3410, p.58. This provided an average radial error of 160 metres.

had to be aware of the dangers of thunderstorms. Penetration of the huge banks of tropical thunderstorms could lead to engine flame-out and severe turbulence.

On a TDP practice sortie for a formation of five Canberras on 29 August the leader, Squadron Leader Doug Harvey, with navigator Flight Lieutenant Bruce Overton[6] (A84-243), was involved in a landing accident, which led to the write-off of their aircraft. After touching down and applying brakes, Harvey found the aircraft swinging to the right, indicating that the port brake had failed.[7] He raised the undercarriage to stop the aircraft, which turned through 180 degrees before coming to rest. The crew evacuated the aircraft safely, but the underside of the fuselage and the starboard mainplane were damaged, and the Canberra was assessed as uneconomic for repair.

Bombing up for a Firedog strike RAAF

The same crew led the first *Firedog* strike, No 2 Squadron's first operational mission since 1945, on 3 September 1958. Both Harvey and Overton had previously completed Lincoln tours in Malaya with No 1 Squadron. The Canberras took off at dawn to strike first in a coordinated attack near Ipoh, south-east of Butterworth, with RAF Canberras and Venoms.[8] The five RAAF Canberras,[9] each carrying a formidable load of six 1,000 lb bombs, were crewed as follows:

6 Wg Cdr B W Overton, 187141/O11260. Nos 1, 10 & 2 Sqns; HQOC, 82 Wg, RAAF Wash, 82 Wg. Regular air force offr of Brisbane; b Brisbane 10 Jul 1930.

7 AA CRS A705/1 32/34/248, A.180 171/58, of 1 Sep 1958. The brake failure was due to a faulty dual relay valve, RAF Form 1669 224G/64/62/58/FS, of 12 Sep 1958.

8 The RAF Canberras were from No 45 Sqn, the Venoms from No 60 Sqn. AP3410, p.166.

9 No 2 Sqn A51 Appendix D (1), of 31 Oct 1958.

A84-233	Sqn Ldr Harvey	Flt Lt Overton
A84-230	Flg Off Collins	Plt Off Simmonds
A84-242	Flg Off Wyatt	Plt Off Hart
A84-246	Flg Off Plummer	Flg Off Morton
A84-240	Flt Lt Turner	Plt Off Little

The notoriously changeable Malayan weather closed in as the Canberras approached the target, a base camp near a clearing on the side of a jungle-clad mountain. As the clouds were quickly building up in the target area, Harvey was unable to bomb visually, so he led the flight at 12,000 feet, under control of No 1 (RAF) TDP.[10] The Canberra's navigation capability allowed for accurately flown and timed TDP runs, which was not possible for the single-seat Venom, and Harvey later commented:

> There were no snags in this operation. There was for the Venoms. They could not go in at low level as planned because of the cloud.[11]

This observation showed the value of the Canberra. In this, the first Australian jet-bomber mission, the well-known capabilities of the Canberra's loitering allowed for the flexibility of carrying out a secondary plan. Also, if the low cloud had been marginally higher, a low-level visual attack could have been carried out due to the Canberra's level bombing capability. This was an advantage over fighters, like the Venom, which relied on dive bombing and therefore demanded a higher cloud base. These were the aircraft's great attributes which would prove the type in No 2 Squadron's next war.

To maintain the visual bombing capability, 2 Squadron practised at the Song Song weapons range to the north of Butterworth. A standard detail was individual practice attacks, carrying up to eight 25 lb smoke-and-flash bombs. However, these skills were again not required on the next *Firedog* on 30 September, when the CO, with navigator Flight Lieutenant Hank Hurley,[12] led five aircraft against another camp area near Grik, 100 kilometres east of Butterworth.[13] Once more, 30 1,000 lb bombs were dropped on the jungle under TDP control. Army security forces which then searched the target area reported that the bombing was "deadly accurate" and the "camp obliterated but nobody found".[14]

Two days later, 2 October, the third *Firedog* was flown, again led by the CO against a camp area north-east of Ipoh.[15] With an overcast cloud base at 600 feet and heavy

10 No 1 (RAF) TDP was equipped with the AA No 3 Mk 7 radar. Another unit was No 2 (Army) TDP, which was deployed to cover southern Malaya. AP 3410, p.58.

11 G Odgers, "RAAF Canberra Jets' First Bombing Mission", *Wings*, Dec 1958, p.27.

12 AirCdre H J Hurley, O33300. Prev RAN. No 86 Wg, 77 Sqn, 30 CU, SAN, DepAir, 2 Sqn, DepAir, SAN, RAAF Wash, DefAir; comd East Sale 1976–77; DefAir. Regular air force offr of Perth; b Middlesex, Eng, 10 Nov 1927.

13 Crews were: Steley/Hurley (A84-242), Harvey/Overton (A84-239), Johnston/Johnson (A84-233), Turner/Little (A84-240) and Collins/Simmonds (A84-246).

14 No 2 Sqn A51 Appendix "D" (2), of 31 Oct 1958.

15 Steley/Hurley (A84-240), Harvey/Overton (A84-239), Johnston/Johnson (A84-236), Wyatt/Hart (A84-242) and Plummer/Morton (A84-233).

rain, the formation of five Canberras bombed under TDP control. Again, ground troops reported the pattern of bombs right on target, with the devastation so great it was impossible to say if the camp was occupied. The fourth *Firedog* required No 2 Squadron to provide 10 single sorties throughout the morning of 8 December. The CO again led, taking off before dawn, against a camp area at Sungei Tinggi, 50 kilometres north-west of Kuala Lumpur.[16] Sixty 500 lb bombs were dropped on TDP passes at 10,000 feet, with inconclusive results.[17]

However, in this conflict the use of offensive air support was nearing its end. An insignificant number of kills had been achieved by air strikes, but sanctuary had been denied by the constant harassing operations, and the main core of the terrorists had retreated to the relative immunity of Thailand. There was, therefore, little point in continuing the air effort, and no strikes took place during the first seven months of 1959.

Apart from operations, other tasks now occupied the Squadron. In November, the Canberras assisted in the ferry of 19 Sabres of No 3 Squadron, by shepherding the fighters on the overwater leg from Labuan, in Borneo (Brunei), to Butterworth. This elaborate ferrying operation from Williamtown, the then biggest peacetime movement of aircraft ever undertaken by the RAAF, involved complex flight planning and detailed safety precautions. Furthermore, it meant pioneering a new air route from Australia to Malaya—via West New Guinea, the Philippines and Borneo.[18] These were the first Australian fighters for the Commonwealth Reserve, and were joined the following year by No 77 Squadron, to constitute No 78 (Fighter) Wing. Other tasks for the Canberras involved participation in air displays at Vientiane, Laos, in November and Manila, in the Philippines, during December. The Manila Air Display was to become almost an annual Squadron commitment at the end of each year, as a demonstration of SEATO unity.

Squadron Leader Doug Harvey and Flight Lieutenant Bruce Overton were then selected in early 1959 for the enviable task of flying a Canberra to England for three weeks for liaison with the RAF. They departed from Butterworth on 19 February, and flew training sorties with No 139 Squadron at Binbrook, in Lincolnshire, to learn the techniques of "Target Marking" attacks. This involved marking the target with incendiaries for the high-level bomber stream, which would release on corrections supplied by the marker aircraft. The RAAF crew returned to Butterworth on 13 March to introduce this new method of attack to the RAAF, and it was this tactic which became standard for the Far Eastern Air Force (FEAF) bomber force that the Squadron practised for the next three years.

16 Steley/Hurley (A84-240), Turner/Little (A84-242), Wyatt/Hart (A84-238), Collins/Simmonds (A84-241), Plummer/Morton (A84-245), Harvey/Overton (A84-237), Johnston/Johnson (A84-239), Hewett/Keenan (A84-238).

17 No 2 Sqn A51 Appendix "C", of 31 Jan 1959.

18 "Operation Sabre Ferry", *Flight*, London, 19 Dec 1958, p.959.

Malaya RAAF

One pilot who flew and developed this form of attack was Flight Lieutenant Arthur Barnes,[19] who recalled the details of "Marker Stream" bombing:

19 Gp Capt A Barnes DFC AFC, O36433. No 10 Sqn, SAN, 1 & 2 Sqns, CFI 1BFTS 1966–68; t/comd 1 OCU 1969–70; 2 Sqn; t/comd 2 Sqn 1971, comd CFS 1973–75; RAAFSC; comd BSqn Amb 1979–80. Regular air force offr of Melbourne; b Boort, Vic, 11 Mar 1931.

The reason why we flew this bomber stream technique was basically the three Canberra squadrons of FEAF, 2 Squadron, plus a New Zealand Air Force squadron and a Royal Air Force squadron which were both based at Singapore, were all quite defenceless against enemy air opposition, and the only hope we had of surviving a bombing raid was to bomb at night. The problems of night bombing were many and major with our lack of equipment. This particularly was so for the Kiwis and the Poms. Their only navigational aid was a radio compass, for practical purposes we can forget astro. We did not believe that the Indonesians would be kind enough to switch on their beacons for us,[20] so that left the two Squadrons from Tengah virtually without any nav aids other than Mark 1 eyeballs. So, a technique was developed, which we called a bomber stream. This technique was devised to make the best use of the Green Satin equipment which was carried in the 2 Squadron aircraft, and we worked out a way to provide vital nav information for all FEAF Canberras.

Briefly, the plan was for 2 Squadron to lead the stream with two "shepherd" aircraft, one at 42,000 feet, "Shepherd 1", and "Shepherd 2", 2000 feet below and two minutes behind at 40,000 feet. The rest of the aircraft in the bomber stream would be staggered two minutes behind, and alternating between 42,000 and 40,000 feet. This meant that we had aircraft staggered out, probably for the space of over half an hour, in the bomber stream. After the aircraft got airborne, they joined into the stream at a given position near their base, and all aircraft were then to fly very accurate airspeeds, headings and heights, until the shepherd, at a predetermined time, worked out a correction of headings to fly. He gave the alteration in code, which was to take place at a certain time, plus two minutes, depending on an aircraft's position in the stream. Basically, the shepherd navigated for all high level aircraft of the three squadrons. Shepherd 2 was the backup, also he was to be alert for significant met changes between 42 and 40,000 feet.

Hopefully, this system would get the bomber stream within visual distance of the commotion going on down below in the target area, they would then take over and bomb visually according to the directions given to them by another aircraft [the master bomber] at low level in the target area. Now the next question is how were these high level aircraft going to pick up, see and identify the target at night? The answer to this gives us the second part of the bomber stream technique. 2 Squadron also provided what was called a "marker force" of two aircraft. These aircraft operated and navigated independently, getting themselves to the target area before the bomber stream was due to arrive. On arriving in the target area, the "Marker 1" then proceeded to an initial point which could be readily seen and identified from low level at night, this normally was something to do with water which stood out quite well—a coastline, a river mouth, a lake. Positioning themselves above this identified point, the marker crew would use their "along and across track" function of the Green Satin to get over the target … at between 3000 and 4000 feet above ground level.

Shortly before expecting to be over target, Marker 1 would release eight parachute flares, these were spaced at about 200 yard intervals. After the last flare cleared the aircraft, the marker would go into an orbit around the flares and, hopefully, be able to pick up the target from the light of the flares he had just dropped. Believe it or not, this did work reasonably successfully, especially if the target was a simple one like an airfield or a bridge. After having identified the target, Marker 1 would then drop the first of his target indicators (TIs).[21] Target indicators were a bomb, about the equivalent size of an 800-pound bomb, which after leaving the aircraft would detonate at 800 feet above ground level, and discharge

20 Indonesia was perceived as a potential adversary by 1960, as its expansionist intentions regarding Dutch New Guinea were against Australia's wishes.

21 This was from a dive bomb attack from 5,000 feet, releasing at 2,000 feet, and recovering below 1,000 feet—at night!

> several hundred small incendiary bombs. On striking the ground, they would form a circle of about 100 yards in diameter, and the bomblets would start igniting.
>
> The bomblets all had their own time delay built into them ... four different lots of time delays, the first igniting immediately and burning for about two or three minutes, by which time they were starting to die out, then the next batch would light up, burn for two to three minutes[22] ... Before the last lot of incendiaries had gone out, the marker force had to roll in and drop another target indicator to keep the target illuminated. Corrections were then given to the bomber stream on their run-in by the marker ... they would call the aircraft running in to bomb 100 yards short of the target indicator, or 50 yards short, left, or right. Hopefully, the bomber stream would be arriving overhead at 42,000 or 40,000 feet and drop their bombs in a nice, orderly fashion. This, of course, never happened!
>
> The second marker was a backup, with both flares and TIs, and would take over progressively marking the target when Marker 1 ran out of TIs, or if something went wrong with the flare run he would have to go in to use his flares. The TIs were placed on target by a dive-bombing technique ... dive bombing in a Canberra on a pitch black tropical night, with basic air driven instruments, initiating the dive at about 4000 feet, releasing the TI at about 2000 feet, and hopefully recovering at around 800–1,000 feet, with bombs from the high level bomber stream falling around you, was a sure way of getting the adrenalin flowing.[23]

Meanwhile, during March 1959, No 2 Squadron had participated in its first SEATO air defence exercise, "Operation Air Express", in Bangkok. Other tasks involved flying Canberras to Hong Kong for radar calibration, or to escort Vampire fighters on the long transit to the colony. From September 1959, No 2 Squadron was able to fly monthly "Lone Ranger" flights to Hong Kong, which were designed to maintain aircrew proficiency in air route procedures for long-distance deployments. There was soon, however, one prerequisite introduced for a "Lone Ranger" to Hong Kong: only those who had been before could go! Because of the tricky approach into Kai Tak airport over the high-rise buildings of neighbouring Kowloon, pilots had to fly their first trip in the spare "jump" seat beside the pilot. (Later, when the two-seat Canberra trainer was received, this was utilised for the Hong Kong Ranger.) Having heard and seen the approach procedures, and dangers, a pilot was then qualified to fly to Hong Kong.

Away from these exotic diversions, the war with the CT in Malaya was still continuing, albeit in its closing phase. During August 1959, several terrorist camps east of Bentong in Northern Pahang were discovered, and two targets were bombed on 13 August by Canberras of Nos 2 (RAAF), 45 (RAF) and 75 (RNZAF) Squadrons. On this, the Squadron's fifth *Firedog*, five Canberra sorties were flown. In addition, Sabres from Butterworth's No 3 and 77 Squadrons strafed the target marked by a British Auster air controller.[24]

Four days later, on 17 August 1959, the Canberras attacked a target on the northern slopes of Bukit Tapah, in Perak. No 2 Squadron's sixth *Firedog* operation, flown by four Canberras, was the last offensive air support provided in the Malayan

---

22 The operational target indicator, for war use, would last for about 18–20 minutes.

23 Arthur Barnes, taped recollections, of Nov 1993.

24 AP3410, p.71.

Emergency, even though the campaign continued for a further eleven months. From the point of view of tangible results:

> The offensive air support provided during the Malayan campaign was hardly worthwhile but the incalculable effects which it had on weakening the terrorists' morale and reducing their ability to mount offensives or withstand security force pressure was considered to be an important factor in preventing the insurgents from progressing beyond the first stage of their campaign to the domination and control of selected areas. In any case, when the terrorists retired to deep jungle areas, air power was frequently the only method of maintaining some pressure against them and was therefore directly instrumental in shortening the duration of the campaign.[25]

The Dakotas of "C" Flight, meanwhile, had begun what was to become a long and vital presence in providing transport support for Australian interests in South-East Asia. In addition to supporting Canberra and Sabre deployments, the Dakotas began transporting service units, and visiting politicians and service chiefs. This VIP transport soon extended to ambassadorial visits around Malaya, Thailand, Laos and South Vietnam. Over February and March 1960 emergency relief missions were flown by the Dakotas to Vientiane, carrying blankets and medical supplies. It was this type of operation which enabled Australia to provide tangible assistance to the region, a commitment held by the Dakotas until their withdrawal from Butterworth in 1980.

The loss of another Canberra at Butterworth occurred on 8 March 1960. Two pilots, Flight Lieutenants Gavin Collins[26] and Bob Wyatt,[27] a recent graduate of the Advanced Navigation Course and acting as navigator, had to abort the take-off of a heavily laden A84-239. Collins later explained the incident to a colleague:

> The aircraft ran off the end of the runway, and the instrument panel collapsed onto his legs. The crash also distorted the throttle linkage to one of the engines holding it in full throttle position, causing the engine to act like a blowtorch.
>
> The fire crew were first on the scene and initially intended to amputate both his legs with an axe to free him from the aircraft, due to the extreme urgency of the situation ... Collins said that he found he had powers of persuasion previously unheard of to convince them they had time to lever the instrument panel off his legs.[28]

A84-239 was a write-off, but both crew members escaped with only minor injuries. The accident necessitated a review of the maximum permissible take-off weight when operating the Canberra in the tropics. This had been a lot more traumatic for Collins than an incident he had earlier experienced with A84-240 the previous September. On that occasion he had been taxying into the operational readiness pan, just prior to take-off, when the port wheel had collapsed, causing undercarriage, mainplane and engine damage.[29] The problem had been stress corrosion in the port undercarriage side stay bracket, which necessitated inspections of all FEAF Canberras.[30]

---

25 AP3410, p.150.

26 Wg Cdr G M Collins, 439209/O34193. Nos 77, 38 & 34 Sqns, 2AD, 2 & 21 Sqns, HQSC, DepAir. Regular air force offr of Melbourne; b Melbourne 14 Jan 1926.

27 Sqn Ldr R F Wyatt, O39336. Nos 6 & 1 Sqns, AAS, 6 & 2 Sqns, SAN, 1BFTS; comd TSF 1968–69; SAN, HQSC. Regular air force offr of Melbourne; b Yallourn, Vic, 30 Dec 1929.

28 WOff G Foran, "Letter to the Editor", *RAAF News*, Vol 35 No 5, Jun 1993, p.2.

29 AA CRS A705/1 32/34/301, A.180 200/59 2SQN 32/34/301, of 1 Sep 1959.

30 RAF Form 1669 224G/C64/63/59/FS, of 22 Sep 1959.

A beautiful sight, the Butterworth Officers' Mess on the shores of the Penang Strait
VIA A WILKINSON

Squadron Leader "Skin" Hewett[31] handed over "B" Flight to Squadron Leader David Evans[32] during June 1960. The new flight commander was to be instrumental in developing tactics for low-level operations as the policy for the employment of the Canberra was changed. Squadron Leader Evans, who would become CO of the Squadron in 1967 and ultimately Chief of the Air Staff, recalled:

> When I arrived at No 2 Squadron in June 1960, the Bomber Stream/Marker system had been in use for about two years. I took on the task of leading the force of four aircraft in the marker team. With the experience of some years the technique was quite good—certainly the marker team was successful on most occasions. The three FEAF squadrons deployed to Darwin in November 1960 for exercises using the Quail Island range. In general, comments by participating crews judged this to be the most successful exercise when using the Bomber Stream technique and using live weapons. I recall on one occasion when Master Bomber, and being out of flares and TIs, I noted my last marker going out. I called Marker 2—Arthur Barnes—to get in there and remark, quickly. By the time "AB" was over the target the

31 Wg Cdr W O K Hewett, 407428/O4405. No 24 Sqn, 11 EFTS, 77 & 23 Sqns, HQEA, 77 Sqn, DepAir, 2 Sqn, HQOC; comd BSqn Darwin 1964–66; DepAir. Regular air force offr of Tintinara, SA; b Bordertown, SA, 6 Sep 1921.

32 AM S D Evans AC DSO AFC, 433900/O22127. No 38 Sqn, RAAF Airlift Sqn, 1BFTS, FTS & CFS RNZAF, Queen's Flt 38 Sqn, 86 Wg, 34 Flt, DepAir, 2 Sqn, Asst AA Wash; comd 2 Sqn 1967–68; DefAir; comd Amberley 1975–77: DefAir CAFOPS 1977–79; DepDef CJOP 1980–81; CAS 1982–85. Regular air force offr of Sydney; b Sydney 3 Jun 1925.

marker had flared out completely, and he did his dive from 5000 feet into the black of Quail Island largely on guess work. However, it was not all that far from the target and I was able to give the remaining stream aircraft appropriate off-set advice.

In all, the technique was about as good as could be used if the Canberras were to be called upon to operate in a non-permissive air environment and, in a SEATO role, this would certainly have been the case. Even so, the survivability of the marker diving from 5000 feet, initiating the dive at low speed, would have been problematical. But it was the only way to go at that time and in such circumstances.[33]

In recognition of the efforts made in becoming such a vital unit to FEAF, No 2 Squadron was presented with the Gloucester Cup for the RAAF's most proficient unit for 1960–61.[34] However, the concept of Marker Bombing in a large force was about to change. From mid 1961, the emphasis was removed from high-level operation, to a more tactical focus. Low-level navigation sorties and attacks became more frequent, with the Squadron flying army cooperation sorties, and high-level bombing disappeared by the end of the year.[35] It had been acknowledged that the complexity and inaccuracy of the bomber stream made it not a viable wartime tactic in the future. As Arthur Barnes recalled:[36] "We were lucky if we could even hit the ground from 40,000 feet!"

Medium-level bombing, from around 8,000 to 10,000 feet, and low-level attacks now became the tactic in 1962, in addition to fighter-escorted training strikes with the Sabres of No 3 Squadron. On several occasions the entire Squadron would fly in formation, escorted by the complete complement of 3 Squadron's 16 Sabres. These tactics, which gave the fighters training, may also have helped the Canberras survive, as some targets, now being selected in Indonesia, were within the range of Butterworth-based Sabres. Squadron Leader David Evans was instrumental in developing medium-level bombing and improving bombing results:

I had been surprised during my operational conversion at the poor accuracy of Canberra bombing and the ready acceptance of this by those in authority at all levels. We bombed from 30,000 feet plus, and a bomb of 100 yards was considered a good bomb—the best that "the system could do".

This was in my mind when I arrived at No 2 Squadron. One of the first things I did was to study the weapons manuals and note the number of sorties required to destroy or neutralise a target at various CEPs.[37] For instance, to put a concrete and steel bridge out of action with a CEP of 100 yards would take about 300 sorties. Looking at some of our potential SEATO targets—in defended areas—the thought of achieving 300 sorties was clearly fanciful. Clearly something had to be done to improve the accuracy attained by the Canberra.

I addressed this in a crewroom briefing, and declared that we must achieve a squadron bombing average of 50 yards. The experienced Canberra crews were as derisive as discipline would allow. We then began intensive training in a sequence of trials based on certain

33 Air Marshal Evans, interview with author, of 21 Dec 1993.

34 The presentation was made by the C-in-C FEAF at Butterworth on 21 May 1962.

35 Arthur Barnes' flying logbook, 1961.

36 Arthur Barnes tape.

37 CEP—Circular Error Probable, a measure of weapons accuracy.

assumptions. One, we could not achieve 50 yards CEP from high level. Two, we could not fly low level at night. Three, in daylight we could only survive in a defended area at low level.

Our trials showed us that target acquisition and bomb run, using the sight, was best achieved at about 8,000–10,000 feet. However, this was clearly the worst of all levels to fly in a heavily defended area—and most of our SEATO targets were in such areas. Finally, we found that a very low level approach to the target, with a pop-up to 8,000–10,000 feet for a 30 second bomb run, was the best compromise available to us. With practice and practice and practice the error steadily decreased and the Squadron did achieve an average bombing error of 50 yards. In the process we received complaints from HQ 224 Group for having sunk the Song Song high level target—twice.

The low level pop-up technique became the standard for the three Canberra squadrons of FEAF.[38]

Another development was in March 1962, when training sorties commenced with a Forward Air Controller (FAC) at the Sungei Patani and Asahan training areas. Artillery would mark the target with smoke, indicating to the bomber crews the point of desired impact, with a ground-based FAC calling a correction on the smoke. Flight Lieutenant Arthur Barnes remembered this as an exercise requiring split-second timing:

We would call as we ran in at low level, at a point where we hoped would give time for the artillery to fire the round, the round to strike the ground, somebody to give us a correction on the smoke, then we would pop-up and bomb the smoke ... altering heading and diving back down to ground level.[39]

A crew that had joined the Squadron from Amberley during these developments was pilot Flight Lieutenant Al Reed[40] and navigator Flight Lieutenant Frank Lonie.[41] This crew had reached the exceptional level of "A" Category flying proficiency, something very rare, which was acknowledged by their award of the Queen's Commendation for Valuable Service in the Air. Frank Lonie remembered:

I was the first 2 Squadron navigator to qualify as a ground FAC. The attacking crews would be tasked and briefed before take-off. In their pull-up from low level, I as a ground FAC, would give a short target brief, talk the aircraft on, and issue final clearance. The aircraft generally operated in threes, and in the training area would drop 25-pound practice bombs. On the first of these missions, the target was defended by 40mm AA gun batteries. The gunners afterwards said that from the low level approach, it was virtually impossible to train their guns on the attacking Canberras.[42]

During 1962, communist insurgency in Thailand's north-eastern provinces, near Laos, had presented a vulnerable "domino". In response to a request from Thailand

38 Air Marshal Evans, interview with author, of 21 Dec 1993.

39 Arthur Barnes tape.

40 AVM A R Reed AO, O52363. Nos 10, 1 & 2 Sqns, RAAF Acad, 1 Sqn, 12th TRS USAF, DefAir; comd 6 Sqn 1971–72; DefAir, HQSC, HQ Amb; comd Point Cook 1982–84; AA Wash 1984–87; AOC HQSC/HQLC 1987–90. Regular air force offr of Perth; b Albany, WA, 16 Dec 1933.

41 Gp Capt F R Lonie, O35101. No 36 Sqn, ARDU, 11, 6, 1 & 2 Sqns, SAN, RAF CAW, 2 Sqn, HQOC, RAAFSC, DefAir, DepDef. Regular air force offr of Ballarat, Vic; b Sydney 29 Sep 1931.

42 Frank Lonie, interview with author, of 29 Nov 1993.

under the SEATO commitment, the RAAF deployed a squadron of Sabre fighters to Ubon. The Dakotas of "C" Flight enabled inspections of this new base to be made, and the requirement to support the Sabre commitment at Ubon led to another "Goony Bird" being added to the flight's strength. The Canberras, too, would be involved in supporting No 79 Squadron, as the unit was to become, by assisting with the changeover of Sabres between Butterworth and Ubon.

Australia was now also becoming militarily involved in Indochina, by a commitment to South Vietnam. The Government announced on 24 May 1962 the dispatch to South Vietnam of 30 officers and NCOs, comprising the Australian Army Training Team Vietnam (AATTV). The main body arrived that August, and became the first Australians to become combatants in the war.[43]

The Canberras continued their low-level training, involving practice bombing at Song Song, and live HE bombing at the China Rock range, north-east of Singapore, and FEAF exercises in Malaya, Singapore and over the South China Sea. During October 1962 the Canberras participated in an annual exercise known as "Bomber Joss Stick", on Okinawa. RAAF involvement in the "Joss Stick" series of USAF air defence exercises had previously been restricted to the Sabres of No 78 Wing. Supported by an RAF Beverley transport and staging through Clark Air Force Base in the Philippines, the CO, Wing Commander Sam Dallywater,[44] led five crews to Kadena:

A84-237	Wg Cdr S W Dallywater	Sqn Ldr K Tongue
A84-238	Flt Lt A R Reed	Flt Lt F R Lonie
A84-241	Sqn Ldr B E Goy	Flt Lt F C Cutler
A84-203	Flt Lt R J Crimmins	Flt Lt E B Bolger
A84-228	Flt Lt A Barnes	Flg Off W Henderson[45]

The deployment gave the RAAF crews the opportunity to fly in the back seat of the USAF F-100 tactical nuclear strike aircraft. Arthur Barnes recalled being given the controls to fly a toss bombing technique attack:

> I did the attack all right, and threw the bomb about 800 yards off the target. I apologised to Captain Woske, and his comments were: "Hell, that's close enough for the bombs we've got!"[46]

During this exercise, the RAAF crews were locked away for a day as the base at Kadena, like United States bases worldwide, was put on full alert status for the Cuban missile crisis. However, working with the USAF had also demonstrated the emphasis for low-level FAC operation in a South-East Asian war. On return to Butterworth, the Canberra crews began immediate bombing sorties at the Asahan Range, under control of a ground FAC. This technique was also demonstrated later in November to the

43 Edwards, p.243.

44 AirCdre S W Dallywater OBE, 415405/O5824. No 8 Sqn, 1 OTU, 9 CU, 36, 30 & 10 Sqns; comd 38 Sqn 1955–56, 2ATU 1958–60, 2 Sqn 1961–63; HQSC, DepAir; comd 82 Wg 1967; RAAF Wash 1967–69; comd Pearce 1971–72, Edinburgh 1973–74; HQSC. Regular air force offr of Perth; b Rajahmundri, India, 1 Jul 1922.

45 The deployment was from 17 Oct until 26 Oct 1962. No 2 Sqn A50.

46 Arthur Barnes tape.

visiting staff from the School of Land Air Warfare. It was obvious to the proponents of air power that a regional war would require such tactics, and when providing close support to ground forces it was essential to know the disposition of all combatants: the enemy and the "friendlies".

Over 1962, the new federation of Malaysia was announced—the merger of Malaya, Singapore, and the British territories of Borneo. This unification was perceived by Indonesia as imperialist and neo-colonialist, and Jakarta had commenced a policy of "Confrontation" against the new state.[47] Small incursions by Indonesian forces into Malaysian territory, mostly in northern Borneo, occurred. This harassment was to build up to an intense period of military activity over the second half of 1964.

During December 1962, as the Valetta transports of No 52 Squadron RAF were to be withdrawn from Kuala Lumpur, the Dakotas of "C" Flight began to survey drop zones in northern Malaya in anticipation of assuming the supply dropping role. By May 1963 the Dakotas were heavily involved in leaflet dropping to the remaining terrorists, and supply dropping to the security forces operating on the Thai-Malayan border. The marked increase in supply dropping activity in the border region to the troops of the Royal Australian Regiment led to a fourth Dakota coming on strength during 1964.

Canberra involvement in SEATO exercises continued in June 1963 with a deployment to Ubon for "Dhanarajata". Although long-range interdiction missions were flown by the five Canberras, there was emphasis on close support with troops of the 28th Commonwealth Brigade, these missions being flown under the control of an airborne FAC. However, the deployment to Ubon had encountered the same problems from the Malay Government that the Sabres had experienced the previous year. As Malaya was not a member of SEATO, only a party to its protocol, the Government did not want to be seen participating as a member of the alliance. Accordingly, when the Canberras deployed to and from Ubon, they had to fly through Singapore.

"Joss Stick" involvement continued in August 1963 with a detachment of five Canberras to Kadena, and in November two aircraft flew to Agra for the air defence exercise "Shiksha" with the Indian Air Force. On this detachment the CO, Wing Commander Frank Griggs, who had temporarily commanded No 2 Squadron a decade previously, took two of his younger pilots, Flying Officers Ian Westmore[48] and Pete Jones.[49] Another SEATO exercise at Ubon in April 1964 was "Air Boon Choo", followed the next month with a "Joss Stick" at Clark, in the Philippines.

Low-level operations to the Butterworth Canberras had now been further developed, and literally meant tree-top height. The belief was that going against high-threat targets in daylight necessitated an ingress and escape right on the deck, with

47 Edwards, p.258.

48 AirCdre I M Westmore AM, O314592. No 2 Sqn, 1BFTS, 442nd TFTS USAF, RAAF Acad, 6 Sqn, HQ Amb, DefAir; comd BSqn Ric 1980; DefAir; comd Amberley 1985–86; DefAir DGDFD 1987–88. Regular air force offr of Melbourne; b Melbourne 1 Jan 1941.

49 Wg Cdr P D Jones, O18752. Nos 2 & 1 Sqns, 2FTS, CFS, HQSC, 2 Sqn; comd 2 Sqn 1976–77; DefAir. Regular air force offr of Brisbane; b Merthyr Tydfil, Wales, 22 Feb 1939.

the attack as low as possible. For the attack, the Canberras would pull up, open the bomb doors passing 300 feet, and bunt to level at 600 feet in a level attack at 350 knots.[50] With only a few seconds for acquisition and tracking the target, it was then a case of avoiding damage from bomb fragments and making an ultra low-level escape. Steep dive bombing attacks were also trialed, with a Lincoln turret gunsight fitted in front of the pilot. The pilot would roll into a 30 degree dive from 9,000 feet, at 180 knots, with speed brakes out and throttles at idle. Tracking the target by the vertical reticle on the sight, the pilot would attempt not to overspeed the 360 knot limit on the bomb doors, and release bombs to then bottom out of the dive at about 1,500 feet.[51]

With virtually all Canberra operations now at high speed and low level, culminating in single-run attacks on simulated targets throughout the Malay peninsula, the probability of birdstrikes became a fact of life. One of the more serious of these emergencies occurred on 6 August 1964, when Pilot Officers Ron Biddell[52] and "Paddy" O'Farrell[53] (A84-237) suffered a severe strike by a large bird through the perspex nose. Ron Biddell related:

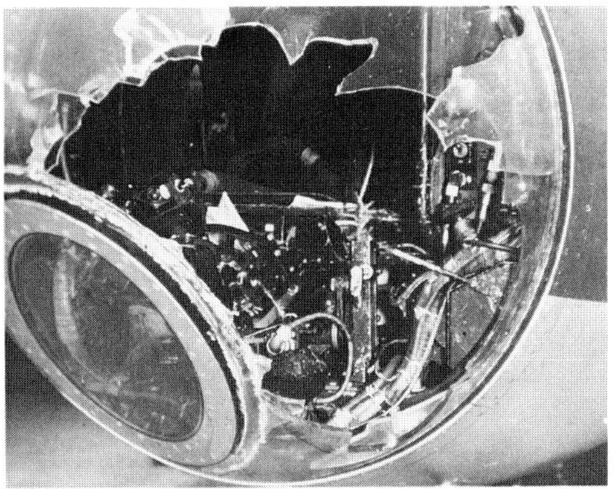

A birdstrike to A84-237 on 6 August 1964 — the bird penetrated the nose, left, where …
VIA R BIDDELL

> We were flying as No 2 to Ian Westmore, and had completed nearly the first half of the low level navigation for a simulated strike on a large rail bridge. Typically, we were at tree-top level and 350 knots, with eyes peeled for obstacles and hazards, particularly birds. If sighted early, then most of the time, birds could be avoided without much difficulty. As we approached the target area, Paddy had moved to the prone position in the nose of the aircraft to prepare the bombsight for the simulated bombing run.
>
> During my scan well ahead of the aircraft, I thought I had sighted two birds just above the horizon, but as I continued I could only find one. I monitored this

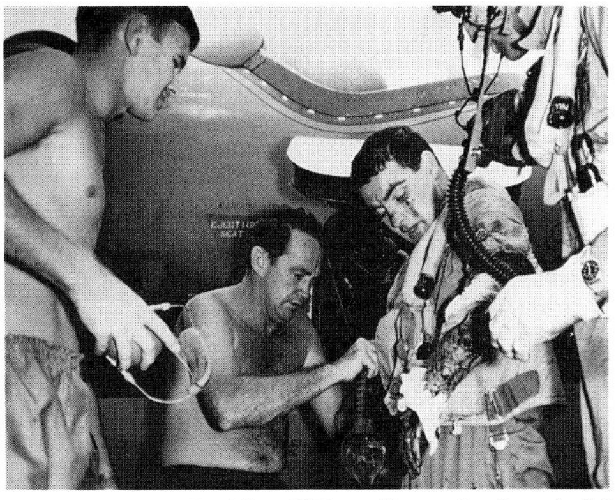

… navigator Paddy O'Farrell received painful injuries
VIA R BIDDELL

50 Ron Biddell, interview with author, of 29 Jan 1994.

51 Ron Biddell, telephone interview with author, of 7 Feb 1994.

52 Wg Cdr R L Biddell AFC, O43796. Nos 6, 2, 6, 2 & 6 Sqns, 1 FTS, CFS, PTS RNZAF, CFS, DefAir; comd CFS 1980–81. Regular air force offr of Adelaide; b Adelaide 1 Dec 1939.

53 Flt Lt J M O'Farrell, O56264. Nos 6 & 2 Sqns, SAN. Regular air force offr of Perth; b Perth 28 Nov 1939.

bird's flight until it was certain to pass the aircraft safely, after which I looked ahead just in time to see a large hawk or eagle directly in front of us. Its reappearance was too late for me to even shout a warning to Paddy. The bird hit the nose cone causing an almighty thud and roar, imparting considerable damage to Paddy and the aircraft.

My first reaction was to place the aircraft into a climbing attitude, apply full power and trim the aircraft well nose-up. Then came the task primarily of trying to find out what had happened to Paddy and attempt to assist him. Aircraft damage and handling checks were to follow some time later. Paddy struggled out of the nose with his oxy mask broken off, face and arm covered with blood, and pieces of bird, feathers and perspex everywhere. Major remnants of the bird somehow finished up between my feet and the rudder pedals. Subsequently we found the aircraft handling characteristics unchanged (except for noise), the bombsight was broken from its mounts and most of the control switches and knobs on the bombsight computer were broken.

Very obviously Paddy was badly hurt but the exact extent was impossible to determine as he was pale from shock and almost passing out. All I could say was "sit down and don't pass out, or I won't know if you are dead or alive". Strapped into my ejection seat and having us on autopilot, with Paddy sitting on the floor next to me, I was able to determine his main injury was to his forearm. My only means of first aid was my handkerchief, which I attempted to tie above his wound to stem the bleeding. This was totally useless and finished up draped around his wrist!

Meanwhile, Westmore was handling all radio procedures for the emergency. We happened to be at our furthest distance on our route from Butterworth, with no certainty that Paddy wasn't more seriously hurt as he was still in shock. After a slower than desirable return to Butterworth (the nearest airfield), we were met by all emergency services and Paddy was treated for a badly gashed forearm, that was broken in two places. He didn't fly again for four months. (Within 24 hours of the accident, mates of Paddy at RAAF Richmond held a wake as they had heard that his arm had been amputated.)[54]

By September 1964, the frequency of Indonesian raids in northern Borneo led to the plan for transferring the Australian infantry battalion from near the Thai border in Malaysia into Borneo.[55] As tensions grew, Australia agreed on 18 December to a British request that RAAF aircraft would be available for offensive action against Indonesia, and Darwin could be used as a strike base.[56] In the event of any major Indonesian attack on Malaysia, RAF and RAAF aircraft would operate against Indonesian air bases in retaliation. This meant No 2 Squadron would strike Indonesia from the north, and the Canberras of 82 Wing from the south.

No 2 Squadron's targets included the Indonesian air bases on Sumatra. Navigator Flying Officer Tony Wilkinson[57] was allocated the MiG-21 base at Medan as his prime target. While doing his mandatory target study and preparation, he noticed a new reconnaissance photograph of the airfield, that had been taken only three days previously. To obtain this intelligence, the RAF had blacked out the Indonesian radars by the electronic jamming from their Victor aircraft, enabling their photo-reconn

---

54 Ron Biddell, letter to author, of 20 Jan 1994.

55 Grant, p.168.

56 Edwards, p.341.

57 Wg Cdr A R Wilkinson, O221023. Nos 6 & 2 Sqns, SAN, 428th TFS USAF, 82 Wg, 1 & 6 Sqns, HQOC, DepDef. Regular air force offr of Sydney; b Sydney 4 Nov 1940.

Canberras to overfly the targets undetected and update their targeting.[58] There was no doubt that although Australia and Indonesia never severed formal diplomatic ties, the nations were effectively at war.[59]

The British reconnaissance was typical of the serious approach that FEAF had to Confrontation. On another occasion, a flypast was planned over the ocean liner departing Singapore with the outgoing C-in-C of FEAF. The RAF briefed a formation overfly of the ship in international waters, which was to include four 2 Squadron Canberras. However, it was pointed out by the RAAF that the ship was not within international airspace, but in airspace controlled by Indonesia. The RAF briefing officer excused himself, and soon returned with the solution. "It's all okay", he stated, "we've armed the Javelins!".[60] The period of Confrontation was eventually concluded with the overthrow in Indonesia of President Sukarno's regime.

A "bare-base" deployment to Gong Kedak, on Malaysia's east coast near Kota Bharu, in May 1965
VIA A WILKINSON

During April and May 1965 a mobility exercise was held by the Squadron, deploying for "bare base" operations at the Gong Kedak airfield, at Kota Bharu, on Malaysia's east coast. Such an exercise involved living in tents, at the edge of the jungle. One evening, late in the night after some celebrating, one member returned to the mess tent to report he'd seen a tiger prowling around the domestic site. This was naturally dismissed by the assembled gathering. The next morning the local police were searching between the tents. They were looking for a wild tiger that had wandered out of the jungle. The prints of enormous tiger paws were in evidence up and down the tent lines![61]

58 Tony Wilkinson, interview with author, of 17 Nov 1993.
59 Grant, p.168.
60 Tony Wilkinson, interview with author, of 17 Nov 1993.
61 Tony Wilkinson, interview with author, of 17 Nov 1993.

The Squadron, now under the command of Wing Commander Leo Britt,[62] was still committed to assist in the changeover of No 79 Squadron's Sabres from Ubon. Because of the Malaysian reticence at being seen to be openly assisting the SEATO pact, the swapping of Sabres between Butterworth and Ubon was a sensitive matter. It was considered, because of the limited aids in the Sabre, that the fighters needed a Canberra navigational escort. Each time a "Sword Move"[63] exercise was necessary, a Canberra would flight plan for a navigational sortie from Butterworth to overhead Korat, in Thailand, and return. The plan was for two replacement Sabres to take off from Butterworth, in radio silence, with the escorting Canberra, and peel off over the Korat navigation beacon. At that point the Canberra executed a 180 degree turn, during which time two Sabres, due for replacement from Ubon, would formate on the Canberra for the return leg. This subterfuge was not attempting to fool the Malays, but it satisfied the diplomatic protocol.

No 2 Squadron officers, Butterworth, 1966, with Commanding Officer Wing Commander Leo Britt (centre front) VIA A WILKINSON

The danger of birdstrikes to the Canberras was always present as the emphasis remained on low flying. During participation in "Exercise Straight Flush" near Singapore on 3 November 1966, Flying Officers Ben Schiemer[64] and "Gus" Gribble[65] (A84-232) had a critical birdstrike on the canopy. Gribble was standing beside the pilot as the canopy completely disintegrated, and was able to crawl back to the navigator's

62 Gp Capt L Britt DFC, 427373/O5844. No 462 Sqn, HQNEA, HQSA, 25 & 6 Sqns; comd Lincoln Flt 1955; 1 Sqn; comd Mel Uni Sqn 1957; HQTC, 82 Wg; comd BSqn Pea 1961–63, 1 Sqn 1963–65, 2 Sqn 1965–66; RAAFSC, HQSC. Regular air force offr of Melbourne; b Melbourne 23 Aug 1921.

63 "Sword" was the popular nickname for the Sabre fighter.

64 Flt Lt B W Schiemer, O219955. 2AD, Nos 1 & 2 Sqns, 2FTS, 1 OCU, 2 Sqn, CFS. Regular air force offr of Mudgee, NSW; b Mudgee 21 Mar 1938.

65 Flt Lt B D Gribble, O56363. Nos 6, 2 & 6 Sqns, 1 OCU. Regular air force offr of Perth; b Perth 13 Oct 1942.

station and strap into his ejection seat. Schiemer elected to return to Butterworth, and the aircraft was escorted back for a safe landing.

The Dakota commitment continued, both for supply dropping, transport support tasks for the forces, and the VIP Ambassadorial tours. As the Australian involvement in Vietnam increased, so did battle casualties mount, and medical evacuation sorties were now added to the roles of "C" Flight. The patients brought down to Butterworth on the Dakota medevac missions were treated at No 4 RAAF Hospital, and then flown back to Australia on RAAF Hercules transports. The importance of the duties carried out by the Dakotas resulted in the formation of the independent Transport Support Flight in February 1967, which would eventually expand to six aircraft. This separation of the Dakotas from No 2 Squadron had, however, been caused by other developments.

No 2 Squadron flight line at Butterworth, early 1967 VIA A WILKINSON

Wing Commander Rolf Aronsen[66] had assumed command of No 2 Squadron on 28 November 1966. For the CO of a flying squadron, his timing was perfect. Within a month, on 23 December, the Government announced the deployment of No 2 Squadron to the increasing Australian presence in South Vietnam.

66 Wg Cdr R B Aronsen, 434360/O12651. No 10 Sqn, 111 ASR Flt, RAAF Coll, CFS, 1 Sqn; comd Qld Uni Sqn 1958–59; RAAF London; comd 1 Sqn 1966, 2 Sqn 1966–67; HQ But, Dept Air. Regular air force offr of Timaru, NZ; b Wananui, NZ, 1 Aug 1921.

Going off to war — No 2 Squadron taxies out of Butterworth on the deployment to Vietnam, 19 April 1967. VIA R BIDDELL

CHAPTER 24

VIETNAM

Canberra bombers from the RAAF's No 2 Squadron have supported most of the major ground operations conducted by the Allied Forces in South Vietnam in its first 10 weeks of combat flying.[1]

The Geneva Accords of July 1954 had divided Vietnam, with most communist forces regrouping in the north and predominantly Catholic refugees moving to the southern zone. A referendum in October 1955 established the Republic of Vietnam, which generally became known as South Vietnam, with Ngo Dinh Diem as President.[2] The Democratic Republic of Vietnam was to be called North Vietnam.

South Vietnam's crescent shaped area of 67,000 square miles was along the south-eastern edge of the Indochinese Peninsula. Due to its long shape, its border with North Vietnam was less than 80 kilometres, with a sea coast of nearly 3,000 kilometres on the South China Sea and the Gulf of Siam. Its land borders to the west were shared with Laos and Cambodia.

What was of international significance was that South Vietnam had been designated under the protocol to the Manila Treaty, as being under SEATO's protection.[3] Despite some assurances, therefore, of some foreign assistance in supporting his regime, Diem's fundamental problem was a lack of any strong support in the countryside. Saigon was perceived by farmers as an imposition upon them for the benefit of Diem's urban supporters.[4] As dissidence grew to conflict against Diem in the late 1950s, Australia publicly committed its support to the survival of South Vietnam.[5] However, a lack of unity within SEATO caused discord between members on what constituted appropriate action in Indochina,[6] and Australia was to act unilaterally to requests from Washington and Saigon to support the United States build-up in Vietnam.[7]

The CO, Wing Commander Rolf Aronsen, welcomed to South Vietnam VIA R BIDDELL

1 "Canberras Strike At Viet Cong", *RAAF News*, Vol 9, No 7, Aug 1967, p.2.

2 Edwards, pp.192–3.

3 Edwards, p.197.

4 Edwards, p.198.

5 Edwards, p.207.

6 Edwards, p.221.

7 Edwards, pp.234–6. At the beginning of 1961, there were 800 American advisers in Vietnam; by late 1962 this had increased to 11,000.

When Australia became involved in 1962 with the dispatch of the AATTV to assist the United States Military Assistance Command Vietnam (MACV), it was considered that a communist victory would threaten the rest of South-East Asia, jeopardising Australia's security. This had been consistent with the "forward defence" policy of the 1950s, and indeed South Vietnam was seen as another "domino". But there was more to it; Australia had to convince the Americans we were more than "paper allies", and the Vietnam involvement became "a sort of life insurance policy".[8] This need for such a firm alliance with the United States by the Menzies Government was a natural consequence of the memory of Japanese advance, as Menzies, after all, had also been Prime Minister in 1939–41.[9]

Hanoi had already deployed some 30,000 regular troops of the North Vietnamese Army (NVA) to the south before the United States committed its combat troops in 1965.[10] In addition to this opposition, Saigon also had to contend with about 100,000 locally recruited part-time guerillas—the Viet Cong. In response to American requests to see "more flags" involved, Australia had doubled the size of the AATTV in 1964, and committed a detachment of RAAF Caribou transports.[11] In April 1965 Australia decided to send an infantry battalion. This force later grew to three battalions, forming the Australian Task Force, supported by a squadron of RAAF battlefield helicopters. Furthermore, Australian Canberra bombers would support ground combat forces.

A84-248, with the 2 Squadron hangar in the background RAAF

Before No 2 Squadron could deploy to Vietnam, the Canberras had to be modified and facilities needed to be prepared. To be compatible with United States tactical aircraft in the area, the Canberras were fitted with better TACAN navigation displays, UHF radios were to be mandatory, and armour plating was fitted to the cockpit floor to afford the crew some measure of protection from groundfire.[12] The make-up of the Squadron was also to change greatly. Even though aircrew strength would remain at around 11 crews, many of whom arrived from No 6 Squadron during March 1967, the ground support had to be dramatically increased. At Butterworth, the Squadron strength had typically numbered 140 personnel. At Phan Rang, without the normal RAAF support of a Base Squadron to provide all domestic facilities, this total of personnel doubled. No 2 Squadron became part of the USAF's 35th Tactical Fighter Wing (35th TFW), which

8 Edwards, p.249.

9 Edwards, p.380.

10 Edwards, pp.293–4, 339.

11 Edwards, pp.297–9.

12 Nitrogen purging fuel systems were also later added, to prevent tank fires if hit by groundfire.

operated F-100 fighters and B-57 dive-bombers (the American variant of the Canberra) from Phan Rang Air Base.

Phan Rang was on the eastern coastal plain, nearly 300 kilometres north-east of South Vietnam's capital, Saigon. The facilities for the new squadron of the 35th TFW had been hurriedly constructed by Detachment B of No 5 Airfield Construction Squadron. Squadron Leader Dick Gurevitch[13] had arrived at Phan Rang at the beginning of January, and with the assistance of the USAF 554th "Red Horse" Civil Engineering Squadron, had designed the 2 Squadron accommodation.[14] The advance element of 5ACS then arrived, and had commenced work on 17 January. Without the efforts of the 5 ACS detachment, under the command of Squadron Leader Graham Anderson,[15] there would have been no living quarters, messing facilities, technical or administrative areas ready by No 2 Squadron's arrival on 19 April. These facilities had also involved construction of a maintenance hangar and annexes, communications and operations sections, recreation areas and a sewerage system. With assistance from the "Red Horse", 5 ACS had built the complete accommodation and infrastructure, which allowed No 2 Squadron to begin immediate bombing operations.

An interesting insight into the enemy intelligence capability was demonstrated immediately prior to Squadron Leader Anderson's party arriving from Australia. North Vietnam's version of "Tokyo Rose"—known as "Hanoi Hannah"—had mentioned Anderson by name, and welcomed his band of engineers to the country. "Feel welcome in our country", she had broadcast, "but the efforts you will put in with US forces at Phan Rang for the Allied cause will be in vain".[16]

Control of No 2 Squadron handed over to the United States 7th Air Force when Wing Commander Rolf Aronsen, with navigator Flying Officer Frank Burtt,[17] led his Canberras from Butterworth on 19 April 1967. The formation for the Squadron's deployment to war comprised:

A84-230	Wg Cdr Aronsen	Flg Off Burtt
A84-236	Plt Off O'Hanlon	Flg Off McKenzie
A84-231	Sqn Ldr Thomson	Flt Lt McIndoe
A84-245	Flg Off Walters	Flg Off Halvorson
A84-234	Sqn Ldr Kilian	Flt Lt Hanigan
A84-240	Flt Lt Montgomery	Flg Off Gribble
A84-237	Flt Lt Squires	Flt Lt Furlonger
A84-242	Flg Off Biddell	Flg Off Waring

13 AirCdre R N Gurevitch AO, O211602. 2ACS, 5ACS, DepAir, 5ACS, DepAir, DepDef; DGAW 1984–87, DGF 1988–91. Regular air force offr of Sydney; b Hong Kong 13 Dec 1936.

14 AirCdre Richard Gurevitch, telephone interview with author, of 9 Feb 1994.

15 Sqn Ldr G P Anderson, 055139. 5ACS, DepAir; comd Det B 5ACS 1967; DefAir. Regular air force offr of Perth; b Wongan Hills, WA, 17 Apr 1938.

16 Graham Anderson, telephone interview with author, of 10 Feb 1994.

17 AirCdre F E Burtt OBE, O17778. Nos 6, 2 & 6 Sqns, 1FTS, DefAir, HQ Amb, RAAFSC, DefAir, HQADF, ADFA; comd Williams 1989–91; DepDef DGSC 1991–93; AHQ. Regular air force offr of Brisbane; b Tully, Qld, 1 Jan 1941.

Other aircrew were transported with the groundcrew by RAAF C-130 Hercules aircraft. An advance party, under Squadron Leader John Chesterfield[18] as the O-i-C Administrative Flight (virtually CO of Base Squadron for 2 Squadron's domestic support), had arrived several days previously. With 5 ACS preparation and construction having been underway for three months, the technical, operational, administrative, messing, sleeping and recreational facilities were virtually complete for the Squadron's arrival in April. It was No 2 Squadron's responsibility to then finish off their accommodation, the painting, adding the comforts, and the vital sandbagging. This also included construction of adequate bunker protection from mortar bombardment, as outside the perimeter wire of the air base, at least initially, the night belonged to the Viet Cong. Although security patrols did much to deter enemy activity near the "wire", for the duration of No 2 Squadron's tour at Phan Rang the base was subjected to intermittent mortar and rocket fire. Personnel became proficient at a quick sprint along the well-worn tracks from their rooms to their protective bunkers.

After two days of training flights to familiarise crews with USAF air traffic procedures and possible diversion airfields, the first missions were flown on 23 April. The first of these Combat Skyspot (CSS) missions—radar-directed, high-level bombing missions, controlled to the extent where the ground controller called the release of the bombs—was flown by the OC Flying, Wing Commander Vin Hill,[19] and Flight Lieutenant Tom Wright (A84-240). The OC Flying position had been founded on the Squadron's move to Vietnam, to take responsibility for flying operations and thereby allow the CO to concentrate on the administrative and technical running of the unit.

For the first five days, eight missions were flown daily over the whole of South Vietnam, until 28 April, when the Squadron settled into the routine of eight sorties each night. The first night mission was flown by the CO and navigation leader, Squadron Leader Dave Ingall[20] (A84-237), and of the eight aircraft "fragged"[21] that night, two aircraft were forced to return to base with their bombs. For the ground radar station to accurately track a Canberra, the aircraft had to be fitted with a radar transponder, known as a "music box", as accurate controlling was not always possible on the reflected radar "skin paint". On this first night, one aircraft was not fitted with a transponder, the other's transponder had gone unserviceable. These teething problems were soon overcome as the crews settled into the routine of the first nightly take-off at about 8 p.m., with the others following at about hourly intervals. The last aircraft would not arrive back until about 6.30 a.m.

---

18 AirCdre J M Chesterfield AM, O32860. No 3 Sqn, 1AFTS, 76 & 2 Sqns; comd 9 Sqn 1972–73, ASU 1973–75, AJWE 1975–77, RAAFSC 1977–78, Townsville 1980–81; DepDef DGPS 1981–83. Regular air force offr of Sydney; b Sydney 9 May 1933.

19 AirCdre V J Hill DFC* AFC, 428313/O33618. Nos 25, 76, 38, 77 & 25 Sqns, ARDU, RCAF exch, DepAir; comd 6 Sqn 1966–67; 2 Sqn; comd 82 Wg 1971–72; HQSC, DefAir. Regular air force offr of Melbourne; b Melbourne 19 Aug 1924.

20 Wg Cdr D Ingall, 173007/O32534. No 1 Sqn, 76 Sqn RAF, 6 Sqn, SAN, DepAir, 2 Sqn, RAAF Wash, RAAFSC. Regular air force offr of Hamilton, NSW; b Newcastle, NSW, 28 May 1931.

21 The fragment of the 24-hourly tasking schedule, issued nightly by Headquarters 7th Air Force.

On 14 May two waves, each of four Canberras, attacked the same target during daylight. The target was an enemy base camp, and the tunnel and bunker fortifications needed the firepower afforded by the penetrating 1,000 lb bombs from the Canberras, and the massive loads of the B-52s.[22] The CO and nav leader (A84-242) led the first flight in a "finger-four" formation, on two runs, each aircraft dropping three bombs per run on concentrated pattern bombing, reminiscent of the Target Direction Post attacks of the Malayan Emergency. Flying number three in this first wave were Squadron Leader Trevor Kilian,[23] whose navigator Flying Officer Frank Burtt (A84-240) recalled:

> The target was a VC base area in III Corps, but we saw nothing but trees as we were flying at medium level so it was an anti-climax. It was, however, impressive to see that number of bombs strike simultaneously, although there was no discernible effect.[24]

This observation was typical for crews flying the Skyspot missions. Although occasionally reports could come later from FACs in the area, for most of these missions it was not possible to obtain accurate bomb damage assessment (BDA).

> This is because of the saturation method of bombing favoured by the Americans. After the Canberras drop their bombs, the Americans pound the area with artillery, send in Iroquois helicopter "gunships" with machine guns and rockets, then sweep the area with ground troops. By the time the target is examined it's anybody's guess who claims the credit.[25]

To meet the daily commitment of eight sorties, No 2 Squadron kept eight Canberras on-line at Phan Rang, which were maintained at exceedingly high serviceability levels. The Squadron maintenance personnel carried out the normal daily flight servicing and rectification, in addition to weekly "A" and monthly "C" services, and over the time at Phan Rang a serviceability rate of 96 per cent was maintained. A further three Canberras would be undergoing second-level "D" (250-hourly) maintenance by Transport Support Flight, back at Butterworth. Ultimately No 478 (Maintenance) Squadron assumed this task. In addition, a twelfth 2 Squadron Canberra was always undergoing depot-level "E" (1250-hourly) maintenance at No 3 Aircraft Depot at Amberley. As the aircraft came out of maintenance, each would be proved by No 1 OCU, before being returned to the war zone.

Initially, the standard bomb load for the Canberras had been six 500 lb bombs, but on the night of 9–10 May six 1,000 lb bombs became the norm.[26] From the night of 15–16 May each aircraft carried an increased load of nine 500 lb bombs.[27] However, the bomb-bay load continued to cause problems, and between 15 and 25

---

22 Charlie Reif, telephone interview with author, of 25 Nov 1993.

23 Sqn Ldr C T Kilian, O52721. Nos 77 & 75 Sqns, 1AFTS, 34, 1 & 2 Sqns. Regular air force offr of Perth; b Perth 9 Aug 1930.

24 AirCdre Frank Burtt, interview with author, of 26 Nov 1993.

25 "Canberras Strike at Viet Cong", p.2.

26 These were the old World War II stocks of a variety of single lug 500 lb bombs, or Mk 1 1,000 lb bombs, all carried in the bomb-bay.

27 The bomb-bay load was two Avro Triple Carriers, each capable of carrying three 1,000 lb bombs, or three Triple Carriers, each capable of carrying three 500 lb bombs.

May there were seven cases of negative releases, or "hang-ups".[28] On the night of 1 June, two Canberras each had double hang-ups, and this frequent occurrence had been traced to an electrical problem with the Avro Triple Carrier.[29] The following evening, Flying Officers Wally Walters[30] and Lance Halvorson[31] (A84-234) had quite a harrowing experience with the bomb carrier. Halvorson related:

> We were flying with a USAF nav in the "jump" seat and experienced a hung bomb. On our return to base and on the descent in turbulent conditions, the bomb dropped off the rack, hit the closed doors and rumbled forward striking the protective plate over the bomb door jack against the cabin bulkhead, making a loud ringing noise. The white-faced USAF nav said "Christ, what was that?" To which we replied "Oh, just the bomb that hung up had dropped off, but it is OK, it is safe! You'll see when we jettison it off Cam Ranh Bay." When the bomb hit the water, it exploded. The USAF nav's face went even whiter!! I met him 3 years later in Tucson when training on the F-4E—he had not forgotten the incident.[32]

Trialling the bomb fit for the Canberra in 1967, in the foreground 1,000 lb bombs, 500 lb bombs in the background RAAF

After this unnerving event, the following night's loads reverted to the 1,000 lb bombs. However, the Triple Carriers continued to cause serious problems. On the night of 17–18 June, after more bombs fell onto the bomb doors and in one case, a bomb was inadvertently released as the doors were opened, the use of these carriers was discontinued:

---

28 Letter 2SQN/511/3/ARM (45) to HQSC, of 30 May 1967.

29 No 2 Sqn A51, of 1 Jun 1967.

30 Sqn Ldr W F Walters, O16897. Nos 6, 2, 6, 2 & 6 Sqns, 1FTS, 6 Sqn, DefAir, 1FTS; comd RAAFSU Pth 1981–82. Regular air force offr of Mareeba, Qld; b Mareeba 27 May 1942.

31 Wg Cdr L J Halvorson, O51382. Nos 1, 2, 6, 2 & 6 Sqns, 2FTS, 6 Sqn, HQ Amb, DefAir, HQOC, HQADF. Regular air force offr of Perth; b Perth 27 May 1939.

32 Lance Halvorson, telephone interview with author, of 28 Nov 1993.

> As a result of the two incidents above, Avro Triple Carriers will not be used until an effective remedy is found.[33]

There can be no doubt that No 2 Squadron was very lucky that these bombing problems did not have tragic consequences. Fortunately, alternate carriers had been held in reserve, but these only allowed for the reduced load of eight 500 lb bombs,[34] or four 1,000 lb bombs, in the bay. Back in Australia, the RAAF's armament store, No 1 Central Reserve at Kingswood, had been modifying the lugs of the old stocks of World War II bombs to ensure better releases from the Canberras. One armourer, Flying Officer "Alex" Alexander,[35] recalled over the period it took to deplete these stocks in 1967–68, that a total of 27,000 bombs were reworked by the armament teams at No 1 Central Reserve, under Operation "Rapid Pack".[36] These events coincided with the adoption of wing-tip bomb carriers, which had been trialed in Australia by the RAAF's Aircraft Research and Development Unit (ARDU). Each carrier had been cleared for a load of up to 370 kg (810 lb)[37] to enable carriage of a 500 lb bomb on each wing tip. These carriers would remain standard for the rest of the Canberra's bombing life. These problems had highlighted the danger of going to war without having established and trained with these configurations in peacetime.

While the Squadron had been plagued with these technical problems, there was also a lot to be done domestically to become established at Phan Rang. Chores included painting and improving Squadron accommodation, and the all-important sandbagging of buildings, including the construction of more bunkers for adequate protection against enemy mortar fire. In addition, on 10 June, the Squadron was again presented with the Duke of Gloucester Cup, the annual squadron proficiency award in the RAAF, for its performance in Malaysia over 1965–66.[38]

The tasking of No 2 Squadron by Headquarters 7th Air Force, in Saigon, had been stipulated by the Australian Government to be wholly within South Vietnam—not over the North or into Laos, nor within 20 kilometres of Cambodia. Not that these restrictions hampered the Canberras' operations; the crews found ways of working within these constraints. The United States had begun bombing North Vietnam in March 1965 with the USAF "Rolling Thunder" campaign, which was maintained until November 1968. However, in the south, typical targets included Viet Cong concentrations, food storage dumps, infiltration routes, gun positions, river crossings, bridges, base camps and military buildings. The military buildings in the south

---

33 No 2 Sqn A51, of 17/18 Jun 1967.

34 These were two Avro 100/1000 Standard Carriers on Multi Adaptors in the bomb-bay, with four bombs on each carrier.

35 AirCdre N Alexander, O316953. Prev RAF. 1CR, 1 OCU, 2 & 11 Sqns, 1CAMD, HQOC, DefAir; comd 1CAMD 1980–81; Austn Ord Council, HQSC, DepDef; President AOC 1987–88. Regular air force offr of Windsor, Berks, Eng; b Burscough, Lancs, 12 Jan 1936.

36 AirCdre Alex Alexander, interview with author, of 8 Dec 1993.

37 The tip pylon incorporated the standard MA4A multi adaptor, which allowed carriage of the twin-lug 500 lb Mk 13 bomb and later the 750 lb American M117 bomb.

38 The Duke of Gloucester Cup had been competed for annually by the flying squadrons of the RAAF since 1946.

amounted to little more than huts which housed the VC and their caches, and were identified by reports received by the USAF from Army reconnaissance patrols, airborne USAF FACs, photographic reconnaissance, and intelligence reports.

On 25 June, the first close air support missions under the direction of a FAC were flown. Wing Commander Vin Hill and Flight Lieutenant Tom Wright (A84-235), with Flight Lieutenants Barry Squires[39] and Charlie Reif (A84-230), were sent as a trial by 7th Air Force to assess the suitability of the Canberra in the low-level visual bombing role. Their target was a suspected VC troop assembly area in II Corps, 30 kilometres south-west of Tuy Hoa airfield. As Squires opened his bomb doors on the first attack, something was seen to fall from the aircraft, so he was forced to return to base.[40] It was discovered that a bomb tail had fallen off. However, the FAC marked the target with a white phosphorous ("willie-pete") smoke rocket, and the other aircraft was able to accurately attack the target.

The view from the nose of the Canberra
VIA R ALLCHIN

The result of this success was that from 20 September a pair of Canberras was sent each day on visual bombing missions (VBM). Flying first on this new schedule were Wing Commander Vin Hill and Flying Officer Gus Gribble (A84-247), with Squadron Leader Jock Thomson[41] and Flight Lieutenant Charlie Furlonger[42] (A84-237). At dawn each morning, two crews departed for close air support missions. Some of these early daylight bombing raids were in support of the Australian Army Task Force in Phuoc Tuy Province. Once again, the raids were directed by a FAC in a Cessna O-1 "Bird-Dog" observation aircraft, marking the target with smoke rockets.

The Canberras, now flying six high-level radar Skyspot and two low-level FAC missions per day, continued to pound enemy concentrations and targets in huge operations throughout South Vietnam. On one operation, "Junction City", 600 VC were killed by Allied forces in one savage battle, supported by the USAF and

39 Wg Cdr B Squires, O216046. 1 & 2 ATU, 1BFTS, 1AFTS, Nos 34, 1 & 2 Sqns, DepAir; t/comd 2 Sqn 1973; HQOC. Regular air force offr of Sydney; b Wollongong, NSW, 29 Aug 1935.

40 Charlie Reif, telephone interview with author, of 25 Nov 1993.

41 AirCdre J I Thomson AM, 160910/O32446. Nos 38, 24 & 75 Sqns, 1BFTS, ARDU, 3AD, 2 Sqn; t/comd 1 Sqn 1969; DepAir; comd ARDU 1975–76; DefAir, HQ Amb; comd Amberley 1983–84; DefAir DGOR. Regular air force offr of Moe, Vic; b Moe 24 Feb 1932.

42 Sqn Ldr C R Furlonger, O28874. No 11 Sqn, 78 Wg, 1, 2 & 6 Sqns, SAN, HQSC, 36 Sqn. Regular air force offr of Lismore, NSW; b Bonalbo, NSW, 18 Jan 1934.

RAAF aircraft. Other major operations were in support for "Task Force Oregon" to the north in I Corps, and operations "Pershing" in Binh Dinh Province and "Francis Marion" in the Pleiku area of II Corps. To the south of Phan Rang, the Squadron flew in "Diamond Head" to the north-west of Saigon and in support of the Australian Task Force (2nd and 7th Battalions) in Phuoc Tuy Province in III Corps, and continued strikes in the Mekong Delta of IV Corps.

During October 1967, a large changeover of aircrew took place to enable the first crews who had served at Phan Rang to prepare for the delivery of the RAAF's new F-111 strike aircraft. Wing Commander "Bill" Hughes[43] then became OC Flying. In late November, Wing Commander David Evans assumed command of the Squadron, and as Hughes was required back in Australia in January to lead No 1 Squadron for the F-111 program, the Wing Commander OC Flying position was disestablished.

The CO, Wing Commander David Evans

RAAF

Headquarters 7th Air Force had, by November, been suitably impressed by the high degree of accuracy that the "Magpies"—No 2 Squadron's radio callsign—could deliver in close air support with the FACs, and four visual strikes became the pattern each day, with four CSS sorties each night.[44] Quite often, however, weather conditions would preclude visual bombing, so the Canberras would be retasked airborne to a Skyspot mission. This would most probably be to another target, and sometimes to another corps. What had been intended as visual bombing in the country's south along the canals of IV Corps, could become a route interdiction task in the mountains of northerly I Corps under Skyspot control. If the urgency of the ground war required immediate air support, the Canberras en route towards a preplanned target could be diverted to a hot troops-in-contact (TIC) "fire fight". Such was the flexibility of the Canberra: its crews made it an

---

43 AVM H A Hughes DFC, O21144. Nos 77, RAF CFS, CFS, 76, 3 & 79 Sqns; comd RAAF Ubon 1965, 1 Sqn 1966–67; 2 Sqn; comd 82 Wg 1969–70; DefAir, HQSC, AA Wash, HQOC; DefAir CAFM 1979–81; DCAS 1981–83. Regular air force offr of Sydney; b Junee, NSW, 25 May 1928.

44 From Oct 1969, No 2 Squadron was being tasked with eight visual sorties a day.

accurate delivery system with a substantial load, and it could remain airborne for over three hours, twice that of other tactical aircraft involved in the fighting.

By late 1967, the North Vietnamese leadership had proposed a major shift in tactics from the guerilla style of conflict to massed assault on prime locations.[45] From the beginning of 1968 intense pressure was placed on the United States fire support bases in northern I Corps, immediately south of the Demilitarised Zone (DMZ), the border with North Vietnam. These assaults were planned to begin on the Vietnamese Buddhist lunar new year, or Tet,[46] which usually falls at the end of January or early February. Even if the attacks would not be successful militarily, the North Vietnamese were confident that they would undermine the American resolve in the war.[47]

On the night of 31 January 1968, more than 100 cities, military installations and transport lines were attacked. The important city of Hue, to the north in I Corps, was taken and held for over three weeks, and heavy fighting took place in Saigon. The Australian Task Force in Phuoc Tuy successfully defeated the enemy's D445 battalion in an attack on the provincial capital of Baria by heavy street and house fighting.

Caught up in the fighting on that first night of Tet was the CO, Wing Commander David Evans, and the Squadron Navigation Officer, Squadron Leader Marc Robin.[48] They had been in the capital overnight, and awoke to the sound of gunfire, just outside their hotel. They thought initially it was fireworks, but found it was the Vietnamese police (the "White Mice") engaging the VC and North Vietnamese Army troops in the streets. The enemy had also been able to breach the defence of the United States Embassy, only a couple of hundred metres up the street. Dodging the fighting as best they could, they were able to return to their Canberra at the Tan Son Nhut air base, only to be prevented from taking off because of the hostile fire around the perimeter. There were dead VC and North Vietnamese troops on the airfield, so the CO assessed he would be safer at his home base. By bluffing that he was "urgently needed back at Phan Rang for tactical operations", he was able to get released. His steep climbing departure ensured they avoided any hostile fire that was directed towards the Canberra.[49]

Perhaps the best known of the Tet offensive attacks was against the United States base at Khe Sanh over 2–4 February. Khe Sanh was an American outpost which defended the northern coastal cities from any enemy advance from Laos. As the North Vietnamese divisions pressed south to capture the base, the United States Commander in Vietnam, General Westmoreland, placed the defence of Khe Sanh above all other operations in the country. Flying on these sorties during operation "Niagara" on 4 February was Wing Commander Evans (A84-235), who recalled:

---

45 Grant, p.365.

46 Abbreviation of *Tet Nguyen Dan.*

47 Grant, p.365.

48 Sqn Ldr M J Robin, O32565. Nos 2 & 1 Sqns, 44 Sqn RAF, 11 Sqn, DepAir, 1BFTS, RAAF Acad, 6 & 2 Sqns, DepAir. Regular air force offr of Adelaide; b Karoonda, SA, 3 Sep 1930.

49 Air Marshal S D Evans, interview with author, of 21 Dec 1993.

> The US Marines had got themselves isolated into a situation at the fire support base at Khe Sanh, similar to the way the French had been in 1954—and the Marine commander later stated he had never heard of Dien Bien Phu! They were supported by a continuous tactical airlift operation, which resupplied the base through constant enemy fire. Our Canberra strikes were part of the enormous air campaign mounted by tactical aircraft to prevent the enemy overrunning the base, which was ultimately saved by air power.[50]

During the next weeks of the siege, the enemy was defeated in his attempts to take the base by round-the-clock close support by tactical aircraft. General Westmoreland later reported:

> Between 22 January 1968 and 31 March 1968, tactical aircraft flew an average of 300 sorties daily, close to one every five minutes and expended 35,000 tons of bombs and rockets. At the same time Strategic Air Command's B-52s flew 2,602 sorties and dropped 75,000 tons of bombs during the siege, preventing the enemy from assembling in large formations.[51]

Integration of No 2 Squadron into the 35th TFW had gone surprisingly smoothly, but inevitably some changes caused problems. Engine fitter Sergeant Peter Ashley[52] explained one of the differences of working within the USAF, away from normal RAAF logistics:

> Our aircraft operated on the United States supplied fuel JP-4, which had less lubricative qualities than our Avtur or Jet A1, and resulted in our engine fuel pumps being put onto ½ life ... they were "beasts" of units to change.[53]

Apart from technical inconveniences such as this, personnel being posted from Australia had to contend with living in a hazardous war zone. The air base at Phan Rang was basically secure, but was subjected to intermittent mortar and rocket fire, normally at night, from VC in the surrounding countryside. Peter Ashley remembered one such attack soon after his arrival at 2 Squadron:

> On the night of 5 May 1968, some 17 82mm mortars, of the M30 type made in Communist China, landed adjacent to the 2 Squadron domestic quarters. One was 50 yards from the Sergeants' living quarters, three more 120 yards away and the remainder up the hill towards the working area and flight line. It was believed that the attempt with the mortars was to fire them over our quarters and the hill and into the flight line.
>
> On another occasion, a sapper attack group was detected trying to breach the wire by the defence perimeter guards. The VC were carrying satchel charges which they intended to place on aircraft to blow them up. The perimeter guards had a truck with a 50-calibre, or .5-inch, machine gun mounted on the back ... they cut this sapper attack to pieces. One of the satchel charges was hit with fire—it blew up and a number of the VC sappers were blown to bits. I think there were 16 or 17 of them; one was identified later as the unit barber and another worked in one of the messes.[54]

---

50 Air Marshal Evans, interview with author, of 21 Dec 1993.

51 G Odgers, *Mission Vietnam*, Department of Defence, Canberra, 1974, pp.73–5.

52 WOff P W Ashley, A51903. No 34 Sqn, ARDU, 2 Sqn, ARDU, 2 Sqn, HQSC, 481 Sqn. Engineer of Bridgetown, WA; b Bridgetown 15 Mar 1936.

53 Peter Ashley, letter to author, of 10 Nov 1993.

54 Ashley, letter to author, of 10 Nov 1993.

The external defence of Phan Rang was provided by the Republic of Korea Army, with internal security by the USAF. Coordinated with these forces, No 2 Squadron also made an active contribution to the base's defence with its own flight of airfield defence guards (ADGs):

> ... the ADGs of No 2 Squadron patrol through the night. They also sweep through the area periodically by day to seek any trace of enemy activity, to familiarise themselves with the ever changing pattern of the countryside, and to let it be known that there is no "no-man's-land" in the vicinity of their base.[55]

The CO, Wing Commander David Evans, remembered the value in having ADGs as part of the Squadron:

> I felt very fortunate to have a flight of ADGs—even more fortunate was that I had as Officer Commanding the flight, one Flight Lieutenant George Foskett,[56] an ex RAF Regiment Wing Commander. He looked at the Ground Defence Plan formulated for the base by the USAF Air Police and declared it a recipe for disaster. I discussed this with the USAF Commander and he asked for Foskett's assistance. In the end, our man virtually wrote the Ground Defence Plan for Phan Rang base. The external defence was, of course, up to the Korean Battalion adjacent to the base.
>
> Foskett pestered me asking that the ADGs be able to carry out night patrols outside the wire. Eventually I agreed to a limited area outside—my memory is uncertain on the exact distance, 400 yards I think. Foskett was always pressing me for more but I refused—the ADGs role was really to guard No 2 Squadron's assets. However, on one occasion while I was there, our patrol intercepted a VC squad setting up for a mortar attack and dispersed them.[57]

The standard bomb loads for the Canberras over late 1967 and 1968 had been eight 500 lb or six 1,000 lb bombs. As the RAAF's stocks of the old Australian bombs were becoming rapidly depleted by the high Squadron usage, alternative weapons were considered. A trial over 31 March to 3 April 1968 saw one aircraft daily dropping six of the American 750 lb M117 general purpose bombs. While awaiting acceptance of this new weapon, the Canberras commenced carrying four 1,000 lb bombs in the bay, and a 500 lb bomb on each tip. Over the period of a week in June 1968, a total of over 250 of the American bombs were dropped on missions, clearing its acceptance for the RAAF. Within two months this would be the Squadron's standard weapon.[58]

A trial in July saw 1,000 lb bombs dropped with variable timing (VT) fuses, with the aim of producing an airburst over the jungle canopy.[59] The advantage of airburst attacks was that soft-skinned targets could be destroyed without cratering. For

55 G J J Beck, *No 2 Squadron—Vietnam 1968*, Squadron annual booklet, Phan Rang, p.30.

56 Sqn Ldr G W Foskett MBE, O223127. Prev Brit Army & RAF. No 2 Sqn, BSqn Wlm, HQOC, 3AD; comd 1RTU 1975–77; BSqn But, 3AD. Regular air force offr of Newton Abbott, Devon, Eng; b Newcastle, Northumberland, 25 May 1925.

57 Air Marshal Evans, interview with author, of 21 Dec 1993.

58 No 2 Sqn A50, of 11 Aug 1968: "Today saw the start of the change over to American M117 750-lb bombs". In late 1969 some stocks of 1,000 lb bombs were again dropped. These were post-WWII Mk 10 HAHS bombs, that had specifically been designed for the Canberra and RAF V-bombers.

59 No 2 Sqn A51, of 27 Jul 1968.

instance, if a helicopter landing zone (LZ) was required in the jungle, this could be achieved by an airburst to clear the trees, without creating a deep crater caused by normal impact or delay fusing. The CO felt that this was unique in Vietnam, and wanted to prove it. On a medium-level Skyspot mission, he released his bombs, then dived to a lower altitude to observe the results. He saw one bomb impact—and the puffs of the others above his aircraft![60] The bombs had detonated prematurely, and fortunately had not damaged the aircraft, but the unreliability of the VT fusing prevented its further use on the Canberra.

By the second half of 1968, the majority of sorties being flown by the Squadron were low-level visual missions, which were more satisfying for the crews, and the Squadron was able to be credited with specific bomb damage assessments (BDA). Wing Commander Evans recalled:

Typical canal line target in the Mekong Delta
RAAF

> I pressed HQ RAAF Vietnam to seek from 7th Air Force all daylight attacks. In the end, the daylight tasking was for six daylight sorties and two night Skyspot sorties. It was not that the night sorties were necessarily ineffective, but rather that results were not always known, or made known to the Squadron.
>
> On one occasion, when working with a FAC on an early morning daylight sortie, the FAC pointed out to me a collection of demolished buildings—a dozen or so large wooden structures—and said: "That was a Viet Cong storage base until you fellas creamed it last night".[61]

The Squadron's accepted minimum height for bombing had been 3,000 feet, which enabled adequate target acquisition and generally kept the aircraft out of small arms fire. It was also above the height of aircraft running the risk of self damage from their bombs. However, if the cloud base did not permit bombing at this height, they were to refuse this and request another. It was felt by the crews this reflected poorly on their ability, so the CO cleared all crews to bomb at the minimum level which the bombsight would function, with a pull to safety height after bomb release. Depending on the aircraft's ground speed, this was between 800 and 1,000 feet. One navigator flying during this stage of the campaign was Pilot Officer Paul Goodwin,[62] who recalled an added advantage of bombing at lower levels:

> By about June we were up to 6 visual missions per day ... much of this work was "canal busting" in the Delta. It didn't take much to work out that line accuracy was critical to these

60 Air Marshal Evans, interview with author, of 21 Dec 1993.

61 Air Marshal Evans, interview with author, of 21 Dec 1993.

62 Wg Cdr P W Goodwin, O223696. Nos 2, 37 & 1 Sqns, SAN, 1 Sqn, DAW RAF Coll, DefAir, HQADF, RAAFSC. Regular air force offr of Toronto, NSW; b King Is, Tas, 7 Aug 1946.

missions and some of us started to drop from 1,000 feet (with a pull to 3,000 feet or so before the bombs went off). In about August 1968 ... a trial period of 1,000-foot bombing clearly showed the advantages of low level bombing.[63]

The lower height minimised left and right (line) errors, and became the accepted technique for bombing targets requiring particular accuracy in line, such as beside the long straight canals. This was just part of the overall aim to achieve an accuracy that would enable the Squadron to operate effectively in tactical operations. And improved bombing results did not just happen. Wing Commander Evans had achieved such results in Malaya, and was intent on achieving better accuracy at Phan Rang:

It was very evident from observation of our bombs dropped in daylight that accuracies well below 50 yards were required. Meticulous attention was paid to the bombsight—it was string aligned after every sortie and given a full calibration the first time a bomb exceeded, say, 50 yards without a logical explanation. The aim was to reduce the Squadron average error to 20 yards—and this was achieved.[64]

After eighteen months of operations in Vietnam, No 2 Squadron had flown 4,200 missions, striking some 6,200 targets by day and night.[65] Pilot Flight Lieutenant Gary Beck[66] related some of the bomb equipment problems and changes the Squadron faced:

... it soon became evident that some equipment, long accepted by the Service, lacked reliability. The aircraft bombing systems, never tested under operational conditions, also developed faults initially. After several worrying months, the unreliable equipment was withdrawn from service, and modifications, an awareness of the trouble areas, and a general "shaking down" of the equipment produced improving results.

As the number of World War II bombs was limited, the problem of replacement equipment (both bombs and bomb carriers) was considered soon after the Squadron was deployed in the area. This task was undertaken by units in Australia, and a trouble-free transition from single lug vintage bombs to the new standard American M117 750-lb bombs was made in August 1968.

The introduction of American bombs and fuses, incorporating safety features not present with the detonator pistol fusing system previously used, allowed the Squadron to adopt American handling and fusing procedures. As more and more American equipment is absorbed into the Service, both at overseas bases and in Australia, this trend toward accepted American practices probably will continue.[67]

Gary Beck's words have been proved correct, as 25 years later the RAAF's air power relies heavily on United States equipment. RAAF adoption of American procedures has facilitated combined operations ever since. The early teething problems with armament systems had driven home the point that becomes evident to every service involved in conflict—train in peacetime for the way you will fight in wartime.

63 Paul Goodwin, letter to author, of 16 Dec 1993.

64 Air Marshal Evans, interview with author, of 21 Dec 1993.

65 Beck, p.11.

66 AVM G J J Beck AO, O19607. SAN, Nos 34,1 & 2 Sqns, 2FTS, SAN, DefAir; comd 2 Sqn 1977–79; HQSC, NDU Wash, RAAF Acad; comd Fairbairn 1986; DGSPP 1987–89; DGPRM 1990; AC PERS 1991–92; AC AUST 1993–. Regular air force offr of Innisfail, Qld; b Bendigo, Vic, 25 Mar 1942.

67 Beck, pp.19–23.

CHAPTER 25

Bombs on Target

Let's face it, the Canberra was built originally as a high level bomber and not as a tactical aircraft. Many people, perhaps with some justification, felt it would not be suitable for close support of ground troops. But even this fear, we have been able to dispel.

Wing Commander David Evans[1]

A concentrated flying effort was mounted over 22–27 October 1968 with the Squadron flying 10 missions a day, seven visual and three Skyspots. The following month, on 18 November, Wing Commander John Whitehead[2] assumed command. His flight commanders were Squadron Leaders Sam Todhunter[3] and Graham Dyke.[4] Bombing leader was Flight Lieutenant Geoff "Zeke" Cramer,[5] who subsequently handed over to Flight Lieutenant Brian Bolger.[6] Both of these navigators were awarded the DFC for the roles they played in the Vietnam air war.

On 25 November, Pilot Officers John Ross[7] and John Reis[8] (A84-232) were hit by groundfire near their target in IV Corps, and the navigator was wounded by the flying plexi-glass from the damaged nose cone. The citation for John Ross's MID read:

> ... while carrying out an attack against a Viet Cong Base Camp, enemy ground fire hit Pilot Officer Ross's aircraft inflicting blinding eye injuries on his navigator who was unable to continue his duties. With coolness and presence of mind, Pilot Officer Ross jettisoned his remaining bombs and alone carried out all necessary procedures to land his aircraft safely at the nearest airfield to obtain medical attention for his injured crewman.[9]

---

1 RAAF PR Release "Fine RAAF Canberra Effort in Vietnam Proved Critics Wrong", of 2 Jan 1969.

2 AirCdre J A Whitehead DSO, O33710. No 77 Sqn, 74 & 92 Sqns RAF, HQHC/HQOC, RAAF Acad; comd Mel Uni Sqn 1962–64; 6 Sqn, HQOC; comd 2 Sqn 1968–69; RAAFSC; comd ARDU 1972–73; Def Adv KL 1974–75, HQOC; comd Point Cook 1977–78; DefAir DGM 1979–81. Regular air force offr of Sydney; b Rockhampton, Qld, 3 Aug 1933.

3 Sqn Ldr S N Todhunter, O53169. No 36 Sqn, 1BFTS, 1AFTS, 2ATU, 6 Sqn, 1 OCU, 2 Sqn; comd 2ATU 1969–70 Regular air force offr of Margaret River, WA; b Sussex, Eng, 9 Apr 1932.

4 AirCdre G Dyke DFC, O14384. No 1 Sqn, SAN, Ant Flt, 1 & 6 Sqns, 1BFTS, 1 OCU, 2 Sqn; t/comd 1 OCU 1970–71; DepAir; comd 2 Sqn 1974–76; DefAir; Def Rep NZ 1978–80; comd East Sale 1981–83; DefAir. Regular air force offr of Ipswich, Qld; b Sydney 22 Nov 1935.

5 Sqn Ldr G M Cramer DFC, O22886. Nos 38, 30 & 11 Sqns, 82 Wg, 1BFTS, 2 Sqn, 1 OCU; comd 24 Sqn 1970–71, 1974–75; 36 Sqn. Regular air force offr of Sydney; b Sydney 11 Apr 1928.

6 Sqn Ldr E B J Bolger DFC, O23741. Nos 10, 6 & 2 Sqns, CFS, 2 Sqn, 1 OCU, 6 & 1 Sqns, DepAir. Regular air force offr of Sydney; b Goulburn, NSW, 7 Aug 1931.

7 Flt Lt J K Ross, O113665. No 2 Sqn, 5 OTU, 6 & 1 Sqns, DefAir. Regular air force offr of Cunnamulla, Qld; b Harden, NSW, 16 Dec 1947.

8 Flt Lt J F Reis, O223697. Nos 2, 36 & 1 Sqns, 430th TFS USAF, 6 Sqn. Regular air force offr of Sydney; b Newcastle, NSW, 11 Nov 1941.

9 Citation for award of MID, of 16 Dec 1968.

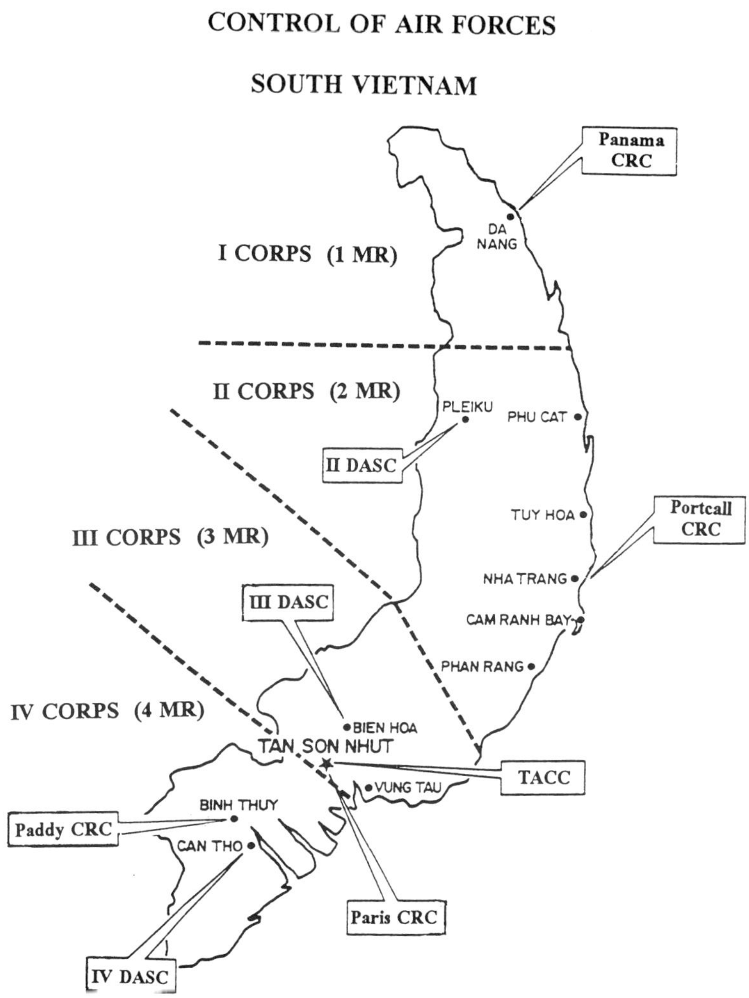

Tactical Air Control Centre (TACC)

The main air operations centre from where all tactical air operations were controlled

Direct Air Support Centre (DASC)

The subordinate operational component for control and direction of close air support operations; the DASC was under the operational control of the TACC and located near the command post of the supported ground unit

Control and Reporting Centre (CRC)

A subordinate air control element of the TACC from where operations in the area of responsibility are controlled

Control of air forces South Vietnam RAAF

A 2 Squadron Canberra unleashing its load D HERSEY

Ross had already dropped five of his bombs on the target, so had to jettison just one remaining bomb and divert to the nearby airfield at Binh Thuy.[10] John Reis soon recovered from his eye injury, and was fit to fly again a week later. John Ross was to feature in another significant strike only four days later, on 29 November. With

10 John Ross, telephone interview with author, of 12 Dec 1993.

navigator Pilot Officer Paul Goodwin (A84-233), he flew in support of a ground operation in the An Xuyen Province, the southernmost region of IV Corps. Although making their attack in marginal weather, with very poor visibility, to the crew it was "just a normal day's work".[11] In a letter of commendation, the FAC on the mission reported:

> Magpie 81 quickly located the FAC and identified the target area. The FAC only had one smoke rocket remaining and Magpie 81 was not able to see the smoke on his first pass so he was forced to make another run and to drop without the benefit of a marking smoke. The professional ability of this crew was evident as they placed their bombs exactly on the desired target ... It is indeed a pleasure to direct airstrikes flown by professionals of this calibre.[12]

Another activity that involved 2 Squadron personnel at Phan Rang was civil action, assistance to the local Vietnamese community. Responsibility for projects fell upon the Squadron chaplain, Squadron Leader Pat McCormick,[13] who was awarded the MBE for his efforts. No 2 Squadron's approach to civil action was to assist the people with projects that they themselves decided they needed most. Such projects took the form of assisting with health, education, agricultural production, social welfare and water supply.

By the beginning of 1969, No 2 Squadron was flying seven visual missions and one Skyspot mission daily. This would have been severely curtailed if an enemy attack on Phan Rang in late January, to coincide with Tet, had met with more success, as a Sydney newspaper reported:

> Viet Cong guerillas attacked the giant American air base at Phan Rang, where Australia's Canberra bomber squadron is based. The Viet Cong opened fire with rocket grenades and small arms, but a US military spokesman said later there were no reports of Australian planes being damaged. The Viet Cong, carrying explosive satchel charges, cut their way through the perimeter wire around the air base ... they were apparently trying to blow up planes at the base. The guerillas were met by rifle-firing US Air Force security police, who called in helicopter gunships to help beat back the Viet Cong. After two and a quarter hours of fighting the Viet Cong withdrew leaving 16 dead. Twelve Americans were wounded, according to the spokesman.
>
> Viet Cong guerillas also sneaked through Saigon's defences to rocket a civil defence post. Military commanders here had warned that the guerillas would launch a series of small attacks on cities when the Paris peace talks began. The raids would be designed to strengthen their negotiators' hand at the conference table.[14]

Delegates from both sides had been meeting in Paris in an attempt to achieve a peaceful solution to the conflict. This proved to be no more than a political forum for charges and counter charges over who was a puppet and represented the people of South Vietnam. Although the American officials believed there had been good

11 Paul Goodwin, letter to author, of 16 Dec 1993.

12 Maj C R Wilson USAF, senior FAC An Xuyen Province, letter to CO 2 Sqn, of 4 Dec 1968.

13 Chaplain (Gp Capt) P J McCormick MBE, O19853. BSqns Wagga, Ric, Butt & Amb, No 2 Sqn, BSqns Lav, Amb & Darwin. Regular air force offr of Brisbane; b Brisbane 25 Jun 1931.

14 "Viet Cong Raid US Air Base", *Daily Telegraph*, Sydney, 27 Jan 1969.

prospects for peace by the end of that year,[15] with at least agreement on the de-escalation of the fighting, the series of meetings produced little.

As the number of missions and bombs dropped mounted, it was decided to publicise the dropping of the Squadron's 40,000th bomb in Vietnam. This event was calculated to occur on 25 April 1969, so a bomb was appropriately inscribed with "Anzac Day Special". About a week before the event, Pilot Officer Pete Salvair[16] and the unit's navigation leader Squadron Leader Bruce Hunt[17] posed with the bomb for publicity shots.[18] Ultimately, it was uncertain who eventually dropped the bomb: neither of these officers flew on Anzac Day.[19] Indeed, it had not been a particularly happy week on the Squadron. One of the Squadron's airfield defence guards, AC Trevor Petith,[20] had been killed in a shooting accident on 21 April.

Mission planning — Flt Lt Al Pearson relaxes with a coffee after a mission, and watches navigator Plt Off John Wilkinson (foreground) and pilot Plt Off Al Blyth prepare for their next strike.
VIA J WILKINSON

15 "Viet Cong Raid US Air Base".

16 Sqn Ldr P T Salvair, O57678. No 2 Sqn, 1 OCU, 6, 1 & 6 Sqns, SAN, 6 Sqn. Regular air force offr of Perth; b Bridgetown, WA, 23 Mar 1947.

17 Sqn Ldr T B Hunt, 167547/O22820. Nos 2 & 1 Sqns, 2 AD, 30, 11, 6, 2 & 1 Sqns, 1 OCU, 82 Wg, HQOC, 2 Sqn, HQOC; comd 23 Sqn 1972–74. Regular air force offr of Coonabarabran, NSW; b Baradine, NSW, 23 Feb 1927.

18 Pete Salvair, telephone interview with author, of 13 Dec 1993.

19 No 2 Sqn A51, of 25 Apr 1969.

20 AC T G Petith, A319010. 3AD, No 2 Sqn. Of Laverton, Vic; b 13 Jul 1946. Killed in accident 21 Apr 1969.

The United States Presidential Citation, earned during the dark days of 1942, was finally awarded to No 2 Squadron on Friday 11 July 1969. It had been awarded by the late President Franklin D Roosevelt, and promulgated in a United States War Department order of 4 January 1943. This was to cause concern in Australia, as the awards to No 2 Squadron and to No 13 Squadron were unique. Unit decorations were not the practice in the Australian or British forces. Accordingly, the Secretary of the Department of Defence, Mr A S Drakeford, wrote to the Prime Minister and Minister for Defence, John Curtin, recommending contact with the Secretary of State for Dominion Affairs in London for advice:

> As these awards do not conform to British practice and as they apparently involve additions to the uniform, it is submitted that the Prime Minister should be advised that in the view of this Department, the whole matter should be referred to the Dominions Office for advice to the King's wishes in this regard.[21]

Although Drakeford had considered that the Presidential streamer from the Squadron standard would be acceptable, he did not favour the wearing of special devices on uniforms. This was, naturally enough, the British view, reflected by the reply of the Dominions Office:

> As regards the question of wearing of a device by all members of the squadrons cited, it has been agreed so far as United Kingdom units are concerned that there should be no acceptance of individual decoration for personnel honoured by USA citations. It is also regarded as undesirable that whole units should wear medal ribbons presented by an Ally. It is hoped that His Majesty's Government in the Commonwealth of Australia will agree that a similar practice should be followed in the case of Australian units. In that event it might be best if refusal were based on inability to return such compliments in kind and on natural reluctance to accept more than can be given when a United States unit is cited in our own orders.[22]

So although His Majesty's Government in the United Kingdom did approve acceptance of the citation, albeit grudgingly, on 20 April 1943,[23] it appears this imperial decree caused the award to be conveniently forgotten. This was indeed a sorry conclusion for such a high foreign acknowledgment, and a shabby treatment of the efforts of the men of No 2 Squadron during 1942. The issue of the citation apparently did not surface for over a decade, until a past CO of the Squadron, Group Captain Garrisson, visited the Squadron in Butterworth in 1959. He instituted inquiries to obtain a copy of the citation, and approval for members to wear the decoration during their period of service with No 2 Squadron.[24]

At last, the citation was presented to the Squadron by the Commanding General, United States 7th Air Force, General George Brown. Armament fitter Warrant Officer

21 AA CRS A5954 228/25, Minister for Defence letter, of 5 Mar 1943.

22 AA CRS A5954 228/25, Secretary for Dominion Affairs cablegram, of 20 Apr 1943.

23 RAAF PR S4324/69, Ministerial Press Release, of 10 Jul 1969.

24 2 Sqn letter 1228/1/P3 (17A) from HQ Butterworth to Dept of Air Melbourne, of 18 May 1959.

Jimmy Lee,[25] and ex-medical orderly Warrant Officer Clem Bolger[26] were flown from Australia to Phan Rang for the ceremony. These men, as serving members from the period covered by the award of 1942, were entitled to wear the decoration in perpetuity of their service. The only other serving members with this entitlement were AVM Frank Headlam and Group Captain Norm Lamb, both of whom were unable to attend the ceremony.[27] Others posted to No 2 Squadron were entitled only to wear the decoration for the period they were on Squadron strength. Pilot Officer Al Blyth[28] carried the RAAF ensign for the attachment of the citation streamer, showing "Banda Sea 1942".

Presentation of the US Presidential Citation for the 1942 Banda Sea campaign — USAF General Brown, Commander 7th Air Force, presents the pennant to standard bearer Pilot Officer Al Blyth, 11 July 1969 RAAF

The Canberras were the only visual level bombers in the country, and in addition to delivering great accuracy,[29] the Canberras could be called in to bomb under low

25 WOff E J Lee, 13319/A31522. Nos 13 & 2 Sqns, 1 OTU, 6464 & 6453 Serv Echs, 2CR, 3CR, 7SD, 1 CAMD. Regular airman of Melbourne; b Colac, Vic, 22 Oct 1921.

26 WOff C P Bolger, 36634/A2235. No 2 Sqn, 6SFTS, HQ Wlm, HQ Amb, BSqns Esl, Amb, Butt, Ric & Amb. Regular airman of Sydney; b Sydney 27 Apr 1922.

27 "Citation for 2 Squadron", *RAAF News*, Vol 11, No 7, Aug 1969, p.1.

28 Gp Capt A T Blyth, O16941. No 2 Sqn, 1BFTS, 6 Sqn, CFS, DefAir; comd 6 Sqn 1986–87; AJWE, DefAir, ADFWC. Regular air force offr of Hervey Bay, Qld; b Monto, Qld, 29 Jan 1945.

29 In the twelve months from Nov 1968 to Nov 1969, No 2 Sqn's weapon delivery accuracy, assessed by photography, was 50% of bombs within 40 metres, 90% within 100 metres.

Preparing for war — No 30 Canberra Conversion Course at Amberley in 1969, training to join No 2 Squadron:
Rear (from left): Plt Off Bob Cuttriss, Sqn Ldr Frank Lonie, Sqn Ldr Blue Sweeney, Plt Off Adrian Slootjes. Front: Plt Off John Bennett, Flg Off Mike Herbert, Plt Off Greg Weekes, Plt Off Des Hyde RAAF

cloud ceilings. This occurred especially during the south-west monsoon season, between May and October, when much of the country was under cloud. Other aircraft were prevented from striking targets under these conditions, as higher cloud bases were needed for their dive attacks. Furthermore, the Canberras could deliver "sticks" of bombs, ideal for targets along the straight canals in the Mekong Delta region in IV Corps, or could loiter to make individual attacks with single bombs against pinpoint targets. This mode of attack was often used in close air support situations, where friendly forces were in close proximity to the enemy, and a "short round" would have had tragic results.

Typical of these close support missions was one on 21 July, in the Mekong Delta. Pilot Officer Dick Allchin[30] and Flying Officer John Bushell[31] (A84-236), flying as Magpie 51, headed south-west out of Phan Rang:

> "Magpie 51, this is Paris," came the voice of Saigon control. "I have you under radar. Your FAC will be Richard 62. Your frequency will be 324.9."
>
> It seemed only a few minutes before "Paris" gave way to "Paddy", the voice of Binh Tuy control, deep in the Delta. Binh Tuy guided Allchin and Bushell to the Delta area, where they contacted the FAC.
>
> "Magpie 51, this is Richard 62. Do you see that river mouth down there that curves to the north?"
>
> "Roger," answered Allchin, surveying the muddy, brown swath through the rice paddies below.
>
> "Okay, there's another river just the other side of it."
>
> "Does it look like a toilet plunger?" asked Allchin, determined to make sure he had the right target.
>
> "Roger that," answered the FAC. "Drop your bombs on the adjacent treeline."
>
> As Allchin was talking with the FAC, Bushell crawled by him and squeezed through a small opening, into the nose of the aircraft. There he lay on a green pad, with just enough room to move his hands and start adjusting the bombsight.
>
> "Go north to south," the FAC told Allchin, "and release your bombs at 50-metre intervals."
>
> "I wonder if we could come in south to north," asked Allchin. "If we came in the other way, the weather would obscure the target."

30 Flt Lt R L Allchin, O317534. Nos 2 & 38 Sqns, SAN, 36 & 34 Sqns. Regular air force offr of Bahrain; b Ewell, Surrey, Eng, 16 Feb 1942.

31 Gp Capt J A Bushell, O221710. Nos 1, 6, 2 Sqns, 1 OCU, 2 Sqn, SAN, 6 Sqn, ARDU, DefAir; comd SAN 1983–85; DefAir; AA Jakarta 1990–91. Regular air force offr of Binnaburra, NSW; b Lismore, NSW, 9 Feb 1945.

"Hold on. There was a troop insertion near the area you would be flying over if you came in from the south. Let me check and see if they're still there."

"Roger."

"Magpie 51, this is Richard 62. You are cleared to come in from south to north."

Allchin brought his ship in for the bomb drop. The Mk.20 is a large airplane. It neither shakes nor shudders when it drops 4,500 pounds of bombs. Only a sudden "Grupp-grupp-grupp-grupp-grupp-grupp" told Allchin that his bombs were away.

Allchin swung the craft around, waiting for the FAC's damage report. The land below looked harmless, peaceful. Miles of rice paddies glowed under the morning sun. The treeline where the bombs had fallen was quiet again.

"Magpie 51, this is Richard 62. Seven bunker complexes were destroyed. Enjoyed working with you."[32]

Another sortie, a week later on 27 July, was flown in III Corps by Flying Officer Bob Sivyer[33] and Pilot Officer Keith "Piggy" Padgett[34] (A84-235), as Magpie 21. Troops from the United States 25th Infantry Division had come under fire while sweeping a village, and called in air support. The FAC (using callsign "Issue 26") marked the target, and in poor visibility and low cloud which made target acquisition difficult, Magpie 21 dropped the first bomb of the stick on the FAC's smoke, with the rest of the bombs rippling through the enemy's position. The FAC and the ground commander were highly complimentary of the precision bombing, adding "we look forward to working with the Magpies because of such pinpoint accuracy".[35]

Three days later, on 30 July, Pilot Officer John Kennedy[36] and Flying Officer Dave "Stumpy" Palmer[37] (A84-231) were hit by groundfire in the starboard engine, but the aircraft was quickly repaired and flew the following day. Kennedy suffered more damage the following week when on 10 August, flying with Flight Lieutenant Nev Duus,[38] A84-236 was hit while bombing a bunker complex in IV Corps. The cloud base at 1,000 feet would normally preclude level bombing attacks, but as Nev Duus later related:

> We were forbidden to bomb at less than 1000 feet above ground level, but 2 Squadron had a reputation for not bringing bombs home. As a result, we decided to bomb from 800 feet. We

---

32 "Flyboys From Down Under", *Typhoon*, Vol III, No 9, Sep 1969, US Forces, APO San Francisco, pp.7–9. This article, for security reasons, altered radio callsigns. The article used the callsign "Magpie 35" but Allchin and Bushell actually flew that day as Magpie 51. The erroneous callsign "Magpie 35" had never been used but, probably as a result of the error in the journal, was included on the Vietnam Memorial in Canberra. The FAC was most probably "David 72".

33 Flt Lt R T Sivyer, O110502. No 2 Sqn, 1 OCU, 1 & 6 Sqns, HQSC. Regular air force offr of Rockhampton, Qld; b Rockhampton 8 Jul 1944.

34 Wg Cdr K W G Padgett, O112430. Nos 2, 6, 1, 37, 6 & 1 Sqns, DefAir, HQOC; comd 23 Sqn 1986–87; DefAir, Air HQ, 82 Wg. Regular air force officer of Brisbane; b Brisbane 26 May 1947.

35 Letter from 25th Inf Div to III DASC, "Magpie 21, Mission 5252, 27 Jul 1969", of 20 Aug 1969.

36 Wg Cdr J M Kennedy, O317782. Nos 2 & 1 Sqns, 1FTS, HQSC, DefAir; comd 25 Sqn 1993–. Regular air force offr of Geelong, Vic; b Geelong 21 Jul 1948.

37 Wg Cdr D K Palmer, O44431. Nos 36, 2, 37 & 6 Sqns, AJWE, DefAir, HQ Amb, HQOC. Regular air force offr of Adelaide; b Adelaide 23 Dec 1944.

38 Sqn Ldr N L Duus, O17899. No 6 Sqn, RAAF Acad, 2 Sqn, 1 OCU, TSF, 34 & 6 Sqns, HQOC, 23 Sqn. Regular air force offr of Brisbane; b Brisbane 14 Dec 1939.

dropped one bomb, and through the perspex window below the bomb sight I watched it fall. I saw it explode, then almost immediately the window disintegrated. We thought we'd been hit by groundfire.[39]

Unbeknown to the crew, in spite of pulling up to a higher altitude after bomb release, the aircraft had been hit by a fragment from one of their bombs, which blew out the bomb aimer's window.[40] After carrying out a slow-speed handling check at a higher level and finding no further apparent damage, the crew elected to continue their attack and dropped the rest of their bombs. On the return leg to base, Duus inspected the damage in the nose:

I had not been wearing flying gloves and received small cuts in my hands. I then glanced up and around for the first time and saw a large piece of shrapnel from our own bomb, which had penetrated through the nose cone and was now protruding through the upper skin of the aircraft. My Dalton nav computer, which had been lying beside me in the nose had been smashed by the shrapnel as it went by.[41]

Crews were more conscious of the risks when operating below bombing safety heights after that. Luck continued to run for the Magpies, when on 26 August Flight Lieutenant Merv Lewis[42] and Flying Officer Bob Molony[43] (A84-247) were hit by groundfire in the starboard fuselage, with no casualties or serious damage.

Bomb strike, the white smoke from the Forward Air Controller's mark is visible RAAF

In spite of the Squadron's good results, sometimes the bomb damage assessment (BDA) suggested the futility in attacking apparently uninhabited jungle. Like the aircrews before them bombing CT targets in the Malayan jungle, a poor BDA made crews question not whether they should be there in the war, but the employment of tactical air power against insignificant targets. Indeed, this very point had been queried by B-25 crews as they operated in the "backwaters" in 1945. However, later feedback during August 1969 showed that an attack of one of these "pointless" targets had borne fruit. On 31 July, Pilot Officer Pete

39 Nev Duus, telephone interview with author, of 28 Nov 1993.

40 John Kennedy, telephone interview with author, of 25 Jan 1994.

41 Nev Duus, telephone interview with author, of 28 Nov 1993.

42 Flt Lt M J Lewis, O216390. SAN, Nos 6, 2, 1 & 2 Sqns, 1 OCU, 6 Sqn, 82 Wg, 9 Sqn, BSqn Amb. Regular air force offr of Bathurst, NSW, b Bathurst 23 Dec 1937.

43 Sqn Ldr R J Molony, O112878. Nos 36, 2 & 6 Sqns, SAN, 6 Sqn, HQ Amb, DefAir, AJWE, 23 Sqn. Regular air force offr of Brisbane, b Brisbane 29 Dec 1946.

Nuske[44] and Flying Officer Lloyd Brown[45] (A84-244) had bombed a VC base camp in the Mekong Delta. Local Viet Cong had been attending a cadre meeting at the time, and 10 of them, including one of the VIPs in the VC infrastructure, had been killed as a result of the Canberra strike.[46] The commander of the 35th TFW, Colonel Frank L Gailer, acknowledged the success of his Australian squadron: " ... their battle damage assessment is the best of any operational unit in Southeast Asia".[47]

Operations Flight Commander, Sqn Ldr Ivan Grove, congratulates Flt Lt Alex Alexander's team of armourers on the 50,000th bomb to be dropped by No 2 Squadron in Vietnam, 28 November 1969.
VIA N ALEXANDER

44 Flt Lt P R Nuske, O224148. Nos 2, 76 & 77 Sqns, 2 OCU. Regular air force offr of Newcastle, NSW; b Toowoomba, Qld, 17 Apr 1943.

45 Wg Cdr L D Brown AM, O110501. Nos 36, 2, 36 Sqns, 509 BW USAF, 36 Sqn, HQOC, DefAir; comd BSqn Wagga 1985–86. Regular air force offr of Gladstone, Qld; b Gladstone 2 Jan 1945.

46 Agent report, No 2 Sqn A50, of 22 Aug 1969.

47 "Flyboys From Down Under", p.7.

One of the more unusual missions carried out by aircrews in Vietnam was visual night bombing, known as "Night Owls". Normally, of course, visual bombing was done by day, but at night it was possible for the target area to be illuminated by parachute flares, fired by an aircraft "flare ship" or by artillery. The FAC would mark the target, as usual, by a smoke rocket. On the night of 28 October, Pilot Officers Pete Salvair and Pete Growder[48] (A84-240) were called in to make a strike to the east of Bien Hoa, in III Corps. With the flare ship, a C-47 Dakota, operating at 4,000 feet, and the FAC at 2,000 feet, the Canberra made its level run between them, at 3,000 feet. The crew decided to leave the Canberra's navigation lights on as an added safety factor, but when enemy tracer rounds started to arc below the aircraft, they decided this might not have been such a good idea.[49] This was one of the few "Night Owl" missions to be flown by the Squadron during the four years in Vietnam.

By the time the new CO, Wing Commander Jack Boast, assumed command on 10 November, the Squadron was being tasked with visual bombing missions, and a very occasional Skyspot.[50] Throughout 1969, No 2 Squadron had successfully flown 2,862 missions, beginning the year with a daily routine of six day visual bombing missions and two night Skyspots, changing in late November to all-day missions assigned to FACs.[51] The Squadron's 50,000th bomb had been dropped on 28 November.[52] At this stage, Squadron Leader Frank Lonie took over as navigation leader from Bruce Hunt, and the flight commanders were Squadron Leaders Ivan Grove[53] and Brian "Blue" Sweeney.[54]

However, despite the operational success of the year, 1969 was to end on a sad note. On 3 December, two of the Squadron's airmen, Corporal Brian Fitzpatrick[55] and LAC Ivan Wooley,[56] were involved in a motor vehicle accident on the road to Cam Ranh Bay, north of Phan Rang. Fitzpatrick, a motor transport driver, was killed instantly. Wooley, an airfield defence guard, died as a result of his head injuries on 11 December, at the USAF 12th Hospital, Cam Ranh Bay.

---

48 Gp Capt P W Growder, O117122. Nos 2, 37 & 1 Sqns, 430th TFS USAF, 6 Sqn, 82 Wg; comd RAAF Coll 1990–91; DefAir DPO. Regular air force offr of Brisbane; b Brisbane 7 May 1946.

49 Pete Salvair, telephone interview with author, of 13 Dec 1993.

50 From 4 Oct 1969 only VBMs were fragged. The author's flying logbook shows that over the period Nov 1969 to Nov 1970, of 260 missions flown, only six were CSS.

51 R Howe, *2 Squadron Vietnam—1969*, Squadron annual booklet, Phan Rang, p.4.

52 Howe, p.17. The 50,000th bomb was dropped by Kennedy/Curr (A84-240). Most of the bombs dropped during 1969 had been the US 750 lb M117 (15,900 dropped), with the Australian Mk10 1,000 lb bombs introduced at the end of the year (1,150 dropped).

53 Sqn Ldr I L Grove DFC AFC, O34222. 30 TU, SAN, 1AFTS; comd Ant Flt 1957–58; 6 & 2 Sqns, 1 OCU, 2 Sqn, DefAir. Regular air force offr of Helidon, Qld; b Brisbane 11 Oct 1930.

54 Sqn Ldr B T Sweeney, O32566. Nos 6 & 1 Sqns, DepAir, ATU, RAAF Coll, OTS, 6 & 30 Sqns; comd 30 Sqn 1967–69; 2 Sqn, DepAir. Regular air force offr of Ararat, Vic; b Ararat 17 Oct 1930.

55 Cpl B F Fitzpatrick, A34167. 2ACS, 1SD, No 2 Sqn. Regular airman of Melbourne; b Melbourne 22 Apr 1932. Killed in ground accident 3 Dec 1969.

56 LAC I A Wooley, A118435. 3AD, BSqn Fbn, No 2 Sqn. Regular airman of Brisbane; b Murwillumbah, NSW, 18 Apr 1948. Died of injuries from ground accident 11 Dec 1969.

RECOMMENDATION FOR HONOURS AND AWARDS

O19853 CHAPLAIN 3rd CLASS PATRICK JOHN McCORMICK

Chaplain McCormick arrived at No 2 Squadron PHAN RANG on 17th October 1967.

Since his arrival Chaplain McCORMICK has worked tirelessly in his capacity as Squadron Chaplain and as Civic Affairs Officer. He has made a particular point of forming a close personal association with all ranks and by his willingness to give assistance, wise counsel and comfort he has won the confidence and respect of all No 2 Squadron. His organisation of Squadron sporting activities and his participation in these and all facets affecting the welfare of squadron personnel has contributed very significantly to the high morale that pertains in this squadron.

In addition to his work within the squadron, Chaplain McCORMICK, as Civic Affairs Officer, has travelled at least once each day to the local towns and hamlets to supervise, and work on civic aid projects undertaken by the squadron. These include the construction of two dispensaries in the hamlets of AP CONG THANH and AP THANH-Y, the building of a playground and laundry at the TAN TAI Orphanage, the distribution of commodities such as baby foods, fruit and other foodstuffs. Chaplain McCORMICK has encouraged squadron members to take part in these activities and has always been the first, and most regular to engage in the hard manual work involved in the construction work undertaken.

It is to Chaplain McCORMICK's credit, and typical of his devotion to the concept of Civil Aid that he has continued his daily visits to the local towns and hamlets during and after the enemy offensive of January/February when security in the local area was extremely precarious and only essential military traffic was permitted to travel outside the Base perimeter.

By tireless devotion to his duties Chaplain McCORMICK has contributed greatly to the high morale of this Squadron. His continued involvement in providing personal and material aid to needy segments of the local population has won the admiration of local authorities and has reflected great credit on the RAAF and AUSTRALIA.

This recommendation was made by Wing Commander S D Evans on 29 May 1968, and resulted in the award of the MBE to Chaplain McCormick.

CHAPTER 26

THE MAGPIES

We gotta get out of this place
If it's the last thing we ever do ... [1]

On the night of 11 February 1970, Corporal Eddie Power,[2] an airfield defence guard (ADG) assigned to No 2 Squadron's Airfield Defence Flight, led his eight-man clearing patrol outside the perimeter of the base to a pre-planned ambush position.[3] In the early hours of the morning, an enemy reconnaissance squad of about six troops attempted to infiltrate the perimeter wire. Two Viet Cong were killed, and others were seen retreating between the Americans, inside the base, and the Australian ADGs, outside the wire. With American fire shooting past the enemy's position, and over the heads of the RAAF squad, Corporal Power moved his section to avoid the friendly fire and cut off any possible route of escape.

Moving his men about a kilometre through the darkness, Power set up an ambush beside two bridges which crossed the perimeter canal. Power recalled:

> We were in the ambush position for about a minute or so, when the M60 gunner sighted enemy movement on the other side of the canal. I obtained authorisation to fire from the US Air Police control centre on the base. We then opened fire on the enemy, and saw at least two in the waist-high grass, who threw a couple of grenades in our direction. After a three minute firefight, we waited for twenty minutes and then completed a sweep of the area. Nothing was found, so we went back to our position near the bridge. At first light, we withdrew to the confines of the base. Later that morning, a clearing patrol went out on another sweep, and picked up an NVA major who had been wounded. He was handed over to US Intelligence for interrogation. Also, maps, a medical kit, an AK-47, a pistol and a cache of crudely-made grenades were found. From the Intelligence interrogation, the prisoner was the only NVA regular and he was being escorted by local VC for reconnaissance, possibly with the view for a Tet offensive, which then never eventuated.[4]

This enemy patrol was most probably reconnoitring entry and escape routes for an attack on the aircraft lines. From their entry on the south-west side of the base, any attacking force would only have had a kilometre to cross to the closest aircraft on the lines—2 Squadron's Canberras. For the success of thwarting this enemy mission, Power was awarded the Military Medal, the first to go to a member of the RAAF since World War II.

---

1 Chorus from a contemporary rock song, remembered by those who were there.

2 WOff N E Power MM, A111932. Prev Army. 1RTU, No 2 Sqn, BSqn Amb, 77 & 482 Sqns, 3AD, HQ Amb, 23 Sqn. Regular airman of Brisbane; b Brisbane 26 Apr 1939.

3 "Award", *RAAF News*, Vol 13, No 2, March 1971, p.4.

4 Eddie Power, interview with author, of 28 Jan 1994.

Visual low-level bombing continued as the prime activity for the Squadron and, accordingly, the Canberras were more susceptible to battle damage from their own bombs. On 11 March, Pilot Officer Barry Carpenter[5] and the bombing leader, Flight Lieutenant Bob Howe[6] (A84-238), received bomb fragment damage to their aircraft during an attack that morning on a VC base camp in IV Corps. The fragment punctured the port integral fuel tank, and lodged in the upper skin of the mainplane. The nitrogen purging system prevented what may have been a devastating fire. Five days later, Squadron Leader Ivan Grove and Pilot Officer Ross Hardcastle[7] (A84-240) received a bomb fragment in the starboard wheel well of their aircraft. The flipper door, tyre and hydraulic lines were damaged.

The 60,000th bomb from No 2 Squadron, with the crew of "Magpie 21", navigator Flt Lt RonAitken (left) and pilot Flg Off Mike Herbert, 29 June 1970 VIA N ALEXANDER

It is testament to the tireless efforts of the groundcrews that both these damaged aircraft remained off-line for little more than a week, and no sorties were lost because of the reduced aircraft availability. As navigator Flight Lieutenant Bob Howe recalled, the professionalism of the groundcrew played a large part in 2 Squadron's success:

5 Sqn Ldr B J Carpenter, O45902. No 2 Sqn, 1FTS, RepSingAF, CFS, 2 Sqn, OTS, DefAir, HQ Ric, 34 Sqn. Regular air force offr of Adelaide; b Adelaide 21 Nov 1948.

6 Wg Cdr R W Howe, O218328. No 10 Sqn, RAAF Acad, 1 OCU, 2 Sqn, 82 Wg, RAAF Wash, DefAir. Regular air force offr of Canberra; b Canberra 10 Jan 1940.

7 Wg Cdr R D Hardcastle, O117506. Nos 2, 37, 2, 1 & 6 Sqns, SAN, DefAir, USAF Space Div, DefAir. Regular air force offr of Brisbane; b Brisbane 22 Aug 1945.

> The enduring impression I have about the 2 Squadron maintenance crews is their fantastic ability to fine tune our aircraft systems so that we virtually flew around the clock carrying out our assigned eight missions a day without fail. The Canberra was an old bird, and I recall one example where makers of our doppler navigation radars were amazed at the serviceability that 2 Squadron was getting from this obsolete equipment.[8]

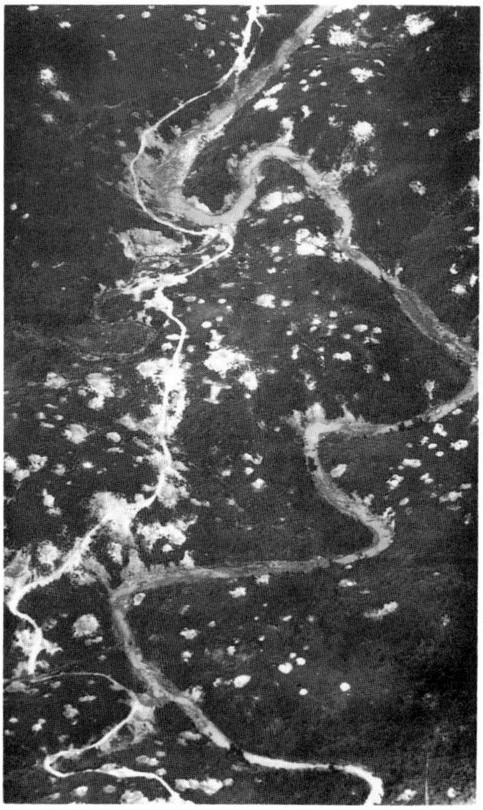

The bomb-scarred trail over Tiger Mountain, in the A Shau Valley of 1 Military Region, the supply route to the south from North Vietnam, May 1970 R HOWE

No 2 Squadron moved into a new field of operations in its air campaign in April 1970, with the interdiction of roads in I Corps, the northernmost region of South Vietnam, immediately south of the Demilitarised Zone (DMZ). The main area of operation during this phase was bombing the road through the A Shau Valley, the principal supply route from the Ho Chi Minh Trail in eastern Laos to VC base camps in northern I Corps, the area surrounding the huge United States base at Da Nang. This corridor to the south provided a conduit of supplies, which was to prove the key to the failure of military strategy in the war.[9] The interdiction of this route proved highly successful while it lasted, and the speed with which the VC tried to repair road damage showed that the Canberra strikes were seriously denying the enemy his supplies.

On one of the first sorties near Tiger Mountain, along the A Shau route, Flight Lieutenant Brian Hammond[10] and Pilot Officer Bob Cuttriss[11] (A84-228) achieved a successful 15 metre cut on the road. Enemy trucks that evening found the road impassable, and 17 were destroyed by USAF AC-119 gunships. The success of cutting these routes by the Canberras saw three missions a day being flown to the A Shau Valley, and soon 7th Air Force requested an increased rate of flying from the Canberras. From the middle of June

8 Bob Howe, letter to author, of 8 Jan 1994.

9 Edwards, p.226.

10 Wg Cdr B G Hammond, O55985. 2 & 1 ATU, Tpt Flt Vn, No 38 Sqn, 1 & 2 ATU, 2 Sqn, 1 OCU, 34 Sqn, DefAir; comd SU Tengah 1981–82. Regular air force offr of Perth; b Perth 2 Sep 1937.

11 Wg Cdr R J Cuttriss, O318065. Nos 2 & 37 Sqns, SAN, 36 Sqn, 509 BW USAF, 36 Sqn, DefAir; comd BSqn Lav 1985–87. Regular air force offr of Port Arlington, Vic; b Geelong, Vic, 21 Jun 1949.

1970, No 2 Squadron was tasked with nine sorties a day, and on the 29th, the Squadron dropped its 60,000th bomb.[12]

Pilot Officers Mick Birks[13] and Al Curr[14] (A84-241), returning from a mission on 11 July, were unable to extend their Canberra's nose wheel. After a well-executed smooth landing on the foam-covered runway with the nosewheel still retracted, Wing Commander Jack Boast had to physically prevent the USAF from bulldozing the damaged Canberra off the edge of the runway. Rightfully, he felt a few diverted aircraft was a small price to pay for a Canberra, especially as he didn't own many! The bright side to this crashed landing was for "Big Al" Curr who, with Pilot Officer Rick O'Ferrall,[15] then returned the aircraft back to Australia for repairs:

> While the skin damage to A84-241 was relatively minor, the 2 Squadron "techos" could only effect a temporary "patch job" at Phan Rang. Rick O'Ferrall and I were tasked to ferry the aircraft back to Amberley for permanent repairs. However, due to the patch, we were limited to 25,000 feet (the 2 Sqn engineers assessed that the pressure differential at higher altitudes would cause the repairs to rupture). On 25 August, we set off in a range-limited Canberra for Butterworth. The next day's flight to Darwin would include a refuelling stop at Djuanda, an Indonesian naval airfield at Surabaya, in Java. While we were on the ground, three decrepit and squeaky Ilyushin Il-28 Beagle bombers taxied out, quite deliberately past us, to the end of the runway. Only two were able to get airborne, and then only for a few circuits, leaving everything down. This was their "show of strength"![16]

The pilot, Rick O'Ferrall, recalled how odd it had been that no aircraft were airborne when they arrived, and as they departed the two old bombers had been ordered to land. The third aircraft had continued to taxy throughout the Canberra's stay on the ground, disappearing in different directions, only to reappear to give the impression of much activity![17] The remainder of the trip back to Amberley was uneventful, as Al Curr remembered:

> On the three sectors to Darwin, we had not detected any leak in the cabin pressurisation. Thus, keen to get to Amberley, we decided to test the integrity of the repairs and climbed to 40,000 feet. The techos had done a great job and we were able to fly direct to Amberley, with no requirement for a refuelling stop at Mt Isa.[18]

The term "Corps" for designating the military division of South Vietnam was discontinued on 2 August 1970. From then, the term "Military Region" was used; however, the boundaries remained the same. Thus, IV Corps, for example, became 4MR. This was in keeping with the "Vietnamisation" of the war, whereby United

12 Magpie 21, Herbert/Aitken A84-241.

13 Gp Capt M B Birks AM, O224235. No 2 Sqn, 1BFTS, RepSingAF, 34 Sqn, DefAir, SAN, HQOC; comd 34 Sqn 1987–90; HQTC. Regular air force offr of Sydney; b Thursday Is 20 Apr 1948. Killed in flying accident 30 Nov 1991.

14 Wg Cdr A J Curr, O117504. Nos 2, 37, 2, 23 & 1 Sqns, SAN, 1 Sqn, SAN; comd BSqn Esl 1988–89; APSC, HQADF. Regular air force offr of Ballandean, Qld; b Stanthorpe, Qld, 10 Feb 1949.

15 Sqn Ldr R E O'Ferrall, O211344. No 2 Sqn, 2 FTS, CFS, 391st TFS USAF, 1, 6 & 1 Sqns, DefAir, 1FTS. Regular air force offr of Sydney; b Sydney 24 Jul 1946.

16 Al Curr, interview with author, of 10 Dec 1993.

17 Rick O'Ferrall, telephone interview with author, of 17 Dec 1993.

18 Al Curr, interview with author, of 10 Dec 1993.

States policy now concentrated on training the Vietnamese to take over the running of operations at all levels.

The greatest impact that Vietnamisation had had on No 2 Squadron was in operating with VNAF FACs. This had begun earlier in the year in the Mekong Delta, with USAF FACs instructing the locals on directing air strikes. The Vietnamese naturally had a limited English vocabulary, so the Australian crews attempted using understandable pidgin-type language. Unfortunately, as soon as the FAC heard the word "smoke", he would fire one of his few rockets at the target, irrespective of the position of the attacking Canberra. However, the Vietnamese took little time adapting to the Canberras' unique level run-ins, which not only required the timely laying of the smoke on the target, but also a time-consuming target brief. Because of the Canberra's fuel reserves, the Australian crews were able to persist with the initial language problems, and ensure that a thorough target brief was understood. The shorter range USAF fighters were unable to hold for these lengthy periods, which made their integration with VNAF FACs more difficult.

A rapport began between the VNAF FACs and the Australian crews by a system of "exchanges" which was implemented from June. This enabled Australian crews to visit the bases of the FACs for several days, and for the Vietnamese to reciprocate by visiting Phan Rang. Not only were RAAF crew members able to fly FAC sorties, and occasionally attack missions in Vietnamese A-37 Dragonfly aircraft, the VNAF FACs were able to fly in the Canberras. There is no doubt that these exchanges improved relations and understanding between the crews of both air forces.

The Squadron's 10,000th mission in Vietnam was flown by Plt Offs Al Curr (left) and Mick Birks, on 3 October 1970 in A84-236. VIA A CURR

The speed with which the VNAF FACs were able to adapt to the Canberra's unusual mode of operation was demonstrated on one support mission, when the FAC had only one "willie-pete" marker rocket remaining. The FAC held the Canberra, while he used his last rocket to mark the target for a formation of VNAF A-1 Skyraiders. After some conversation in Vietnamese, the FAC then called in the Canberra. As the Canberra approached the target, one of the A-1s dive bombed with his last Mark 82 500 lb bomb. The FAC directed the Canberra's aim point in relation to this bomb impact: an impressive piece of ingenuity.

With the increased effort of flying nine sorties a day, No 2 Squadron was sometimes even called to fly a 10th. On the afternoon of 3 September, Pilot Officers Adrian Slootjes[19] and John Bennett[20] (A84-234) had just returned from a mission in

19 Wg Cdr A Slootjes, O317895. No 2 Sqn, 1BFTS, 2 Sqn, HQSC, 2 Sqn; t/comd 2 Sqn 1982; SRO Syd, HQSC, HQ 1 Div, OTS, HQ IADS, Air HQ. Regular air force offr of Benalla, Vic, b Rotterdam, Holland, 24 Jul 1944.

20 Sqn Ldr J W Bennett, O224236. No 2 Sqn, 1 OCU, 2, 1 & 36 Sqns, 15 Sqn & 237 OCU RAF, DefAir, HQADF. Regular air force offr of Sydney; b Sydney 27 Jan 1949.

2MR. During their debrief, a request came from the Tactical Air Coordinating Centre to launch an immediate mission to 3MR in support of troops in contact (TIC) with the enemy. Departing as soon as an aircraft was available (A84-237), they arrived at the target at nightfall, and received the brief from the FAC for attacking their target under "Night Owl" illumination flares, fired by nearby artillery. Although friendly forces were only 200 metres from the enemy, and the extinguishing flares could easily be confused for the FAC's marking smoke, the attack was successful. Having "friendlies" so close to the target was not the only anxious aspect of the mission; having the flares exploding at the aircraft's bombing altitude was somewhat disconcerting.

The following week, 12 September, Pilot Officer Barry Carpenter and Flying Officer Tom Morrissy[21] (A84-236) were also instrumental in a TIC action. They were called in by a VNAF FAC to support ARVN troops in 4MR. As the friendly troops pulled back to 300 metres, the Canberra accurately dropped its bombs killing the enemy troops in a treeline.

Handover of No 2 Squadron by CO Wg Cdr Jack Boast to Wg Cdr John Downing, November 1970
VIA G BICKLE

Wing Commander John Downing[22] took command of the Squadron on 1 November and immediately was confronted with bad news. On a Skyspot mission in 1MR on the night of 3 November, a Canberra flown by Flying Officer Mike Herbert[23] and Pilot Officer Bob Carver[24] (A84-231) was lost:

21 Wg Cdr T M Morrissy, O317411. Nos 10, 2, 11 & 1 Sqns, SAN, 6 & 1 Sqns; comd SU Adl 1984–85, BSqn Lav 1988–89; 23 Sqn. Regular air force offr of Melbourne; b Melbourne 13 Jan 1943.

22 Wg Cdr F J L Downing DFC, O34496. Nos 1 & 6 Sqns, RAAF Acad, 1 OCU, HQSC, 1 OCU, DepAir; comd 1 OCU 1970, 2 Sqn 1970–71; DepAir, HQSC, DefAir. Regular air force offr of Albany, WA; b Perth 19 Apr 1934.

23 Flg Off M P J Herbert, O44310. No 2 Sqn. Regular air force offr of Adelaide; b Freeling, SA, 6 Sep 1946. Missing in action 3 Nov 1970.

24 Plt Off R C Carver, O119223. No 2 Sqn. Regular air force offr of Ipswich, Qld; b Toowoomba, Qld, 2 Jul 1946. Missing in action 3 Nov 1970.

> The aircraft was reported missing in the Da Nang area, and a search began at dawn today. It is the first Canberra jet bomber of No 2 Squadron reported missing or lost in Vietnam since the Squadron began operations in that country in 1967.[25]

The Canberra, Magpie 91, had last reported flying at 22,000 feet, at 8.22 p.m., having dropped its bombs under the direction of the United States Marine "Milky" Skyspot controller.[26] The Canberra had acknowledged the heading given by "Milky" that would take the aircraft on a direct track back to Phan Rang. That was the last heard from Magpie 91. After the fuel endurance of the Canberra had elapsed, and all inquiries to other agencies and bases failed to reveal any knowledge of the aircraft, search and rescue procedures were initiated.

Intensive searches by all the Squadron's Canberras and other Allied aircraft in the area, under the coordination of the United States 3rd Air Rescue and Recovery Group, were conducted for days after the loss. Over three days No 2 Squadron flew 38 sorties, amounting to 113 hours, all of which failed to reveal any trace of the missing A84-231. Mike Herbert and Bob Carver are remembered by plaques for those Missing in Action at the Vietnam Memorial, on Anzac Parade, in Canberra.

Inspection of No 2 Squadron by the Commander of the RAAF in Vietnam (COMRAAFV), Air Commodore "Spud" Spurgeon, accompanied by the Squadron's senior engineering officer, Squadron Leader Graham Bickle, Australia Day 1971 VIA G BICKLE

If this loss was attributable to enemy action, then perhaps it could have been avoided if the Canberras had been equipped with electronic warfare equipment. Such systems alert crews to enemy radar activity for directing surface-to-air missiles (SAM) and anti-aircraft artillery (AAA). The Canberra in the late 1960s was always "about to be replaced". Accordingly, the crews were to persist for over another decade with inadequate Martin Baker Mark 1 ejection seats, and no electronic warfare equipment.

The crews were, however, subjected to another form of electronic warfare, known as MIJI—Meaconing, Intrusion, Jamming and Interference. Meaconing occurred when the enemy attempted to interfere with the signals transmitted by navigation beacons, such as the ground-based non-directional beacons and TACAN (Tactical Air Navigation) stations. Corrupting the beacon signals would cause Allied aircraft to have

25 RAAF PR S4460/70, of 4 Nov 1970.

26 No 2 Sqn A51, of 3 Nov 1970.

wrong navigational information, which might tempt them to stray into more hostile environments. For instance, the Pleiku TACAN in 2MR was repeatedly meaconed by enemy transmissions emanating from Cambodia. During October 1969 and March 1970, there had been 16 reported incidents of this activity. Intrusion was when the radio was used by English-speaking VC or NVA, in an attempt to redirect aircraft over enemy territory. The enemy used correct radio procedures and callsigns to direct aircraft to hold at a certain position off a certain TACAN, which when plotted would put the aircraft over North Vietnam. There had been 31 recorded cases of intrusion over October 1969 and March 1970.[27] Jamming and Interference normally occurred at a critical point for an aircraft, such as approaching the drop point on a Skyspot. The Allied radio frequencies would be monitored by the enemy, who would jam the communications by yelling into the microphone, or by playing oriental music. On other occasions the NVA directed AAA fire at the aircraft using the same frequency.

Despite the inexplicable loss of Herbert and Carver, the Squadron continued to fly nine or 10 sorties per day, and this remained the level of activity for the remainder of No 2 Squadron's operations in Vietnam. On the night of 23 December members of another crew thought that they would be the Squadron's next casualties. Flying Officers Dave Smith[28] and Greg Weekes[29] (A84-234) were flying a Skyspot in 2MR close to the Laotian border, where the national boundary loops back on itself to the north-east. As Australian crews were not permitted to violate international borders, the attack necessitated a run-in from the east, a fast turn onto south for bomb release, and an even faster 180 degree turn back to the north-east to clear the border. They were at 14,000 feet and, just as they turned left for the attack, the sky to their immediate 2 o'clock position lit up with angry orange fireballs of flak. This turn had probably saved the aircraft. A hurried call to their United States ground radar controller that they were taking heavy AAA solicited a query: "Roger Magpie, confirm that your ECM gear is on?". The navigator quickly countered: "What ECM? We're a bloody Canberra!"[30] ECM—Electronic Counter Measures—was the smart electronic equipment which would respond to a hostile radar, and by jamming the enemy display would deny him the accuracy for firing his guns and missiles. Fortunately, bomb release point was only a matter of seconds away, and before the enemy 37 mm AAA site could re-acquire the Canberra, the bombs were dropped, and the Canberra hastily turned back to the north-east.

On several occasions in February 1971, the Canberras were called upon to fly eleven missions per day. On the morning of one such day, 9 February, the CO and the nav leader, Squadron Leader Bernie Johnson[31] (A84-234), were diverted to fly a most

---

27 Bob Howe, notes taken in Vietnam, 1969–70.

28 Flt Lt D R Smith, O224754. Previously Army, Nos 2, 34 & 2 Sqns. Regular air force offr of Sydney; b Sydney 11 Oct 1945.

29 Wg Cdr G D Weekes, O118068. Nos 2 & 38 Sqns, HQSC, 9 Sqn, HQOC; comd 22 Sqn 1984–85, 38 Sqn 1986–88; HQOC, 1 Div, DefAir. Regular air force offr of Dalby, Qld; b Charleville, Qld, 12 Jul 1949.

30 Greg Weekes, letter to author, of 30 Dec 1993.

31 Wg Cdr B A Johnson, O35022. 2 Sqn, SAN, RAAF London, RAAF Wash, 82 Wq, 2 Sqn, DefAir. Regular air force offr of Geelong, Vic; b London, Eng, 30 Jun 1934.

hazardous attack in support of troops in contact in 2MR. Flying below low cloud that was shrouding the high rugged peaks around which they would have to work, a USAF fighter pilot flying in the O-2 FAC aircraft reported:

> The FAC indicated the seriousness of the engagement on the ground and, having diverted Wing Commander Downing and Squadron Leader Johnson from a preplanned strike for immediate assistance, was especially concerned to see if they could operate in such an environment. With no hesitation they maneuvered under the overcast and initiated preparations for attack ... they were determined to make a maximum number of passes, exposing themselves on five successive deliveries to intense ground fire, in order to enhance the effectiveness of their 750-pound bombs and provide some respite for the beleaguered friendlies.
>
> Flying just 100 feet above their minimum altitude over the bomb impact area, grazing the overcast ceiling, they repeatedly flew across a ridge line from which they were receiving intense ground fire, clearing it by only 800 feet. Their restricted run-in placed them in this extremely vulnerable position and their requirement of wings-level stabilized speed/heading deliveries made them sitting ducks. The target could not be acquired until the last possible moment due to the enemy held ridge line, the ragged cloud bases, haze, smoke and thick triple canopy jungle which obscured the FAC's mark. The accuracy of the bombs dropped by the crew of the Canberra drew excited praise from the friendly ground troops who were located 800 metres from the impact area on a mountain top. Other friendly forces had set up ambush positions within 500 metres of the target; these locations could not be marked without revealing their emplacement to the enemy.
>
> On their last pass Wing Commander Downing and Squadron Leader Johnson sustained serious battle damage to their aircraft. A massive chunk of bomb fragment ripped through the nose section where Squadron Leader Johnson was lying prone ... The fragment passed by him and went up into the pilot's compartment striking the control wheel column and partially severing elevator controls; it continued with high energy six inches past Wing Commander Downing's leg and smashed into the IFF panel.[32]

Subsequent damage assessment of the target area by ground forces indicated the enemy had been planning a major assault. The Canberra had destroyed the main body of their forces, a radio station and a large part of their ammunition cache. The USAF report continued: "... the friendlies might not have survived the assault had it not been for the efforts of Wing Commander Downing and Squadron Leader Johnson". For Bernie Johnson it was second time around for flying the Canberra in combat. He had flown 2 Squadron's *Firedog* missions in Malaya. Vietnam was proving to be a little hotter.

By now No 2 Squadron had become heavily involved in operations in the northern part of 1MR. One pilot, Flight Lieutenant Dave Godfrey,[33] recalled the intensity of the campaign:

> With the launch of *Operation Lam Son 719*—the ARVN push along Route 9 into Laos to try and capture the town of Tchepone, which was a dismal failure—the 2 Squadron sortie rate

32 Capt L D Bruce, 612th TFS, letter "Outstanding Air Strike Performance" of 9 Feb 1971, to Commander 35th TFW, RAAF Personnel Records, Canberra.

33 Wg Cdr D J H Godfrey, O316600. Prev RCAF, No 1BFTS, 1 OCU, 2 Sqn, 1FTS, 1 Sqn, HQOC; comd CPE 1981–83; HQ Amb. Regular air force offr of Toronto, Ontario, Canada; b Weston, Ontario, 11 Dec 1938.

was increased. In late February, the United States Marine Corps Air Support Radar Team (ASRT) at Quang Tri was relocated to Khe Sanh. Between 23 February and 15 March, Ron Aitken[34] and I flew 17 missions directed by Khe Sanh ASRT, mostly at night. The fighting was pretty intense around Khe Sanh and the border. A lot of the night sorties were in direct support of the marine controllers at Khe Sanh, bombing within 200 metres of the base perimeter. I remember one night at about 0200, as we were being vectored on the bomb run, the controller asked what it was like where we were. I told him we were above a layer of cloud but otherwise a great night. He said he was in a bunker 18 feet underground and they had been taking about 100 rounds of incoming rocket fire an hour through the night, and it would be real fine if our bombs could quiet things down a bit. They did, for a while. The Magpies had a mission to Khe Sanh every couple of hours throughout the night, every night, for about a week.

One of the few day sorties to Khe Sanh was on Sunday 14 March [A84-248]. Ron and I were on the "frag" following John Downing and Al Pinches.[35] The ASRT controller told us he had lost contact with the Magpie ahead of us and that he had probably been hit. He advised that there was a lot of 57mm and 100mm activity in the area. He didn't mention SAMs. The pucker factor was just a bit greater that afternoon ...[36]

That afternoon, Magpie 41, the CO with Flight Lieutenant Al Pinches, had indeed been hit. They had been flying at 14,000 feet near the DMZ. Fortunately the crew survived by ejecting; this was the first time that an RAAF aircraft had been downed by a missile. The starboard wing of their Canberra, A84-228, had virtually been blown off by a Soviet type SA-2 SAM, and Wing Commander Downing was able to relate:

Then the bloody sky fell in. A terrific explosion seemed to envelop the aircraft ... I called to Al that it was a SAM and that we were ejecting ...[37]

Navigator Al Pinches recalled that they had just been turning on to their target run, when they were rocked by the explosion:

... my nav's station was immediately a great turmoil of blasting airstream, flapping canvas, flying papers and charts, dirt, dust and noise. It had also got suddenly darker. John saw another SAM go by but it didn't explode. A second or so later we felt the aircraft beginning to break up and John ordered me to eject. I blasted my hatch off and pulled the face blind to fire the ejection seat, but nothing happened. A second desperate tug was successful. I was barely clear of the aircraft when I saw John blasting through his pilot's canopy.[38]

Although Downing's parachute worked, as advertised, by opening automatically at 12,000 feet, the navigator's did not. Al Pinches was about to deploy it manually as finally the canopy blossomed open:

I remember being annoyed ... annoyed because of the noise, annoyed because my watch had been ripped off, and because John's parachute opened and mine would not.[39]

---

34 Sqn Ldr R W Aitken, O219684. Nos 36, 34, 2 & 37 Sqns, Surv Regt, 37 & 22 Sqns. Regular air force offr of Sydney; b 29 May 1939.

35 Flt Lt A J Pinches, 160765/O32334. Nos 38, 36, 10, 1, 6 & 11 Sqns, 1BFTS, 2 & 23 Sqns, TSF. Regular air force offr of Melbourne; b Tumut, NSW, 9 Oct 1929.

36 Dave Godfrey, letter to author, of 21 Nov 1993.

37 C Black, "Flyers Nearly Bought It From An Angry SAM", *RAAF News*, Vol 13, No 5, Jun 1971, p.5.

38 Al Pinches, letter to author, of 25 Feb 1994.

39 Black, p.5.

"Magpie" R HOWE

They had parachuted through cloud, and in rain and near-zero visibility were floating down into the mountainous jungle. John Downing crashed through trees on the side of a steep ridge, 3,000 feet high, and landed on his knees, fracturing a kneecap. Al Pinches fortunately had his fall broken by the thick jungle canopy, which snagged his 'chute, leaving him dangling a metre above the boulder-strewn ground. He had received back injuries on ejection, but falling on the boulders would most probably have killed him. He recalled:

> As I lowered myself, I noticed pain in my back but thought that I had merely bruised or sprained it. I managed to cut some equipment and a dinghy from the seat pack, and scurried from the scene in case there were any "unfriendlies" nearby.[40]

Now in enemy territory, Downing also collected his survival kit. He headed up the mountain ridge line to evade, and secure a likely rescue site. His radio calls from this vantage point, however, remained unanswered. Al Pinches, a long 400 metres away, could hear his pilot's radio calls, but was prevented from climbing the ridge line by his back injury. He could not bend over, and was exhausted after covering a few hundred metres in a couple of hours, so found shelter for the night. After the long night, with pain, hunger, discomfort and jungle noises preventing sleep, the dawn brought renewed efforts to make contact on their radios. About midday they established contact with one another, quite a boost to their morale. They then alternated their distress calls to conserve battery power.

Their radio calls were picked up that afternoon by a USAF C-130 Hercules—"King 26"—from the 39th Aerospace Rescue and Recovery Squadron, which coordinated the rescue. No Search and Rescue (SAR) "Jolly Green Giant" helicopters were available, and their positions were pinpointed by a USAF O-2 FAC aircraft. A rescue "dustoff" UH-1 Iroquois helicopter was called in to attempt hoist extractions.

Not long after the rescue attempt started, the cloud descended all around them and they had to temporarily call it off. "King" began arranging for artillery or B-52 strikes to be laid on in the area during the night in case "Charlie" came looking for the downed airmen. As all these arrangements were being made, the weather suddenly improved and the rescue resumed:

> "Dustoff" had trouble finding his way north through the cloud-enshrouded valleys, but persevered, periodically firing bursts from his guns to find out if he was close enough for us to hear. After the longest wait in my life he finally broke through into our valley. Never make comments about the funny "wok-wok" noise choppers make. It is the most beautiful sound in the world.[41]

40 Al Pinches, letter to author, of 25 Feb 1994.

41 Al Pinches, letter to author, of 25 Feb 1994.

Three hours after making radio contact, the downed crew were picked up by a skilled helicopter crew, whom Wing Commander Downing praised:

> That pilot was fantastic. In heavy cloud he flew up and down the sides of those mountains trying to locate us. They were stinking weather conditions and we almost had to call off the rescue at one stage.[42]

Al Pinches had dragged himself through heavy jungle to the clearing of a bomb crater:

> When the pilot came in to pick me up he actually chopped away tree branches with his helicopter's rotors. He was really sticking his neck out.[43]

The extraction system used by rescue helicopters was known as the "jungle penetrator", which could be lowered by cable through the jungle canopy to lift out downed airmen. Pinches had been expecting a more rudimentary harness type hoist:

> Rather than waste any precious seconds working out how to unfold it (I now know it only takes a second), I slid one leg through a canvas webbing loop and hung on. Immediately that I was clear of the tree tops, the chopper swung out into the valley so that within seconds I was suspended about a thousand feet above the ground. All that was between me and a big drop was a loop of canvas held in place with a press-stud and eyelet, and my grip on the cable. That's how I remained until the winch operator was able to get me into the cabin.[44]

The survivors were flown, after a refuelling stop at a United States fire support base, to the United States 18th Surgical Hospital at Quang Tri. Here they were examined: Downing had a fractured knee cap, which required plaster, and Pinches had fractured vertebrae in the lower back.[45] Meanwhile, south at Phan Rang, the Canberras had continued bombing strikes, leaving the search for the missing aircraft to the USAF SAR experts. At the precise time of the rescue, Padre Bill Wheeler[46] had been conducting a prayer service. As the two crew members were being extracted, a running commentary was relayed back to the gathering, and with its successful conclusion, the service developed into a party.

Early the next morning, the survivors were moved again, as Al Pinches recalled:

> We were cleared from the hospital to make way for heavy casualties coming in from the Laotian push, but our fun and adventure days still hadn't finished. The medevac Caribou lost both engines on take-off. Fortunately, the pilot did a skilful emergency landing and after some repairs, we were off.[47]

Downing and Pinches were transferred to the Cam Ranh Bay hospital, near Phan Rang, where they were visited by Squadron Leader Arthur Barnes, the Squadron's temporary CO. John Downing remembers Arthur Barnes arriving with a large briefcase:

---

42 Black, p.5.

43 Black, p.5.

44 Al Pinches, letter to author, of 25 Feb 1994.

45 No 2 Sqn Signal A184, of 160330Z Mar 1971.

46 Chaplain (AirCdre) W T Wheeler, O218505. No 11 Sqn, BSqns Lav, Pce, Wag, 2 Sqn, BSqns Edn, Butt, Edn, Amb, Butt, AHQ. Regular air force offr of Sydney; b Cowra, NSW, 29 Mar 1937.

47 Al Pinches, letter to author, of 25 Feb 1994.

> He said, "I have some confidential papers for you." Then he opened the case. It was crammed full with cans of beer ...[48]

On 7 April, Flight Lieutenant Stan Fenton[49] and Flying Officer Pete Murphy[50] (A84-236) conducted one of the most successful strikes flown by the Squadron for the entire war. Flying as Magpie 81, they rendezvoused with their FAC, Herb 52, in the mountains near the Cambodia, Laos, Vietnam border:

> American ground troops on Fire Base 6 were under heavy pressure from the enemy from a distance of only 200 metres. The friendly forces commander in spite of the potential danger cleared the Canberra to bomb. Herb 52 dived in and marked the target ... here with adversaries fighting it out in close contact friendly lives could be lost by a misdirected bomb. The strike completely disrupted the enemy attack and relieved the heavy pressure on the friendlies on Fire Base 6.[51]

The ground forces later confirmed 80 enemy troops had been killed as a result of the strike.[52] The same day of this successful mission the new CO had arrived. After Wing Commander Downing had been shot down, Squadron Leader Arthur Barnes had taken command of the Squadron, followed by Squadron Leader Brian O'Shea,[53] while the new commander, Wing Commander Tom Thorpe,[54] hurriedly completed his training in Australia. Thorpe then rushed to Vietnam, but his tour of operations was to be short-lived; it had already been announced that No 2 Squadron was to be withdrawn. It was ironic that the tactics of high-level Skyspot missions, flown by the Squadron on arrival in 1967, had turned full circle: through low-level visual bombing, back to the upper air in 1 Military Region. Once again the priority was to cause roadslides to hinder enemy supply routes to the south.

The last Magpie missions were tasked for 31 May 1971. The crew selected to drop the last bombs, and sever No 2 Squadron's involvement in the conflict, was the crew that had been serving the longest at Phan Rang. Flying Officers Dave Smith and Pete Murphy (A84-244) flew the last Magpie war mission in support of the 2nd Brigade of the United States 101st Division in 1MR's A Shau Valley. On the last bomb to fall from Magpie 61 was the message: "76,389 and last bomb Compliments to 'Charlie' from No 2 Squadron RAAF Uc Dai Loi".[55] As Smith returned south to Phan Rang, he transmitted the following message:

---

48 Black, p.5.

49 Wg Cdr S J Fenton, O218272. 2AD, 82 Wg, BSqn Darwin, Nos 30, 10, 2, 76, 6 & 1 Sqns, DefAir, 1 Sqn, HQ Amb, HQSC; comd 6 Sqn 1983–85. Regular air force offr of Lord Howe Is, NSW; b Lord Howe Is 4 Aug 1941.

50 Sqn Ldr P V Murphy, O117342. Nos 37, 2 & 34 Sqns, DefAir, 1, 6 & 23 Sqns. Regular air force offr of Brisbane; b Brisbane 25 May 1948.

51 Odgers, *Mission Vietnam*, p.115.

52 No 2 Sqn A51, of 7 Apr 1971.

53 Sqn Ldr B E O'Shea, O116226. Nos 10 & 6 Sqns, 1BFTS, 1 OCU, 2 Sqn; t/comd 2 Sqn 1971; DefAir, 37 Sqn, HQ Ric. Regular air force offr of Rockhampton, Qld; b Rockhampton 8 Feb 1937.

54 Gp Capt T H Thorpe AFC, O14501. Nos 1 & 6 Sqns, DepAir, 1AFTS, DepAir; comd 1 OCU 1971, 2 Sqn 1971–73; DefAir; comd BSqn Butt 1975–77, DefAir. Regular air force offr of Perth; b Moe, Vic, 11 Mar 1930.

55 *Uc Dai Loi* is Vietnamese for "Australia".

Attention all aircraft in the Vietnam area. This is Magpie 61 with a message from the last Magpie mission in Vietnam.

On behalf of No 2 Squadron, Royal Australian Air Force, I would like to thank everybody for their assistance, cooperation and friendship over the past four years and we wish you all a safe tour and all the best for the future.

Good luck and goodbye, Vietnam.[56]

The crew for the last mission, Flg Offs Dave Smith (left) and Pete Murphy, with the "76,389th and last bomb, compliments to 'Charlie' from No 2 Squadron RAAF Uc Dai Loi" RAAF

One of the Squadron's young pilots, Pilot Officer John Wood,[57] remembered the fun farewells from Vietnam, that Wing Commander Tom Thorpe and the Australians were to experience:

> The USAF invited the officers up the hill [to the USAF Officers' Mess] about a week before departure—the fighter vs bomber rivalry appeared towards the end of the night, when the USAF determined a competition was in order. Out came a set of darts and a dart board hanging from a pole—it had about a 6" bull and a larger outer. The USAF colonel chief fighter pilot stood behind the line and obviously threw his 3 bulls—then it was "Uncle Tom's" turn. The USAF then blind-folded Uncle Tom, turned him around three times, put one foot in a bucket of water, the other up on a chair, reversed the dart board with its normal minuscule bull, gave it a swing—and we all had a laugh, although we obviously didn't like being taken advantage off ...

56 D Ekstrom, "Of Grey Hairs and Sharp Teeth—The Canberra", *Australian Aviation & Defence Review*, Aerospace Publications, Canberra, Dec 1980, p.84.

57 Wg Cdr R J Wood, O58068. No 2 Sqn, 2FTS, 1 & 6 Sqns, 2FTS, DefAir; comd 2FTS 1987–89. Regular air force offr of Bunbury, WA; b Perth 20 Dec 1949.

> We began plotting for the return match—the engineers' expertise was called upon. The 2 Squadron Officers' Mess, a couple of days later, and late into the night the return match began. The boss was dressed in the navigators' parachute harness with D-ring fitted to the back, and hoisted up and connected to a bomb carrier fixed to the ceiling. He was handed his three darts, and from his horizontal position was able to eye them into the enlarged bull below. The Colonel's turn was sweet revenge. The parachute harness changed bodies, a blind-fold was affixed, he was turned around three times and hoisted onto a flying fox leading 100 metres down the track to the swimming pool—this area had been covered by curtains and had not previously been visible to the party-goers. The scream, when this guy realised half way down the fox and tore the blind-fold off, was blood-curdling. Needless to say, when he arrived a few seconds later there was a "gunny" with a long pole, who obligingly released him from the bomb carrier—a perfect dive bomb … into the water![58]

The aircraft were prepared with their wing tip fuel tanks for the ferry back to Australia, and No 2 Squadron aircrew flew their Canberras from Phan Rang on 4 June 1971, arriving at Amberley the following day. The Squadron had been absent from Australia since 29 June 1958, the longest continuous period for an RAAF squadron. In recognition for its service during its four years in Vietnam, No 2 Squadron was again awarded the Duke of Gloucester's Cup, this time for 1970–71.[59] The citation read:

> No 2 Squadron moved to Vietnam in April, 1967, and was employed in support of allied operations until June, 1971, when the Squadron returned to Australia. During that period, squadron personnel maintained an extremely high standard of proficiency in every respect, culminating in this award of the Cup.

This was not the final award for No 2 Squadron's service in Vietnam. The Republic of Vietnam Unit Citation, the Gallantry Cross with Palm, was awarded to the 35th TFW, which included No 2 Squadron members who had served from 1 September 1968 to 9 October 1970.[60] In addition, the USAF awarded its Outstanding Unit Award, for "exceptionally meritorious service in support of military operations against opposing armed forces in South-east Asia from 19 April 1967 to 31 May 1971".[61] The USAF honour was gazetted on 11 January 1973, and presented to the Squadron on 9 May that year.

The young men who had served with No 2 Squadron in Vietnam had carried out their duties with the same determination of Australian youths in previous wars. With the RAAF's entry into the Vietnam War, the pilot training rate had been greatly increased, to the extent that four courses, each of 45 trainees, were commencing training each year at Point Cook's No 1 Basic Flying Training School (1 BFTS). This meant that 1 BFTS had a massive flying effort of 22,000 flying hours per year over the Vietnam period.[62] A significant observation made by the Operations Flight Commander, Squadron Leader Arthur Barnes, was one of being proud and impressed

58 John Wood, letter to author, of 15 Dec 1993.

59 The presentation was made at Amberley on 6 Jul 1972.

60 RAAF PR S4550/72 Ministerial Press Release, of 1 Mar 1972. The award had been approved in *Commonwealth of Australia Gazette* No.87, of Thursday 9 Sep 1971.

61 Odgers, *Mission Vietnam*, p.122.

62 Arthur Barnes, telephone interview with author, of 3 Dec 1993.

by the young aircrew. Having served as the Chief Flying Instructor at 1 BFTS, and then at No 1 Operational Conversion Unit, Arthur Barnes had seen these young aircrews develop throughout their training, and then had supervised them on combat operations.[63] This was the lasting impression too of CO, Wing Commander Jack Boast.[64]

The Squadron's aircrew ride in on the wing of A84-244, 31 May 1971 VIA G BICKLE

Australia's involvement in the Vietnam War has always been controversial and, as the war continued, and our commitment had grown, the tide of public opinion had turned towards the withdrawal of Australia from the war. The returning veterans were not feted, as their predecessors of previous conflicts had been. Our military contribution was perceived by the general public as ill-conceived, and Vietnam veterans became unpopular accordingly. Yet these volunteers, who served their country well, did not doubt the role they had to perform. It was duty, and No 2 Squadron served with distinction. If the returning personnel were not favoured by the populace, they could hold their heads high with past members of No 2 Squadron; the highest traditions of Australian airmen had, as expected, been maintained.

It was only after 1973, by which time all USAF tactical squadrons had also left South Vietnam, that the North was finally able to mount a decisive campaign.[65] Without that Allied air campaign, albeit mounted with the incredible political constraints from Washington, the fall of the South may well have happened a decade earlier. And then, as the domino fell in the 1960s, the question would have been asked: "Why did Australia not help an ally?" Australia would have had no right to expect any assistance in the future.

63 Arthur Barnes, telephone interview with author, of 19 Nov 1993.

64 AirCdre Boast, telephone interview with author, of 6 Dec 1993.

65 Stephens, p.125.

CHAPTER 27
Bird of Paradise

The Canberra's effectiveness and the soundness of its basic design is well illustrated by the fact that more than four decades after its first flight, the type is still in service ... in roles other than the bomber/strike mission for which the aircraft was originally designed.[1]

Wing Commander Tom Thorpe retained command of No 2 Squadron as it settled into its more peaceful surroundings back at Amberley. Bombing was stopped immediately on the Squadron's return, as photographic reconnaissance (PR) was the new role proposed for the unit. However, in spite of a tactical reconnaissance course being run on the Squadron during July 1971, and consideration of developing the Canberra Mark 20 into a dedicated PR aircraft by the addition of a camera pack, this was not to be the fate of 2 Squadron.

Presentation of the new Squadron Standard at Amberley in 1971 — Governor-General Sir Paul Hasluck and Pilot Officer Rod Scotland RAAF

However, during 1972, one aircraft (A84-235) was fitted with a CSIRO multi-spectral camera to assist the Mineral Physics Section in surveys of mineral deposits and pastoral resources. Photography of land surface features using the camera's infra-red and colour filters would be compared with images taken by NASA's Earth Resources Technology Satellite (ERTS), low altitude photography, and ground surveys. The camera modification was test flown by Flight Lieutenant John Holt[2] and Flying Officer Greg Weekes in late May, after which they deployed to Alice Springs over the period 13–24 June, to photograph areas around Tennant Creek, Mt Isa and Alice Springs.[3] The photography had to be flown at 45,000 feet, and sorties averaged 4 1/2 hours duration, most of which was spent by the navigator prone over the camera sight in the aircraft's nose. Apart from the extreme cold experienced at these levels as the aircraft heating was so inadequate, the navigator recalled having slight attacks of the bends with fiery pains in the knees and elbows. Such discomfort readily dissipated on descent, and in those days was accepted as one of the hazards of high-level jet operations. The CSIRO support

1 Wilson, p.79.

2 Sqn Ldr J F Holt, O315918. Nos 11 & 2 Sqns, 2FTS, 1 & 6 Sqns. Regular air force offr of Robinvale, Vic; b Melbourne 4 Jan 1943. Killed in flying accident 29 Sep 1977.

3 Greg Weekes, letter to author, of 30 Dec 1993.

continued over several months, with different 2 Squadron crews flying sorties in August and September 1972.

The roles of air defence training—target towing and high-level radar targeting—which had previously been carried out by the training unit, No 1 OCU, were also taken over by No 2 Squadron, together with routine photography of strategic significance around Australia. The target-towing duties also involved biannual deployments to the RAAF fighter squadrons in Butterworth, known as "TugBut". In all, 24 of these six-week TugBut deployments were to be conducted at Butterworth.[4]

Support duties, such as air defence training, were supplemented by more rewarding flying for the aircrews in 1973, when aircraft were converted under Modification 769 to dedicated cartographic survey photographic aircraft with the addition of the large format Wild RC-10 survey camera[5] in the bomb-bay, with the bomb doors modified so that they closed around the lens. The RC-10's associated NF-2 periscopic sighting and tracking equipment was fitted for the navigator to use in the cabin. A84-245 was the first Canberra modified in early 1973, followed by a further three aircraft.[6]

The first overseas deployment for one of the newly modified aircraft was on 31 March 1973 when the CO, Wing Commander Tom Thorpe, and Flying Officer John Bennett (A84-245) flew to Port Moresby to begin the *Skai Piksa*[7] survey of Papua New Guinea (PNG). Since 1963, periodic aerial surveys of the country had been carried out by United States Navy reconnaissance aircraft operating from Townsville. This had been done in collaboration with the RAAF, and the results had been made available to the Australian Services.[8] However, these efforts had not been enough to adequately cover the mammoth task of complete, accurate mapping of the region. Most maps were still wildly inaccurate topographically, and out of date with marked cultural features, such as "Japanese base camp"!

At the time, there were only three airfields in PNG that could be used to conduct Canberra operations—Port Moresby, Lae and Nadzab. However, due to the speed, endurance and flexibility of the aircraft, it appeared that all operations could be conducted from Port Moresby. One early exception to this rule related to condensation problems encountered with A84-245, the first RC-10 aircraft. The survey camera had been hung from the bomb beam, and a semicircular aperture had been cut in each bomb door, so that photography could be taken with the doors closed. The circular aperture was closed by a horizontal sliding door with its own rudimentary actuator. The camera bay was not sealed or pressurised, nor was it heated. This arrangement was satisfactory for the normal mission of climb to height (40,000 feet), transit and photography at high altitude, but was inadequate when a high-level

4 The last of these operations, TugBut 24, was conducted over Jan–Mar 1982.

5 The RC-10 produced a negative of 10" x 10" (25.4 cm^2).

6 Modified Canberras were A84-233, -234, -238 and -245. As these became time expired, the modified components were fitted to A84- 230 and -232.

7 Pidgin for "Sky Picture".

8 "Aerial Survey", *RAAF News*, Vol 11, No 7, Aug 1969, p.1.

transit was followed by lower level photography. The camera and periscopic sight would "cold soak" during transit and be subjected to heavy condensation on descent, which would obscure the lens. This difficulty was remedied on later aircraft by blowing warm air around the camera and bleeding nitrogen from the fuel tank purge system through the periscopic sight.

Mission planning for Skai Piksa *at Lae, Flt Lts Al Stephens and Mike Rawlinson* VIA M RAWLINSON

In the early detachments, condensation remained a problem for any lower level photography to the north of the Owen Stanley Ranges, as the short trip over the mountains was sufficient to cool the camera and produce condensation on the lens filter from the very moist tropical air. A solution to the lower level survey of the Trobriand Islands in the Solomon Sea to the north of PNG was to operate from the north side of the Owen Stanleys and transit below photographic height. This formula was followed during May by Flight Lieutenants Al Stephens[9] and Mike Rawlinson[10] who successfully operated A84-245 out of Lae for two days to capture low-level island photography. As far as can be determined, this was the only Canberra ever to operate from Lae. This is not surprising as the runway at Lae was shorter than Port Moresby, and marginal for a Canberra with full fuel. Mike Rawlinson recalled:

9 Wg Cdr A W Stephens, O43976. No 10 Sqn, BSupt Flt Viet, 2 ATU, 2 & 38 Sqns, DefAir, RAAFSC; comd 2 Sqn 1979–1981; DepDef, APSC. Regular air force offr of Adelaide; b Peterborough, SA, 17 Jul 1944.

10 Gp Capt M J Rawlinson, O56545. Nos 36 & 2 Sqns, RAAF Acad, DefAir, RAAFSC; comd OTS 1983–84; ADFA; comd Laverton 1988, Commandant RAAF College 1989–90; HQTC. Regular air force offr of Perth; b Clacton, Essex, Eng, 18 Nov 1944.

> While we were confident of the early morning take-off on the first day (9 May), this confidence was apparently not shared by all, as it seemed as if the whole population of Lae turned out to see us depart. What did they know that we didn't? It was a little tight. I can remember looking through the nav window and seeing the water of the Huon Gulf appear immediately following rotation.[11]

On 2 July, Flying Officers Geoff Carter[12] and John Bennett deployed with a second aircraft (A84-238). The results during 1973 for this first phase of *Skai Piksa* were most successful and, when completed in September, had exceeded the most optimistic of expectations. The large amount of "X-grade" photography[13] obtained was once again due to the amazing flexibility of the Canberra. It could transit quickly, conduct photography at very high level, and had the range to complete areas in different regions of the country within the same sortie. Moreover, the task was not commercially driven, so the RAAF crews were able to check actual weather conditions, and not rely on meteorological forecasts, thereby obtaining photography of areas which had previously never been photographed. This was critical in New Guinea, where weather would change dramatically over a short time, and timing was critical for the operation, as *Skai Piksa* had a deadline to meet: the independence of PNG in September 1975.

Wing Commander Graham Dyke returned to 2 Squadron to take command in January 1974, as the unit began flying in the *Gading* operation. This was a joint Australian-Indonesian program, which had commenced in 1970, to survey and map Kalimantan, Sumatra and Irian Jaya. No 2 Squadron commenced its involvement in *Gading 4*, with photography of Sumatra, initially operating from Butterworth. Two Canberra crews, Flight Lieutenants John Gosper[14] and Gordon French[15] (A84-238), and Flying Officer Geoff Carter and Flight Lieutenant Mike Rawlinson (A84-233), began the *Gading* survey and, in addition to three Canberras on target duties at Butterworth, No 2 Squadron had a total of five aircraft operating again from Butterworth. Survey operations were also conducted from Tengah Air Base in Singapore for the *Gading 4* task.

Skai Piksa photography in PNG commenced again from Port Moresby in May 1974, with the CO and Flight Lieutenant Tony Stankevicius (A84-233) opening Phase Two of the operation. This was also a momentous month for the Squadron, as it coincided

---

11 Mike Rawlinson, letter to author, of 17 Dec 1993.

12 Flt Lt G N Carter, O318546. Nos 34 & 2 Sqns. Regular air force offr of Auckland, New Zealand; b Auckland 16 Jul 1949.

13 Survey photography requirements classified "X", "Y" and "Z" grade results. "X" photography was desirable, being of high quality on required track and completely useable. "Y" grade had cloud coverage or shadow below 10 per cent, but was suitable, and "Z" indicated cloud coverage was excessive and, therefore, was generally unsuitable. Results of Phase One *Skai Piksa* were so successful that by the end of Aug 1973, 90 per cent of PNG coverage had been obtained. No 2 Sqn A50, of Aug 1973.

14 Flt Lt J R Gosper, O222923. Nos 36, 5, 2 & 37 Sqns. Regular air force offr of Sydney; b Brisbane 19 Mar 1947.

15 Wg Cdr G C French, O219605. No 10 Sqn, 2ATU, Qld ATC, 37, 2 & 1 Sqns; comd 23 Sqn 1979–81; DefAir, 23 Sqn. Regular air force offr of Sydney; b Sydney 18 Oct 1938.

with the 21st anniversary of the first flight of an Australian Canberra, A84-201. "Old 201" was still serving on 2 Squadron, and the celebrations on 29 May featured a handling display of the aircraft by Flight Lieutenants Stu Cooper[16] and Pete Murphy. Two dignitaries who participated in the celebrations were Mr John Miles, the GAF test pilot who had been the first to fly A84-201, and Air Commodore Jel Cuming, the first RAAF pilot to fly the aircraft.

Bad weather in PNG restricted *Skai Piksa* operations, so the effort was diverted to obtaining photography of the Cape York Peninsula area. Over 4 and 5 August, Flight Lieutenants Stu Cooper and Al Curr (A84-233) flew four sorties that stretched the Canberra's fuel and film capacity to the absolute limit.[17] In almost 18 hours of flying out of Port Moresby, they obtained more than 6,000 kilometres of X-grade photography—the Squadron's most productive survey sortie to date.

By the following month, September 1974, there appeared little prospect of obtaining good photography due to the onset of the wet season, so the detachment was withdrawn to Amberley. The end of the year saw the devastation of Darwin once more, this time by Cyclone "Tracey", so the Squadron's Canberras photographed the stricken area to assist the relief efforts of the Disaster Committee.

Phase Three of *Skai Piksa* commenced in May 1975 with the deployment by the CO and Flight Lieutenant Mike Rawlinson (A84-233), and Flight Lieutenant Adrian Slootjes and Flying Officer Pete Campbell[18] (A84-238). Better weather allowed completion of the operation this year, and when Flight Lieutenant Noel Williamson[19] and Flying Officer Pete Zawojski[20] (A84-233) returned to Amberley on 29 July, *Skai Piksa* had been successfully concluded on time. No 2 Squadron did return to Port Moresby, however, when three crews flew in on 11 September to take part in the PNG Independence Day celebrations.[21]

As Wing Commander Pete Jones took over the Squadron in January 1976, the next survey deployment was being prepared. *Gading 5* was planned to complete the Sumatra survey, flying from Kemayoran Air Base, in Jakarta, on Java. The new CO with the first crew—Flight Lieutenants Dick Waddy[22] and Doug Ireland[23] (A84-233)—

---

16 Wg Cdr S D Cooper, O222452. Nos 38 & 35 Sqns, 2FTS, 2 & 1 Sqns, Regular air force offr of Sydney; b Sydney.

17 Al Curr, interview with author, of 16 Dec 1993.

18 Sqn Ldr P A Campbell, O121627. Nos 10, 2 & 9 Sqns, 2FTS, SRO Brisbane, CFS, DefAir. Regular air force offr of Rockhampton, Qld; b Rockhampton 23 Feb 1950.

19 Sqn Ldr J N Williamson, O223430. BSqn Pck, No 2 Sqn, 2FTS, Nos 2 & 1 Sqns, HQSC, 2 Sqn; comd 23 Sqn 1982–84; 6 Sqn; comd 23 Sqn 1985–86. Regular air force offr of Sydney; b Condoblin, NSW, 8 Jul 1942.

20 Flt Lt P J Zawojski, O58385. Nos 37 & 2 Sqns, TSF, 34 Sqn. Regular air force offr of Perth; b Katanning, WA, 18 May 1951.

21 Crews: Sqn Ldr Jones/Sqn Ldr Kirby (A84-232), Flt Lt Gosper/Flg Off Phillips (A84-235), and Flt Lt Jacobs/Flg Off Campbell (A84-240).

22 Sqn Ldr R I Waddy, O46802. Prev Army. No 9 Sqn, HQOC, 2 Sqn, 1FTS, 2 Sqn, HQ Amb, 1FTS, Vic Sqn ATC. Regular air force offr of Adelaide; b Adelaide 30 Oct 1945.

23 Wg Cdr D C Ireland, O119228. Nos 10 & 2 Sqns, SAN, 2 & 6 Sqns, DefAir; comd BSWg Tvl 1989–93. Regular air force offr of Brisbane; b Brisbane 22 May 1950.

commenced operations in March. Successful results were not obtained until August, when the Squadron deployed to Mokmer airfield, on Biak, an island some 100 kilometres north of Irian Jaya, to commence a new phase of No 2 Squadron's survey work—Operation *Cenderawasih.*

Former COs of No 2 Squadron at Amberley in September 1976 to commemorate the Squadron's 60th Anniversary:
Wing Commander Smokey Douglas DFC (1945–46), Wing Commander John Downing DFC (1970–71), Group Captain Leo Britt DFC (1965–66), Group Captain Frank Griggs DFC DFM (1953, 1963–65), Air Vice-Marshal Alan Charlesworth CBE AFC (1939), Air Commodore Bob Dalkin DFC (1949), Air Commodore David Evans DSO AFC (1967–68), Group Captain Jack Boast DSO AFC (1969–70), Group Captain John Whitehead DSO (1968–69), Wing Commander Pete Jones (1976–77), Wing Commander Graham Dyke DFC (1974–76), Group Captain Joe Fitzgerald (1955–56), Group Captain Jim Graney AFC (1949–51).* RAAF

Cenderawasih, Indonesian for "Bird of Paradise", was an Australian defence cooperation program with Indonesia, for the survey and mapping of Irian Jaya. The project was conducted jointly by the Army's 2 Field Squadron, with air support from No 173 (General Support) Squadron, and the Indonesian Survey Corps. No 2 Squadron again undertook the high-level aerial photography and provided logistic support, and the RAN's Landing Craft Heavy (LCH) were used to position stores and fuel.[24] In this rugged terrain, there were four separate operations necessary to compile a topographic map:

> Initially, mapping photography is carried out by 2 Squadron RAAF using RC-10 cameras.
>
> Secondly, geoceivers are used to fix accurate positions. American-designed and produced, this equipment has revolutionised surveying. Costing more than $90,000 apiece, the machines are able to utilise information transmitted from navigation satellites, and make it possible to pinpoint a position within 1½ metres on the earth's surface. The field squadron

24 "Operation Cenderawasih", *TRIAD*, No 7, Department of Defence, Canberra, 1977, p.21.

is equipped with six geoceivers, enabling five to be used continuously in the field at any one time. Surveying Irian Jaya would have been almost impossible by traditional methods. The geoceivers enable surveyors to pinpoint their position without reference to any other known point. Altogether, 128 survey points are used throughout Irian Jaya by Army survey teams.

Thirdly, a [Beechcraft] Queenair is used to measure the height of terrain on set tracks. It carries laser height-measuring equipment (aerial profile-recorder) designed and built by Australia's Weapon Research Establishment.

Finally, an Army Pilatus Porter aircraft supplements photos taken by Canberra aircraft to establish the exact spot where the geoceiver points are located.

Information is collected in Jakarta and survey establishments in Australia before the maps are finally printed at Bendigo, Victoria.[25]

An example of the survey photography carried out by No 2 Squadron over 1973–1982, this frame is from Skai Piksa *survey of New Ireland in 1973.* VIA A STEPHENS

Cenderawasih 76 continued until the onset of the wet season in November, and the withdrawal allowed other photographic tasks to be completed, such as the survey of Christmas Island and Cocos/Keeling Islands, in the Indian Ocean, during December. The CO, Wing Commander Pete Jones, made the following comments about the survey success of 1976:

> The Squadron's achievements over the past year in this task have been excellent. However, results just don't happen, they are an accumulation of hard work, dedication and a high degree of professionalism. The airmen and officers of this Squadron have, throughout the year, demonstrated these qualities.[26]

Cenderawasih 77 commenced from Biak in May 1977, with the detachment of Squadron Leader Terry Jacobs[27] and Flight Lieutenant Greg Fitzgerald[28] (A84-230), and Flight Lieutenant Stu Hawker[29] and Pilot Officer Steve Clarke[30] (A84-245). The Australian effort on Biak for this operation amounted to 140 Army and RAAF personnel. In addition to the 2 Squadron effort, the RAAF also provided logistic

25 "Operation Cenderawasih", p.21.

26 Wg Cdr Pete Jones, "Christmas Cheer from 2 Sqn", *Strike*, RAAF Amberley magazine, Christmas Issue, 1976.

27 Wg Cdr T B Jacobs, O224393. Nos 77, 3 & 2 Sqns, RAAF Acad, HQ 1 Div, 6 Sqn. Regular air force offr of Sydney; b Sydney 26 Sep 1945.

28 Wg Cdr G J Fitzgerald, O225197. Nos 2 & 1 Sqns, HQ Amb, 1 Sqn, 82 Wg, HQ 6 Bde, HQ 1 Div; comd 1 Sqn 1991–93; HQ SRG. Regular air force offr of Sydney; b Sydney 18 Oct 1950.

29 Flt Lt S D Hawker, O227325. Nos 38 & 2 Sqns. Regular air force offr of Sydney; b Christchurch, New Zealand, 30 Jul 1951.

30 Sqn Ldr S W Clarke, O123856. Nos 2, 6 & 1 Sqns, 2FTS, 38, 1, 6, 25 & 1 Sqns. Regular air force offr of Brisbane; b Newcastle upon Tyne, Eng, 15 Oct 1956.

support from Iroquois helicopters of No 9 Squadron, for transporting the survey teams, and Caribou transports of No 38 Squadron, to resupply stores and fuel. By the time the new CO of the Squadron, Wing Commander Gary Beck, assumed command in December, *Cenderawasih 77* had achieved successful results and was completed prior to the "wet".

During 1978, satellite coverage of Irian Jaya provided more accurate cloud cover information, assisting crews in heading for the clear regions for photography. The operation commenced at the end of March 1978, with the detachment under the command of Squadron Leader Dick Strudwick,[31] who deployed to Mokmer through Townsville with Flying Officer Geoff Garley[32] (A84-233). A chartered Beech Queenair aircraft, equipped with a profile recorder, was again used by the army to assist in the aerial survey, but no ground survey work was required during the 1978 operation.[33]

Flying his first survey mission in the area on 9 May 1978 was Pilot Officer Mike "Boggy" Smith,[34] flying with Flying Officer Bruce Davidson[35] (A84-230), who gave details of 2 Squadron's operations:

> A standard day's flying commenced in the early hours of the morning with an extensive met briefing to determine which area to attempt to survey. Based on the weather report, the crew then made an assessment or guess on which area of operations to conduct survey, flight planned, then took-off to try and beat the cloud build-ups. Sorties were flown as early as possible simply because the cloud build-up began very early in the tropics. Weather was of particular concern to operations; firstly because it ultimately determined mission success, secondly the weather played a big factor in navigation especially with limited navigation aids (Mark 1 eyeball!), and finally because the Canberra was not fitted with an airborne radar, tropical thunderstorms posed a significant threat.
>
> Once airborne, the Canberra was normally climbed to about 40,000 feet (in less than 10 minutes!) and the pilot and navigator began visually searching for cloud free areas. Air Traffic Control (ATC) was a real problem,[36] simply because Biak radio did not answer your HF transmissions ... Irian Jaya was generally clear of international air routes or restricted airspace, therefore crews normally tracked randomly throughout Indonesia as required. When an area was deemed suitable for aerial photography, the navigator had to quickly pinpoint the aircraft's position then check and see whether the particular photo run had previously been completed.[37] If not, the pilot then flew the photo run whilst the navigator controlled the tracking and operated the RC-10 camera.

---

31 Sqn Ldr R G Strudwick, O223103. No 2 Sqn, 2ATU, 2FTS, CFS, 97th Trg Sqn USAF, 2FTS, 2 Sqn, DefAir. Regular air force offr of Sydney; b Sydney 13 Mar 1947.

32 Sqn Ldr G K Garley, O320234. Nos 36 & 2 Sqns, SAN, 6 Sqn, DefAir, SRO Brisbane, DefAir, HQADF. Regular air force offr of Melbourne; b Merridian, WA, 20 Feb 1953.

33 "Completion of Irian Jaya Mapping", Defence PR Release 139/78, of Friday 8 Sep 1978.

34 Sqn Ldr M G Smith, O124770. Nos 2 & 1 Sqns, HQ Amb, 1 Sqn, 208 Sqn RAF, AJMWC, 6 Sqn, 82 Wg. Regular air force offr of Brisbane; b Port Moresby, TPNG, 30 Jan 1958.

35 Flt Lt B Davidson, O228256. SAN, No 2 Sqn, 2FTS, 2 Sqn, CFS. Regular air force offr of Cowra, NSW; b Cowra 21 Aug 1953.

36 Each detachment was supported by a qualified ATC officer, whose primary responsibility was to flight follow and hold SAR whenever the Canberras were airborne.

37 Photo runs were normally planned along east/west headings, and annotated on a series of charts carried by the navigator. The maps were continually updated by the Photo Interpreters (PIs) when successful imagery was obtained, therefore the navigator always knew if the run needed to be flown or not.

Post mission the film was downloaded by the photographers, processed and interpreted by the PIs to assess suitability for future map making. Results were therefore available first hand and it quickly became apparent if the mission flown was a success or not.

My first mission out of Biak was eventful in that we luckily obtained imagery over the "spine" of eastern Irian Jaya near the PNG border. That particular morning, I was flying with "Davo", who'd been there for several weeks and knew the ropes. We headed off towards the PNG border, an area notorious for incomplete maps and seemingly eternally shrouded in cloud, not surprisingly because the mountain ranges soared to over 16,000 feet. Navigation was a bit difficult, especially with large white areas indelibly marked "RELIEF DATA INCOMPLETE" on certain sections of the navigational charts. By sheer luck, Davo found a clear area in the vicinity of the "spine" early that morning, and after determining that the photo runs had not been completed, the Canberra flew two east/west runs before the cumulus cloud started to form.

My excitement after Day One's mission was short-lived. I was quickly brought down to earth when the PIs plotted my two photo runs and assessed that each photo run showed a marked tendency to drift south at the completion of each run. After much soul searching, excuses, etc etc, it was finally realised the reason this occurred was that I had forgotten to leave zero degrees magnetic variation on the compass, and that we should have been flying a true track instead of magnetic! Fortunately, the magnetic variation was only 4 degrees east, and most of the imagery was useable, albeit slightly off track. Several other attempts during the following seven weeks failed to obtain any useful photography of the same area due to cloud. The learning curve for a "bograt" navigator in a Canberra over Indonesia trying to take aerial photographs was steep, to say the least![38]

For the high level-photographic coverage of southern Irian Jaya, airfields in Australia's north were able to be used during *Cenderawasih 78.* During July, Wing Commander Beck and Squadron Leader John Riches[39] (A84-241) had, therefore, carried out an airfield inspection of Gove, which had not been used by 2 Squadron since World War II. The airfield was found to meet the Canberra's requirements and, from November, Phase Two of *Cenderawasih 78* was concluded by flying from Gove and Darwin.

Areas in Irian Jaya, with continual cloud coverage, had caused some degree of frustration. Although large areas had been successfully covered, the illusive "socked-in" areas were to require constant visits in the following year's operation. Results were obtained with great difficulty, as the Canberras were operating with no wing tip tanks, due to the limited runway length at Gove.[40] As Wing Commander Gary Beck reminded his Squadron:

Patience is still the main virtue required of our survey teams anywhere, but an extra large serving is required in Irian Jaya.[41]

---

38 Boggy Smith, letter to author, of 21 Dec 1993.

39 Wg Cdr J D Riches, O57015. Nos 11 & 2 Sqns, SAN, RAAF SU Brisbane, 2 Sqn, DefAir, HQSC; comd BSqn Tvl 1988–89; HQ OSG, Air HQ. Regular air force offr of Perth; b Climpy, Scotland, 17 Aug 1946.

40 AVM Beck, telephone interview with author, of 1 Feb 1994.

41 Wg Cdr Gary Beck, "Greetings from 2 Sqn", *Strike*, RAAF Amberley magazine, Christmas Issue, 1978, p.10.

2 Squadron Canberras during a regular deployment to Butterworth in 1973 RAAF

The following year's goal was to complete the survey by "gap-filling" during *Cenderawasih 79*, which commenced in March with operations flown from Darwin. This final phase of the year's operation was terminated early in June as the stubborn clouded areas refused to clear. The timing of this decision was fortunate. In July, all Canberras were grounded with the onset of fatigue problems. "The Queen of the Sky", which had served No 2 Squadron so admirably for nearly three decades, was beginning to show her age. The grounding kept the Squadron out of the air until October. By this stage, the army found that some more "gap" photography was required, and this was completed over the last two months of 1979, as Wing Commander Al Stephens took over as the Squadron's commander.

As part of the Squadron's air defence support role, the Canberras involved in target-towing duties at Butterworth would at times be tasked as "bombers" for an air defence exercise (Adex). The Adexes, under the five-power Integrated Air Defence System in Malaysia and Singapore, provided the crews with a bit more challenge and interest than their normal mundane target flying. On such an occasion during IADS Adex 1/80 during March 1980, which coincided with the TugBut 20 deployment, one crew was introduced to the dangers of thunderstorms in the tropics. Flying Officers

Kingsley Abbott[42] and Dave Harrison[43] (A84-232) were on 17 March to fly a simulated attack on Singapore out of Butterworth. The navigator, Dave Harrison, has never forgotten the experience:

> We flew down to Singapore on a high-low profile for a dusk strike on Tengah airfield. After a successful attack, we headed north for Butterworth to be confronted by unforecast thunderstorms and lightning around Kuala Lumpur. Believing Butterworth to be OK, we picked our way through them but were eventually trapped in a fairly serious storm during the descent. Butterworth weather was "red", and with insufficient fuel for a diversion, we requested a GCA.[44] The noise from the storm and the hammering of rain on the aircraft made communication almost impossible—we had to shout over the intercom to each other. Radar lost us in heavy precipitation on finals so we went around. On the second approach we were not visual at the minima. Around we went for a min fuel GCA, but radar lost us in heavy rain again. Fuel was now critical and a fourth approach had to be successful, or it was out over the Straits of Penang for a night ejection. By now Air Traffic had roused their best GCA operator, who gave us the most wonderful GCA to touchdown. We saw the runway lights in the flare. The rain was so heavy that it was impossible to taxy safely, so we stopped on the runway—no-one else was going to use it! Just to cap off the evening we were struck by lightning, which caused our hair to literally stand on end. We learnt about flying in thunderstorms from that![45]

Although target flying was a necessary duty, it was the survey task that brought No 2 Squadron the tangible results, and the accolades. Pilot Officer Rick Owen[46] remembered the satisfaction in knowing what No 2 Squadron's efforts were achieving during *Cenderawasih*:

> Survey operations in Indonesia always gave me a sense of achievement. Flying from Biak, an island in the "Birds Head" area of Irian Jaya, involved early morning take-offs in order to beat the inevitable midday cloud build up. A typical survey mission would be a hunt for as many miles of open sky as possible. Some areas of Irian Jaya had not had sky clear days for over seven years, so returning to Biak with a full magazine of photography was somewhat of an event.
>
> Whilst flying at 45,000 feet over Irian Jaya I often thought about the early Dutch map makers whose information was used in the maps I was now using. Although the Canberra was not the most comfortable aircraft to conduct survey from, the danger and hardships these men endured made my job seem snug in comparison. Perhaps 50 or 60 years hence, when updated maps of the region are being produced by images from space, the survey crews and men from 2 Squadron will be remembered in the same way.[47]

---

42 Flt Lt F K Abbott, O320777. Nos 34 & 2 Sqns. Regular air force offr of Melbourne; b Melbourne 21 Oct 1954.

43 Sqn Ldr D K Harrison, O229976. Nos 2, 34 & 37 Sqns, Air HQ; comd 22 Sqn 1990–92; 37 Sqn. Regular air force offr of Canberra; b Sydney 13 Aug 1958.

44 A radar Ground Controlled Approach.

45 Dave Harrison, letter to author, of 2 Feb 1994.

46 Sqn Ldr R D Owen, O125059. Nos 2, 1 & 6 Sqns, RAF Coll, HQADF, 6 Sqn. Regular air force offr of Brisbane; b Lincoln, Eng, 20 Feb 1956.

47 Rick Owen, letter to author, of 9 Dec 1993.

Cenderawasih 80 was conducted over May and June 1980 from Biak, with the operation mounted again during November and December, flown by the CO and Flight Lieutenant John Harvey[48] (A84-245). Attempts to finalise the task during the early months of 1981 continued to be plagued by weather, as some areas were never clear of cloud. The effort was switched back to New Guinea in May, when Squadron Leader Adrian Slootjes and Flying Officer Dave Harrison (A84-232) carried out additional survey operations from Port Moresby for *Skai Piksa 81*.

In addition to the demanding air defence and survey commitments, during the late 1970s No 2 Squadron also developed a cadre of tactical reconnaissance experience. For these low-level operations, the Canberra could carry the old F24 and F52 aerial cameras, or the specialist KS87 reconnaissance camera, in the rear camera bay.[49] Some navigators conducted training in the United Kingdom with the RAF, and several crews underwent tactical "recce" training with No 77 Squadron at Williamtown, New South Wales. For its efforts over almost a decade in survey and air defence support, No 2 Squadron was again awarded the Gloucester Cup, this time for 1980–81.

The Canberra's fatigue problems had not disappeared, but operations were not inhibited. Wing Commander Alan Stephens explained the corrosion problems that were now occurring with the Canberra:

> The trouble we had was with exfoliate corrosion, which was particularly bad in the Canberra because of the type of aluminium that had been used in the wings. This was in the centre section forging, which basically held the wings on. With my engineering officer, Flight Lieutenant Des McGrory,[50] I paid a great deal of attention to the hours servicing stagger, to keep flying as many aircraft as possible. We had the flexibility of a 10% extension of flying hours before a major servicing, and in that way we projected that we could still generate nearly 3000 hours a year from an active fleet of nine aircraft. Operations were not affected, but there were plenty of maintenance headaches. On one occasion, we took a good wing off an aircraft that had previously been grounded, and fitted it to one which had a condemned wing, thereby resurrecting an aircraft.[51]

December 1981 saw Wing Commander Jules Wills[52] assume command of No 2 Squadron. He had served on the Squadron as a Pilot Officer at Phan Rang, and became the first navigator CO of an RAAF flying squadron. No 2 Squadron continued air defence support flying through 1982, until finally the decision was made in May to cease Canberra flying because the corrosion problems had become insurmountable. The suitability of the Canberra as a survey platform had been well and truly proved. With its great range, the aircraft was limited only by the amount of film it could carry. Even then, No 2 Squadron produced quality film at a rate far greater than could be

---

48 Wg Cdr J P Harvey, O229962. No 2 Sqn, DepDef, 1 Sqn, DepDef, F-111 AUP Office USA, DepDef, NZ MOD, SDSC. Regular air force offr of Sydney; b Sydney 4 Sep 1954.

49 Geoff Garley, interview with author, of 15 Dec 1993.

50 Sqn Ldr D D McGrory, O12512. 78 Wg, 3AD, 78 Wg, No 2 Sqn, 82 Wg, 75 Sqn, HQSC, 2 Sqn, Sch Tech Trg, 2 & 492 Sqns. Regular air force offr of Brisbane; b Brisbane 6 Aug 1932.

51 Al Stephens, interview with author, of 13 Dec 1993.

52 Wg Cdr J A Wills, O56872. Nos 6, 2, 6 & 34 Sqns, 430th TFS & 442nd TFTS USAF, 6 Sqn, RAF College, DefAir; comd 2 Sqn 1982; DefAir. Regular air force offr of Perth; b Sydney 23 Jan 1946.

absorbed by the Army Survey Corps. Small business jets then attempted to replace the Canberra in the survey role, as navigator Squadron Leader Geoff Garley recalled:

> After the disbandment of 2 Squadron, specially modified Lear Jets were used as part of 6 Squadron Survey Flight. I flew in these for a year to support the survey effort. It turned out that all the Lear photography has been rejected by Survey Corps, who claimed that the images were being irregularly distorted by the optical flat glass.[53] There can be no doubt that the Canberra was, despite some limitations, the most flexible photo platform available.[54]

Squadron Leader Adrian Slootjes had become the temporary commander for the last two months, until No 2 Squadron was disbanded at the end of July 1982. He recalled the dramatic change No 2 Squadron and its personnel went through when it was announced the unit was to be deactivated:

The last CO, Squadron Leader Adrian Slootjes, with a plaque presented by Army Survey Corps, to commemorate the contribution of No 2 Squadron to the mapping of the region to Australia's north RAAF

> Until it became general knowledge that the Squadron was to cease operations and become part of the history of the RAAF, the effectiveness of No 2 Squadron's air operations appeared to be at a peak. The tasking board was full and projected work out into the next few years. The Squadron was under pressure to produce. Each available airframe was working either from Amberley or a deployed location—it was common to have three deployments in progress concurrently. The Squadron had a near-full complement of personnel and everyone had a positive outlook to the future.
>
> The external support for continued operations seemed willingly provided by Base Headquarters, Operational Command and Air Force Office. The sense of security gained from such support fostered the positive unit attitude and ability for which No 2 was known; it provided the basis for making commitments to the many tasks on our plate. The collective attitude was professional, the Squadron operations room was a no-nonsense place, the flight line seemed to work without a hitch, the administrators were busy keeping people happy, and aircraft unserviceabilities were fixed with a minimum of fuss. If a question was asked on any aspect of the Squadron, someone knew the answer, and if he or she did not, then they would know someone who did. No 2 Squadron was a complete and autonomous unit, employed in the unique roles of aerial photography, air defence targeting, and training. Our customers wanted what we produced.

53 More recently, Army-sponsored Lear Jet operations have recommenced, using the same door modification with optical glass and Wild RC-10 combination. One of the 6 Squadron Lear navigators later commented: "There is some conjecture that the rejection, if not 'political' by the Army, was due to computer mismatchings of survey tie-in points and the appropriate earth models". Sqn Ldr "Doc" Millar, interview with author, of 27 Jan 1994.

54 Geoff Garley, interview with author, of 15 Dec 1993.

Then arrived the thunderous news that the Squadron was to be deactivated—that we were to wrap it up and quickly too! We were collectively stunned. There was disbelief, then followed months of a process not unlike that of grieving. The professional attitude gave way to pride, frequently reflected in a question such as: could anyone else do a better job than us? The inevitability of the demise of the Squadron was reiterated each time a question was confirmed, and with it came an end to the certainty about the future of work—the certainty that was the basis for commitments. People started to feel threatened by the prospect that lay ahead: what was to become of them, what should they be concentrating on now? Management tried to anticipate questions to prop up the once positive attitude we had. There was anger about not knowing the answers to questions of individual survival, but there was enough respect, loyalty and discipline to keep this anger from showing. Nevertheless, the men now had other things than work on their mind. The attitude changed, irreversibly.

An air of resignation was carefully steered into an activity that nearly everyone could accept—a farewell party for No 2 Squadron. Planning time was short, but we would do our best. There were momentoes to be made and procured, a final parade to be planned, important people and past members to be invited, and much more. The tempo gathered with the ever increasing pace of the approaching final day, and somehow this recaptured the imagination that was part of the unit character of the past. It had a magic that flows from the histories of once proud units such as you find in photo albums, except we were living that magic in making the last day of the Squadron one to remember. As it turned out, not all the planned activities would fit into one day, so the party raged for several. There were final dinners in the messes, there were farewell drinks in the hangar, there was a ceremony outside the front gate, we had our parade and laid up the proud colours, and then there was the final fly past. The whole thing took days to do, and it was done right—in typical No 2 Squadron fashion. It was a fitting end for a Squadron with such a distinguished and meritorious history. And if history repeats itself, as so often it does, then I am as confident as I am hopeful that we shall see the colours of No 2 Squadron fly again as an active unit in our Air Force.[55]

At the time of the disbandment, No 2 Squadron's personnel had been planning for tomorrow and working to get the job done today. Some might say they did it with yesterday's equipment—the Canberra had become old. Airmen of World War I remembered the SE-5A, and those of World War II recollected the Hudson or Mitchell. To the postwar airman, No 2 Squadron *was* the Canberra. She was the *Queen of the Skies*, and she had served No 2 Squadron well, for almost 30 years. Vale the Canberra—but the spirit of the Squadron's traditions lives on.

55 Adrian Slootjes, letter to author, of 7 Feb 1994.

EPILOGUE

The rapid disbandment of No 2 Squadron, with what appeared as indecent haste, did cause concern for past members of the unit. The Squadron Standard was held for many years at Amberley, the home of No 82 Wing, but was finally laid to rest at Point Cook on 1 November 1993.

Many cannot believe that this can be the end of such a notable unit, the only Australian squadron that has served in four wars. The purchase of fifteen additional F-111 bombers, the F-111G model, was announced by the Government in October 1992. These aircraft were ferried from the United States to join No 82 Wing over 1993–94. In these days of shrinking defence spending, it is unlikely another regular bomber squadron bearing the number 2 will be formed. However, as the RAAF pursues the use of Reserve forces, perhaps Reserve aircrew may be allocated more flying hours, and be held in No 2 Squadron. To be equipped with such a formidable exponent of air power as the F-111 would be a most suitable new chapter for the Squadron with the highest traditions.

The laying up of No 2 Squadron's Colours, at Point Cook in November 1993 RAAF

Appendix 1

BATTLE HONOURS

No 2 Squadron

Date	*Commanding Officer*	*Base*	*Aircraft*
SQUADRON FORMED KANTARA, EGYPT			
20.9.16	Capt W O Watt	Kantara	----
23.12.16	Maj W O Watt		
30.1.17		Harlaxton	Pup, Strutter, 504
7.17			DH-5
23.9.17		Baizieux	
7.12.17			SE-5A
6.1.18		Auchel	
22.1.18		Savy	
16.2.18	Maj W Sheldon		
2.4.18		Bertangles	
4.4.18		La Bellevue	
22.5.18	Maj A M Jones MC		
4.6.18		Foquerolles	
21.6.18		Liettres	
1.7.18		Reclinghem	
1.10.18		Serny	
20.10.18		Auchel	
26.10.18		Pont-a-Marcq	
10.1.19		Hellemmes	
1.2.19		Serny	
4.3.19		Hurdcott	----
6.5.19	SQUADRON SAILED FOR AUSTRALIA, DISBANDED		
	REFORMED CADRE BASIS		
10.1.22	Flt Lt A H Cobby DSO DFC*	Point Cook	SE-5A
1.7.22	SQUADRON DISBANDED		
	SQUADRON REFORMED		
3.5.37	Sqn Ldr J H Summers	Laverton	Demon
28.10.37			Demon, Bulldog
29.11.37			Demon, Bulldog, Anson
1.2.38	Wg Cdr J H Summers		
1.3.39	Wg Cdr A M Charlesworth AFC		
24.7.39			Anson
3.12.39	Sqn Ldr F W Thomas		
12.6.40			Hudson
15.4.41	Sqn Ldr F Headlam		
1.7.41	Wg Cdr F Headlam		
7.12.41		Penfoei, Namlea	
10.1.42		Penfoei	
19.2.42		Darwin	

20.2.42		Darwin, Daly Waters	
20.4.42	Sqn Ldr A B McFarlane	Darwin	
14.5.42	Wg Cdr A B McFarlane		
3.7.42		Darwin/Rapid Creek	
16.8.42		Batchelor	
24.10.42	Sqn Ldr R H Moran		
1.12.42	Wg Cdr R H Moran		
5.4.43	Wg Cdr J M Whyte		
12.4.43		Hughes	
7.7.43	Sqn Ldr C C Williams		
10.7.43	Sqn Ldr K M Kemp DFC (Temp)		
5.8.43	Wg Cdr K S Hennock		
2.12.43			Hudson, Beaufort
23.3.44	Wg Cdr L A Ingram		
1.5.44			Beaufort, Mitchell
22.5.44			Mitchell
7.7.44	Sqn Ldr J S Lee (Temp)		
20.7.44	Wg Cdr DW I Campbell DFC		
29.11.44	Wg Cdr T S Ingledew		
31.5.45	Wg Cdr L A Douglas DFC		
22.8.45		Balikpapan	
21.12.45		Laverton	----
15.5.46	SQUADRON DISBANDED		
	SQUADRON REFORMED From No 21 SQN		
23.2.48	Wg Cdr A E Cross AFC	Amberley	Lincoln, Liberator
6.9.48	Flt Lt F B M Wilson DFC (Temp)		
29.10.48	Sqn Ldr A B Jay AFC		
1.49			Lincoln
18.4.49	Sqn Ldr R N Dalkin DFC		
17.11.49	Flt Lt J P Graney AFC (Temp)		
3.3.51	Sqn Ldr J M Sutherland		
1.8.52	Sqn Ldr W M Dixon DSO DFC (RAF)		
30.1.53	Sqn Ldr P F Raw DFC		
23.2.53	Flt Lt F J Montgomery AFC (Temp)		
15.6.53	Flt Lt F M Griggs DFC* DFM (Temp)		
27.7.53	Wg Cdr A D J Garrisson		
18.12.53	Sqn Ldr P F Raw DFC AFC		Canberra
13.5.55	Maj R K Stoner USAF (Temp)		
11.7.55	Wg Cdr J B Fitzgerald		
5.12.55	Wg Cdr A E Cross AFC		
30.3.57	Wg Cdr R A Hosking DFC AFC		
21.4.58	Wg Cdr C W Steley DFC		
1.7.58		Butterworth	
1.11.60	Wg Cdr H W Connolly DFC* (Temp)		
16.1.61	Wg Cdr S W Dallywater		
14.1.63	Wg Cdr F M Griggs DFC* DFM		
14.12.64	Sqn Ldr L R Watkin (Temp)		
15.1.65	Wg Cdr L Britt DFC		
28.11.66	Wg Cdr R B Aronsen		
19.4.67		Phan Rang	

24.11.67	Wg Cdr S D Evans AFC	
18.11.68	Wg Cdr J A Whitehead	
10.11.69	Wg Cdr J R Boast AFC	
1.11.70	Wg Cdr F J L Downing	
14.3.71	Sqn Ldr A Barnes AFC (Temp)	
24.3.71	Sqn Ldr B E O'Shea (Temp)	
7.4.71	Wg Cdr T H Thorpe AFC	
9.6.71		Amberley
24.7.73	Sqn Ldr B Squires (Temp)	
18.12.73	Wg Cdr G Dyke DFC	
12.1.76	Wg Cdr P D Jones	
1.12.77	Wg Cdr G J J Beck	
1.11.79	Wg Cdr A W Stephens	
21.12.81	Wg Cdr A J Wills	
7.6.82	Sqn Ldr A Slootjes (Temp)	
31.7.82	SQUADRON DISBANDED	

Appendix 2
HONOUR ROLL
No 68 (Australian) Squadron RFC
No 2 Squadron AFC
World War I

Captain S K Muir	Killed in Accident	12 Sep 1917
Lieutenant D G Morrison	Died of Wounds	29 Oct 1917
Lieutenant D G Clark	Killed in Action	22 Nov 1917
Lieutenant S W Ayers	Died of Wounds	23 Nov 1917
Lieutenant A Griggs	Died of Wounds	23 Nov 1917
Lieutenant G H Cornell	Killed in Accident	11 Dec 1918
Captain J Bell	Died of Wounds	27 Dec 1917
2/Lieutenant P H Lawson	Killed in Accident	6 Jan 1918
Captain R W Howard MC	Died of Wounds	22 Mar 1918
Lieutenant T Hosking	Killed in Action	28 Mar 1918
Lieutenant L J Primrose	Killed in Accident	4 Jun 1918
Lieutenant T J Hammond	Killed in Action	12 Jun 1918
Lieutenant C R Ebeling	Died of Injuries	23 Aug 1918
Lieutenant J A McKeown	Killed in Action	14 Oct 1918
Lieutenant F Howard	Killed in Action	27 Oct 1918
Lieutenant F C Markham-Mills	Died of Illness	18 Dec 1918
2 A/M G L Taylor	Died of Illness	11 Feb 1919
Lieutenant G H Cowan	Died of Illness	22 Feb 1919

Appendix 3
AWARDS
No 68 (Australian) Squadron RFC
No 2 Squadron AFC
World War I

MILITARY CROSS (MC)

Captain G C Wilson DCM
Captain R C Phillipps
Captain F G Huxley
Lieutenant H Taylor MM
Lieutenant L H Holden
Lieutenant R W Howard

BAR to MILITARY CROSS (MC*)[1]

Captain R C Phillipps MC

DISTINGUISHED FLYING CROSS (DFC)[2]

Captain R L Manuel
Captain H G Forrest
Major A M Jones MC
Captain R C Phillipps MC*
Captain A T Cole MC
Captain E D Cummings
Captain F R Smith MC

BAR to DISTINGUISHED FLYING CROSS (DFC*)

Captain R L Manuel DFC
Major A M Jones MC DFC

MILITARY MEDAL (MM)

Sergeant R Lonsdale
Sergeant B F Jones
Corporal W B Campbell
1 A/M H S Raphael

MERITORIOUS SERVICE MEDAL (MSM)

Warrant Officer H P Hansen

1 * Indicates a Bar to the award.

2 The DFC was introduced on 3 Jun 1918, to replace the MC for gallantry in the air in combat.

Appendix 4

HONOUR ROLL

No 2 Squadron RAAF

World War II

Flt Lt R E Hitchcock	Killed in Accident	13 Aug 1940
Plt Off R F Wiesener	Killed in Accident	13 Aug 1940
Cpl J F Palmer	Killed in Accident	13 Aug 1940
AC C J Crosdale	Killed in Accident	13 Aug 1940
LAC C R Liebke	Died of Illness	28 Aug 1940
Flg Off P C Gorrie	Killed in Action	12 Jan 1942
Sgt W L Morgan	Killed in Action	12 Jan 1942
Sgt J E Farrington	Killed in Action	12 Jan 1942
Sgt K R T McCrackan	Killed in Action	12 Jan 1942
Flt Lt P H R Hodge	Killed in Action	12 Jan 1942
Sgt H C Shore	Killed in Action	12 Jan 1942
Sgt J Mawdsley	Killed in Action	12 Jan 1942
Flt Lt R W B Cuming	Killed in Action	20 Jan 1942
Plt Off R H Martin	Killed in Action	20 Jan 1942
Plt Off P H Richards	Killed in Action	20 Jan 1942
Sgt J Bessell-Browne	Killed in Action	20 Jan 1942
Sgt A E Blake	Killed in Action	20 Jan 1942
Sgt H J Dean	Killed in Action	20 Jan 1942
Sgt K H Kelaher	Killed in Action	20 Jan 1942
LAC O F E Crawford	Killed in Action	20 Jan 1942
LAC E G Mears	Killed in Action	20 Jan 1942
LAC V H Smith	Killed in Action	20 Jan 1942
AC F Morris	Killed in Action	20 Jan 1942
Sgt N K Rayner	Killed in Action	14 Feb 1942
Flt Lt W V D White	Executed by Japanese	20 Feb 1942
Sgt J Baker	Executed by Japanese	20 Feb 1942
Sgt I W Read	Executed by Japanese	20 Feb 1942
F Sgt L P Schroeter	Died of Illness	9 Mar 1942
Sgt J G Wright	Killed in Action	23 Mar 1942
Sgt D G J Horsburgh	Killed in Action	23 Mar 1942
Sgt J R Maddern	Killed in Action	23 Mar 1942
Flg Off P H A Thompson	Died of Illness	31 Mar 1942
Cpl R E Andrews	Died of Illness	13 Apr 1942
Sgt D A Miller	Died of Injuries	14 Apr 1942
Sgt H F Hearle	Executed by Japanese	8 May 1942
Sgt W D Witham	Executed by Japanese	8 May 1942
Flg Off J H Venn	Killed in Action	13 May 1942
Sgt W J James	Killed in Action	13 May 1942
Sgt B A Kilpatrick	Killed in Action	13 May 1942
Sgt L A Omsby	Executed by Japanese	13 May 1942
Plt Off K I Kersten	Killed in Action	22 May 1942

Plt Off W C McLay	Killed in Action	22 May 1942
Flg Off R C Muecke	Killed in Action	30 Jul 1942
Flg Off F S Moss	Killed in Action	30 Jul 1942
Plt Off L M Miners	Killed in Action	30 Jul 1942
Sgt J E Laman	Killed in Action	30 Jul 1942
Flg Off R D J Tregonning	Killed in Action	30 Jul 1942
Plt Off S W Faull	Killed in Action	21 Aug 1942
Sgt W R Edeson	Killed in Action	21 Aug 1942
Sgt F M O'Reilly	Killed in Action	21 Aug 1942
Sgt W H Gould	Killed in Action	21 Aug 1942
Flg Off K L McDonnell	Killed in Action	15 Sep 1942
Flg Off K J Ahern	Killed in Action	15 Sep 1942
Flg Off V Treloar	Killed in Action	15 Sep 1942
Sgt G S Dobbs	Killed in Action	15 Sep 1942
Sgt G Ward-Smith	Killed in Action	15 Sep 1942
Sqn Ldr S J Fraser	Killed in Accident	11 Oct 1942
Plt Off R K R Dunning	Killed in Action	25 Oct 1942
Sgt M L Kenihan	Killed in Action	25 Oct 1942
Sgt L H S Fraser	Killed in Action	25 Oct 1942
Sgt D W Francis	Killed in Action	25 Oct 1942
Flg Off G W C Allchin	Executed by Japanese	22 Nov 1942
Sgt L J Montgomerie	Executed by Japanese	22 Nov 1942
Sgt G J Thame	Killed in Action	4 Dec 1942
Flg Off M W C Johns	Killed in Action	27 Dec 1942
Plt Off J F Nixon	Killed in Action	27 Dec 1942
Sgt J D Horseman	Killed in Action	27 Dec 1942
Sgt D L Tyler	Killed in Action	27 Dec 1942
Sgt K Ross	Killed in Action	27 Dec 1942
Sqn Ldr M C Burns	Killed in Action	3 Apr 1943
Plt Off L R Jope	Killed in Action	3 Apr 1943
Sgt K Jones	Killed in Action	3 Apr 1943
Sgt R J Johnstone	Killed in Action	3 Apr 1943
F Sgt A A Dean	Killed in Action	3 Apr 1943
Flt Lt K E Daniel	Killed in Action	20 Apr 1943
Flg Off J Laws	Killed in Action	20 Apr 1943
Plt Off W J H Gove	Killed in Action	20 Apr 1943
Plt Off J G Butler	Killed in Action	20 Apr 1943
F Sgt T W Wilson	Killed in Action	20 Apr 1943
Flg Off K R Mills	Killed in Action	20 Apr 1943
Flg Off N E John	Killed in Action	20 Apr 1943
Plt Off J L Barnes	Killed in Action	20 Apr 1943
Plt Off T W Bassan	Killed in Action	20 Apr 1943
F Sgt H H K Hadley	Killed in Action	20 Apr 1943
Flg Off R R James	Killed in Action	28 Apr 1943
Plt Off R D Ryan	Killed in Action	28 Apr 1943
W Off H E Cutten	Killed in Action	28 Apr 1943
F Sgt K G Keech	Killed in Action	28 Apr 1943
F Sgt P S Reen	Killed in Action	28 Apr 1943
Sgt R E Norris	Killed in Action	28 Apr 1943
Flg Off D R Hicks	Killed in Action	7 May 1943
W Off V L Jackman	Killed in Action	7 May 1943

F Sgt E Quinn	Killed in Action	7 May 1943
F Sgt I W MacKenzie	Killed in Action	7 May 1943
Sgt D A Emery	Killed in Action	7 May 1943
Flt Lt J F Mason	Killed in Accident	12 Jun 1943
Flt Lt J R Wood	Killed in Accident	12 Jun 1943
Flg Off L G Knauer	Killed in Accident	12 Jun 1943
Flg Off T W Swann	Killed in Accident	12 Jun 1943
Sgt G R Rowland	Killed in Accident	12 Jun 1943
Sqn Ldr C C Williams	Killed in Action	9 Jul 1943
Flt Lt A McP Fowler	Killed in Action	9 Jul 1943
Flg Off W A Hepworth	Killed in Action	9 Jul 1943
Flg Off R C Lindsey	Killed in Action	9 Jul 1943
F Sgt D C Farmer	Killed in Action	9 Jul 1943
F Sgt J B Lamb	Killed in Action	15 Aug 1943
F Sgt C A Ikin	Killed in Action	15 Aug 1943
F Sgt M H Jackson	Killed in Action	15 Aug 1943
F Sgt N D McLean	Killed in Action	15 Aug 1943
F Sgt T W Wise	Killed in Action	15 Aug 1943
Flg Off L A Martin	Killed in Action	8 Sep 1943
Plt Off M C Wettenhall	Killed in Action	8 Sep 1943
Plt Off R H Williams	Killed in Action	8 Sep 1943
Flg Off J P Oldridge	Killed in Action	15 Oct 1943
F Sgt D N Rumble	Killed in Action	15 Oct 1943
F Sgt J F G Cullen	Killed in Action	15 Oct 1943
Sgt F R Drake-Brockman	Killed in Action	15 Oct 1943
Sgt D W Petch	Killed in Action	15 Oct 1943
F Sgt J W R Boanas	Executed by Japanese	31 Oct 1943
F Sgt M A Graham	Executed by Japanese	31 Oct 1943
Flt Lt J S Austin	Died of Illness	9 Nov 1943
Plt Off C D Brockhurst	Killed in Action	3 Apr 1944
Plt Off R J Philpot	Killed in Action	3 Apr 1944
Plt Off R L Black	Killed in Action	3 Apr 1944
Plt Off H J Gillies	Killed in Action	3 Apr 1944
F Sgt R G Frew	Killed in Action	3 Apr 1944
Flg Off D C Beddoe	Killed in Captivity	24 May 1944
W Off I S McCallum	Killed in Action	19 Jul 1944
W Off J S McC Campbell	Killed in Accident	6 Aug 1944
W Off A K Griesbach	Killed in Accident	6 Aug 1944
Sgt F H Conaghan	Killed in Accident	6 Aug 1944
Flg Off D G Lane	Died of Injuries	8 Aug 1944
Sqn Ldr J L Scott	Executed by Japanese	16 Aug 1944
F Sgt R I M King	Executed by Japanese	16 Aug 1944
F Sgt B E Wallace	Executed by Japanese	16 Aug 1944
F Sgt K C Wright	Executed by Japanese	16 Aug 1944
Plt Off A K Buckland	Died of Injuries	19 Aug 1944
Flt Lt H C Easton	Killed in Accident	2 Sep 1944
W Off G S J King	Killed in Accident	2 Sep 1944
Flg Off A W Slater	Killed in Action	22 Sep 1944
F Sgt K R Philipson	Killed in Action	22 Sep 1944
Flg Off M S Millett	Killed in Action	22 Sep 1944
Flg Off J F Daggett	Killed in Action	22 Sep 1944

Flg Off B A Wisniewski	Killed in Action	22 Sep 1944
Sgt D F Harberger	Killed in Action	22 Sep 1944
Flt Lt J H Selway	Killed in Action	4 Nov 1944
Plt Off H B Worman	Killed in Action	4 Nov 1944
Flg Off A E Pott	Killed in Action	4 Nov 1944
W Off J F Stormon	Killed in Action	4 Nov 1944
W Off A E Hawkins	Killed in Action	4 Nov 1944
F Sgt R C Palfreyman	Killed in Action	4 Nov 1944
Flg Off W F E Thompson	Killed in Action	20 Dec 1944
Flg Off L T Forsyth	Killed in Action	20 Dec 1944
F Sgt J A Rolfe	Killed in Action	20 Dec 1944
F Sgt T H Rowlands	Killed in Action	20 Dec 1944
W Off J E S Thompson	Killed in Action	20 Dec 1944
W Off F H Mathews	Killed in Action	20 Dec 1944
Flt Lt E M White	Killed in Accident	14 Aug 1945
Flg Off A K Morrell	Killed in Accident	14 Aug 1945
F Sgt B M O'Brien	Killed in Accident	14 Aug 1945
F Sgt R A MacGregor	Killed in Accident	14 Aug 1945
Cpl W J C Maxwell	Killed in Accident	14 Aug 1945
LAC I S Coleman	Killed in Accident	14 Aug 1945
LAC B A Marshall	Killed in Accident	14 Aug 1945
LAC V A Morgan	Killed in Accident	14 Aug 1945
Flt Lt L A Kirk	Killed in Accident	15 Sep 1945
Flg Off L Bishop	Killed in Accident	15 Sep 1945
Flg Off P A Taylor	Killed in Accident	15 Sep 1945
WOff C R M Ricketts	Killed in Accident	15 Sep 1945
F Sgt F J Stolweather	Killed in Accident	15 Sep 1945
Cpl R O Byrne	Killed in Accident	15 Sep 1945
LAC M S White	Killed in Accident	15 Sep 1945
LAC E A Zosky	Died of Illness	30 Jan 1946
LAC E P Kenny	Died of Illness	18 Feb 1946

Appendix 5
AWARDS
No 2 Squadron RAAF
World War II

MEMBER OF THE ORDER OF THE BRITISH EMPIRE (MBE)

Flt Lt J L MacAlister
Flg Off A F Cole
Flt Lt B Rofe

DISTINGUISHED FLYING CROSS (DFC)

Wg Cdr K S Hennock
Wg Cdr A B McFarlane
Wg Cdr J M Whyte
Sqn Ldr S Austin
Sqn Ldr S J Fraser
Sqn Ldr K M Kemp
Sqn Ldr J S Lee
Flt Lt J S Austin
Flt Lt N T Badger
Flt Lt A L Greenaway
Flt Lt G Hodges
Flt Lt J L MacAlister
Flt Lt H D Mackenzie
Flt Lt I A McCombe
Flt Lt F K Morcombe
Flt Lt R F Overheu
Flt Lt J C Simpson
Flg Off M M Helsham
Flg Off W V D White
Plt Off A W Hawkesford
Plt Off A J Sharp

DISTINGUISHED FLYING MEDAL (DFM)

FSgt W L Allen
FSgt M L Franklin
FSgt B M Hocking
FSgt E A Keith
FSgt H D H Learmonth
FSgt E R Oldfield
Sgt A J Crowe
Sgt R C Lindsey

AIR FORCE MEDAL (AFM)

WOff G Wiburd

BRITISH EMPIRE MEDAL (BEM)

LAC L R Borgelt

MENTIONED IN DISPATCHES (MID)

Flt Lt H D Gillett
Flt Lt I H A Hay
Flt Lt P T Sanders
Flt Lt J P Squires
Flg Off A A Cambridge
Flg Off K E Coughlan
Flg Off F Moriarty
Flg Off A B Weston
Plt Off R R James
Sgt M A McDonald
Sgt P S Reen
Sgt H Reilly
Sgt F V Webster
LAC C R H Francis

Appendix 6
HONOUR ROLL
No 2 Squadron RAAF
Vietnam

AC T G Petith	Killed in Accident	21 Apr 1969
Flg Off R M O'Hanlon	Died of Illness	3 Jul 1969
Corporal B F Fitzpatrick	Killed in Accident	3 Dec 1969
LAC I A Wooley	Killed in Accident	11 Dec 1969
Flying Officer M P Herbert	Missing in Action	3 Nov 1970
Pilot Officer R C Carver	Missing in Action	3 Nov 1970
AC R D Hewitt	Died of Illness	21 Jun 1972

Appendix 7
AWARDS
No 2 Squadron RAAF
Vietnam

DISTINGUISHED SERVICE ORDER

Gp Capt S D Evans AFC
Wg Cdr J R Boast AFC
Wg Cdr J A Whitehead

MEMBER OF THE ORDER OF THE BRITISH EMPIRE (MBE)

Chaplain P J McCormick
Flt Lt G W Foskett

BAR to DISTINGUISHED FLYING CROSS (DFC*)

Wg Cdr V J Hill DFC

DISTINGUISHED FLYING CROSS (DFC)

Wg Cdr F J L Downing
Sqn Ldr A Barnes AFC
Sqn Ldr R D Crump
Sqn Ldr G Dyke
Sqn Ldr I L Grove AFC
Flt Lt E B J Bolger
Flt Lt G M Cramer

MILITARY MEDAL (MM)

Sgt N E Power

BRITISH EMPIRE MEDAL (BEM)

Flt Sgt C M Robb
Sgt A W Rudd
Cpl J B Vanderkyl

MENTIONED IN DISPATCHES (MID)

Wg Cdr H A Hughes
Wg Cdr F R Lonie
Sqn Ldr J M Chesterfield
Sqn Ldr B Squires
Sqn Ldr S N Todhunter
Sqn Ldr J I Thomson
Sqn Ldr B E O'Shea
Sqn Ldr B A Johnson
Sqn Ldr T D Wright
Chaplain S B Ford
Chaplain D T Pritchard
Sqn Ldr L T Winn
Flt Lt L D Brown
Flt Lt G N Goddard
Flt Lt B G Hammond
Flt Lt R W Howe
Flt Lt M F Linden
Flt Lt D K Palmer
Flt Lt J A Power
Flt Lt B E S Williamson
Flt Lt A J Taylor
Flg Off P V Murphy
Flg Off P R Nuske
Flg Off J J Wilkinson
Plt Off J K Ross
WOff L Alderton
FSgt J M Hayes
FSgt J Penn
FSgt H K Rieck
Sgt R O'Neill
Sgt K W Marks
Cpl A D Gale
Cpl C J R Gray
Cpl M D Johnson
Cpl A E Matthews
LAC D J Dubber

APPENDIX 8

DH-5 AIRCRAFT OF NO 68 (AUSTRALIAN) SQUADRON RFC

Serial	*Date On Sqn*	*Code*	*Date Off Sqn*	*Remarks*
A9197	7.17	–	Unknown	named "NSW No 15, The Upper Hunter", presumed returned to RFC
A9224	9.17	5	17.11.17	damaged engine fail Arras 20.10.17 (Taylor), engine failure Warloy (Taylor), to 2 ASD
A9226	9.17	E	11.10.17	damaged engine fail Roequinqy (James), 2 ASD
A9242	9.17	W	1.10.17	named "Women's Battleplane", engine failure Honnecourt (Morrison)
A9245	9.17	C	11.10.17	named "Upper Hunter Battleplane", damaged when crashed landing (Holden), to 2 ASD
A9255	22.11.17	V	19.12.17	to 2 ASD
A9263	22.9.17	6	23.11.17	shot down Bourlon Wood (Ayers killed)
A9265	9.17	2	22.11.17	shot down by groundfire (Pratt wounded)
A9271	22.9.17		2.10.17	forced down behind lines (Agnew POW)
A9273	9.17	B	16.10.17	engine failure Gouy (Bartlam), to 2 ASD
A9275	.17	–	12.9.17	crashed Harlaxton (Muir killed)
A9277	3.10.17	W	13.10.17	shot down Queant (Morrison killed)
A9278	13.10.17	C	23.11.17	damaged by EA nr Bapaume 20.11.17 (Holden), to 2 ASD
A9279	14.10.17	W	7.12.17	damaged by EA Bourlon Wood 6.12.17 (Johnson), to 2 ASD
A9284	9.17	A	16.10.17	damaged by EA Gouy (Howard) written-off
A9288	9.17	3/D?	7.12.17	to 2 ASD
A9292	14.10.17	E	15.12.17	to 2 ASD
A9294	17.10.17	A?	22.11.17	damaged by EA, crashed landing (Howard), to 2 ASD
A9324	1.10.17	–	1.10.17	damaged, engine fail on delivery (Huxley), to 2 ASD
	1.12.17	F	11.12.17	stalled at Baizeux (Cornell killed)
A9326	22.11.17	C	23.11.17	damaged by EA Bourlon Wood (Holden)
A9336	23.11.17	5	4.12.17	damaged by crashed landing (Taylor), to 2 ASD
A9338	1.12.17	2	15.12.17	to 2 ASD
A9341	23.11.17	X?	1.12.17	damaged by EA Wagholien (Benjamin), to 2 ASD
A9344	24.11.17	C	15.12.17	to 2 ASD
A9363	9.17	–	9.17	not deployed to France
A9378	18.11.17	5	20.11.17	shot down by groundfire (Taylor)
A9395	9.17	–	9.17	not deployed to France
A9399	3.10.17	F	20.11.17	shot down by groundfire Marcoing (Ward POW)
A9402	9.17	–	9.17	not deployed to France

Left to right from top:

A9242 at Harlaxton, September 1917, before proceeding to France, and becoming coded "W" in "C" Flight AWM C1858

A 9245 at Harlaxton, the "Upper Hunter Battleplane" AWM C1855

A9449 at Baizieux, with Capt G C Wilson in the cockpit, showing the white rear fuselage band (indicating 68 Squadron), the number "1" (indicating "B" Flight), and the flight commander's pennant *AWM E1444*

A9432 at Harlaxton — it appears that this aircraft was not flown by 68 Squadron in France AWM C2800

A9395 at Harlaxton, again a training machine in England not deployed to Baizieux AWM C1856

A9363, devoid of any presentation inscriptions and squadron markings, only served a short time on 68 Squadron at Harlaxton RAAF

A9428	24.10.17	U	23.11.17	shot down by groundfire Bourlon (Griggs killed)
A9432	9.17	–	9.17	not deployed to France
A9445	5.12.17	X?	19.12.17	to 2 ASD
A9449	10.10.17	1	19.12.17	to 2 ASD
A9451	22.11.17	Z?	10.12.17	WFS due to unsatisfactory condition
A9457	9.17	X	20.11.17	shot down by groundfire (Sheppard wounded)
A9459	9.17	D,3	19.12.17	to 2 ASD
A9461	3.10.17	4	7.12.17	damaged by EA Bourlon Wood (Robertson), to 2 ASD
A9462	9.17	4	1.10.17	to 2 ASD
A9464	9.17	1	14.10.17	damaged engine fail 9.10.17 (Wilson), to 2 ASD
A9466	21.11.17	Y	1.12.17	damaged by groundfire (Robertson), to 2 ASD
A9469	9.17	U	24.10.17	damaged engine failure Baizeux 23.10.17 (Matthews)
A9473	9.17	V	22.11.17	shot down groundfire 20.11.17 (Bell killed), to 2 ASD
A9477	17.10.17	B	22.11.17	failed to return from Bourlon (D Clark killed)
A9483	9.17	Y	20.11.17	damaged by EA nr Bapaume (Robertson), to 2 ASD
A9495	13.12.17		15.12.17	to 2 ASD
A9517	23.11.17		15.12.17	damaged by EA Bourlon 29.11.17 (Howard), 2 ASD
A9530	24.11.17		30.11.17	forced land Bapaume 29.11.17 (Grant), to 2 ASD
A9531	23.11.17	6	19.12.17	to 2 ASD
A9532	23.11.17		30.11.17	shot down by EA Bourlon Wood (Cornell)
A9535	24.11.17		7.12.17	to 2 ASD
A9536	23.11.17	–	23.11.17	forced landed on delivery at Couin (Grant)
A9537	24.11.17		15.12.17	to 2 ASD
A9544	1.12.17	Z	7.12.17	damaged by EA, crashed Fremlicourt 6.12.17 (McKenzie), to 2 ASD
B377	22.9.17	Z	20.11.17	damaged by EA Queant (McKenzie)
B4938	4.12.17	Y	19.12.17	to 2 ASD

Notes:

Aircraft shown as on Squadron strength at 9.17 are the original 15 machines flown to France on 21.9.17, and as records do not exist prior to that date, it is possible these aircraft could have been on strength as early as 7.17.

The use of Squadron code letters is described in Chapter 4. The symbol "?" beside a code letter indicates this is unconfirmed, but interpolated from the pilot who flew the aircraft.

Most aircraft that were damaged were returned to 2 ASD where they were struck off charge as being uneconomic for repair, most probably influenced by the type's imminent withdrawal from service.

Abbreviations:

ASD Aeroplane Supply Depot
EA enemy aircraft
WFS withdrawn from service

Appendix 9

SE-5A Aircraft of No 2 Squadron AFC

Serial	*Date On*	*Code*	*Date Off*	*Remarks*
A4856[1]	27.10.17	–	8.12.17	first SE-5 for No 68 Sqn, to 24 Sqn RFC
A8906	19.12.17		9.1.18	forced landing Yvrench (Truscott), 2 ASD—repaired
	21.3.18		18.4.18	damaged mid-air collision 17.4.18 (Davies), to 1 ASD
A8913	9.3.18		28.3.18	failed to return from patrol (Hosking killed)
A8936	15.12.17	C	22.1.18	wrecked, returned to 1 ASD
B12	19.12.17		8.3.18	ex-56 Sqn, forced landing Marieux (Truscott) written-off
B55	7.12.17		6.1.18	crashed on landing Auchel (Lawson killed)
B61	7.12.17		16.2.18	crashed on landing (Paxton) written-off
B64	7.12.17		27.12.17	crashed on landing (A Clark) written-off
B70	6.12.17		4.6.18	crashed take-off Foquerolles (Hamilton) written-off
B74	15.12.17		3.1.18	ex-64 Sqn,crashed landing (Taylor), to AD
B102	21.2.18		28.3.18	shot down by EA at Etaing (Flight POW)
B149	30.3.18		7.4.18	damaged landing Bertangles 30.3.18 (Hamilton), 1 ASD
B150	7.4.18		20.4.18	damaged landing La Bellevue (Cox), to 1 ASD
B181	23.3.18		28.3.18	damaged by EA (Holden), to 1AD
B184	30.3.18		4.7.18	forced landing Tancqued 1.7.18 (Alberry), to 1 ASD
B185	18.3.18		6.4.18	forced landing nr Doullens 5.4.18 (Sexton), to 1 ASD
B187	30.3.18		5.8.18	to 2 ASD
B188	25.4.18		3.5.18	shot down by EA nr Merville (Cummings)
B194	12.4.18		30.5.18	forced landing Bertangles 28.5.18 (Primrose), to 2 ASD
B195	4.5.18		10.6.18	forced landing 8.6.18 Cachy le Grand (Cummings) to 1 ASD
B504	15.12.17		24.1.18	ex-56 Sqn, forced landing (Power), to 1 ASD
B505	19.12.17		14.3.18	ex-56 Sqn, forced landing 12.3.18 (Benjamin), 1 ASD
B513	23.2.18		13.3.18	crashed landing 11.3.18 (Follett), to 1 ASD
B525	28.3.18		1.6.18	forced down behind lines Bourchoir (Rackett POW)

1 A4895 was an SE-5 model, powered by a 150hp Hispano-Suiza, provided to the Squadron for training. Other machines were SE-5A models powered by the 200hp Hispano-Suiza, and later by the 200hp Wolseley Viper.

Left to right from top:

A8936, "C" of "A" Flight, was flown by Lt Holden, before being wrecked on take-off at Auchel on 22 January 1918. RAAF

Unidentified SE-5A, "5" of "B" Flight, later in 1918, showing the "5" also marked on the upper wing and the 2 Squadron boomerang marking replaced by a white bar RAAF

SE-5A "W" of "C" Flight (possibly C9489), damaged in a landing mishap, early 1918 RAAF

A later "W" of "C" Flight, C1948, which joined 2 Squadron in July 1918 to see six months service with the unit RAAF

B535	9.1.18		5.3.18	damaged by EA 21.2.18 (Lang), time expired to 1 ASD
B571	15.3.18		17.5.18	crashed landing (Jones), to 2 ASD
B579	19.12.17	Y?	18.3.18	forced landing 16.3.18 (Sexton), to 1 ASD
B619	5.1.18	E	21.2.18	failed to return from patrol (Logan POW)
B699	7.12.17		12.12.17	returned to AD
B4858	17.12.17	A	5.2.18	damaged landing (Phillipps), to 1 ASD
B4859	15.12.17	E	3.1.18	crashed landing Baizieux (R Clark), written-off
B4895	7.12.17		26.6.18	WFS war worn after 145 hours over lines, to 1 ASD
B8237	5.3.18		14.3.18	forced landing 12.3.18 (Rackett), to 1 ASD
B8392	24.8.18	1	9.10.18	damaged landing 27.8.18 (Currie), forced landing 7.10.18 (Long), to 1 ASD
B8400	11.5.18		11.6.18	landed with seized engine 2.6.18 (Cox), to AD
C1060	21.1.18	U?	6.4.18	forced landing nr Marieux 5.4.18 (Truscott), to 1 ASD
C1125	28.6.18		29.11.18	damaged Reclinghem (Follett), to 6 ASD
C1129	17.10.18		19.2.19	to No 32 Squadron RAF
C1429	.2.18		18.2.18	forced landing (Moore), written-off
C1837	13.6.18		25.6.18	forced landing Eggewaertsceapple (Manuel), written-off
C1934	23.6.18		14.8.18	crashed landing Allonville 13.8.18 (Cole), to 2 ASD
C1948	7.7.18	W	31.1.19	last sortie on sqn 16.1.19, WFS
C5309	13.3.18		2.4.18	forced landing 29.3.18 (Follett), to 1 ASD
C5312	15.12.17	T	22.1.18	damaged landing Savy (Power), to 1 ASD
C5320	15.12.17		6.2.18	crashed Roclincourt (Brettingham-Moore), to 1 AD
C5328	29.12.17		18.2.18	forced landing Estres-Cauchee (Adams), to 1 ASD
C5347	22.1.18		18.3.18	forced landing (Rackett), to 1 ASD
C5349	24.1.18		6.4.18	damaged on transit to La Bellevue, to 1 ASD
C5380	16.2.18		8.3.18	crashed landing (Hosking) written-off
C5382	12.3.18		24.4.18	damaged landing La Bellevue 23.4.18 (Studdert), 2 ASD
C5387	21.2.18		15.3.18	crashed, engine fail Savy (J Adam), to 1 ASD
C5440	2.4.18		28.7.18	forced landing 25.7.18 Rubecourt (Disney), to 1 ASD
C5441	4.5.18		4.6.18	crashed low flying nr Clermont (Primrose killed)
C5442	2.4.18		17.6.18	forced landing Verrines 16.6.18 (Cummings), to 1 ASD
C5443	4.5.18		20.5.18	forced landing Savy 19.5.18 (Cameron), to AD
C6403	4.6.18		9.11.18	shot down by groundfire Enghien (F Smith)
C6473	18.6.18		15.11.18	WFS war worn after 219 hours, to 1 ASD
C8728	26.6.18		17.7.18	crashed taking off (Forrest), written-off
C8901	19.11.18		20.2.19	to No 32 Squadron RAF
C9489	19.12.17	W?	18.2.18	wrecked, to 1 ASD
C9496	9.1.18		27.6.18	WFS war worn after 181 hours
C9539	19.12.17	V	26.6.18	WFS war worn after 137 hours over lines, to 1 ASD

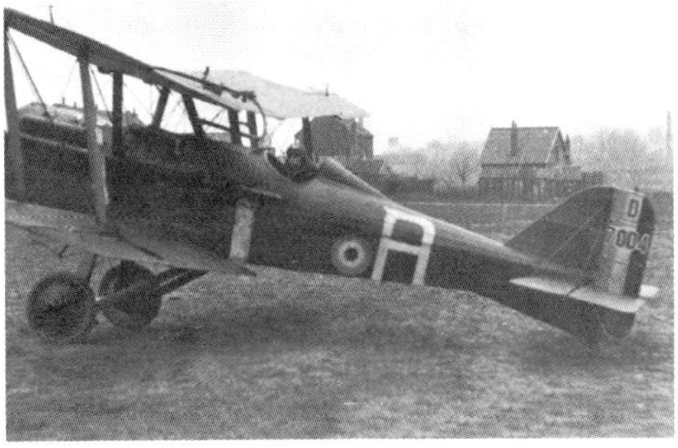

From top:

D7004, "A" of "A" Flight, joined 2 Squadron after the war's end, and was written off in an accident on 10 February 1919. RAAF

"C" Flight at Reclinghem August/September 1918 — visible aircraft are D6950 "V", D6903 "Z", and E5965 "X" RAAF

This photo of E5965 was most probably taken at the same time as the photo above. RAAF

C9541	6.2.18		5.8.18	WFS war worn after 142 hours 26.7.18, to 1 ASD
C9619	19.2.18		27.3.18	forced landing at La Gorgue (Smith), to 1 ASD
D204	15.3.18		7.4.18	forced landing Villers Bocage 6.4.18 (Benjamin), 1 ASD
D212	27.1.18		22.3.18	shot down over lines by EA (R Howard killed)
D214	26.1.18		11.4.18	forced landing Amiens 4.4.18 (Primrose), to 1 ASD
D379	29.10.18		31.1.19	WFS war worn
D406	28.7.18		26.10.18	to 1 ASD
D3429	30.3.18		10.6.18	damaged by groundfire 9.6.18, crashed landing (W Adams), to 1 ASD
D3431	7.4.18		25.7.18	forced landing 24.7.18 Reclingham (Alberry), to 1 ASD
D3502	16.3.18		27.3.18	shot down by groundfire Bozincourt (B-Moore)
D3512	20.4.18		10.5.18	forced landing Famecon (Studdert), to 2 ASD
D3514	7.4.18	T	30.5.18	forced landing La Bellevue 29.5.18 (Hammond), 2 ASD
D3535	19.4.18		4.5.18	forced landing Tangry 3.5.18 (Phillipps), to AD
D3923	24.4.18		5.5.18	damaged landing Auchel 3.5.18 (Cameron), to AD
D3924	7.4.18		24.4.18	forced landing Treizennes 22.4.18 (Cummings), 2 ASD
D3960	20.5.18		12.6.18	failed to return from patrol (Hammond killed)
D3962	17.5.18		10.7.18	crashed landing 8.7.18 (Knight), to 1 ASD
D6187	6.6.18		10.6.18	forced landing Courperque 8.6.18 (Simonson) to 1 ASD
D6190	6.6.18		10.7.18	forced landing 8.7.18 (Smith), to 1 ASD
D6191	6.6.18		9.6.18	forced landing St Laurent (Sexton), to 1 ASD
D6194	11.6.18		5.8.18	to 2 ASD
D6196	11.6.18		9.7.18	crashed landing 7.7.18 (Cox), to 1 ASD
D6860	31.5.18		21.10.18	forced landing Fleur-Paix 18.10.18 (Davies), to 1 ASD
D6869	30.5.18		8.10.18	damaged on landing 15.7.18 (Copp), to 1 ASD
D6891	12.7.18		7.9.18	to 1 ASD
D6903	26.6.18	Z	14.10.18	to 1 ASD
D6912	7.7.18		20.2.19	to No 32 Squadron RAF
D6913	26.6.18		26.8.18	ex-24 Sqn, crashed take-off Marquise 23.8.18 (Ebeling killed), to ASD
D6919	10.7.18		18.2.19	to 1 ASD
D6941	6.8.18		7.9.18	crashed landing 3.9.18 (Ross), to 1 ASD
D6948	5.8.18		13.8.18	forced landing in trenches 12.8.18 (Alberry), to ASD
D6950	26.7.18	V	27.10.18	failed to return from patrol (F Howard killed)
D6964	14.8.18		19.2.19	to No 32 Squadron RAF
D6968	6.8.18	Z	14.10.18	failed to return from patrol (McKeown killed)
D6995	13.8.18	Y	20.2.19	to No 32 Squadron RAF
D7004	29.11.18	A	10.2.19	crashed nr Fruges (Dibbs)
D7006	17.11.18		18.2.18	forced landing nr Marquise (Wellwood)
E1265	11.6.18		23.6.18	forced landing 22.6.18 Droinville (Simonson), to 2 ASD

E1351	2.7.18		9.7.18	crashed landing 7.7.18 (Jones), to 1 ASD
E3154	11.6.18		18.8.18	damaged 5.8.18 (Disney), to 1 ASD
E5659	8.10.18		20.2.19	to No 32 Squadron RAF
E5757	9.10.18	1	20.2.19	to No 32 Squadron RAF
E5765	21.10.18	A	15.2.19	crashed Aire, nr Marquise (Coy)
E5782	30.10.18	V	20.2.19	damaged landing 23.11.18 (Kay), 32 Squadron RAF
E5965	18.7.18	X	21.9.18	crashed behind lines (Cox POW)
E5989	6.8.18		14.10.18	failed to return from patrol (Cornish POW)
F860	22.9.18		20.2.19	to No 32 Squadron RAF
F930	29.11.18		20.2.19	to No 32 Squadron RAF
F940	29.11.18		20.2.19	to No 32 Squadron RAF
F5457	14.10.18	2	18.11.18	believed damaged 14.11.18 (Stuttert), to 1 ASD
F5465	7.9.18		26.10.18	crashed transit Auchel-Pont a Marcq 26.10.18 (Marshall), to 1 ASD
F5466	7.9.18		13.2.19	crashed nr Marquise (Denton)
F5511	17.10.18		20.2.19	to No 32 Squadron RAF
F8955	14.11.18		26.11.18	to 11th Salvage Section for 1 ASD
H7165	28.10.18		26.11.18	crashed landing 23.11.18 (Coy), 11th Salv Sect, 1 ASD

Notes

Quite often aircraft were wrecked in crashes and forced landings, but the action to write-off the machine was not taken until the wreckage was returned to Depot.

Some ADs not identified—records sometimes incorrectly abbreviate ASD to AD.

Abbreviations

AD Aircraft Depot
ASD Aeroplane Supply depot
WFS withdrawn from service

Appendix 10

CLAIMS BY No 2 SQUADRON AFC

World War I

Date	*Pilot*	*Aircraft No*	*EA Type*	*Location*	*Combat Rpt No*	*Remarks*
FLYING DH-5 AIRCRAFT						
22 Nov 17	Lt F G Huxley	A9461	Albatros D.V	Bourlon Wood	11	destroyed
22 Nov 17	Lt R W Howard	A9294	Albatros D.V	Graincourt	12	driven-down
			DFW C	Graincourt	12	driven-down
22 Nov 17	Capt R Phillipps	A9288	two-seater	Cambrai	13	driven-down
26 Nov 17	Lt H Taylor	A9336	DFW C	Bourlon Wood	14	destroyed
29 Nov 17	Lt R W Howard	A9517	DFW C	Bourlon Wood	15	driven-down
30 Nov 17	Capt G C Wilson	A9449	DFW C	Bantouzelle	18	destroyed
1 Dec 17	Lt F G Huxley	A9461	Aviatik C	Bourlon	20	destroyed
1 Dec 17	Lt R W McKenzie	A9451	Albatros D.V	Villers Guislain	22	destroyed
6 Dec 17	Lt F G Huxley	A9461	DFW C	nr Honnecourt	25	destroyed
FLYING SE-5A AIRCRAFT						
18 Feb 18	Capt F G Huxley	B4895	Albatros D.V	Lille	27	OOC
18 Feb 18	Capt R W Howard	D212	Albatros D.V	nr Carvin	29	OOC
19 Feb 18	Lt R W McKenzie	C9539	Albatros D.V	Pont a Vendin	31	destroyed
21 Feb 18	Lt A G Clark	B4895	Albatros D.V	Brebieres	35	OOC
21 Feb 18	Lt L Benjamin	C1060	Albatros D.V	Brebieres	35	OOC
8 Mar 18	Capt R W Howard	D212	two-seater	Beaumont	36	driven-down
8 Mar 18	Capt R W Howard	D212	two-seater	Henin Lietard	38	OOC shared Paxton
	Lt Paxton	C9496	two-seater	Henin Lietard	38	OOC shared Howard
8 Mar 18	Capt R W Howard	D212	Albatros D.V	Brebieres	40	OOC
8 Mar 18	Lt L H Holden	D214	Albatros D.V	nr Illies	41	OOC
9 Mar 18	Capt R W Howard	D212	Albatros D.V	Illies	42	OOC
			Albatros D.V	Illies	42	OOC
10 Mar 18	Capt R W Howard	D212	Albatros D.V	Pont a Vendin	43	OOC
11 Mar 18	Lt L H Holden	C9696	two-seater	Illies	44	OOC
12 Mar 18	Lt L H Holden	B8237	Albatros D.V	Wingles	45	destroyed
12 Mar 18	Lt R W McKenzie	D214	two-seater	Wingles	45	destroyed
13 Mar 18	Lt R W McKenzie	B579	Albatros D.V	Wingles	46	destroyed
18 Mar 18	Capt R W Howard	D212	Rumpler C	Haubourdin	47	destroyed
22 Mar 18	Capt H G Forrest	C9539	two-seater	Bullecourt	48	destroyed
			Albatros D.V	Bullecourt	48	OOC
			Albatros D.V	Bullecourt	48	OOC
22 Mar 18	Lt R W McKenzie	C5382	Albatros D.V	Omissy	49	OOC
22 Mar 18	Capt R Phillipps	C9541	Fokker Dr.I	Marquion	50	OOC
22 Mar 18	Lt L H Holden	D214	Albatros D.V	Queant	51	OOC
22 Mar 18	Lt A R Rackett	B70	two-seater	Honnecourt	53	OOC
23 Mar 18	Lt R W McKenzie	C5382	two-seater	Brebieres/Vitry	52	destroyed
23 Mar 18	Capt R Phillipps	C9541	Albatros D.V	Bourlon Wood	54	OOC

23 Mar 18	Capt H G Forrest	C9539	two-seater	nr Bapaume	55	destroyed
24 Mar 18	Capt R Phillipps	C9541	two-seater	nr Vaux	57	destroyed
24 Mar 18	Lt A G Clark	B4895	Albatros D.V	nr Vaux	58	OOC
26 Mar 18	Lt L Benjamin	C5382	Albatros D.V	nr Pozieres	59	destroyed
27 Mar 18	Capt R Phillipps	C9541	Fokker Dr.I	Suzanne	60	destroyed
			scout	Meault	61	destroyed "new type scout" poss Pfalz D.XII
27 Mar 18	Lt Hosking	A8906	FokkerDr.I	Fricourt	62	destroyed
30 Mar 18	Capt H G Forrest	C9539	two-seater	Queshel/Beaucourt	62	OOC
2 Apr 18	Capt H G Forrest	C9539	DFW C	Demuin	63	destroyed, shared Manuel
	Lt R L Manuel	B184	DFW C	Demuin	63	destroyed, shared Forrest
2 Apr 18	Lt A G Clark	D3429	two-seater	Amiens/Corbie	64	destroyed, shared Blaxland
	2Lt G H Blaxland	B571	two-seater	Amiens/Corbie	64	destroyed, shared Clark
4 Apr 18	Lt L J Primrose	B187	Albatros	Villers-Bretonneux	65	destroyed
6 Apr 18	formation	–	two-seater	Lamotte	66	destroyed, shared by Lt Clark (D3429), Lt Paxton (B525), Lt Primrose (B187), Lt Blaxland (B571)
12 Apr 18	Capt A G Clark	B4895	Albatros D.V	Lorgies	67	OOC
12 Apr 18	Capt H G Forrest	C9539	two-seater	Vieille Chapelle	68	destroyed
3 May 18	Lt E D Cummings	B168	Fokker Dr.I	Meteren	69	destroyed
8 May 18	Lt J A Adam	B4895	Fokker Dr.I	Houplines	70	OOC
9 May 18	Lt J A Adam	B4895	Pfalz D.III	Cambrai/Marcoing	71	destroyed, shared Smith
	Lt F R Smith	C9496	Pfalz D.III	Cambrai/Marcoing	71	destroyed, shared Adam
9 May 18	Capt H G Forrest	C9539	two-seater	Ervillers	72	OOC
10 May 18	Lt G H Blaxland	B571	two-seater	Morcourt	73	OOC
16 May 18	Capt R Phillipps	C5441	Pfalz D.III	Beaumetz	74	destroyed
17 May 18	Capt A M Jones	C9496	Fokker Dr.I	Bapaume	75	OOC
30 May 18	2Lt G J Cox	D3431	Pfalz D.III	Bapaume	76	destroyed
1 Jun 18	Lt W Q Adams	D3429	Pfalz D.III	Cappy	77	destroyed
1 Jun 18	Lt E D Cummings	B195	Pfalz D.III	Pozieres	78	destroyed
1 Jun 18	Lt R L Manuel	B184	Pfalz D.III	Villers-Bretonneux	79	OOC
1 Jun 18	Lt L J Primrose	D6860	Pfalz D.III	Chuignes	81	OOC
1 Jun 18	Lt G J Cox	D3431	Pfalz D.III	Chuignes	82	OOC
1 Jun 18	Capt H G Forrest	C9539	Fokker Dr.I	Chuignes	83	destroyed
1 Jun 18	Capt A T Cole	D3962	Fokker Dr.I	Estrees	85	destroyed
2 Jun 18	Lt R L Manuel	B184	Pfalz D.III	Villers-Bretonneux	80	destroyed
			Pfalz D.III	Villers-Bretonneux	80	OOC
2 Jun 18	Capt H G Forrest	C9539	Pfalz D.III	Estrees	84	OOC
			Pfalz D.III	Albert	86	OOC
2 Jun 18	Lt W Q Adams	C9496	Pfalz D.III	Albert	86	OOC

2 Jun 18	Lt J A Adam	D3962	Pfalz D.III	Peronne	86	OOC
2 Jun 18	Lt H E Hamilton	B4895	Pfalz D.III	Peronne	87	OOC
2 Jun 18	Capt A T Cole	D3429	Pfalz D.III	Peronne	88	OOC
2 Jun 18	formation	–	Pfalz D.III	Clery	89	OOC by formation
				Lt J Adam (B4895), Capt Cole (D3962), Lt Smith (D3429), Lt Hamilton (C9496)		
11 Jun 18	Lt T J Hammond	C9541	two-seater	Cuvilly	90	destroyed
12 Jun 18	Capt R Phillipps	D6860	Fokker Dr.I	Gury	91	destroyed
			Fokker Dr.I	Cannectancourt	91	destroyed
			LVG C	Marqueglise	93	destroyed
			Fokker D.VII	Gournay	94	destroyed
12 Jun 18	Lt R L Manuel	B184	Pfalz D.III	Bussy	92	destroyed
4 Jul 18	Lt F R Smith	D6190	Fokker D.VII	Armentieres	95	destroyed
17 Jul 18	Capt A T Cole	C1934	Fokker Dr.I	Armentieres	96	destroyed
22 Jul 18	Lt R L Manuel	C1948	Pfalz D.III	nr Lens	97	OOC
			Fokker D.VII	nr Lens	97	OOC
25 Jul 18	Capt R Phillipps	D6860	Fokker D.VII	Houplines	98	destroyed
29 Jul 18	Capt A T Cole	C1934	Hannovarin	Foret de Nieppe	99	driven-down
31 Jul 18	Lt F W Follett	C1125	Fokker D.VII	Laventie	100	destroyed
31 Jul 18	Lt E D Cummings	C6473	LVG C	Laventie	101	destroyed
			LVG C	Laventie	101	OOC
31 Jul 18	Capt R L Manuel	C1948	Albatros D.V	Estaires	102	OOC
31 Jul 18	Lt J J Wellwood	D6913	Rumpler C	Armentieres	103	driven-down
31 Jul 18	Capt R Phillipps	D6860	DFW C	La Bassee	104	destroyed
			Fokker D.VII	La Bassee	104	OOC
1 Aug 18	Lt J J Wellwood	D6913	Rumpler C	Lille	103A	destroyed
7 Aug 18	Lt F R Smith	C6403	Fokker D.VII	Wavrin	105	OOC
7 Aug 18	Capt A T Cole	C1934	Fokker D.VII	Herrin	106	OOC
12 Aug 18	Lt Markham-Mills	E5989	Fokker D.VII	Licourt	107	OOC
12 Aug 18	Capt R Phillipps	D6860	Fokker D.VII	Licourt	108	destroyed
12 Aug 18	Lt J J Wellwood	D6968	Fokker D.VII	Licourt	109	destroyed
13 Aug 18	Lt F W Follett	C1125	Fokker D.VII	Erchen	110	OOC
13 Aug 18	Lt Markham-Mills	E5989	Fokker D.VII	Erchen	111	OOC
14 Aug 18	Lt Markham-Mills	E5989	biplane scout	Steenwerck	112	destroyed
14 Aug 18	Capt R L Manuel	C1948	Fokker D.VII	Lampret	113	destroyed
14 Aug 18	formation	–	two-seater	Wavrin	114	destroyed EA broke up without shots fired
16 Aug 18	Lt F W Follett	C1125	Albatros C	Harbourdin	115	destroyed
19 Aug 18	Capt A T Cole	D6948	Fokker D.VII	Harbourdin	116	OOC
			Fokker D.VII	Harbourdin	116	OOC
25 Aug 18	Capt A T Cole	D6964	DFW C	Epinoy	117	destroyed, shared Wellwood
	Lt J J Wellwood	D6968	DFW C	Epinoy	117	destroyed, shared Cole
27 Aug 18	Lt E E Davies	D6860	Fokker D.VII	Lecluse	118	destroyed
27 Aug 18	Lt G J Cox	E5965	Fokker D.VII	Sains	119	destroyed
			Fokker D.VII	Sains	119	OOC
			Pfalz D.III	Sains	119	OOC

3 Sep 18	Lt C O Stone	D6919	Fokker D.VII	Cambrai	120	OOC
3 Sep 18	Lt F R Smith	C6403	Fokker D.VII	Cambrai	121	OOC
4 Sep 18	Lt V E Knight	C1125	Fokker D.VII	Douai	122	OOC
6 Sep 18	Lt J J Wellwood	D6968	Fokker D.VII	Douai	123	OOC
6 Sep 18	Lt F R Smith	C6403	Fokker D.VII	Lille	124	destroyed
15 Sep 18	Capt E Cummings	C6473	Albatros C	Macquart	133	destroyed, shared Davies
	Lt E E Davies	D6860	Albatros C	Macquart	133	destroyed, shared Cummings
16 Sep 18	Lt F R Smith	C6403	Fokker D.VII	Lille	126	OOC
			Fokker D.VII	Lille	126	OOC
16 Sep 18	Lt F Alberry	D6995	Fokker D.VII	Lille	127	destroyed
16 Sep 18	Lt C H Copp	D6869	Fokker D.VII	Steenvoorde	128	OOC
16 Sep 18	Lt G E Holroyde	D6903	Fokker D.VII	Lille	130	OOC
16 Sep 18	Capt R L Manuel	C1948	Fokker D.VII	Harbourdin	131	OOC
			Fokker D.VII	La Bassee	132	destroyed
17 Sep 18	Lt F Alberry	D6995	Fokker Dr.I	Lille	134	OOC
			Fokker Dr.I	Lille	134	OOC
20 Sep 18	Capt R L Manuel	C1948	Fokker D.VII	Quesnoy	135	driven-down
24 Sep 18	Capt R L Manuel	C1948	Albatros C	La Bassee	136	OOC, shared Smith
	Lt F R Smith	C6403	Albatros C	La Bassee	136	OOC, shared Manuel
24 Sep 18	Capt A T Cole	D6964	Pfalz D.III	Harbourdin	137	destroyed
24 Sep 18	Lt F R Smith	C6403	Fokker D.VII	Seclin	138	destroyed
			Pfalz D.III	Seclin	139	driven-down
24 Sep 18	Lt L Franks	F5465	Fokker D.VII	Harbourdin	140	destroyed
24 Sep 18	Lt F L Roberts	D6995	Fokker D.VII	Harbourdin	141	OOC
24 Sep 18	Lt J J Wellwood	D6968	Pfalz D.III	Perenchies	142	destroyed
24 Sep 18	Lt Markham-Mills	E5989	Fokker D.VII	Harbourdin	144	OOC
24 Sep 18	Capt E Simonson	D406	Fokker D.VII	Harbourdin	143	destroyed
			Pfalz D.III	Harbourdin	143	OOC
1 Oct 18	Lt E E Davies	D6860	DFW C	Ligny	144	OOC
4 Oct 18	Lt G H Blaxland	D6903	Fokker D.VII	Lille	145	OOC
9 Oct 18	Lt F R Smith	C6403	Fokker D.VII	Lille/Armentieres	146	OOC
			Fokker D.VII	Lille/Armentieres	146	OOC
9 Oct 18	Lt L Franks	F5465	Fokker D.VII	Lille/Armentieres	147	driven-down
9 Oct 18	Lt J McKeown	D6968	Fokker D.VII	Lille/Armentieres	148	driven-down
9 Oct 18	Capt R L Manuel	C1948	Halberstadt	Pont a Vendin	149	driven-down
14 Oct 18	Lt G H Blaxland	C1125	Fokker D.VII	Grand Ennettieres	150	destroyed
			Fokker D.VII	Grand Ennettieres	150	OOC
14 Oct 18	Lt F R Smith	C6403	Fokker D.VII	Pont a Marcq	151	destroyed
			Fokker D.VII	Pont a Marcq	151	destroyed
14 Oct 18	Lt L Franks	F5465	Fokker D.VII	Mouyeaux	152	destroyed
14 Oct 18	Capt E Cummings	C6473	Fokker D.VII	Cysoing	153	destroyed
			Fokker D.VII	Gruson	153	OOC
			Fokker D.VII	Hertain	153	OOC
14 Oct 18	Lt F L Roberts	C1125	Fokker D.VII	Tournai	154	OOC
14 Oct 18	Lt C O Stone	D6919	Fokker D.VII	Tournai	155	destroyed
			Fokker D.VII	Tournai	155	OOC
14 Oct 18	Lt F R Smith	C6403	Fokker D.VII	Tournai	156	destroyed

			Fokker D.VII	Tournai	156	OOC
18 Oct 18	Lt F Alberry	D6995	Fokker D.VII	Tournai	157	OOC
18 Oct 18	Lt G H Blaxland	F5457	Fokker D.VII	Tournai	158	destroyed
27 Oct 18	Lt C H Copp	E5659	Fokker D.VII	Tournai	159	OOC
			Fokker D.VII	Tournai	159	OOC
28 Oct 18	Lt G H Blaxland	F5457	Fokker D.VII	Peruwelz/Blaton	160	destroyed
28 Oct 18	Capt A T Cole	D6964	Fokker D.VII	Peruwelz	161	destroyed, pilot jumped from EA by parachute
28 Oct 18	Lt F Alberry	D6995	Fokker D.VII	Bandour	162	destroyed
			Fokker D.VII	Bandour	162	OOC
28 Oct 18	Capt E Simonson	C1125	Fokker D.VII	La Louviere	163	destroyed
28 Oct 18	Lt C O Stone	D6919	Fokker D.VII	Mons	164	OOC
28 Oct 18	Lt J A Egan	F860	LVG C	La Louviere	165	destroyed
30 Oct 18	Capt E Simonson	C1125	Fokker D.VII	Tournai	166	OOC
1 Nov 18	Capt E Cummings	C6473	LVG C	Antoing	167	OOC, shared Davies
	Lt E E Davies	E5765	LVG C	Antoing	167	OOC, shared Cummings
4 Nov 18	Lt J J Wellwood	F5611	Fokker D.VII	Renaix	169	destroyed
4 Nov 18	Lt E E Davies	E5765	LVG C	Renaix	170	destroyed
			Fokker D.VII	Renaix	170	OOC
4 Nov 18	Capt E Simonson	C1129	Fokker D.VII	Renaix	171	destroyed
4 Nov 18	Lt C O Stone	D6919	Fokker D.VII	Renaix	172	destroyed
4 Nov 18	Lt G H Blaxland	F5457	Fokker D.VII	Houtaing	173	driven-down
4 Nov 18	Capt E Simonson	C1129	Fokker D.VII	Houtaing	174	driven-down
4 Nov 18	Lt E E Davies	E5765	Fokker D.VII	Houtaing	175	driven-down
			Fokker D.VII	Houtaing	175	driven-down
4 Nov 18	Lt C O Stone	D6919	Fokker D.VII	Houtaing	176	driven-down
4 Nov 18	Lt F Alberry	D6995	Fokker D.VII	Houtaing	177	driven-down
4 Nov 18	Lt J J Wellwood	D6964	Fokker D.VII	Houtaing	178	driven-down

Appendix 11
TRAINING AIRCRAFT OF No 2 SQUADRON (1937-39)

HAWKER DEMON

Sixty-four Hawker Demon Mark I and II two-seat general-purpose aircraft were delivered to the RAAF from 1934. Two provided the initial equipment for No 2 Squadron on its reformation at Laverton in May 1937. From then until February 1939, the "B" Flight was equipped with three or four Demons.

Serial	*Date On Sqn*	*Date Off*	*Remarks*
Demon I			
A1-19	3.5.37	21.7.38	damaged 14.7.38 landing Laverton (Symonds), to 1AD
	21.9.38	7.2.39	to No 12 Sqn, Laverton
A1-38	3.5.37	7.2.39	damaged 17.1.39 Laverton (Lewis), to No 12 Sqn, Laverton
Demon II			
A1-57	18.1.38	23.3.38	damaged 22.3.38 heavy landing (Green) Laverton, to 1AD
A1-58	2.3.38	7.2.39	to No 12 Sqn, Laverton
A1-59	23.3.38	26.9.38	CR 3.9.38 stalled in gusty winds, to 1AD

BRISTOL 105A BULLDOG II

The RAAF received eight Bulldog single-seat fighters in 1930. They were flown primarily by Fighter Squadron of No 1 FTS at Point Cook, but the survivors were made available for No 2 Squadron in late 1937, and were operated by "C" Flight. The last three in RAAF service saw out their days as training airframes for the Engineering School at Melbourne Showgrounds in 1940.

Serial	*C/N*	*Date On Sqn*	*Date Off*	*Remarks*
A12-1	7389	28.10.37	5.39	to Comm Flt, Laverton
A12-2	7390	21.3.38	22.4.38	crashed Richmond (Sutherland)
A12-3	7391	21.3.38	5.39	to Comm Flight
A12-6	7394	28.10.37	2.4.38	damaged landing Laverton (Newbound), 1AD
		7.38	5.39	to Comm Flt, Laverton

NORTH AMERICAN NA-16

Two North American trainers were obtained from the United States in 1937 as pattern aircraft for the Australian Wirraway, built by the Commonwealth Aircraft Corporation. The first United States aircraft, an NA-16-1A, model NA-32, was flown by No 2 Squadron's "C" Flight over the first half of 1938. The second aircraft, a model NA-33 with retractable undercarriage and three-bladed propeller, also known as an NA-16-2K, operated only briefly during April that year when the Squadron participated in the Flemington and Richmond Air Pageants.

Serial	*C/N*	*Date On Sqn*	*Date Off*	*Remarks*
A20-1	32-387	7.1.38	24.6.38	to No 1 FTS, Point Cook
A20-2	33-388	4.4.38	28.4.38	to CAC for further trials

AVRO 643 MK.II CADET

34 Avro Cadet trainers were operated by the RAAF from 1935. Two were flown briefly by No 2 Squadron in conjunction with the Ansons and Bulldogs in 1939.

Serial	*C/N*	*Date On Sqn*	*Date Off*	*Remarks*
A6-25	1060	10.2.39	24.7.39	to 1 SFTS, Point Cook
A6-26	1061	10.2.39	18.7.39	to 21 Sqn, Laverton

The remaining Bulldogs from 2 Squadron ended their days at the Engineering School at Ascot Vale in Melbourne, in 1940. AWM P449/212/036

APPENDIX 12

AVRO ANSON AIRCRAFT OF No 2 SQUADRON

Prior to the war, the RAAF ordered 48 Ansons as general reconnaissance bombers. These were delivered from 1936 and serialled A4-1 to A4-48. To make good RAAF expansion plans, another 40 Ansons were leased from the RAF, and delivered from 1938. With the advent of war and the Empire Air Training Scheme (EATS), a further 937 Ansons were received for service at RAAF Service Flying Training Schools (SFTS) throughout Australia. These were delivered from 1940, and also retained RAF serial numbers. Ansons were initially delivered in silver doped colour scheme, but with war were camouflaged. No 2 Squadron Ansons carried the squadron code letter "B".[1]

AVRO 652A ANSON Mark I

Serial	*C/N*	*Date On Sqn*	*Date Off Sqn*	*Remarks*
Ansons from original RAAF orders				
A4-9	975	18.4.40	8.7.40	to No 1 SFTS, Point Cook
A4-12	978	14.11.38	5.4.39	to No 25 Sqn, Pearce
A4-26	1021	12.5.38	16.6.40	damaged 8.2.39 landing Point Cook (Collings), to No 1 SFTS, Point Cook
A4-29	1024	22.2.38	10.8.38	crashed Arthur's Seat, Vic (Gillespie), written-off
A4-33	1028	29.11.37	1.7.40	to No 1 SFTS, Point Cook
A4-34	1053	30.11.37	9.5.38	to Training Depot 1AD, Laverton
A4-36	1055	14.2.38	21.2.38	to 2AD, Richmond, for No 22 Sqn
A4-37	1056	25.2.38	27.4.39	to 1AD, Laverton
		21.2.40	1.7.40	to No 1 SFTS, Point Cook
A4-39	1079	30.8.38	16.6.40	to No 1 SFTS, Point Cook
A4-42	1082	28.8.38	16.6.40	to No 1 SFTS, Point Cook
A4-44	1084	12.9.38	8.7.40	to No 1 SFTS, Point Cook
A4-46	1086	21.9.38	19.4.39	to Trg Depot 1AD for Survey Flt
Ansons leased prewar from RAF				
N1332	–	27.1.39	5.4.39	damaged 14.2.39 landing Laverton (Anderson), to No 25 Sqn, Pearce
N1333	–	10.7.39	16.6.40	to No 1 SFTS, Point Cook
N4883	–	20.7.39	8.7.40	to No 1 SFTS, Point Cook
N4926	–	30.3.39	10.7.39	to No 12 Sqn, Laverton
N4936	-	21.4.39	25.8.39	to No 1 FTS, Point Cook
N4946	-	5.4.39	15.4.40	to No 21 Sqn, Laverton
N4955	-	19.2.40	9.7.40	to No 1 SFTS, Point Cook
N4960	-	5.4.39	4.7.40	to No 1 SFTS, Point Cook

Note: No 1 FTS became No 1 SFTS 1.5.40

1 As directed by AGI No.C11 of 22 Sep 1939, the code letter "B" was allocated to No 2 Squadron to distinguish this unit's aircraft. Similarly, "A" was allocated to No 1 Sqn, "C" No 3 Sqn, "D" No 4 Sqn, "E" No 5 Sqn, etc.

Three 2 Squadron Ansons during 1938–39, in the original silver scheme applied to the aircraft when delivered to the RAAF

From top: A4-29, which crashed at Arthur's Seat on 10 August 1938 RAAF

A4-33, which was passed to 1 SFTS when 2 Squadron received Hudsons in June 1940 F JEFFERIES

Appendix 13

LOCKHEED HUDSON AIRCRAFT OF No 2 SQUADRON

The first 50 RAAF Hudsons were ordered in 1938, and were basically similar to the RAF Mark IV, but were known in the RAAF as the Hudson Mark I. The Hudson began to replace the Anson on No 2 Squadron at Laverton during June 1940.

During operations from Darwin over the NEI, Hudsons from No 13 Squadron were flown by No 2 Squadron crews, and 2 Squadron aircraft were sometimes flown by 13 Squadron crews. This is not documented in the sparse Squadron records over January–April 1942. However, where known, Hudsons belonging to No 13 Squadron being flown by 2 Squadron crews are marked below by "(13)". Similarly, A16-73, which was allotted to 2 Squadron, was flown mainly by 13 Squadron.

The Lockheed constructor's number (c/n) was painted on each Hudson nose, immediately aft of the perspex nose. This is useful in identifying individual aircraft from photographs. On the early aircraft this number, possibly for security reasons, was changed: for instance c/n 1885 became "285".

From 1939 RAAF operational aircraft had specified fuselage code letters applied to indicate the squadron. Accordingly, the side letter "B" was carried by No 2 Squadron's early Hudsons.[1]

Serial	*C/N*	*Previous Identity*	*Date On Sqn*	*Code*	*Date Off Sqn*	*Remarks*
LOCKHEED Model B14S HUDSON Mark I						
A16-5	1856	–	3.1.42	B	4.4.42	damaged at Drysdale 6.3.42 (Badger), written-off
A16-6	1857	–	24.6.40	B	19.2.42	destroyed in air raid at Darwin
A16-9	1860	–	5.7.40	B	30.1.42	lost in strafing attack Mina River (Bonython)
A16-12	1863	–	25.8.40	B	12.1.42	shot down by enemy fighters over sea near Menado (Gorrie)
A16-18	1869	–	25.8.40	B	10.5.42	to 1 AD, to 1 OTU
A16-32	1883	–	24.6.40	B	8.5.42	to 1 AD, to 1 OTU
A16-34	1885	–	24.6.40	B	4.2.42	to 1 AD
			13.3.42	B	19.4.42	to 1 AD, to 1 OTU
A16-46	1897	–	3.1.42	B	12.1.42	shot down by enemy fighters over sea near Menado (Hodge)

A second batch of 50 Hudson was ordered by the RAAF in 1939, and differed from the first 50 by having Hamilton Standard Hydromatic constant-speed propellers. Again, these were basically the RAF Mark IV variant but in the RAAF, although sometimes unofficially referred to as the "Mark II", the type was officially still known as the Mark I.[2] Both these batches were

1 RAAF Aircraft General Instruction No C.11 of 22 September 1939.
2 RAAF Pub No 80C, "Operating Instructions for Australian Hudsons Marks I, III & IV", dated 1942, courtesy of David Vincent.

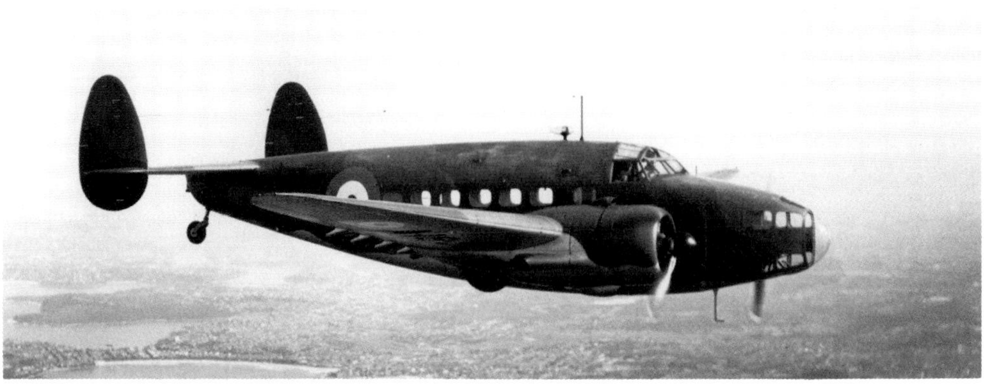

From top:

One of 2 Squadron's first Hudsons, with no dorsal gun position A McEGAN

A16-12, flown by 2 Squadron pilots Nev Hemsworth and Dave Campbell — a classic shot by RAAF photographer John Harrison AWM AC-64

A16-18, with the interim "bath tub" dorsal turret, the code "B" indicating 2 Squadron ownership M LAWSON powered by the Pratt & Whitney Twin Wasp R-1830-SC3G engine, which had been specified by the RAAF for commonality with the Australian-built Beaufort.

A16-57	1910	–	6.1.42	B	20.3.42	apparently damaged by enemy action, withdrawn from service
A16-65	1918	–	12.6.40	B	4.3.42	destroyed in air raid at Darwin
A16-68(13)	1921	–	5.42	N	6.42	13 Sqn a/c, crashed 8.6.42, to 1 AD
A16-73	1926	–	3.1.42	B	26.3.42	to 1 AD, to 1 OTU
A16-74	1927	–	6.1.42	B	6.4.42	to 5 AD, to 1 OTU
A16-77	1931	–	5.7.40	B	26.5.42	to 13 Sqn Hughes
			8.8.42	B	5.11.42	to 5 AD, to 1 OTU
A16-78	1932	–	5.7.40	B	19.2.42	destroyed in air raid at Darwin
A16-79	1933	–	23.7.40	B	20.1.42	crashed on take-off Penfoei (Cuming)
A16-80	1934	–	23.7.40	B	26.5.42	to 13 Sqn Hughes
A16-95	1949	–	23.12.40	B	15.6.41	to 1 AD, to 14 Sqn
A16-96	1950	–	23.12.40	B	15.6.41	to 1 AD, to 13 Sqn
A16-97	1951	–	12.8.40	B	13.8.40	crashed at Canberra (Hitchcock)
A16-98	1952	–	25.8.40	B	29.8.40	to 1 AD, to 14 Sqn
A16-99	1953	–	3.1.42	B	26.3.42	to 1 AD, to 1 OTU
A16-100	1955	–	15.9.40	B	18.3.42	damaged enemy action ditched off Bathurst Is (Venn), written-off

The next 52 Hudsons for the RAAF (A16-101 to A16-152), ordered in 1941, were powered by the Pratt & Whitney SC3G Twin Wasp, and were diverted from a USAAF contact, where they had been designated the A-28. In RAAF service these were referred to by the RAF designation of Hudson Mark IV.

LOCKHEED 414-08 HUDSON Mark IVA

A16-109(13)	6038	41–23179	3.42		23.3.42	shot down by enemy fighter Oepoela, Timor (MacAlister)
A16-121	6050	41–23191	25.1.42		30.1.42	abandoned unserviceable on ground at Laha (White)
A16-129(13)	6058	41–23199	5.42		25.6.42	13 Sqn a/c to 5 AD, to 14 Sqn
A16-133	6062	41–23203	1.2.42		9.2.42	to 13 Sqn Daly Waters
A16-135	6064	41–23205	25.1.42		19.2.42	destroyed in air raid at Darwin
A16-142	6071	41–23212	9.2.42		20.2.42	crash landed at Lake Woods (Lamb)
A16-143(13)	6072	41–23213	5.42		3.6.42	13 Sqn a/c damaged landing, 1RSU

The previous order for 52 aircraft was increased to 146 in 1941,[3] and with the acquisition of a single RAF aircraft,[4] the remaining RAAF Hudsons (A16-153 to A16-247) all came from RAF orders, and were Mark IIIA versions powered by the Wright Cyclone R-1820. (These had been diverted through lend-lease from the USAAF, where the model was known as the A-29.)

The system of squadron code letters was subsequently changed in 1943 and "KO" was allotted to 2 Squadron, followed by another letter to identify individual aircraft.[5] This KO code was only carried by a few aircraft from May 1943 onwards; often only the individual letter was marked on the fuselage. (One Mark IV aircraft, A16-112, although it never served on 2 Squadron, was re-engined with Wright Cyclone engines postwar and has since been restored to represent a typical Mark III of the 1943–44 period carrying the identity KO-Y. Hudson Mark IIIs served with No 2 Squadron longer than any other RAAF squadron in the South West Pacific Area.)

LOCKHEED 414-56 HUDSON Mark IIIA

A16-154	6415	41–23598,BW736	20.8.43		11.3.44	damaged by enemy fighter nr Aroe Is 9.2.44 (Northover), damaged 3.3.44, to 2 AD written-off
A16-156	6417	41–23600,BW738	6.4.42		23.4.42	to 1 AD, to 13 Sqn
			14.4.43		29.10.43	to 3 AD, to 1 OTU
A16-159	6420	41–23603,BW741	6.4.42		13.4.42	shot down in sea Koepang (Cook)
A16-160	6421	41–23604,BW742	7.9.42	KO-X	26.12.43	forced landing Cape Ford 25.10.42 (Mills), crashed Drysdale 21.12.43 (Coleman), 4RSU
			18.1.44		26.3.44	to 2 AD, to 1 OTU
A16-161	6422	41–23605,BW743	6.4.42		6.5.43	to 5 AD, to 1 OTU
A16-169	6098	41–23281, BW419	20.2.43		3.4.43	shot down by groundfire strafing Timoeka airfield (Burns)
A16-170	6424	41–23607,BW745	5.9.42		25.10.42	failed to return from ops Koepang (Dunning)
A16-171	6100	41–23283, BW421	3.5.43		7.5.43	shot down by Rufe, Maikoor Aroe Islands (Jackman)
A16-172	6425	41–23608,BW746	6.4.42		15.9.42	shot down in ship attack at Saumlaki (McDonnell)
A16-174	6427	41–23610,BW748	13.4.42		22.5.42	shot down by Zeros Halong Bay near Ambon (Brooks/ 13Sqn)
A16-175	6347	41–23530,BW668	19.4.42		16.6.42	ditched Port Keats (Sharp), written-off
A16-176	6348	41–23531,BW669	11.4.42		23.5.42	crashed on take-off 7.5.42, to 1RSU written-off

3 Gillison, p.138.

4 A16-222, ex AE488 of the RAF, was a survivor of the Singapore campaign and evacuated to Australia in March 1942. S Wilson, *Anson, Hudson & Sunderland in Australian Service*, Aerospace Publications, Canberra, 1992, p.97.

5 Air Force Confidential Order AFCO A.3/43.

From top:

Two Hudson of 2 Squadron (the front aircraft is A16-161) at Batchelor, 28 October 1942
AWM 27619

A formation of 2 Squadron Hudsons in 1943, prior to applying the "KO" squadron codes; the nearest aircraft "N" is A16-233
AWM NWA 0452

A16-211, "The Tojo Busters", which crash landed on 7 May 1943 after being damaged by Japanese Rufe fighters
AWM P269/93/58

A16-160, KO-X, which crashed at Drysdale Mission, WA, on 21 December 1943
RAAF

A16-177	6349	41–23532,BW670	21.4.42		18.10.42	to 5 AD, to 6 Sqn
			29.3.43		9.1.44	to 4 RSU, to 2 AAU
A16-178	6350	41–23533,BW671	17.4.42		19.6.42	to 1 RSU, to 13 Sqn
			18.7.42		15.10.43	failed to return from ops over Arafura Sea from Millingimbi (Oldridge)
A16-180	6352	41–23535,BW673	12.10.43		25.10.43	to 4 RSU, to 6 CU
A16-181	6353	41–23536,BW674	21.4.42		15.8.43	failed to return from ops to Keaukwa, prob shot down by Zeros (Lamb)
A16-182	6354	41–23537,BW675	11.4.42		23.4.42	failed to return from ops to Koepang (Blanchard/13Sqn)
A16-183	6355	41–23538,BW676	23.4.42		20.4.43	collided after take-off Millingimbi with A16-197 at Rabuma Is (Mills), w-o
A16-184	6340	41–23523,BW661	18.5.42		30.5.42	crashed on take-off Darwin (Hawkesford), written-off
A16-185	6341	41–23524,BW662	3.6.43		19.8.43	damaged, to 4 RSU, w-o
A16-186	6342	41–23525,BW663	4.4.43		12.6.43	crashed landing on test flight Hughes (Mason), written-off
A16-187	6343	41–23521,BW664	11.5.42		22.5.42	shot down by Zeros Halong Bay nr Ambon (Allchin)
A16-189	6345	41–23528,BW666	15.5.43		24.7.43	crash landed Millingimbi 14.7.43 (Brodie), to 4 RSU, to 6 CU
A16-192	6357	41–23540,BW678	4.4.43	KO-U	19.9.43	damaged 11.9.43, to 5 AD, to 4 CU (became VH-BIB)
A16-195	6459	41–36970,FH169	15.1.43		9.7.43	crashed on ops Mt Kekneno, Timor, (Williams)
A16-196	6461	41–36972,FH171	30.4.42		13.5.42	shot down attacking shipping Ambon Harbour (Venn)
A16-197	6462	41–36973,FH172	27.7.42		4.2.43	to 1 RSU, to 13 Sqn
			4.4.43		20.4.43	collided after take-off Millingimbi with A16-183, crashed in sea (Daniel), written-off
A16-199	6464	41–36975,FH174	4.4.43		16.4.44	to 3 CU (to VH-SMM,AGJ), display RAF Museum
A16-200	6467	41–36978,FH177	22.7.42		22.10.42	to 5AD, to 6 Sqn
A16-202	6469	41–36980,FH179	10.4.43		6.10.43	crashed taking-off Hughes (Dennis), written-off
A16-203	6470	41–36981,FH180	1.10.43		3.4.44	failed to return from ops over Arafura Sea (Brockhurst), found Adelaide River 1946
A16-204	6472	41–36983,FH182	10.4.43		28.9.43	to 5 AD
A16-207	6476	41–36987,FH186	11.5.42		19.3.44	to 2 AD, to 1 OTU
A16-208	6433	41–23616,BW754	4.5.42		23.5.42	crashed Millingimbi 22.5.42, to 1 RSU written-off
A16-209	6485	41–36996,FH195	1.6.42		21.8.42	shot down Timor by Zero (Wadey)

A16-210	6431	41–23614,BW752	11.5.42		27.12.42	shot down in sea off Timor by Zeros (Johns)
A16-211	6430	41–23613,BW751	3.4.43		7.5.43	"The Tojo Busters", damaged by Rufes Maikoor, crash landed Millingimbi (Hornby), written-off
A16-213	6432	41–23615,BW753	27.3.43	KO-H	11.10.43	damaged by groundfire Langgoer, crash landed Hughes (Helsham), written-off
A16-215	6429	41–23612,BW750	22.12.43		3.3.44	to 4 RSU, 2 AAU (to VH-AGO)
A16-217	6482	41–36993,FH192	12.6.42	KO-L	17.9.43	damaged by enemy a/c Dobo 24.4.43 (Kilgariff), crashed landing Corunna Downs 29.5.43 (Davies), to 7 AD written-off
A16-219	6486	41–36997,FH196	1.4.43	KO-M	11.2.44	damaged by enemy a/c Timoeka 25.4.43 (Davies), 1 AD, to 2AAU (to VH-BLA, -AGG)
A16-221	6428	41–23611,BW749	30.10.42		18.2.43	to 5 AD written-off
A16-226	6477	41–36988,FH187	10.8.42		12.8.42	to 13 Sqn
			26.6.43		19.1.44	"Houdini", to 4 RSU, to 2 AAU (to VH-BIA)
A16-227	6478	41–36889,FH188	15.4.43		20.7.43	crash landed Millingimbi 1.7.43 (Coleman), to 4 RSU, written-off
A16-229	6490	41–37001,FH200	25.9.43		13.10.43	damaged 4.10.43, to 2 AD, to 1 OTU
A16-230	6491	41–37002,FH201	12.4.43		27.3.44	damaged by groundfire Kai Is 24.4.43 (James), "Jake", shot down nr Loeang Is (Scott)
A16-232	6480	41–36991,FH190	16.6.42		5.12.42	damaged by Zero nr Laga, crash landed Batchelor 4.12.42 (Cambridge), 1 RSU written-off
A16-233	6483	41–36994,FH193	4.4.43	-N	19.8.43	crashed on landing Hughes 18.7.43, to 4 RSU
			7.9.43		8.9.43	shot down nr Tanimbar Is (Martin)
A16-234	6502	41–37013,FH212	8.7.42		30.7.42	collided with enemy aircraft near Kai Is (Muecke)
A16-235	6501	41–37012,FH211	4.4.43	-W?	4.10.43	"Foo", to 3 AD, to 4 CU
A16-236	6503	41–37014,FH213	27.3.43	KO-Y	3.10.43	damaged 15.9.43, to 5 AD, written-off
A16-237	6504	41–37015,FH214	5.7.42		16.3.43	exploded on refuel Batchelor, written-off
A16-238	6488	41–36999,FH198	24.1.43		14.4.43	damaged, to 1 RSU, written-off

A16-240	6493	41–37004,FH203	11.2.43		17.3.43	damaged on take-off Batchelor 15.3.43, to 1 RSU
			10.4.43		31.8.43	to 4 RSU
A16-241	6494	41–37005,FH204	22.7.42		28.4.43	lost engine, crashed in sea north of Millingimbi (James)
A16-242	6495	41–37006,FH205	9.7.42		17.3.43	crash landed Coburg Peninsula 14.3.43 (Graham), to 1 RSU
A16-244	6497	41–37008,FH207	19.5.43	KO-A	7.1.44	crash landed Drysdale 4.11.43 (Reed), 14 ARD
A16-247	6500	41–37011,FH210	27.3.43		9.4.43	damaged by groundfire Timoeka crash landed Millingimbi (O'Neil) 5.4.43, 1RSU written-off

The last Hudson mission was flown by the Squadron in A16-199 on 8 April 1944.

Abbreviations:

AAU Air Ambulance Unit
AD Aircraft Depot
ARD Aircraft Repair Depot
CU Communications Unit
OTU Operational Training Unit
RSU Repair & Salvage Unit
w-o written-off

Appendix 14

DAP Beaufort Aircraft of No 2 Squadron

The intention in late 1943 was to replace the Hudson in No 2 Squadron service with the Lockheed B-34 Ventura. One aircraft, A59-9, was received on the Squadron on 22 Nov 1943.[1] During December this re-equipment decision was reversed, with No 13 Squadron becoming the RAAF's sole Ventura squadron, relinquishing its Beauforts to No 2 Squadron.

700 Bristol Beaufort bombers were manufactured in Australia by the Department of Aircraft Production (DAP). A flight of nine Beauforts was planned for operation by No 2 Squadron, in conjunction with Hudsons. The initial batch of eight Beauforts was received at Hughes on 24 Dec 1943. The ninth aircraft, A9-380 (SF-H), operated by No 13 Squadron, crashed at Canberra on 11 December 1943, prior to its planned ferry north.

The first operational sorties were carried out by No 2 Squadron Beauforts on 12 January 1944. The Squadron built up to a strength of 11 Beauforts over March/April 1944, before being re-equipped with the Mitchell.

Serial	*Date On Sqn*	*Code*	*Date Off Sqn*	*Remarks*
BEAUFORT Mark VIII				
A9-361	24.12.43	KO-	25.4.44	to 14 ARD for 1 Sqn, Gould
A9-363	24.12.43	KO-	24.5.44	to 14 ARD for 1 Sqn, Gould, converted to Mk.IX A9-745
A9-373	24.12.43	KO-	20.5.44	to 14 ARD for 1 Sqn, Gould
A9-375	24.12.43	KO-R	4.4.44	to 14 ARD for 1 Sqn, Gould
A9-378	24.12.43	KO-	24.5.44	to 1 Sqn, Gould
A9-379	24.12.43	KO-	12.5.44	to 7AD, converted to Mk.IX A9-709
A9-386	24.12.43	KO-	15.2.44	to 4 RSU for 1 Sqn, Gould
A9-388	24.12.43	KO-	25.4.44	to 1 Sqn, Gould
A9-392	14.1.44	KO-	27.2.44	to 4 RSU for 1 Sqn, Gould
A9-496	17.1.44	KO-	10.4.44	to 4 RSU for 1 Sqn, Gould
A9-508	17.1.44	KO-	7.4.44	to 14 ARD for 1 APU, Laverton
A9-519	26.1.44	KO-Y	25.5.44	to 1 Sqn, Gould
A9-540	12.3.44	KO-	27.5.44	to 1 Sqn, Gould
A9-560	8.4.44	KO-V	22.5.44	CR 18.5.44 on take-off at Hughes (Kuring), to 4 RSU
A9-562	13.4.44	KO-F	25.5.44	to 14 ARD for 1 Sqn, Gould
A9-564	8.4.44	KO-D	24.5.44	to 1 Sqn, Gould
A9-576	10.3.44	KO-Z	29.5.44	to 14 ARD for 1 Sqn, Gould
A9-578	20.3.44	KO-T	26.5.44	to 1 Sqn, Gould

1 The other Venturas allotted to No 2 Sqn, but not delivered, were: A59-1, A59-3 to A59-8, A59-10 and A59-14. Order HQQO 902 of 16 Dec 1943 cancelled the plan to equip No 2 Sqn with Venturas.

A9-379 was fitted with dual controls. On 12 May 1944 it was ferried by FSgt Jim Henderson's crew to 7AD at Tocumwal, via Alice Springs and Broken Hill, and was converted to Beaufort Mark IX ("Beaufreighter") transport configuration.

As the Mitchell was received on No 2 Squadron from late April 1944, the last Beaufort sorties were flown on 22 May, and then most of the remaining 15 were delivered to No 1 Squadron at Gould.

During Beaufort operations, none was lost to enemy action.

APPENDIX 15

NORTH AMERICAN B-25 MITCHELL AIRCRAFT OF No 2 SQUADRON

The first RAAF Mitchells were received from the Dutch when they were unable to form a second squadron due to lack of crews. No 2 Squadron operated Mitchells alongside No 18 (NEI) Squadron as part of 79 Wing. The first four B-25 aircraft arrived on No 2 Squadron at Hughes in late April 1944, to begin conversion on 1 May of crews under the instruction of NEI pilots. The first Mitchell operations were flown on 11 June 1944.

Aircraft were delivered in standard USAAF olive drab camouflage, but several were stripped to bare metal (giving a speed increase of some 6 knots), and these were often flown by formation leaders to be conspicuous to the rest of the flight. Such B-25D aircraft were A47-3, A47-11, A47-36 and A47-37.

Serial	*C/N*	*Previous Identity*	*Date On Sqn*	*Code*	*Date Off Sqn*	*Remarks*
B-25D MITCHELL						
A47-1	100-24190	N5-183, 42-87607	28.6.44	KO-Q	25.8.45	to 3 AD Amberley for storage
A47-2	100-20748	N5-187, 43-3422	6.8.44	KO-Z	30.12.44	"Stinka", CR 26.12.44 Hughes (Ekert)
A47-3	100-20749	N5-181, 43-3423	18.6.44	KO-	22.9.44	CR into sea on ops off Manatuto (Slater)
A47-4	100-20750	N5-189, 43-3424	13.6.45	KO-	23.1.46	to 12 ARD for storage at 3 AD
A47-5	100-20752	N5-192, 43-3426	20.5.44	KO-N	1.1.46	to 3 AD Amberley for storage
A47-6	100-20753	N5-193, 43-3427	3.6.44	KO-	2.9.44	CR in sea near Peron Island (Simpson)
A47-7	100-23433	N5-194, 43-3607	6.6.44	KO-S	17.12.45	damaged 20.7.44 hit mast of ship (Cherry), damaged 1.9.4 by AA fire (Coughlan), to 3 AD for storage
A47-8	100-23439	N5-195, 43-3613	18.7.44	KO-	4.11.44	damaged 1.9.44 by AA fire (Hocking), shot down in flames into sea Atapoepoe (Selway)
A47-9	100-23447	N5-196, 43-3621	22.5.44	KO-	26.12.44	CR take-off Hughes, bombs exploded (Avery)
A47-10	100-23449	N5-197, 43-3623	26.5.44	KO-W	22.11.45	damaged 25.8.44 by AA fire over Langgoer (Simpson), to 3 AD for storage
A47-11	100-23450	N5-198, 43-3624	6.7.44	KO-C	5.12.44	"Petunia", CR Bathurst Is after being hit by AA fire (Norriss), to 18 RSU for components
A47-12	100-23451	N5-199, 43-3625	12.6.44	KO-	1.9.44	ditched at Peron Island (Carter)
A47-13	100-23452	N5-200, 43-3636	3.6.44	KO-	6.8.44	CR while practice bombing at Pell near Adelaide River (Davies)

All B-25Ds, left to right from top:

A47-25 KO-J	J Henderson
A47-34 KO-E, "My Favourite"	RAAF
A47-36 KO-Z, one of the silver Mitchells	L Power
KO-Q, the RAAF's first Mitchell, A47-1	R Jones
A47-17 KO-C in 1945	R Jones
KO-G, possibly A47-19	RAAF

A47-14	100-23592	N5-201, 43-3766	27.5.44	KO-F	6.12.45	damaged 18.8.44 by AA fire off Timor (Slater), to 3 AD for storage
A47-15	100-23593	N5-202, 43-3767	20.5.44	KO-	25.11.45	damaged 19.7.44 by AA (Pierce), to 3 AD storage
A47-16	100-23594	N5-203, 43-3768	25.4.44	KO-L	23.2.45	"Ell for Leather", damaged 19.7.44 by AA fire Dili (Hocking), to 14 ARD
A47-17	100-23595	N5-204. 43-3769	25.4.44	KO-C	19.5.44	damaged 9.5.44 at Hughes, 4RSU & 14ARD repair
			10.8.44	KO-	19.3.45	to 3 AD storage
A47-18	100-23596	N5-205, 43-3770	6.6.44	KO-Y	25.9.45	to 3 AD Amberley for storage
A47-19	100-23616	N5-206, 43-3790	26.4.44	KO-G?	15.9.45	damaged 31.8.44 by AA fire Selaroe Island (Simpson), CR Bandjermasin (Kirk)
A47-20	100-23617	N5-207, 43-3791	25.4.44	KO-	19.5.44	damaged belly-landing 17.5.44 Hughes (Hill), to 4 RSU for components
A47-21	100-23615	N5-213, 43-3789	20.8.44	KO-L	17.12.45	to 3 AD Amberley for storage
A47-22	100-23656	N5-190, 43-3830	24.7.44	KO-	16.7.45	damaged 10.10.44 by AA fire (Ingram), damaged 10.5.45 during test flight (McQuoid), to 14 ARD for storage at 3 AD
A47-23	100-23658	N5-212, 43-3832	18.7.44	KO-	19.10.45	damaged by mine 25.7.44 (Campbell), damaged 23.9.44 by AA fire (Fraser), damaged on ground at Balikpapan, to components
A47-25	100-23695	N5-215, 43-3869	29.7.44	KO-J	18.12.45	to 3 AD Amberley for storage

The following B-25D aircraft were received from No 18 Squadron, at the end of 1944 and early 1945.

A47-33	100-23842	N5-175, 42-87259	1.12.44	KO-	20.12.44	shot down by AA Saumlaki (Thompson)
A47-34	100-24191	N5-186, 42-87608	8.11.44	KO-E	30.10.45	"My Favourite", to 3 AD for storage
A47-35	100-23999	N5-168, 42-87416	12.5.45	KO-R	8.1.46	ex 18 Sqn GM-P, forced landed 27.8.45 Sepinggang, to 3 AD
A47-36	100-23838	N5-171, 42-87255	22.4.45	KO-Z	28.2.46	to 3 AD Amberley for storage
A47-37	100-23841	N5-174, 42-87258	22.3.45	KO-V	14.8.45	ex-18 Sqn GM-C, lost on transit Hughes-Biak (White)

The B-25D was modified by moving the dorsal turret forward to the navigator's compartment, and repositioning this station to the glasshouse nose, resulting in the B-25J. Both RAAF models were equipped with .5-inch fixed gun pods on either side of the nose and upgraded engine cowls, with individual cylinder exhaust stacks, not normally fitted to the USAAF B-25D.

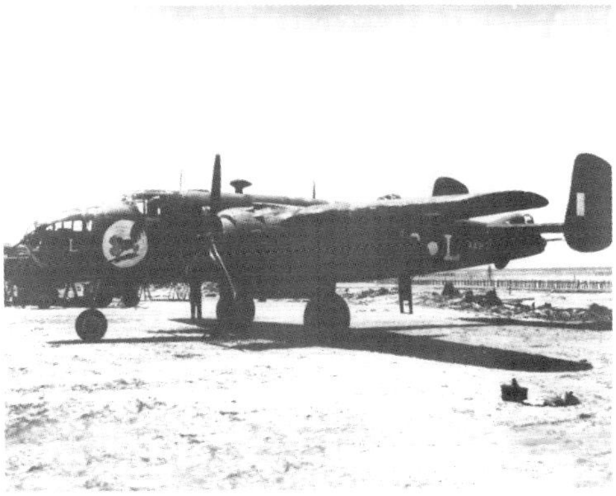

Left to right from top:

A47-29 KO-D, a B-25J, showing its sharkmouth marking D VINCENT

A close-up of "sharkmouth" at Sepinggang, October 1945 RAAF

B-25D A47-21 KO-L D VINCENT

Close-up of KO-L's "bird and gun" marking RAAF

A47-16, an earlier KO-L, "Ell For Leather" (the previous Dutch name, "Kaladjati" is still visible) A HENRY

A47-18 KO-Y, with the reclining "Bitova-Menace" A HENRY

B-25J MITCHELL

A47-26	108-32202	N5-220, 43-27689	29.6.44	KO-K	9.11.45	damaged by AA 6.1.45 at Watoeata (Coward), to 3 AD Amberley for storage
A47-27	108-32204	N5-219, 43-27691	29.6.44	KO-A	21.12.45	damaged 31.8.44 by AA fire (Avery), to 3 AD storage
A47-28	108-32440	N5-224, 43-27927	7.9.44	KO-	1.1.46	damaged 19.2.45 taxying Hughes, to 3 AD storage
A47-29	108-32441	N5-225, 43-27928	20.8.44	KO-D	21.12.45	to 3 AD Amberley for storage
A47-31	108-32696	N5-231, 43-28183	3.9.44	KO-P	21.12.45	to 3 AD Amberley for storage
A47-32	108-32694	N5-227, 43-28181	19.10.44	KO-	16.7.45	damaged practice bombing 29.10.44 (Clifton), to 14 ARD for storage at 3 AD
A47-39	108-33349	N5-235, 44-29024	6.12.44	KO-	19.5.45	damaged 28.4.45 by AA fire (Legge), to 14 ARD
A47-41	108-35214	44-30889	27.6.45	KO-U	1.1.46	to 3 AD Amberley for storage
A47-43	108-35220	44-30895	26.6.45	KO-B	21.12.45	to 3 AD Amberley for storage

Note that aircraft A47-24, A47-30, A47-38, A47-40, A47-42, and A47-44 to A47-50 did not serve with No 2 Squadron.

Appendix 16

GAF LINCOLN B.30 AIRCRAFT OF No 2 SQUADRON

The Avro 694 Lincoln B.1 was manufactured under licence in Australia by the Department of Aircraft Production (DAP), later the Government Aircraft Factories (GAF), as the Lincoln B.30. The first five aircraft (A73-1 to A73-5) were assembled from components shipped out from the United Kingdom, with the first aircraft flying at Fishermen's Bend on 12 March 1946. The first all-Australian aircraft, A73-6, was delivered to the RAAF in November 1946. The following year the three squadrons of No 82 Wing, Nos 12, 21 and 23 Squadrons, began converting to the type from the B-24 Liberator. In February 1948, these units were retitled Nos 1, 2 and 6 Squadrons.[1]

Total Australian production extended to 73 aircraft. A73-1 to A73-50 were the B.30, powered by four 1635bhp British-built Rolls Royce Merlin 85Bs,[2] while aircraft subsequent to A73-50 were the B.30A variant, with four 1650bhp Australian CAC-built Merlin 102s and structural modifications to allow an increase in all-up weight.[3] Earlier aircraft were subsequently brought up to this standard. Nineteen aircraft were converted with a two-metre nose extension for the general reconnaissance role, as the Lincoln GR.31,[4] with ten of these subsequently modified for anti-submarine duties with No 10 Squadron as the MR.31.

The Lincolns at Amberley remained in a Wing "pool", and were flown by all squadrons. In July 1950 No 1 Squadron deployed to Tengah, in Singapore, while Nos 2 and 6 Squadrons continued to fly Wing aircraft, with No 6 Squadron responsible for conversion of crews to the Lincoln. No 2 Squadron flew the Lincoln until re-equipment with the Canberra in 1955. The list below consists of No 82 Wing Lincolns that were flown by No 2 Squadron, with the dates reflecting the Squadron's operations from 23 Feb 1948 to 18 Dec 1953.

Serial	*Date On Sqn*	*Date Off Sqn*	*Fate*
A73-1	23.2.48	1.3.48	to Air Nav School, East Sale
A73-3	23.2.48	20.4.48	to DAP for mods
	25.8.49	14.4.50	to Air Armt School, East Sale
A73-4	12.8.48	29.9.49	to 10 (GR) Sqn, Townsville
A73-5	20.5.48	28.10.48	to ARDU, Mallala
A73-10	23.2.48	24.6.49	to 3 AD; then GAF for mods; 10 Sqn
A73-12	5.11.48	13.9.49	to 10 (GR) Sqn, Townsville
A73-13	20.1.49	12.1.50	to 3 AD; then East Sale
A73-14	23.2.48	22.7.48	to DAP and 1 AD for mods
	4.5.49	12.2.53	to School Air Nav, East Sale

1 In addition to flying the Lincoln, No 2 Squadron also flew the Liberator up until Jan 1949. These B-24 Liberators flown by the Squadron were A73-172, A72-189, A72-193 and A72-369.

2 Some B.30s also flew with Merlin 66s (taken from the RAAF's stocks of Spitfire LF.VIIIs) on the outboard positions, with Merlin 85Bs inboard.

3 Wilson, p.29.

4 Modified to Mk.31 standard: A73-48 (prototype), A73-28, A73-55, A73-57, and A73-59 to -73. A73-62 onwards were delivered, from early 1953, as Mk.31; previous aircraft were retrofitted.

From top:

82 Wing Lincolns at Pearce, WA, to participate in an aerial pageant at Crawley, 9 February 1948 (note aircraft are still marked with RAAF Pacific roundels) RAAF

A73-31, which crashed at Amberley 9 April 1953 RAAF

A73-31 in better times, 10 May 1949, before serving a short tour with 1 Squadron in Singapore RAAF

A73-35 crashed while landing at Amberley 10 March 1949. RAAF

A73-16	23.2.48	24.1.52	to 10 (GR) Sqn, Townsville
	25.3.53	9.4.53	struck on ground by A73-51 at Cloncurry (Isaacs, 6 Sqn)
A73-17	23.2.48	23.9.49	to 10 (GR) Sqn, Townsville
	11.8.53	18.12.53	retained on 82 Wing
A73-18	26.7.49	10.6.52	damaged by hail storm 12.2.53, 3 AD; East Sale
A73-19	23.2.48	15.12.48	to DAP; Air Nav School, East Sale
	12.2.53	4.12.53	to 3 AD; GAF for mods
A73-21	23.2.48	16.12.48	to DAP for mods; Air Armt School
	16.4.53	18.12.53	retained on 82 Wing
A73-22	23.2.48	13.9.49	to 10 (GR) Sqn, Townsville
A73-23	23.2.48	23.3.50	to 10 (GR) Sqn, Townsville
A73-25	16.5.52	22.3.54	nuclear experiment 16.10.53, to 3 AD
A73-26	23.2.48	1.8.49	to DAP for mods; 11 (GR) Sqn, Pearce
	26.3.50	12.11.50	to 11 (GR) Sqn, Pearce
	18.6.53	18.12.53	retained on 82 Wing
A73-27	22.5.50	12.11.50	to 11 (GR) Sqn, Pearce
	18.8.53	4.12.53	taxy accident 13.11.53 with A73-54, 3 AD repairs; cockpit held Camden Avn Museum
A73-28	8.4.48	15.12.49	to DAP for mods; to 10 Sqn; became Mk.31
A73-29	21.9.50	14.6.51	to 1 (B) Sqn, Tengah
A73-30	20.5.48	14.11.49	to 1 AD; GAF; Air Armt School, East Sale
	26.2.53	16.6.53	to 1 (B) Sqn, Tengah
A73-31	22.7.48	29.8.49	to GAF for LRN mod
	17.11.49	15.7.50	to 1 (B) Sqn, Tengah
	12.2.53	8.5.53	damaged when undercarriage collapsed at Amberley 9.4.53 (Trewin, 6 Sqn), 3 AD written-off
A73-32	24.7.48	10.8.49	to GAF for LRN mod
	31.10.49	30.6.53	to 1 (B) Sqn, Tengah
A73-33	1.10.48	1.8.49	to GAF for LRN mod
	17.10.49	20.7.50	to 1 (B) Sqn, Tengah
A73-34	5.11.48	8.11.51	to 1 (B) Sqn, Tengah
A73-35	21.10.48	10.3.49	forced landing Amberley (Davie), became instructional airframe
A73-36	16.12.48	22.7.49	to DAP for LRN mod
	22.9.49	15.7.50	to 1 (B) Sqn, Tengah
	28.8.52	13.11.52	to 1 (B) Sqn, Tengah
A73-37	17.12.48	27.6.49	to DAP for LRN mod
	8.9.49	28.9.50	damaged landing Amberley 21.2.50 (Ride), to 1 (B) Sqn, Tengah
	20.7.53	18.12.53	retained on 82 Wing
A73-38	4.8.49	11.11.53	to 3 AD; then 1 (B) Sqn, Tengah
A73-39	10.8.49	15.7.50	damaged landing Kalgoorlie 3.2.50 (Davie), to 1 (B) Sqn, Tengah
A73-40	15.8.49	18.12.53	retained on 82 Wing
A73-42	24.8.49	15.7.52	to 1 (B) Sqn, Tengah
A73-43	22.8.49	23.11.50	to 1 (B) Sqn, Tengah
	2.8.51	10.1.52	to 1 (B) Sqn, Tengah
A73-44	9.9.49	7.3.50	dived into ground Amberley (Lynch, 1 Sqn)

A73-45	15.9.49	31.5.51	to 1 (B) Sqn, Tengah
	14.9.53	18.12.53	retained on 82 Wing
A73-46	10.11.49	28.12.50	to UK during 1950, to 1 (B) Sqn
	2.11.53	18.12.53	retained on 82 Wing
A73-49	2.3.53	30.6.53	to 3 AD; to East Sale
A73-50	12.7.50	23.9.50	to 1 (B) Sqn, Tengah
A73-51	1.8.50	9.4.53	crashed into A73-16 at Cloncurry (Winchcombe, 6 Sqn), written-off
A73-52	24.8.50	18.12.53	retained on 82 Wing
A73-53	16.11.50	9.12.53	to 3 AD; to 82 Wing
A73-54	10.11.50	11.12.53	damaged taxying into A73-27 13.11.53, to 3 AD; written-off
A73-55	5.51	9.51	became Mk.31
A73-56	22.12.50	13.2.52	to GAF for mods
	11.6.52	16.3.53	to 3 AD; to GAF
	24.9.53	18.12.53	retained on 82 Wing
A73-57	2.3.51	unknown	became Mk.31
A73-59	7.6.51	31.10.52	to 3 AD; to GAF became, Mk.31
A73-60	23.1.52	6.2.53	to GAF, became Mk.31
A73-61	27.8.51	27.10.52	to GAF, became Mk.31

Although maintained in the 82 Wing "pool", some Lincolns were painted with coloured propeller spinners signifying squadron ownership. No 1 Squadron Lincolns had dark blue spinners, No 2 yellow, and No 6 red.[5] These colours differed from those later adopted as tail lightning flashes on 82 Wing's Canberras.

5 Eyre & White, p.6.

APPENDIX 17

GAF CANBERRA AIRCRAFT OF No 2 SQUADRON

The Canberra had been selected as the replacement for the RAAF's Lincolns, and this type was also manufactured in Australia under licence by the Government Aircraft Factories (GAF). No 2 Squadron was the first Canberra bomber squadron in the RAAF. The Australian Canberra Mark 20 was powered by the RA.3 Avon Mark 101 engines (commonly referred to as the Mark 1 engine), similar to the English Electric Canberra B.2. Two British-built Canberra B.2 aircraft, A84-125 and A84-307, were brought to Australia to serve as pattern aircraft to assist Australian production, and the first was received on No 2 Squadron at Amberley in December 1953.

Serial	*Date On Sqn*	*Date Off Sqn*	*Remarks*

CANBERRA B.2

Two British-built bombers, WD939 and WD983, were obtained as pattern aircraft for the Australian production of the Canberra Mark 20. These aircraft were operated by ARDU and 82 Wing's Canberra Flight prior to receipt on No 2 Squadron.

A84-125	14.1.54	11.7.55	to 6 Sqn, to GAF, converted to Mk 21
A84-307	8.12.53	21.7.55	damaged landing Amberley (Turnnidge) 30.6.55, to GAF, converted to Mk 21

Both these aircraft were converted to Mark 21 trainers (see below).

CANBERRA T.4

Two British-built 2-seat trainers, WT491 and WT492, were obtained for instruction on 82 Wing in 1956. The first became A84-501 on 6 Squadron. The second, A84-502, was flown out from England by a 6 Squadron crew for use on No 2 Squadron. Eventually, with the availability of the Australian Mark 21 trainer, both T.4 aircraft served with CFS at East Sale.

A84-502	28.5.56	4.57	to 6 Squadron; to CFS; RSTT Wagga; to Historical Aircraft Restoration Society at St Marys, nr Sydney

CANBERRA Mark 20

A84-201	18.7.54	31.8.55	to GAF, converted to Mk.21, to ARDU
A84-202	19.12.53	16.6.54	crashed Amberley (Davis/Nicholls/Adler)
A84-204	31.7.54	3.3.55	damaged 17.1.55 Stanthorpe (Stewart/ Martin), to ARDU, to GAF Mk.21 conversion
A84-206	10.9.54	21.2.56	damaged 12.1.56 to 3 AD, to GAF Mk.21 conv
A84-207	18.9.54	21.10.57	to GAF for modifications; Channel 7 Perth display
A84-208	1.12.54	11.7.55	to 6 Squadron; display RAAF Museum Point Cook
A84-209	17.12.54	11.7.55	to 6 Squadron
A84-210	16.12.54	31.10.55	to GAF for modifications, to 6 Sqn
A84-211	27.1.55	24.7.57	to 3 AD, to 1 Squadron
A84-212	21.4.55	20.8.57	to GAF for modifications

A84-214	17.3.55	21.10.57	to 3 AD, to 82 Wing
A84-215	9.4.55	23.1.58	to 3 AD, to 82 Wing
A84-216	28.4.55	24.3.58	to 3 AD, to 82 Wing
A84-217	10.6.55	27.5.58	to 3 AD, to 6 Squadron
A84-218	19.7.55	12.55	to 6 Squadron
A84-219	19.7.55	10.10.56	to 3 AD, to ARDU; display Toowoomba Aero Museum
A84-224	29.4.56	12.5.58	to 6 Squadron
A84-225	11.5.56	21.2.58	to GAF for modifications, to ARDU
A84-226	29.4.56	.2.57	to 6 Squadron
	13.7.71	.2.72	approved for disposal 19.11.71, RSTT Wagga as instructional airframe no.2

Subsequent to the first 27 production aircraft (A84-201 to A84-227), an improved version came off the GAF production line. This model, still known as the Mark 20, was similar to the British Canberra B.6, with the more powerful RA.7 Avon Mark 109 engines and triple breech starters, integral fuel tanks in the wings and updated navigation equipment. The first, A84-228, was received by the RAAF on 23 April 1956, with the first (A84-230) to 2 Squadron on 19 August 1957. Nine of this type equipped No 2 Squadron for its move to Butterworth in 1958. From March 1959, a Canberra Mark 21 trainer was also on 2 Squadron strength at Butterworth, but this did not deploy to Vietnam when the unit moved in April 1967.

During its four years in Vietnam, No 2 Squadron had a strength of 12 Mark 20s: eight on-line at Phan Rang, three undergoing minor servicing with No 478 (Maintenance) Squadron in Butterworth, and one on major servicing with No 3 AD at Amberley. (The following table does not record an aircraft's movements for servicing when the aircraft was then returned immediately to the Squadron.)

The following eight aircraft deployed with the Squadron moved from Butterworth to Phan Rang on 19 April 1967: A84-230, A84-231, A84-234, A84-236, A84-237, A84-240, A84-242 and A84-245.

CANBERRA Mark 20 (109)

A84-228	5.5.61	25.1.67	to ARW Parafield for mods, then 82 Wing
	25.5.67	15.4.70	to 3 AD, then 1 OCU
	11.11.70	14.3.71	shot down by SAM in Vietnam (Downing/ Pinches)
A84-229	7.78	14.10.82	to 3 AD, to USA, airworthy as N229CA.
A84-230	19.8.57	8.10.58	to 82 Wing Amberley
	24.3.67	22.8.67	to ARW Parafield for mods, to 82 Wg, 1 OCU
	26.2.71	12.12.74	to 3 AD storage
	16.11.76	14.10.82	to 3 AD; WA Air Force Assoc Perth for display
A84-231	2.5.58	1.9.58	to 82 Wing Amberley
	14.12.61	20.1.66	to ARW Parafield for mods, then 82 Wing
	3.3.67	26.9.67	to ARW Parafield for mods, to 82 Wg, 1 OCU
	18.9.68	4.8.69	to 3 AD, then 1 OCU
	11.2.70	3.11.70	missing in action in Vietnam (Herbert/ Carver)

Left to right from top:

A84-240, which crashed while taxying at Butterworth, 1 September 1959 RAAF

A84-243, written-off in a landing accident at Butterworth, 1 May 1959 RAAF

A84-244, on the way to a target in South Vietnam in 1969 RAAF

Back home in Australia — the "Queen of the Skies" RAAF

A84-232	2.5.58	1.9.58	to 82 Wing Amberley
	9.7.65	15.2.66	to ARW Parafield for mods, then 82 Wing
	22.10.66	26.2.67	to 3 AD, then 82 Wing
	29.8.67	7.11.69	to 3 AD, then 1 OCU
	21.7.70	29.8.82	to 3 AD; then GAF/ASTA Avalon for display
A84-233	13.6.58	15.10.58	to 82 Wing Amberley
	25.3.66	20.7.66	to 82 Wing, ARW Parafield for mods
	14.9.67	13.6.69	to 3 AD, then 1 OCU
	5.6.70	8.9.80	to 3 AD storage, Karinga trials Woomera
A84-234	27.8.65	6.7.67	to ARW Parafield for mods, then 3 AD
	14.9.67	27.5.68	to 3 AD, then 1 OCU
	6.5.69	1.4.81	to 3 AD storage, disposal approved 16.1.84
A84-235	9.4.60	29.4.65	to 3 AD, then 82 Wing
	26.1.66	23.7.70	to 3 AD, then 1 OCU
	16.2.72	18.12.79	to ARDU Edinburgh
	22.4.80	5.7.82	to 3 AD, RSTT Wagga; display RAAF Wagga
A84-236	13.6.58	23.12.60	to 3 AD, mods at GAF, 3 AD storage
	19.2.64	20.1.66	to ARW Parafield for mods, then 82 Wing
	6.66	8.7.66	to 82 Wing Amberley
	30.3.67	11.9.67	to ARW Parafield for mods, then 82 Wg, 1 OCU
	2.10.68	6.10.69	to 3 AD, then 1 OCU
	7.8.70	17.9.74	to 3 AD, then ARDU
	26.8.80	2.8.82	to RSTT Wagga; display RAAF Museum Point Cook
A84-237	2.9.58	15.1.63	to 3 AD, mods at GAF, 3 AD
	8.5.63	27.1.65	to 3 AD, mods at GAF, 82 Wing
	6.3.67	30.9.68	to 3 AD, then 1 OCU
	10.9.69	26.2.71	to 3 AD, ARDU, 11.6.82 Karinga trials Woomera
A84-238	29.7.58	15.10.64	to 3 AD, mods at GAF, 82 Wing
	14.1.66	27.2.67	to ARW Parafield, then 82 Wing
	22.10.67	16.8.68	to 3 AD, then 1 OCU
	23.7.69	12.3.82	to 3 AD storage; sold 6.85, display 8.86 near Amberley
A84-239	23.9.58	8.3.60	crashed on take-off Butterworth (Collins/Wyatt), converted to components 22.6.60
A84-240	1.6.58	11.6.64	damaged taxying 1.9.59 Butterworth (Collins/Keenan) and repaired by GAF, to 3 AD, mods at GAF, 3 AD
	23.1.65	18.12.67	to 3 AD, then 1 OCU
	29.5.68	9.6.70	to 3 AD, then 1 OCU
	22.6.71	14.4.78	to ARDU Edinburgh, then 3 AD
	21.8.80	14.10.82	to 3 AD; 22.8.84 display RNZAF Museum Wigram
A84-241	28.10.58	10.2.64	to 3 AD, mods at GAF, to 6 Squadron
	17.4.65	24.3.67	to ARW Parafield for mods, to 82 Wing, 1 OCU
	20.6.68	25.8.70	damaged landing 11.7.70 Phan Rang (Birks/Curr), to 3 AD, then 1 OCU
	22.6.71	22.9.78	to ARDU, 3 AD; 11.6.82 Karinga trials

A84-242	4.6.58	1.9.65	to 3 AD, mods at ARW, 82 Wing
	1.12.66	12.9.68	to 3 AD, then 1 OCU
	16.1.70	27.7.77	to 3 AD storage, 15.2.83 training aid Amberley
A84-243	13.6.58	29.8.58	crashed landing Butterworth (Harvey/ Overton), converted to components 1.5.59
A84-244	13.6.58	4.8.58	to 6 Squadron
	12.2.65	23.2.66	to ARW Parafield, 82 Wing, 1 OCU
	10.6.68	22.1.70	to 3 AD, 1 OCU
	29.1.71	28.3.78	to 3 AD storage; 6.1.81 Karinga trials Woomera
A84-245	13.6.58	18.12.61	to No 1 Squadron
	1.3.66	15.6.67	to ARW Parafield, then 82 Wing, 1 OCU
	29.1.71	12.3.82	storage at 3 AD; 27.7.83 to ARL for display
A84-246	13.6.58	12.12.58	to 82 Wing Amberley
	20.3.67	13.6.69	to 3 AD, 1 OCU
	7.11.69	2.2.73	to ARDU Edinburgh
	14.11.73	1.5.79	to ARDU
	27.11.79	3.9.80	to 2 AD for destructive testing
A84-247	19.12.58	4.2.65	mods at GAF, 3 AD, 82 Wing
	10.2.66	17.6.68	to 3 AD, 1 OCU
	14.7.69	20.7.73	to 3 AD for repairs
	4.2.75	16.8.77	to 3 AD storage; 4.8.82 to AWM for display
A84-248	5.6.64	1.12.66	to ARW Parafield, then 3 AD
	16.6.67	21.4.69	to 3 AD, then 1 OCU
	9.10.69	15.7.71	to 3 AD for repairs
	6.7.72	12.3.82	to 3 AD storage for Karinga trials; 18.12.86 intended for East Sale display, displayed nr Amberley

Canberra bombers were originally delivered in an all-over natural metal scheme, later having a reflective gloss-white upper fuselage in an attempt to improve crew comfort. With the aircraft's re-roling to low-level in the early 1960s, an olive drab (green) and dark admiralty grey camouflage pattern was applied. The aircraft undersurfaces remained aluminium lacquer.[1] The first aircraft to arrive on 2 Squadron with the new camouflage paint was A84-231, which was received from No 3 AD on 26 Sep 1963. From the experience gained with this aircraft, the camouflage was changed from gloss to matt, and the 84-inch diameter roundels on the mainplanes were reduced to 33-inch diameter. This amended scheme was adopted on 10 October 1963. The dark grey was introduced for undersurfaces on 2 Squadron aircraft in October 1964, and in February 1966 this became the standard for all RAAF Canberras.[2] The large white fuselage serial numbers were also toned down to black.

In December 1967, after the Squadron's move to Vietnam, roundels were removed from the wings and the fuselage roundels were reduced from 33-inch diameter to 18-inch.[3] The familiar red lightning flash was changed after the Squadron's return to Amberley, to include the piping shrike from the unit badge.

1 This scheme was adopted on 17 Jun 1963. File 414/1/21, RAAF Historical.

2 File 579/3/255(114), of 26 Mar 1973, RAAF Historical.

3 File 579/3/255(114), of 26 Mar 1973, RAAF Historical.

The five earliest Mark 20 aircraft (A84-201, and A84-203 to A84-206) were converted to trainer configuration by GAF as the Mark 21. In addition, the two B.2 aircraft received from the United Kingdom, A84-125 and A84-307, were also converted to Mark 21 configuration. These aircraft were primarily for use by the conversion unit, No 1 OCU at Amberley, but one was kept on 2 Squadron strength during its time at Butterworth. After Vietnam, when 1 OCU was disbanded, the Mark 21 was again operated by the Squadron.

CANBERRA Mark 21

A84-125	3.72	11.76	to 3 AD servicing and storage
	11.78	30.6.80	to ARDU Edinburgh
	9.81	29.4.82	to ARDU Edinburgh, held Amberley for National Aerospace Museum
A84-307	20.5.63	6.3.66	to ARW Parafield for mods, to 1 OCU
	30.6.71	12.9.73	to 3 AD; display 1 CAMD RAAF Kingswood
A84-201	2.3.72	20.12.77	to 3 AD, static display at Amberley
A84-203	3.59	8.5.63	to 3 AD, to 1 OCU, later ARDU
	7.72	11.73	to 3 AD storage
	12.73	11.79	to 3 AD
	4.82	7.82	allocated for PNG display
A84-204	24.4.66	9.5.67	not deployed to Vietnam 19.4.67, to TSF/478 Sqn, to ARDU
	22.9.78	21.5.82	to 3 AD for disposal

Canberra Mark 21 trainers retained the silver/white scheme for the duration of their service in the RAAF.

APPENDIX 18

DOUGLAS C-47B DAKOTA AIRCRAFT OF No 2 SQUADRON

The first two Dakotas for the new "C" Flight joined No 2 Squadron at Amberley on 2 June 1958, before deploying to Butterworth. A third Dakota was added in 1963 to take on a higher supply dropping commitment to security forces working on the Thai border. The strength was increased to four in July 1964 as the RAAF involvement in South-East Asia escalated.

With No 2 Squadron's move to Vietnam in April 1967, "C" Flight became an independent unit, known as Transport Support Flight (TSF), based at Butterworth, to provide support to Australian military forces and diplomatic missions throughout the region. Note that aircraft which subsequently served with TSF have not been included in this listing.

Serial	*C/n*[1]	*ex-USAAF*	*Date On Sqn*	*Date Off Sqn*	*Fate*
A65-63	14556/26001	43-48740	21.2.66	23.2.67	to TSF, Butterworth, 1981 to PNGDF as P2-005, 1993 to VH-PWN
A65-69	15682/27127	43-49866	25.1.66	23.2.67	to TSF Butterworth, 1980 became ZD215 for delivery to Berlin, preserved Berlin-Gatow
A65-70	15684/27129	43-49868	31.5.64	24.2.66	to ARW Parafield, the 2 ATU, 1968 became N16892, PK-JDE, PK-EHC
A65-71	15690/27131	43-49870	23.7.63	20.1.66	to ARW, 10 Sqn, TSF Butterworth, 1980 to AWM as "VH-CIN" coded OM-N airworthy as the "heritage" Dakota
A65-73	15923/32671	44-76339	6.6.62	18.12.65	to ARW Parafield, ARDU & 10 Sqn, sold 1973, museum display Townsville
A65-82	15683/27128	43-49867	2.6.58	31.3.61	to ARW Parafield, 1 AFTS Pearce, 1970 to Indon AF as P504
A65-91	16136/32884	44-76552	23.9.59	3.3.60	to 86 Wing Richmond, 1 AFTS Pearce, CFS, sold 1986
A65-94	16358/33106	44-76774	2.6.58	1.4.62	to ARW Parafield, disposal 1985 to WA branch AFA Museum

1 Most C-47 aircraft have two constructor's numbers. This arose as the same c/n blocks were allocated to both Douglas C-47/C-117 and A-26 production lines. In July 1946, aircraft manufactured at Douglas' Oklahoma plant were issued with a corrected c/n, which is the second number shown in each sequence.

From top:

A65-98, with 2 Squadron 1962–66. Note 2 Squadron's red flash on the fin, which was later moved to the rudder, as shown on the aircraft below RAAF

A65-69, which served on 2 Squadron 1966–67, before being passed to TSF RAAF

A65-98	16354/33102	44-76770	14.4.62	14.3.66	to ARW Parafield, ARDU, 1980 to RSTT Wagga, sold to VH-OVM
A65-119	16715/33463	44-77131	10.4.61	17.7.63	to ARW Parafield, disposal 1968 to N16896, VH-EQO, given as aid to Laos 1971
A65-122	16707/33455	44-77123	10.3.66	23.2.67	to TSF Butterworth, 1980 given to PNGDF Lae as a training aid
A65-124	16960/34220	45-957	13.12.65	23.2.67	to TSF Butterworth, 2AD, ARDU, 1981 to AFA Museum Perth.

BIBLIOGRAPHY

PRIMARY SOURCES

A.108(B), RAAF Attack Reports, RAAF Historical Section, Canberra

A.108(D), RAAF Combat Reports, RAAF Historical Section, Canberra

AA CRS A2 series, Prime Minister's Office/Dept, Australian Archives, Canberra

AA CRS A705 series, Air Services Branch/Dept of Air, Australian Archives, Canberra

AA CRS A1196 series, Air Services Branch/Dept of Air, Australian Archives, Canberra

AA CRS A1952 series, Dept of Defence, Australian Archives, Canberra

AA CRS A1969 series, Australian Archives, Canberra

AA CRS A2023 series, Dept of Defence, Australian Archives, Canberra

AA CRS A5954 series, Dept of Defence, Australian Archives, Canberra

AA CRS MP84/1 series, Dept of Defence, Australian Archives, Melbourne

Air Board Agenda, RAAF Historical Section, Canberra

Air Board AMSE correspondence, Australian Archives, Canberra

Air Reconnaissance Reports, RAAF Historical Section, Canberra

Aircraft Crash Records, RAAF Historical Section, Canberra

Aircraft Status Cards, RAAF Historical Section, Canberra

Alexander, Air Cdre N, interview with author, 8 Dec 1993

Ashley, P W, letter to author, 10 Nov 1993

Australian Military Order No 570, of 22 Oct 1912, and *No 381*, of 14 Jul 1914, RAAF Historical Section, Canberra

AWM 4 series, AIF Unit War Diaries 1914–1918 War, AWM, Canberra

AWM 8 series, AIF Embarkation Rolls, AWM, Canberra

AWM 10 series, AIF Administrative HQ Registry, AWM, Canberra

AWM 22 series, AIF HQ (Egypt) Central Registry Files, AWM, Canberra

AWM 25 series, Strength Returns, AWM, Canberra

AWM 30 series, POW Statements 1914–1918 War, AWM Canberra

AWM 38 series, Official History, CEW Bean Records, AWM, Canberra

AWM 54 series, Written Records 1939–1945 War, AWM, Canberra

AWM 131 series, Roll of Honour 1914–1918, AWM, Canberra

AWM 133 series, Nominal Roll of AIF 1914–1918, AWM, Canberra

AWM 140 series, Biographical Details 1914–1918 War, AWM, Canberra

AWM 224 series, Unit Manuscript Histories, AWM, Canberra

AWM PR Private Records series, AWM, Canberra

Barnes, A, taped recollections, Nov 1993

Barnes, A, telephone interviews with author, 19 Nov 1993 and 3 Dec 1993

Batten, A, taped recollections, No 2 Squadron Association, Melbourne

Batten, A, telephone interview with author, 26 Oct 1993

Beck, AVM G J J, telephone interview with author, 1 Feb 1994

Bice, J, letter to author, 11 Jul 1993

Biddell, R L, interviews with author, 29 Jan 1994 and 7 Feb 1994, and letter to author, 20 Jan 1994

Boast, Air Cdre J R, telephone interview with author, 6 Dec 1993

Bonython, K, letter to author, of 13 Oct 1993

Burtt, Air Cdre F, interview with author, 26 Nov 1993

CD.102 series, RAAF records of lost aircraft and crews, RAAF Historical section, Canberra

Cope, H, letter to author, 29 Jun 1993

Curr, Wg Cdr A J, interview with author, 10 Dec 1993

Daymon, R, taped recollections, Jun 1993, No 2 Squadron Association, Melbourne

Dodd, K, interview and letter to author, 16 Sep 1993

Dorsett, R J, notes on Raid on Darwin 19 Feb 1942, RAAF Hudson Squadrons Association, Adelaide

Dunne, A, telephone interview with author, of 1 Nov 1993

Duus, N L, telephone interview with author, 28 Nov 1993

Evans, Air Marshal S D, interview with author, 21 Dec 1993

Findings of RAAF Inquiry, and *Findings of Judicial Court of Inquiry*, crash of A16–97, courtesy of A B McFarlane, Canberra

Freeman, R, interview with author, 10 Nov 1993

Freeman, R, letter to author, 18 Nov 1993

Garley, Sqn Ldr G K, interview with author, 15 Dec 1993

Garrett, Sir R, taped recollections, No 2 Squadron Association, Melbourne

Garrisson, Air Cdre A D J, flying logbook

Godfrey, D J, letter to author, 21 Nov 1993

Goodwin, P W, letter to author, 16 Dec 1991

Green, R, letter to No 2 Squadron Association, 20 May 1993

Green, R, telephone interview with author, 29 Oct 1993

Grey, Wg Cdr R, interview with author, 18 Nov 1993

Halvorson, L J, telephone interview with author, 28 Nov 1993

Harrison, Sqn Ldr D K, letter to author, 2 Feb 1994

Hawthorne, Sqn Ldr B D, personal diary

Henderson, J, letters to author, 20 Mar 1993 and 20 Jul 1993

Henry, A, interview with author, 8 Oct 1993

Howe, R W, letter to author, 8 Jan 1994

Humphris, C B, letter to Alexander Downer MP, 21 May 1985

Humphris, C B, letter to author, 21 Oct 1993

Individual service records, RAAF Personnel Records Section, Queanbeyan

Interview with Wg Cdr K S Hennock DFC, 12 Feb 1945, RAAF Historical Section, Canberra

Interview with WOff A H McEgan, Evacuation of Laha and Initial Raid on Darwin, undated, courtesy of A H McEgan

IWM Documents Collection, Imperial War Museum, London

Jefferies, F, *Statement of June 1993*, "The Crash of A16–97 at Canberra on 13th August 1940", No 2 Squadron Association, Melbourne

Jellis, G, *Statement of June 1993*, "The Loss of Aircraft A47–6", No 2 Squadron Association, Melbourne

Jones, R, letter to author, 30 Sep 1993

Kerr, W, telephone interview with author, 10 Nov 1993

Lonie, F R, interview with author, 29 Nov 1993

MacAlister, J, telephone interview with author, of 11 Nov 1993

Martin, R H, telephone interview with author, 24 Dec 1993

McFarlane, Gp Capt A B, interview with author, 1 Nov 1993

MFC 76/1/293, Microfiche Collection, RAF Museum, Hendon

Millar, Sqn Ldr D G, interview with author, 27 Jan 1994

MS 5020, Manuscript Collection, National Library, Canberra

No 1 Squadron Operations Record Book (Form A50), RAAF Historical Section, Canberra

No 2 Squadron Operations Record Book (Form A50), RAAF Historical Section, Canberra

No 2 Squadron Record of Operations (Form A51), RAAF Historical Section, Canberra

No 6 Squadron Operations Record Book (Form A50), RAAF Historical Section, Canberra

No 13 Squadron Record of Operations (Form A51), RAAF Historical Section, Canberra

No 24 Squadron Operations Record Book (Form A50), RAAF Historical Section, Canberra

No 68 Squadron Routine Orders No 1, AWM, Canberra

No 548 Squadron RAF Operations Record Book (Form A50), RAAF Historical Section, Canberra

NWA Interviews Book 1, RAAF Historical Section, Canberra

NWA Brief, Interview with Flg Off E B Stacy, 17 Nov 1944, RAAF Historical Section, Canberra

NWA Signals, MS 5020 Box 1, National Library, Canberra

O'Ferrall, R E, telephone interview with author, 17 Dec 1993

Owen, Sqn Ldr R D, letter to author, 9 Dec 1993

Owen, Air Cdre T C, telephone interview with author, 28 Jan 1994

Perkins, E, taped recollections of 5 Jul 1993, No 2 Squadron Association, Melbourne

Pinches, A J, letter to author, 25 Feb 1994

Power, N E, interview with author, 28 Jan 1994

PRO Air 1 series, Air Historic Branch WWI Records, Public Record Office, Kew, England

PRO Air 8 series, CAS Records, Public Records Office, Kew

PRO Air 10 series, Air Publications, Public Records Office, Kew

RAAF Crash Critique No 40, RAAF DFS, Melbourne, RAAF Historical Section, Canberra

RAAF General Instruction No C.11, 22 Sep 1939, RAAF Historical Section, Canberra

RAAF HQ Routine Orders, RAAF Historical Section, Canberra

RAAF PR Release, "Fine RAAF Canberra Effort in Vietnam Proved Critics Wrong", 2 Jan 1969, Dept of Defence, Canberra, RAAF Historical Section, Canberra

RAAF PR Releases: Ministerial release S4324/69, 10 Jul 1969; S4460/70, 4 Nov 1970; and S4550/72, 1 Mar 1972, Dept of Defence, Canberra, RAAF Historical Section, Canberra

RAF Form 1669, 22 Sep 1959, in A705 series, Australian Archives, Canberra

Rawlinson, Gp Capt M J, letter to author, 17 Dec 1993

Rayner, W R, telephone interview with author, 23 Dec 1993

Reif, C W, telephone interviews with author, 16 Nov 1991 and 25 Nov 1993

Report on Operations, RAAF WWII operations, RAAF Historical Section, Canberra

Riggs, F, Flying Logbook, No 2 Squadron Association, Melbourne

Ross, J K, telephone interview with author, 12 Dec 1993

Salvair, P T, telephone interview with author, 13 Dec 1991

Sandow, D, letter to author, 27 Sep 1993

Scott, W, letter to No 2 Squadron Association, Melbourne, 22 Oct 1992

Shattock, D K, letter to No 2 Squadron Association, Melbourne, 18 Jun 1993

Slootjes, Wg Cdr A, letter to author, 7 Feb 1994

Smith, Sqn Ldr M G, letter to author, 21 Dec 1993

Stephens, A W, interview with author, 13 Dec 1993

Taylor, R, Record of RAAF Service, Jun 1993, No 2 Squadron Association, Melbourne

Transcript of 3AR ABC Broadcast, 10 Sep 1944, courtesy of P Hocking, 2 Squadron Association, Melbourne

Wallace, B, personal diary, courtesy of Mr J Wallace, Melbourne

Weekes, Wg Cdr G D, letter to author, 30 Dec 1993

White, W M, recollections on service with No 2 Squadron, Nov 1944–Oct 1945, No 2 Squadron Association, Melbourne

Wilkinson, A R, interview with author, 17 Nov 1993

Wilson, Maj C R, USAF, letter to CO 2 Squadron, 4 Dec 1968, courtesy of Wg Cdr K W G Padgett

Wood, R J, letter to author, 15 Dec 1993

Secondary Sources

Alford, R N, *Darwin's Air War*, Aviation Historical Society of NT, Darwin, 1991.

Andrade, J M, *US Military Aircraft Designations and Serials*, Midland Counties, Earl Shilton, Leics, 1979.

Army News, "Unconditional Surrender by Japan: World at Peace", Darwin, 16 Aug 1945.

Balfe, J D, *War Without Glory*, MacMillan, Melbourne, 1984.

Baring, M, *Flying Corps Headquarters 1914–1918*, Buchan & Enright, London, 1985.

Barnes, C H, *Shorts Aircraft Since 1900*, Putnam, London, 1989.

Beck, G J J, *No 2 Squadron — Vietnam 1968*, Squadron annual booklet, Phan Rang.

Billings, H D, My Own Story, Glen Iris, 1989, unpublished manuscript, RAAF Museum, Point Cook.

Bonython, K, *Ladies' Legs and Lemonade*, Rigby, Adelaide, 1979.

Chapman, F S, *The Jungle is Neutral*, Chatto & Windus, London, 1949.

Cobby, A H, *High Adventure*, Kookaburra, Melbourne, 1981.

Coulthard–Clark, C D, *The Third Brother*, Allen & Unwin, Sydney, 1991.

Cox, F R, The First World War Memoirs of Fergus Cox, unpublished manuscript, Museum of Army Aviation and Flying, Oakey.

Cranston, F, "Fairbairn Hoped to Pilot New Bomber", in *The Canberra Times,* 12 Aug 1990, Canberra.

Cutlack, F M, *The Official History of Australia in the War of 1914–1918, Vol VIII,* The Australian Flying Corps, AWM, Canberra, 1923.

Daily Telegraph, "Viet Cong Raid US Air Base", Sydney, 27 Jan 1969.

Defense News, Washington, 26 Oct 1992.

Edwards, P, *Crises and Commitments,* Allen & Unwin, Sydney, 1992.

Ekstrom, D, "Of Grey Hairs and Sharp Teeth— The Canberra", in *Australian Aviation & Defence Review,* Aerospace Publications, Canberra, Dec 1980.

Emme, E M, *The Impact of Air Power,* Van Nostrand, Princeton, NJ, 1959.

Eyre, D, & White, B, "The Avro Lincoln in Australia", in *AHSA Journal,* Vol IX, No 1, Jan–Feb 1968.

Flight, "The DH–5", London, 9 Jan 1919.

Flight, "Operation Sabre Ferry", London, 19 Dec 1958.

Flying, "Training the Military Flyer", Aeronautical Society, London, 14 Feb 1917.

Francillon, R J, *Japanese Aircraft of the Pacific War,* Putnam, London, 1990.

Garbett, M, & Goulding, B, *Lincoln at War 1944–1966,* Ian Allan, London, 1979.

Gillison, D, *Royal Australian Air Force 1939–1942,* AWM, Canberra, 1962.

Graham, B, & Smyth, F, *A Nation Grew Wings,* Winterset House, Melbourne, 1946.

Grant, I, *Dictionary of Australian Military History,* Random House, Melbourne, 1992.

Grantham, S R, *The 13 Squadron Story,* self-published, Sydney, 1991.

Hata, I, & Izawa, Y, *Japanese Naval Aces and Fighter Units in World War II,* Airlife, Shrewsbury, Shropshire, 1989.

Hayward, R, *The Beaufort File,* Air-Britain, Tonbridge, Kent, 1990.

Howe, R W, *No 2 Squadron Vietnam — 1969,* Squadron annual booklet, Phan Rang.

Humphris, C B, *Trapped on Timor,* self-published, Victor Harbor, 1991.

Isaacs, Gp Capt K, *Military Aircraft of Australia 1909–1918,* AWM, Canberra, 1971.

Jackson, R, *Aces' Twilight,* Sphere Books, London, 1988.

Jackson, R, *Canberra — The Operational Record,* Airlife, Shrewsbury, Shropshire, 1988.

Jones, G P, *Attacker — The Hudson and its Flyers,* William Kimber, London, 1980.

Jones, H A, *The War in the Air, Vol IV,* Clarendon Press, Oxford, 1934.

Jones, H A, *The War in the Air, Vol V,* Clarendon Press, Oxford, 1935.

Jones, H A, *The War in the Air, Vol VI,* Clarendon Press, Oxford, 1937.

Jones, R V, *Most Secret War,* Hamish Hamilton, London, 1978.

Kerr, W, "In the Last Great Air Race from London to Christchurch", in *The Canberra Times,* Canberra, 9 Oct 1991.

Laffin, J, *Guide to Australian Battlefields of the Western Front 1916–1918,* Kangaroo Press, Sydney, 1992.

MacCloskey, Brig Gen M, *The United States Air Force,* Praeger, New York, 1967.

MacFarling, Wg Cdr I, "Airborne Anti–Submarine Warfare", in Stephens, A W (ed.) *Defending the Air/Sea Gap,* ADSC, Canberra, 1992.

McCarthy, D, *South-West Pacific Area — First Year,* AWM, Canberra, 1959.

Montgomery, Maj Gen Sir A, *The Story of the Fourth Army*, Hodder & Stoughton, London, 1919.

Morison, S E, *History of the United States Naval Operations in World War II, Vol III*, Little & Brown, Boston, 1948.

Morison, S E, *History of the United States Naval Operations in World War II, Vol VI*, Little & Brown, Boston, 1950.

Moss, P W, "BOAC At War", in *Aeroplane Monthly*, August 1975, London.

Odgers, G, *Air War Against Japan 1943–1945*, AWM, Canberra, 1957.

Odgers, G, *Mission Vietnam*, Department of Defence, Canberra, 1974.

Odgers, G, *The Royal Australian Air Force: An Illustrated History*, Ure Smith, Sydney, 1977.

Ottaway, N, "Head-On Clashes in Timor", in *The Sun*, Sydney, 10 Jul 1943.

Paley, T, *The Sparrows*, Self-Publishing Association, Worcester, Hereford, 1992.

Paris, M, *Winged Warfare*, Manchester University Press, Manchester, 1992.

Prime, A, "Australia's Pearl Harbor", in *Australian Aviation*, Aerospace Publications, Canberra, December 1991.

RAAF Hudson Squadrons Association Newsletter, RAAF Hudson Squadrons Association, Adelaide.

RAAF Log, AWM, Canberra, 1954.

RAAF News, editions since 1961, RAAF Historical Section, Canberra.

Raleigh, Sir W, *The War in the Air, Vol I*, Clarendon Press, Oxford, 1922.

Richards, E J, *Australian Airmen*, Bruce & Co, Melbourne.

Robertson, B, *Air Aces of the 1914–18 War*, Harleyford, Letchworth, Herts, 1964.

Roylance, D, *Air Base Richmond*, 1991.

Shores, C, Franks, N, & Guest, R, *Above The Trenches*, Grub Street, London, 1990.

Sims, E H, *Fighter Tactics and Strategy 1914–1970*, Cassell, London, 1972.

Smith's Weekly, "Knights in the Air", Sydney, 29 Mar 1919.

Stephens, A W, *Power Plus Attitude*, AGPS Press, Canberra, 1992.

Stevens, LCdr D M, "The War Cruise of I–6", in *Australian Defence Force Journal*, No 102, Sep/Oct 1993, Canberra.

Strike, RAAF Amberley magazine, various editions.

Sturtivant, R C, *The Anson File*, Air-Britain, Tonbridge, Kent, 1988.

Sutherland, L W, *Aces and Kings*, Hamilton, London.

Taylor, Sir G, *Sopwith Scout 7309*, Cassell, London, 1968.

The Golden Years, AGPS, Canberra, 1971.

The Malayan Emergency 1948–1960, AP3410, Ministry of Defence, London, 1970.

The Sydney Morning Herald, "State Aviation School", of 20 Jul 1917.

Thetford, O G, & Riding, E J, *Aircraft of the 1914–1918 War*, Harleyford Publications, Marlow, Bucks, 1954.

TRIAD, No 7, "Operation Cenderawasih", Dept of Defence, Canberra, 1977.

Typhoon, Vol III, No 9, "Flyboys From Down Under", APO San Francisco, Sep 1969.

Ure Smith, S, Stevens, B, & Watt, E, *Oswald Watt the Airman*, Art in Australia Ltd, Sydney, 1921.

Wadey, S, The Operation Order for the Day Read, unpublished manuscript, courtesy of Mrs M Wadey, via RAAF Hudson Squadrons Association, Adelaide.

Wallace, G, *Up in Darwin With the Dutch*, Melbourne, 1983.

Werrell, K P, *Archie, Flak, AAA and SAM*, Air University Press, Maxwell AFB, Alabama, 1988.

White, T W, *Guests of the Unspeakable*, Little Hills Press, Sydney, 1990.

Williams, Air Marshal Sir R, *These Are Facts*, AWM, Canberra, 1977.

Wilson, S, *Anson, Hudson & Sunderland in Australian Service*, Aerospace Publications, Canberra, 1992.

Wilson, S, *Lincoln, Canberra & F–111 in Australian Service*, Aerospace Publications, Canberra, 1989.

Wings, Air Force Association magazine, several editions, Melbourne and Sydney.

Wray, C C H, *Timor 1942*, Mandarin, Melbourne, 1990.

Wrigley, Wg Cdr H N, *The Battle Below*, Errol G Knox, Sydney, 1935.

Yeates, V M, *Winged Victory*, Mayflower, St Albans, Herts, 1975.

INDEX

A page reference in italics indicates an illustration not referred to in the discussion on that page. An *n* following a reference indicates that the person or subject referred to is discussed only in a footnote and not in the text.

Indexer: Michael Harrington